AF900323

Translational Medicine Research

Series editors
Zhu Chen
Shanghai, China

Xiaoming Shen
Shanghai, China

Saijuan Chen
Shanghai, China

Kerong Dai
Shanghai, China

Translational medicine converts promising laboratory discoveries into clinical applications and elucidates clinical questions with the use of bench work, aiming to facilitate the prediction, prevention, diagnosis and treatment of diseases. The development of translational medicine will accelerate disease control and the process of finding solutions to key health problems. It is a multidisciplinary endeavor that integrates research from the medical sciences, basic sciences and social sciences, with the aim of optimizing patient care and preventive measures that may extend beyond health care services. Therefore, close and international collaboration between all parties involved is essential to the advancement of translational medicine. To enhance the aforementioned international collaboration as well as to provide a forum for communication and cross-pollenation between basic, translational and clinical research practitioners from all relevant established and emerging disciplines, the book series "Translational Medicine Research" features original and observational investigations in the broad fields of laboratory, clinical and public health research, aiming to provide practical and up-to-date information on significant research from all subspecialties of medicine and to broaden readers' horizons, from bench to bed and bed to bench. Produced in close collaboration with National Infrastructures for Translational Medicine (Shanghai), the largest translational medicine research center in China, the book series offers a state-of-the-art resource for physicians and researchers alike who are interested in the rapidly evolving field of translational medicine. Prof. Zhu Chen, the Editor-in-Chief of the series, is a hematologist at Shanghai Jiao Tong University, China's former Minister of Health, and chairman of the center's scientific advisory board.

More information about this series at http://www.springer.com/series/13024

Dong-Qing Wei • Yilong Ma • William C.S. Cho
Qin Xu • Fengfeng Zhou
Editors

Translational Bioinformatics and Its Application

上海交通大學出版社
SHANGHAI JIAO TONG UNIVERSITY PRESS

Springer

Editors
Dong-Qing Wei
State Key Laboratory of Microbial Metabolism and School of Life Sciences and Biotechnology
Shanghai Jiao Tong University
Shanghai, China

William C.S. Cho
Department of Clinical Oncology
Queen Elizabeth Hospital
Kowloon, Hong Kong, China

Fengfeng Zhou
College of Computer Science & Technology
Jilin University
Changchun, Jilin, China

Yilong Ma
Center for Neurosciences
The Feinstein Institute for Medical Research
New York, USA

Department of Molecular Medicine
Hofstra Northwell School of Medicine
New York, USA

Qin Xu
State Key Laboratory of Microbial Metabolism and School of Life Sciences and Biotechnology
Shanghai Jiao Tong University
Shanghai, China

ISSN 2451-991X ISSN 2451-9928 (electronic)
Translational Medicine Research
ISBN 978-94-024-1043-3 ISBN 978-94-024-1045-7 (eBook)
DOI 10.1007/978-94-024-1045-7

The print edition is not for sale in China Mainland. Customers from China Mainland please order the print book from: Shanghai Jiao Tong University Press.

Library of Congress Control Number: 2017932746

© Shanghai Jiao Tong University Press, Shanghai and Springer Science+Business Media Dordrecht 2017
This work is subject to copyright. All rights are reserved by the Publisher, whether the whole or part of the material is concerned, specifically the rights of translation, reprinting, reuse of illustrations, recitation, broadcasting, reproduction on microfilms or in any other physical way, and transmission or information storage and retrieval, electronic adaptation, computer software, or by similar or dissimilar methodology now known or hereafter developed.
The use of general descriptive names, registered names, trademarks, service marks, etc. in this publication does not imply, even in the absence of a specific statement, that such names are exempt from the relevant protective laws and regulations and therefore free for general use.
The publisher, the authors and the editors are safe to assume that the advice and information in this book are believed to be true and accurate at the date of publication. Neither the publisher nor the authors or the editors give a warranty, express or implied, with respect to the material contained herein or for any errors or omissions that may have been made. The publisher remains neutral with regard to jurisdictional claims in published maps and institutional affiliations.

Printed on acid-free paper

This Springer imprint is published by Springer Nature
The registered company is Springer Science+Business Media B.V.
The registered company address is: Van Godewijckstraat 30, 3311 GX Dordrecht, The Netherlands

Preface

It was May 2015 when I was invited to join the editorial team of the "Translational Medicine Publication Project." I proposed to edit a book entitled *Translational Bioinformatics (TBI)*. I was happy to have invited a few colleagues from China and the USA who are experts in the field to join me as coeditors, Profs. Yilong Ma, William C.S. Cho, and Fengfeng Zhou. Prof. Qin Xu from my research team and my PhD student Huiyuan Zhang spent much time in managing the project. It has been many years since I started to collaborate with Springer. Our proposal was approved quickly as a collaboration project with the Shanghai Jiao Tong University Press.

TBI is an emerging field in the study of health informatics, focused on the convergence of molecular bioinformatics, biostatistics, statistical genetics, medical imaging, and clinical or medical informatics. Its focus is on applying sound informatics methodology to the increasing amount of biomedical and genomic data to formulate knowledge, disease models, and medical tools, which can be utilized by scientists, clinicians, and patients. TBI employs data mining and analytical biomedical informatics in order to generate clinical knowledge for a wide array of applications. Furthermore, it involves cross-disciplinary biomedical research to improve human health through the use of computer-based information systems. This new field has achieved great success in the recent decade by synergic integration of the molecular and genetic footprints in tissue cultures, animal models, and patients with various diseases.

Our book tries to cover, but not limited to, the following topics:

Biomedical knowledge integration
Data-driven view of disease biology
Biological knowledge assembly and interpretation
Human microbiome analysis
Pharmacogenomics
Mining electronic health records in the genomics era
Small molecules and disease

Protein interactions and disease
Network biology approach to complex diseases
Structural variation and medical genomics
Analyses using disease ontologies
Mining genome-wide genetic markers
Genome-wide association studies
Cancer genome analysis
Medical bioinformatics: biomarkers and medical imaging
Neuroinformatics of neurological and psychiatric disorders
Neuroimaging genetics

It is a challenging task that these topics are quite diversified and involved scientists with various expertise. Finally, we tried our best to summarize these diverse topics into five Parts, as in the Contents, with the chapters 2, 6, 10, 14, 16 and 17 edited by Yilong Ma, the chapters 3, 8, 11 and 13 edited by William C.S. Cho, the chapters 5, 6 and 7 edited by Qin Xu, as well as the chapters 1, 4, 9, 11, 12, 14, 15 and 16 edited by Fengfeng Zhou. My assistants Mrs. Ruili Zhao and Ms. Qiuyuan Hu made great efforts in soliciting manuscripts. Mrs. Becky Jinan Zhao from Springer and Mrs. Min Xu and Zhufeng Zhou from the Shanghai Jiao Tong University Press give us a lot of help in formulating this book and applying for funding.

In 2015, we enter the era of "precision medicine," which integrates two major contemporary developments including various omics (e.g., genomics, proteomics and metabolomics) and Big Data. I believe the TBI would play an important role in the endeavor for precision and personalized medicine.

Shanghai, China Dong-Qing Wei
2016-11-13

Contents

Part I
Computer-Aided Drug Discovery

Chapter 1
Drug Discovery

Geetha Ramakrishnan

Abstract An understanding of the process of drug discovery is necessary for the development of new drugs and put into clinical practice, to alleviate the diseases prevalent in modern era. This chapter covers the basic principles of how new drugs can be discovered with emphasis on target identification, lead optimization based on computer-aided drug design methods and clinical trials. The drug design principles in the pharmaceutical industry are explained based on the target and chosen ligand using molecular docking, pharmacophore modelling and virtual screening methods. The drug design is illustrated with specific examples. The clinical trials are necessary to introduce the drugs into market after due validation.

Keywords Lead compound • Computer-aided drug design • Molecular docking • Scoring functions • Virtual screening • Pharmacophore modelling • Quantitative structure-activity relationship (QSAR) • Clinical trials

1.1 Introduction

Drug discovery process deals with the root cause of the disease finding relevant genetic/biological components (i.e. drug targets) to discover lead compounds. Currently specialists in various fields, such as medicine, biochemistry, chemistry, computerized molecular modelling, pharmacology, microbiology, toxicology, physiology and pathology, contribute their research capability to achieve this goal. The drug discovery process (Fig. 1.1) in general is divided into three parts, namely, target identification, lead discovery and clinical trials.

The target identification will normally require a detailed assessment of the pathology of the disease and in some cases basic biochemical research such as study of the basic processes of life, body biochemistry and the use of metabolic analogues; study and exploitation of differences in molecular biology, differential

G. Ramakrishnan (✉)
Department of Chemistry, Sathyabama University, Rajiv Gandhi Salai, Chennai 600 119, Tamil Nadu, India
e-mail: icget2011@gmail.com

© Shanghai Jiao Tong University Press, Shanghai and Springer Science+Business Media Dordrecht 2017

D.-Q. Wei et al. (eds.), *Translational Bioinformatics and Its Application*, Translational Medicine Research, DOI 10.1007/978-94-024-1045-7_1

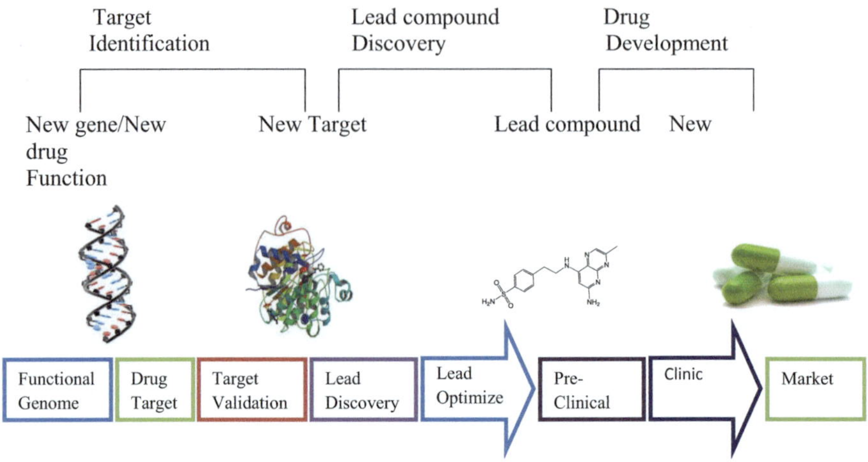

Fig. 1.1 Drug discovery process

cytology, biochemistry and endocrinology; and study of the biochemistry of diseases which will be necessary before initiating a drug design investigation.

The *lead compound* design is the most decisive step in the process of drug discovery. Methods used in lead compound design include folk/ethno-pharmacy and therapeutics, massive pharmacological screening, modification of bioactive natural products, exploitation of secondary or side effects of drugs, an approach through the molecular mechanism of drug action, drug metabolism and chemical delivery systems (Drews 1999, Bodor 1982, 1987). Numerous methods have been invented for the quantification of electronic, hydrophobic and steric effects of functional groups (Franke 1984). Statistical methods, pattern recognition/principal components analysis and cluster analysis can lead to the prediction and optimization of activity and ultimately to the design of newer drugs.

The structure of the proposed lead compound allows the medicinal/organic chemist to prepare the sample by synthetic route, and the lead compound undergoes initial pharmacological and toxicological testing. The selected lead compounds are given to animals for preclinical trials. When the lead compound has been found to be effective and safe in animal testing, it is used for human clinical trials. The lead compound is required to pass three phase clinical trials in human beings. In phase I, studies on healthy subjects are conducted to confirm safety. In phase II, studies are conducted on patients to confirm efficacy. Finally in phase III, large studies on patients are conducted to gather information about safety and efficacy at the population level.

The results of these tests enable the team to decide whether it is profitable to continue development by preparing a series of analogues, measure their activity and correlate the results to determine the drug with optimum activity.

Because of the strict prerequisites of drug authorities, which are becoming ever more demanding, the cost of drug discovery is steadily increasing. Thus, rational drug design becomes the main objective of medicinal chemistry today. Based on rational design, new structures can be developed with a high probability of possessing the required properties and biological activity.

1.1.1 Need for Drug Design

Drug discovery is a time-consuming and costly process. The process takes 12–15 years to release a new drug into market, and average cost for the development of a new drug is about 600–800 million dollars (Adams and Brantner 2006). Among 10,000 drugs that are applied on animals, only ten of them are tested for human clinical trials, in which one or two of the drugs only are put into the market (Hughes 2009). In order to reduce the research timeline and cost, various computational methods were used. The computer-aided drug design process is fast, automatic and less expensive with high success rate and fruitful with respect to intellectual property rights. The problems encountered for this procedure with possible solutions (Kubinyi 1999) are given in Table 1.1.

The strategies to be followed in the drug design include structure-based design of ligands with affinity and selectivity using molecular docking, virtual screening of favourable drug properties and bioavailability and pharmacophore modelling.

1.2 Target Identification

This process involves identification of relevant molecular target based on the known pathology of the disease due to an enzyme, receptor, ion channel or transporter. The next step is to determine the responsible DNA and protein sequence with their function and its mechanism of action (Ryan et al. 2000; Silverman 2004). The mechanism of action can be obtained by the earlier study done on animals as proof and a suitable choice for the target from earlier investigations. Based on the mechanism of drug action, the associated disease and status of the drug are given in Table 1.2.

1.3 Computer-Aided Drug Design

Computer-aided drug design (CADD) is a specialized discipline that uses computational knowledge-based methods to aid the drug discovery process. It is estimated that the computational methods could save up to 2–3 years and $300 million (Price waterhouse coopers 2005). There are several areas where CADD plays an important role in the traditional drug discovery. Genomics and bioinformatics support genetic methods of target identification and validation. Cheminformatics enables

Table 1.1 Problems faced by drug industry with its possible solutions

Sl.No.	Problems	Possible solutions
1.	Target search	Genome information
2.	Target validation	Knockouts, RNA silencing
3.	Lead search	In vitro test models, high-throughput screening
4.	Lead optimization	Parallel syntheses, chemogenomics
5.	Absorption, permeability	Lipinski rules, Caco cells, prodrugs
6.	Metabolism	Liver microsomes
7.	Toxicity	Ames test, hERG models
8.	Drug-drug interactions	CYP inhibition/induction

Table 1.2 Targets with their mechanism, associated disease and status of the drug

Sl. No.	Drug targets	Mechanisms of drug action	Disease	Status of the drug
1.	Enzymes	Reversible and irreversible inhibitors		
	Angiotensin-converting enzyme	Renin-Ang system	Hypertension	Launched
	Tryptase	Phagocytosis	Inflammation, asthma	Clinical phase III
	Cathepsin K	Bone resorption	Osteoporosis	Clinical phase I
2.	Receptors	Agonists and antagonists	Chronic pain	Dopamine, epinephrine, morphine-known drugs
3.	Ion channels	Blocker and opener Ca^{+2}, Na^{+} and K^{+} channel blockers, K^{+} channel openers	Renal Problems	Cyclosporine – launched
4.	Transporters	Uptake inhibitors	H^{+}/K^{+}-ATPase (proton pump)	Omeprazole – as known drug
5.	DNA	Alkylating agents, minor groove binders, intercalating agents	DNA duplication, tumours	Distamycin A, netropsin as known drugs

researchers to process virtual screening for selection of lead compounds for synthesis and screening. This allows researchers to make fast decision on lead compound identification and optimization. In silico ADMET (absorption, distribution, metabolism, excretion and toxicology) modelling aids researchers to identify a bioavailable drug with suitable drug metabolism properties.

CADD methods offer significant benefits for drug discovery. One of them is time and cost savings for lead identification, optimization and ADMET predictions for implementing experimental research. Only the most promising drug candidates will be tested based on the results of CADD. CADD provides deep insight to drug-receptor interactions. Molecular models of drug compounds can reveal intrinsic, atomic scale binding properties that are difficult to envisage. It is classified as structure-based drug design and ligand-based drug design.

1.3.1 Structure-Based Drug Design (SBDD)

The preliminary step in structure-based drug design is to determine the three-dimensional structure of a target molecule (usually protein). This can be achieved by X-ray crystallography or NMR spectroscopy experiments or by approximated computational methods such as comparative modelling (homology modelling uses previously solvated structure as starting point to determine the three-dimensional structure of protein) and ab initio modelling (this method seeks to build three-dimensional protein models based on physical principles rather than previously solved model). The next step in this process is to identify the location of the binding site of a target molecule (receptor). The actual binding site can be located by comparing with known protein-ligand complexes or homology comparisons to related complexes. With well-defined binding site, a ligand (lead) can be determined. Usually, leads can be determined either through de novo design or through large database search for a molecule that matches the binding site. Docking methods are then used to evaluate the quality of ligand.

The molecular docking process mainly involves three steps:

Characterizing the binding site
Positioning the ligand into the binding site
Evaluating the strength of interaction for a specific ligand-receptor complex

Structure-based drug design includes molecular docking methods as a main tool, and certain researchers employ molecular dynamics also, if drug action is known.

1.3.1.1 Molecular Docking

When the structure of protein and its binding site are available, molecular docking techniques are used to identify lead compound. This technique is also used in lead optimization, when modification to known active molecule structure can quickly be tested by CADD before compound synthesis.

Molecular docking is useful in the identification of low-energy binding mode of a molecule or ligand in the active binding site of protein or receptor. A molecule or ligand which binds strongly through hydrogen bonds, van der Waal bonds or any possible electrostatic attractions with receptor or protein associated with disease may inhibit the function and thus acts as a drug. Hydrogen bonds are local electrostatic interaction between the atoms which plays a significant role in recognition of ligand binding with the target. Calculating the accurate protein-ligand interactions is the key principle behind structure-based drug discovery (Cramer et al. 1988).

1.3.1.2 Types of Docking

Three options for docking are available.

Rigid docking – where a suitable position for the ligand in receptor environment is obtained while maintaining its rigidity

Flexible docking – where a favoured geometry for receptor-ligand interaction is obtained by changing internal torsions of ligand into the active site while receptor remains fixed

Full flexible docking – where the ligand is freely rotated via its torsion angles and the side chain of active site residues (selected active site residues within a user-specified radius around the ligand) is freely rotatable.

Most of the docking methods used at the present moment in academic and industrial research employ a rigid target/protein. The algorithms used in docking are given in Appendix I.

The two components of molecular docking are:

(i) Prediction of binding conformation of the ligand in the binding site
(ii) Binding free energy prediction of the ligand (Leach A.R. and Gillet V.J., 2003)

1.3.1.3 Scoring Functions

There are mathematical methods used to predict the strength of the non-covalent interaction called binding affinity between two molecules after docking. The scoring functions have also been developed to predict the intermolecular interaction between two proteins, protein-DNA and protein-drug. The objective of any scoring function is to estimate the free energy change of binding for a ligand in a given binding pose. This can be expressed by the fundamental thermodynamic Eq. (1.1):

$$\Delta G = \Delta H - T\Delta S \tag{1.1}$$

where ΔG is the free energy change of binding, ΔH is the enthalpy change, T is the temperature of the system in Kelvin and ΔS is the entropy change.

Scoring functions are categorized into (i) force field and (ii) empirical (Stahl and Rarey 2001; Perola et al. 2004) (Table 1.3).

Force field scoring functions rely on the molecular mechanics methods. In this method it calculates both the protein-ligand interaction energy and ligand internal energy by van der Waals energy and electrostatic interactions. Advantages of force field-based scoring functions include accounting of solvent, and disadvantages include overestimation of binding affinity and arbitrarily choosing of non-bonded cutoff terms (Kitchen et al. 2004; Moitessier et al. 2008).

Empirical scoring functions – Empirical scoring functions weigh contributions from the different energetic terms in order to make a binding affinity prediction. These terms may include hydrogen bonding using geometric measures as well as force field-based physical potentials. However, the linear weighing of the terms is

Table 1.3 Major docking tools utilized in industrial and academic research institutes

Docking tool	Algorithm/method (Appendix I)	Scoring function
FlexX	Incremental construction	Boehm empirical scoring function
FlexX-Pharm	Incremental construction	Boehm empirical scoring function
Auto Dock	Genetic algorithm	Force filed-based empirical scoring
Dock	Incremental construction	Force filed-based scoring
ICM	Simulated annealing	Force filed-based scoring
GOLD	Genetic algorithm	Empirical knowledge-based scoring
Surflex-Dock	Incremental construction	Empirical Hammerhead scoring
Glide	Simulated annealing/incremental search	Empirical knowledge-based scoring
LigandFit	Shape matching	Empirical knowledge-based scoring

derived from regression methods that fit binding affinity terms to experimental affinities using experimental data and structural information (Teramoto and Fukunishi 2007).

1.3.1.4 Limitations and Challenges

Some key challenges in molecular docking and scoring are discussed based on protein flexibility and role of solvent and scoring function.

Protein flexibility: Docking programmes usually use protein as rigid and ligand as flexible; in this case receptor has one conformation, while the ligands have different conformations. The fundamental goal of virtual screening is to identify molecules with the proper complement of shape, hydrogen bonding and electrostatic and hydrophobic interactions for the target receptor; the complexity of the problem is far greater in reality. For example, the ligand and receptor may exist in different conformations when in free solution, which is different from the conformation when ligand is bound to protein (Koh 2003).

Role of solvent and scoring function: Protein and ligands are surrounded by solvent molecules, usually water. If the water mediation is ignored during docking, then the calculated interaction energy may be low, and favourable interactions with water may be lost (Moitessier et al. 2008). Several methods are now available to predict the binding energy accurately by accounting entropic and solvation effects (Reynolds et al. 1992; Zhang et al. 2001). These methods need greater amount of computational time and inappropriate to use in screening large databases. The molecular docking process is shown in Fig. 1.2.

1.3.2 Ligand-Based Drug Design (LBDD)

The ligand-based drug design starts with a database containing set of ligands with known activity interaction with the same receptor. The first step in this process is to

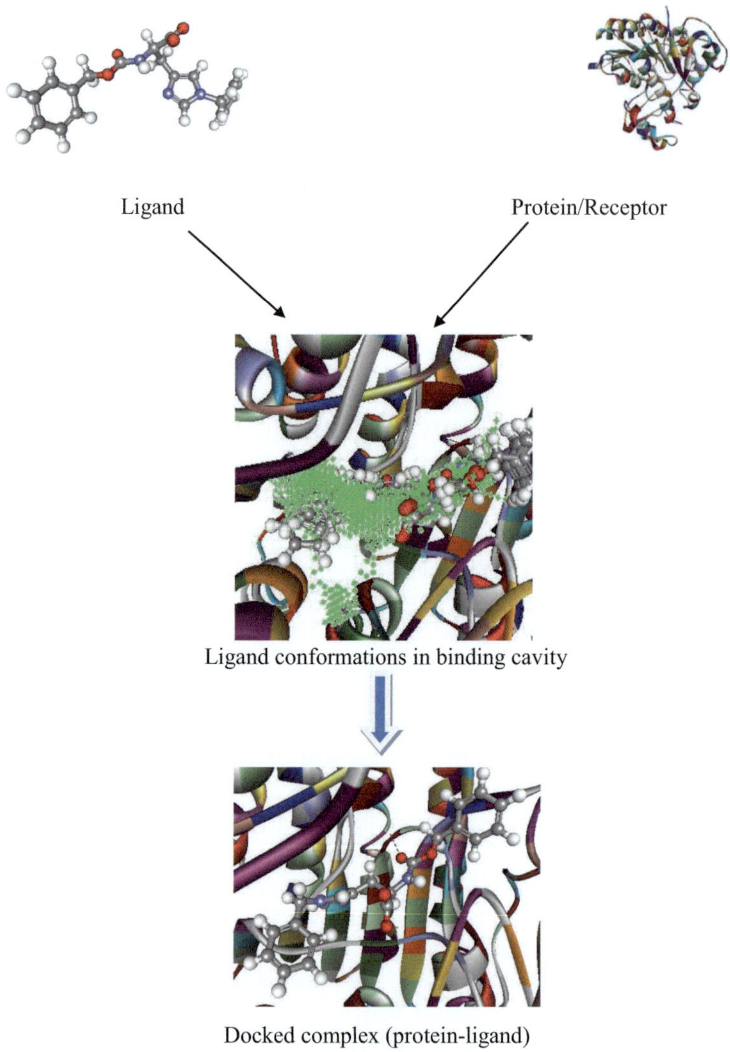

Fig. 1.2 Molecular docking flow chart using a benzamide derivative (MS-275) with HDAC2 protein (Naresh Kandakatla and Geetha Ramakrishnan 2014a, b)

divide the set of ligands into training and test set, and the second step in this process is molecular modelling. Ligand-based approach commonly considers descriptors based on chemistry, shape and electrostatic and interaction points (e.g. pharmacophore points) to assess similarity. A pharmacophore is an explicit geometric hypothesis of the critical features of a ligand (Leach and Gillet 2003). Features usually include hydrogen-bond donors and acceptors, charged groups and hydrophobic patterns. The hypothesis can be used to screen databases for candidate compounds and also can be used to refine existing leads. Another method in ligand-based drug design is quantitative structure-activity relationship (QSAR) modelling method and used for identifying a lead molecule and optimization. The concept of QSAR is based on the fact that the biological properties of a compound can be expressed as functions of its physicochemical parameters. The goal of the QSAR model is to predict the activity of the new molecules (optimized leads). The third step in ligand-based design involves identification of the most promising molecule as lead compound for further experimental investigation.

1.3.2.1 Pharmacophore Modelling

A *pharmacophore* describes a set of interactions required to bind given receptor. The pharmacophore is usually derived from three-dimensional computed conformations of a molecule and is an abstract representation of the molecule.

Common pharmacophore feature types are hydrophobic, hydrogen-bond acceptor, hydrogen-bond donor, aromatic rings and positively ionizable and negatively ionizable groups. The pharmacophore features describe the target binding site, e.g. a hydrophobic feature corresponds to hydrophobic region in the protein and hydrogen-bond acceptor feature as hydrogen bond donating counterpart in the protein. Hydrogen-bond acceptor and donor features usually have direction as parameter. The spatial relationship between the pharmacophore features is defined by interpoint distances between the features.

Pharmacophore modelling is widely used in drug design for identifying novel scaffolds or leads for various targets. Pharmacophore model is classified into two categories as (i) structure-based pharmacophore modelling and (ii) ligand-based pharmacophore modelling.

Structure-Based Pharmacophore Modelling

Structure-based pharmacophore modelling uses a 3D structure of protein co-crystallized with ligand or 3D structure of protein. The structure-based pharmacophore model is further subdivided into two types as protein-ligand complex and protein/receptor without ligand contribution. The protein-ligand-based approach locates the ligand binding sites of the protein target and determines the key interaction points between the protein and ligand. Automated tools for the jobs are LigandScot, Pocket v.2 and GBPM (Wolber and Langer 2005; Chen and Lai 2006; Ortuso et al. 2006). For protein-based approach, Discovery Studio (LUDI) was employed, where LUDI converts the interaction points in the binding site into catalyst pharmacophore features such as H-bond acceptors, H-bond donors and hydrophobe (Bohm 1992). In general structure-based pharmacophore, the generated interaction points consist of a large number of unprioritized pharmacophore features, which complicate further virtual screening process. To overcome this problem, a fast knowledge-based approach, hotspot-guided receptor-based pharmacophores (HS-Pharm) and Apo protein-based approach were used. Hotspot analysis is employed to identify the binding sites, where the ligand forms strong interactions (Barillari et al. 2008). In the second approach, the binding cavity embedded in a GRID and molecular interaction fields of GRID node and protein is calculated using a set of probes; the minimum energy found can be converted into pharmacophore feature (Tintori 2008; Goodford 1985).

Ligand-Based Pharmacophore Modelling

Ligand-based pharmacophore modelling is a key computational strategy in drug discovery in the absence of 3D structure of protein. Pharmacophore model generation extracts common chemical feature from a set of known molecules (usually training set) as a representative of essential interaction between the ligand and target protein of interest. This method involves two steps: the first step involves conformational analysis of training set molecules that allows conformational flexibility of each molecule, and the second step is alignment – aligning of training set molecules to determine the essential common chemical feature to construct pharmacophore models. Currently various commercial and academic computational softwares are available for pharmacophore model development – such as Hip Hop (Barnum et al. 1996), HypoGen (Li et al. 2000) (Accelrys Inc., http://www.accelrys.com), PHASE (Dixon et al. 2006) (Schordinger Inc., http://www.schrodinger.com), MOE (Chemical Computing Group, http://www.chemcomp.com), DISCO (Martin 2000), GASP (Jones and Willet 2000) and GALAHAD (Tripos Inc., http://www.tripos.com). Challenges to overcome are conformational ligand flexibility and molecular alignment. Conformational ligand flexibility problem is solved by computing multiple conformers for each molecule and creating a database. The second method is on-the-fly method, in which the conformational analysis is carried out in the pharmacophore modelling process; it does not need mass storage but requires higher CPU time (Poptodorov et al. 2006). A good conformer should satisfy low-energy configuration which interacts with the receptor. Molecular alignment is another challenging issue in ligand-based pharmacophore modelling. Alignment method can be classified into two categories as point-based and property-based approaches (Wolber et al. 2008). In point-based approach, pair of atoms or fragments or chemical feature points is superimposed using least square fitting. The biggest problem in this approach is to identify anchor points in dissimilar ligands. Property-based approach makes use of molecular descriptors to generate alignment.

Once pharmacophore model is generated, it can be used for virtual screening of small or large databases. Many tools such as ligand-based pharmacophore mapping, search 3D database (Accelrys Inc., http://www.accelrys.com), PHASE (Schordinger Inc., http://www.schrodinger.com), ChemDBS (VLife MDS., http://www.vlifesciences.com/), etc. are available for virtual screening. The full framework of pharmacophore modelling is illustrated in Fig. 1.3.

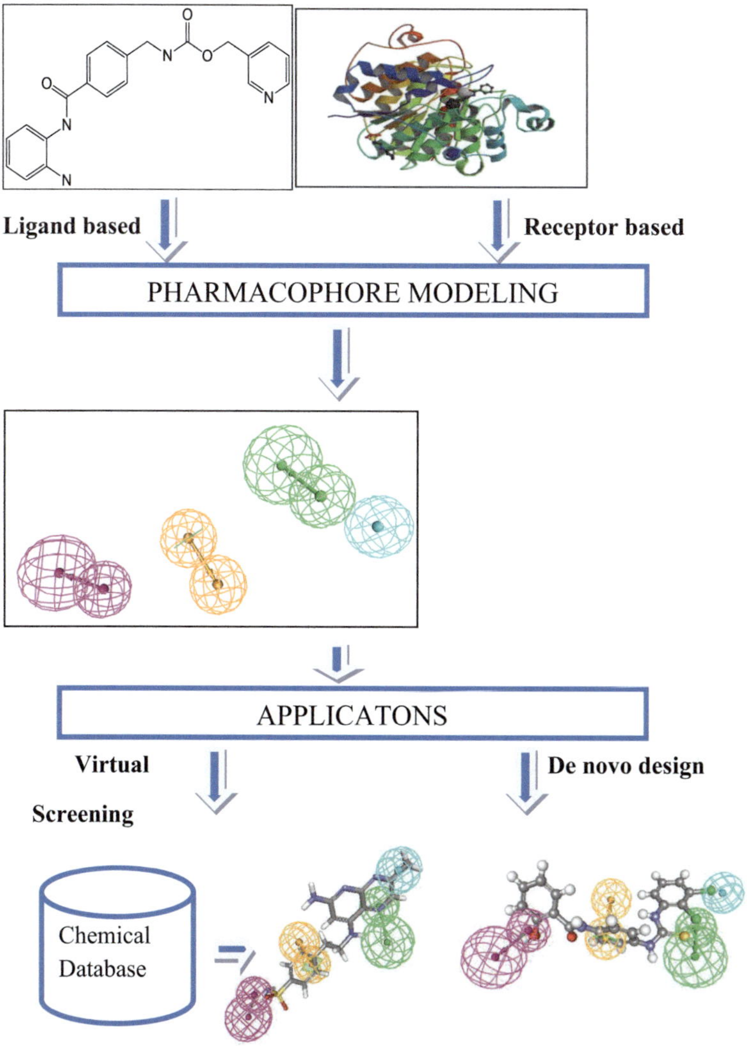

Fig. 1.3 The full framework of pharmacophore modelling

1.3.2.2 Virtual Screening

In silico screening of chemical compound database for identification of novel chemotype is termed as virtual screening. Virtual screening is generally performed on the commercial, public or privately available 2D/3D chemical structural databases. Virtual screening is employed to reduce the number of compounds to be tested in experimental laboratories, thereby focussing on more reliable entities for lead discovery and lead optimization (Rester 2008). The costs and time associated with virtual screening of chemical compounds are significantly lower when compared to screening of compounds in experimental laboratories. Thus virtual screening reduces the size of the haystack by selecting compounds or libraries that are either lead-like or drug-like properties with the potential of oral bioavailability. Virtual screening is divided into two types as (a) ligand-based virtual screening (LBVS) and (b) structure-based virtual screening (SBVS) (refer to Appendix II).

Lipinski Rule

The selection criteria of lead compounds using the rule are referred to as Lipinski analysis (Lipinski et al. 1997).The use of upper and/or lower bounds on quantities such as molecular weight (MW) or logP helps to vary the in vivo properties of drugs. The rule of *5* developed by Lipinski predicts that good cell permeation or intestinal absorption is more probable when there are less than *5* H-bond donors, 10 H-bond acceptors, MW is less than 500 and the calculated logP is lower than *5*. Property ranges for lead-like compounds can be defined: 1–5 rings, 2–15 rotatable bonds, MW less than 400, up to 8 acceptors, up to 2 donors and a logP range of 0.0 to 3.0. The average differences in comparisons between drugs and leads include 2 less rotatable bonds, MW 100 lower and a reduction in logP of *0.5* to 1.0 log units. Thus, one of the key objectives in the identification of lead-like compounds for screening, either by deriving subsets of corporate, or commercial, compound banks or through the design of libraries, is the need for smaller, less lipophilic compounds that, upon optimization, will yield compounds that still have drug-like properties. Figure 1.4 gives the different approaches used in virtual screening process. Further using Lipinski bioavailability rules, neural nets (e.g. drug-like character), pharmacophore analyses, similarity analyses, scaffold hopping and docking and scoring functions, lead compounds can be selected. The example given for selecting the compounds based on the virtual screening method of data bases is illustrated in Sect. 1.3.3.

1.3.2.3 Quantitative Structure-Activity Relationship (QSAR)

In ligand-based drug design, a computational model is needed for further identification of promising molecule as a lead molecule for further experimental investigation. QSAR modelling techniques are used for further lead optimization. It is a mathematical relationship between a biological activity of a molecular system and its geometric and chemical characteristics. QSAR attempts to find consistent relationship between biological activity and molecular properties, so that these "rules" can be used to evaluate the activity of new compounds.

The concept of QSAR was first introduced in 1968 (Selassie et al. 2003), and the model of QSAR is related by the following equation (Crum-Brown and Fraser 1968):

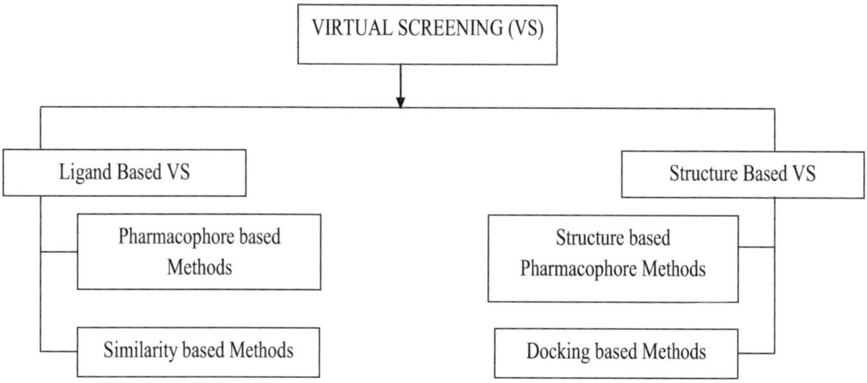

Fig. 1.4 Different approaches to virtual screening process

$$\delta = f(C) \tag{1.2}$$

where the physiological activity δ was expressed as a function of the chemical structure.

Later quantitative approaches combine different physicochemical parameters in a linear additive manner. Free and Wilson proposed structure-activity dependencies by equation

$$AB = u + \Sigma^i a_i x_i \tag{1.3}$$

where AB is the biological activity, u is the average contribution of the unsaturated parent molecule of a particular series (training set compounds), the a_i values are contributions of various structural features and the x_i values denote the presence or absence of particular fragments (Free and Wilson 1964). Since then QSAR has remained a thriving research area in drug design.

More recently developed QSAR modelling approaches include HQSAR (Lowis 1997), inverse QSAR (Cho et al. 1998) and binary QSAR (Gao et al. 1999). The accuracy of QSAR modelling is greatly improved by using sophisticated statistical and machine learning methods, for example, partial least square (PLS) (Dunn and Rogers 1996) and support vector machines (SVM).

QSAR models are regression models used in the chemical and biological sciences; QSAR regression relates a set of physicochemical properties or theoretical molecular descriptors of chemicals to the potency of the biological activity (most often expressed by logarithms of equipotent molar activities) of chemicals. It is a technique that quantifies the relationship between structure and biological data and is useful for optimizing the groups that modulate the potency of the molecule and also predict the activity of newly designed molecules (Hansch 1990).

There are different types of computational methods in QSAR depending upon the data complexity. They are two-dimensional (2D), three-dimensional (3D) and higher methods (Livingstone 2004). 2D QSAR is insensitive to the conformational

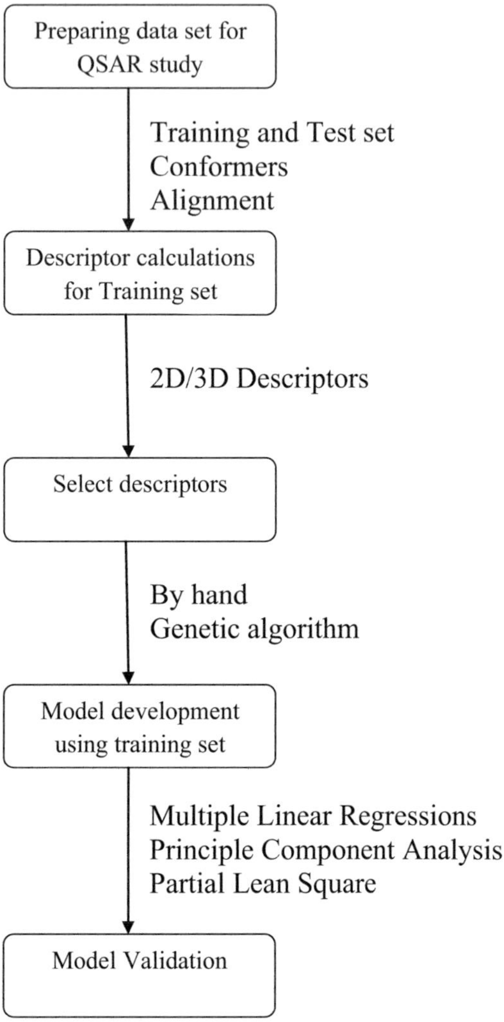

Fig. 1.5 Various stages of QSAR model development

arrangement of atoms in space, while in 3D QSAR needs information on the position of the atoms in three spatial dimensions. In 4D QSAR for each molecule, a set of automatically docked orientations and conformations are developed by genetic algorithms. Induced-fit scenarios of ligands upon binding to the active site and solvation models can be thought of as the fifth (protein flexibility) and sixth (entropy) dimensions in 5D and 6D QSAR, respectively.

The QSAR model development generally is divided into three stages: data preparation, data analysis and model validation. The development of good quality QSAR model depends on many factors like data set and their biological data, selection of descriptors, statistical methods and model validation. The process of QSAR development was given in the flow chart (Fig. 1.5).

The developed models were useful in prediction of untested compounds. In QSAR model development, the main challenge is the selection of data set and group of descriptors, which describes structural physicochemical features associated with the biological activity. The developed QSAR models were validated by (i) cross-validation, (ii) randomization, (iii) bootstrapping and (iv) external validation. The validation methods are needed to establish the predictiveness of a model on unseen data and to help determine the complexity of an equation that the amount of data justifies. The internal validation uses data set that creates model and a separate data set for external validation. Internal methods for validation of models are least square fit (R^2), cross-validation (Q^2), adjusted R^2 (R^2adj), root mean-squared error (RMSE), bootstrapping and scrambling (Y-randomization). The external validation is a best method to validate the model, such as evaluating QSAR model on a test set of compounds. These are statistical methods used to select the best QSAR model.

1.3.3 Illustrated Examples Using CADD

HDAC proteins have been associated with basic cellular events and disease states, including cell growth, differentiation and cancer formation because of their role in gene expression. Several HDAC inhibitors (HDACi) are in clinical trials, namely, benzamide derivatives (Fig. 1.6), hydroxamic acids, cyclic peptides and short-chain fatty acids (Wagner et al. 2010). SAHA (suberoylanilide hydroxamic acid or vorinostat (Zolinza®)) which is structurally similar to trichostatin A (TSA) was the first HDACi approved for the treatment of refractory cutaneous T-cell lymphoma by the Food and Drug Administration (FDA) in October 2006 (Walkinshaw and Yang 2008). SAHA compound inhibits all zinc-dependent HDACs in the low nanomolar range, and recent studies suggested that it has weak inhibitory effect on the class IIa HDACs (Bradley et al. 2009).

Entinostat (SNDX-275, MS-275) belongs to benzamide class HDACi and inhibits HDAC1 and 2, 3 and 9 and has low effect against HDAC4, 6, 7 and 8 (Khan et al. 2007). Entinostat is in phase II clinical trial for treatment of Hodgkin's lymphoma and advanced breast cancer (in combination with aromatase inhibitors) and metastatic lung cancer (in combination with erlotinib). Mocetinostat (MGCD0103) is class I selective HDAC inhibitor and is undergoing phase I and II clinical trials for hematologic malignancies and solid tumours (Blum et al. 2009).

The crystal structure of the HDAC2 protein (PDB ID: 3 MAX) was downloaded from the protein data bank (http://www.rcsb.org/pdb). The crystal structure of histone deacetylase 2 (HDAC2) protein has three chains, which are A, B and C. The reference compounds SAHA and MS-275 (Entinostat) were docked into active sites of all three chains using LigandFit programme in Discovery Studio; out of three chains, chain A has given the best docking score and higher H-bond interactions than chains B and C. The docking score of all three chains with SAHA and Entinostat was shown in Table 1.4. Chain A was selected as active

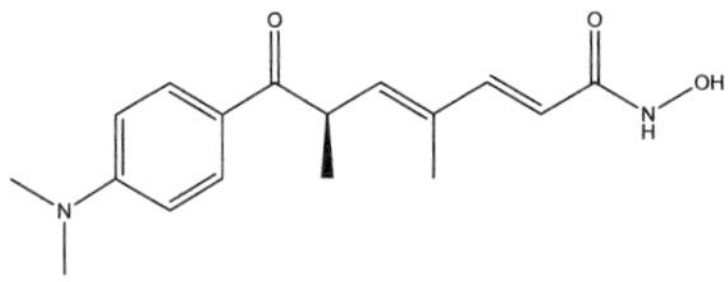

Trichostatin A (TSA)

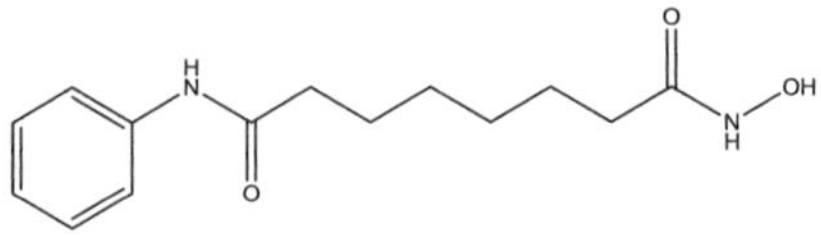

Suberoyl anilide hydroxamic acid (SAHA)

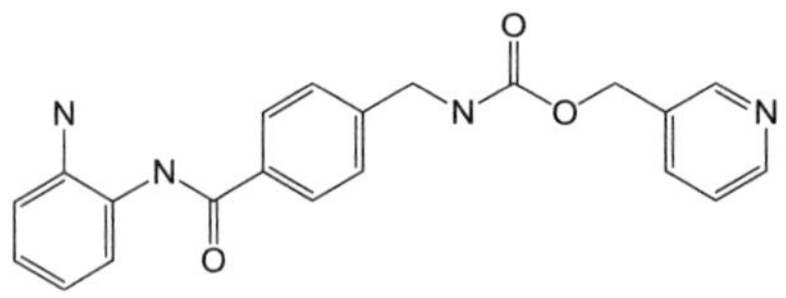

Entinostat (MS-275)

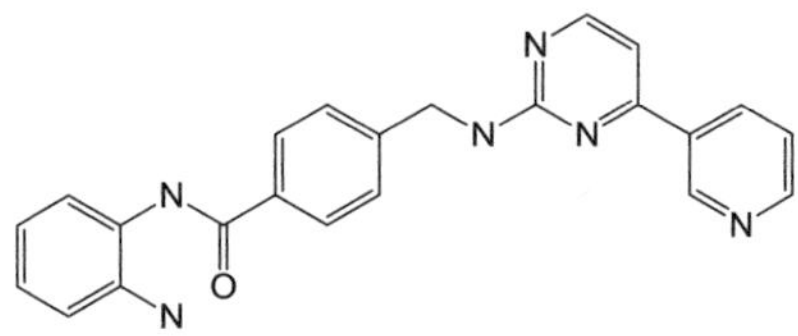

Mocetinostat

Fig. 1.6 Chemical structures of benzamide HDACi

chain, and the optimized benzamide compounds were docked into active site of 3MAX-A. The docking score along with binding orientations and hydrogen bonds were considered for choosing the best pose of the docked compounds. The docking score of the SAHA compound was 40.8 with three hydrogen-bonding interactions with Arg39(2), Gly305 and Gly142(2), and for Entinostat the docking score was 42.6, with four hydrogen-bonding interaction with Arg39, Cys156, Gly305 and His183 and the configurations are given in Fig. 1.7. The designed compounds that scored docking score above than reference compounds with greater interaction with the crucial amino acids were considered as effective HDAC2 inhibitors.

Virtual screening studies were used to find potential lead molecules with increased inhibitory activity against HDAC2 inhibitors. The pharmacophore model Hypo1 (Fig. 1.8) from benzamide compounds was used as 3D query in database screening of the National Cancer Institute (NCI) database containing 265,242 molecules and Maybridge database containing 58,723 molecules. Ligand pharmacophore mapping protocol was used with flexible search option to screen the database. Hit compounds from the database with estimated activity less than 0.1 μM were selected, and further screening of compounds using Lipinski rule of five compounds has (i) molecular weight less than 500, (ii) hydrogen donors less than 5, (iii) hydrogen acceptors less than 10 and (iv) an octanol/water partition coefficient (Log P) value less than 5.

The pharmacophore model development was performed with Discovery Studio (DS) and Schrodinger softwares. Benzamide pharmacophore model was developed by HypoGen algorithm in DS. Hypo1 of HBD, HBA, RA and HY pharmacophore features were selected based on cost difference and correlation coefficient

Table 1.4 The docking score of SAHA and MS-275 with HDAC2 protein

HDAC2 (3MAX)	Chain A		Chain B		Chain C	
	Docking score	H-bond interaction	Docking score	H-bond interaction	Docking score	H-bond interaction
SAHA	40.8	ARG39(2), GLY305, GLY142(2)	22.66	Tyr308, His146, Gly142, Ala141	39.96	Arg39, Gly142
MS-275 (Entinostat)	42.65	Arg39, Cys156, Gly305, His183	39.07	Tyr308, tyr29	36.9	Tyr308, tyr29

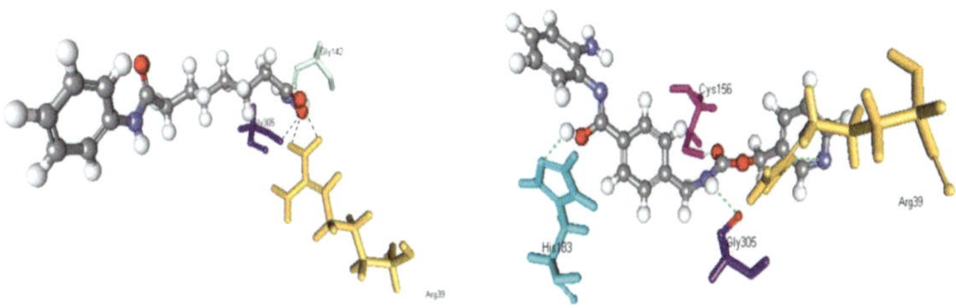

Fig. 1.7 Binding mode of reference compounds SAHA and MS-275

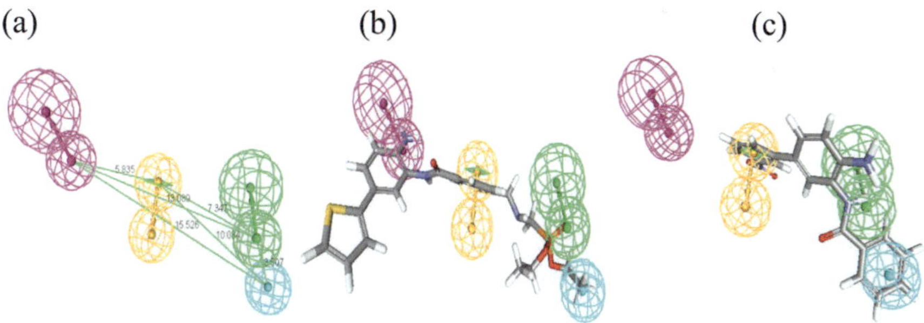

Fig. 1.8 The best pharmacophore model (Hypo1) of HDAC2 inhibitors generated by the HypoGen module: (**a**) the best pharmacophore model Hypo1 represented with distance constraints (Å), (**b**) Hypo1 mapping with one of the active compounds, and (**c**) Hypo1 mapping with one of the least active compound. Pharmacophoric features are coloured as follows: hydrogen-bond acceptor (*green*), hydrogen-bond donor (*magenta*), hydrophobic (*cyan*) and ring aromatic (*orange*) (Naresh Kandakatla and Geetha Ramakrishnan 2014a, b)

(Fig. 1.8). The pharmacophore model can be validated by three methods, such as cost analysis, test set prediction and Fisher's randomization test.

A total of 6130 compounds from NCI and 1379 from Maybridge were mapped using the features of Hypo1. The biological activity IC_{50} (inhibitory concentration

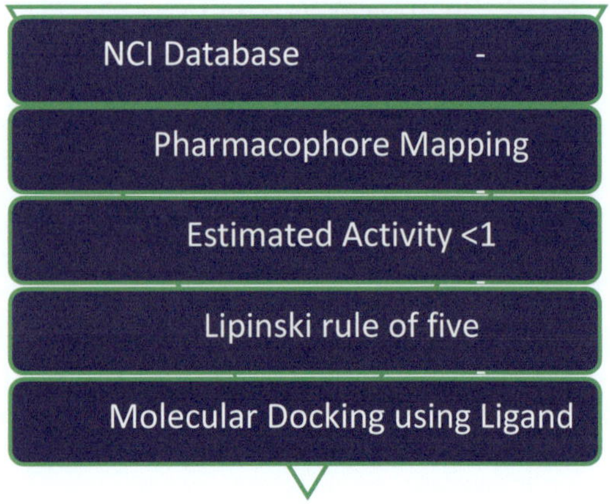

Fig. 1.9 Schematic representation of virtual screening process implemented in the identification of HDAC2 inhibitors

for 50% in μM) was converted to negative logarithmic dose in moles (pIC_{50}) for analysis. The pIC_{50} values of the molecules spanned a wide range from 5 to 8. A total of 1198 and 440 compounds from NCI and Maybridge showed HypoGen estimated value of less than 1 μM for their biological activity and were considered for further studies, and these compounds were screened for Lipinski rule of 5. A total of 625 (382 NCI, 243 Maybridge) compounds obeyed the rule and were subjected to molecular docking studies. The flow chart in Fig. 1.9 was a schematic representation of virtual screening process.

A total of 625 compounds with estimated activity less than 1 μM and favourable Lipinski rule were chosen from NCI and Maybridge databases, and 571 compounds from natural database were subjected to molecular docking studies using LigandFit and LibDock docking programmes. Based on docking score and H-bond interactions, 30 hits were selected from three databases (Naresh Kandakatla and Geetha Ramakrishnan 2014b), and the structure of few of the lead compounds with the respective codes (NSC108392, NSC127064, MFCD01935795, MFCD00830779, ZINC4089202, ZINC4000330) was selected based on structural diversity and stability. These novel compounds can be used for experimental studies for the inhibition of HDAC2 with suitable pharmaceutical formulation.

1.4 Clinical Trials

For a bioactive compound to succeed as a drug, it should pass many selective filters during development like toxicity and in the body including metabolism, uptake, excretion and distribution.

1.4.1 Preclinical Trials

After a lead compound is identified, the medicinal chemist/organic chemist has due interest to prepare them and put into clinical trials. The ability to predict absorption, distribution, metabolism, excretion and toxicology (ADMET) properties from molecular structure has a tremendous impact on the drug discovery process both in terms of cost and the amount of time required to bring a new compound to market. For example, different stereoisomers will exhibit differences in physiochemical properties, such as absorption, metabolism and elimination.

Toxicologists use experimental animals to identify hazardous substances for humans. The main disadvantage is the need for large amounts of substance, several years for the animal studies and relatively expensive. This type of study is of limited value in mechanistic understanding of toxicity. This type of research accounts for 60–65% of the total cost of introduction of a drug into the market. In a nut shell the preclinical activities in the order follows six different sequences as listed below.

Synthesis and purification of the new drug
Pharmacology of the new drug
Pharmacokinetics: absorption, distribution, metabolism, excretion and half-life
Pharmacodynamics: mechanism of action and estimates of therapeutic effects
Toxicology including carcinogenicity, mutagenicity and teratogenicity
Efficacy studies on animals

1.4.2 Human Clinical Trials

To be able to estimate the hazardous risk of humans, additional studies on the mechanism of action, species extrapolation and effects in the low and human-relevant dose range need to be followed. Generally, dose-dependent studies are done for production volume greater than 1000 tons per year in the chemical industry. But drug safety evaluation of pharmaceutical agents is complex as drug exposure to humans is intentional and mechanism of toxicity should be pursued.

An assessment of toxicity requires a broad and interdisciplinary research and development strategy, which includes system biology and case studies on the liver, kidney, cardiovascular, endocrine and in vitro teratogenicity. Further haemotoxicity and peripheral blood cell studies and investigations are done to find their consequences in the drug-induced toxicity (Jurger Borlak 2005).

1.4.3 Types of Clinical Trials

Phase I Trial
In this procedure, how well a drug or procedure can be tolerated in humans acting as healthy volunteers, aged between 18 and 55 years, males and females (however, no females who could be or could become pregnant) of normal weight, no smokers and no alcohol (ab)use will be assessed. The volunteers are given the drug taken with 150 ml water accompanied by standard food, no other therapy and no intake of fruit juices or illegal drugs. The outcome will be to determine a reasonable dose or technique.

Phase II Trial
The phase II trial includes estimation of biological activity or effect (efficacy) and to assess rate of adverse events (toxicity).

Phase III Trial
The phase III trial finds out the effectiveness in comparison to standard treatment or placebo.

Phase IV Trial
Phase IV trial includes long-term surveillance (monitoring) and assesses long-term morbidity and mortality.
Clinical trials provide a systematic framework within which scientific research in human subjects can be carried out efficiently and ethically.

Experimental conclusions are reached in a manner that is statistically defensible.

1.5 Conclusions

Drug discovery process involves target identification, lead compound design and clinical trials. Target identification involves identification of the root cause of the disease. In the case of lead compound selection, virtual screening is a powerful tool to enrich libraries and compound collections. A proper preprocessing of the compound database is of utmost importance in drug design. Further experimental data and theoretical investigations are needed for better p*K*a estimations and better scoring functions. Stepwise procedures (filters, pharmacophore searches, docking and scoring, visual inspection) are most efficient in drug designing. Fragment-based approaches are a promising new strategy in lead structure search and optimization.

The new opportunities in medicinal formulations include genotyping of drug targets and metabolic enzymes which enables cost savings in drug development through better design of clinical trials. The selection of the best drug for a certain patient with individual dose ranges (variance in target sensitivity reduced or increased metabolism) and fewer toxic side effects and drug-drug interactions.

Acknowledgements I acknowledge my institute management and chemistry department faculty for the help rendered during the manuscript preparations. I am also very happy to acknowledge the help given by my student Dr. Naresh Kandakatla in the preparation. I would like to thank my family especially my husband, B. Ramakrishnan, and friends for their kind help.

Appendices

Appendix I

Docking Algorithms

Prediction of correct bound conformation of both protein and ligand is challenging, and this can be achieved by giving proper bound conformation of the protein and prediction of proper bound conformation of the ligand and complex. The problem is the focus of the large majority of docking algorithms though a few incorporate a sampling of receptor conformation as well as optimize the predicted complex coordinates.

Docking algorithms of SBDD have been classified into three types as (a) searching the conformation space during docking, (b) searching conformation space before docking and (c) incremental docking.

The first type of algorithm performs conformation of small molecules and its orientation in the active site. For large chemical databases, it is difficult to do; hence, stochastic algorithms are employed (Taylor et al. 2002).

Monte Carlo (MC) – This method is widely used in stochastic optimization techniques, and it uses sampling technique to generate low-energy conformations. MC simulation makes the ligand position within the binding site through a number of random translational and rotational changes. The standard MC methods generate configuration of system through random Cartesian changes. Each change to the system is evaluated and then rejected or accepted based on a Boltzmann probability. Molecular docking programmes using MC method are AutoDock, ProDock, ICM, MCDOCK, DockVision, QXP and Affinity (Metropolis et al. 1953).

(a) Genetic Algorithm (GA) – GA is one example of evolutionary programming (EP) algorithm. EP is a computational model that takes name and concept from biological process. GA and EP are quite suitable for solving the docking problems because of their usefulness in solving complex optimization problems. In GA, each binding pose of the ligand including its conformations is expressed as a string of values (termed chromosomes). Crossovers are used to generate the new chromosomes, and a complex set of scoring functions are then used to select members within each round of selection. DOCK, GOLD, AutoDock, DIVALI and DARWIN programmes use GA algorithm (Ziemys et al. 2004).
(b) Second-class algorithm – In this method a conformational analysis is carried out first, and all relevant low-energy conformational are then placed in the

binding site. Only the remaining six rotational and translational degrees of freedom of the rigid conformer must be considered. Slide and Fred docking programmes use this docking methodology.

(c) Incremental construction algorithms – The ligand is split in rigid fragments by cutting its rotatable bonds. One of these fragments is termed base fragment, and these fragments are docked rigidly at various positions in the binding site. The largest section is usually selected as the starting fragment and is docked to the receptor. The docked orientation of this fragment is kept and other fragments are added at various orientations and scored. This process is repeated until the entire ligand is assembled. DOCK, FlexX, Hammerhead and HOOK docking programmes use this algorithm.

Taylor RD, Jewsbury PJ, Essex JW. J Comput Aided Mol Des. 2002;16:151–66.
Metropolis N, Rosenbluth AW, Rosenbluth MN, Teller AH, Teller EJ. Chem Phys. 1953;21:1087–92.
Ziemys A, Rimkute L, Kulys J. Nonlinear analysis: modelling and control. 2004;9:373–83.

Appendix II

Virtual Screening Methods: Appendix II

(a) *Ligand-Based Virtual Screening (LBVS)* – In the absence of the 3D structure of receptor information and when one or more active molecules are available, ligand-based virtual screening is used. A common assumption in drug design is that two compounds with similar chemical property also exhibit similar biological effect. This is the main principle and motivation of ligand-based virtual screening. Different methods of LBVS include:

(i) Pharmacophore-based virtual screening (PBVS): When one or more bioactive molecules (usually training set) are available, pharmacophore virtual screening is performed. Developed pharmacophore model from active molecules is taken as template to screen chemical database of unknown compounds for finding compounds with similar chemical features that interact with the target. The hits from the VS are similar to known active molecules, but some might be entirely novel scaffold. The screening process involves two steps as conformational flexibility of molecules and identification of pharmacophore pattern. The conformational flexibility of molecules is handled by either pre-enumerating or on-the-fly method similar to those used in pharmacophore modelling.

(ii) Similarity search: Similarity search is performed when single bioactive compound is available. The basic principle behind this search is that similar molecules have similar bioactivities. Similarity search uses one-, two- and three-dimensional chemical and physical descriptors of molecule to screen chemical database.

LBVS are more limited than SBVS since it uses the properties of known molecule for a given target.

(b) *Structure-Based Virtual Screening (SBVS)* – In the presence of structural information of the target protein, structure-based method is a widely used method to screen the chemical databases. SBVS uses the knowledge of the target protein structure to select the lead compound with which it is likely to interact.

The SBVS workflow involves the following steps:

Step 1 Selection: Selection of the target protein and availability of X-ray crystal structure or NMR structure, if not homology model, chemical compound database and molecular docking software

Step 2 Preparation of target: If the selected target protein is bound with ligand, then it requires preparing binding site of protein by taking ~8–10 Å from the co-crystallized ligand, taking care of significant amino acids for the activity that are included in the binding site.

Step 3 Screening: Screening of chemical databases using molecular docking studies.

Step 4 Results analysis: Results based on the docking score and binding mode of the compound inside the binding cavity.

Step 5 Selection: Visualization of interesting protein-ligand complexes and final selection of compounds for experimental testing.

Glossary

Ligand Any molecule that binds to a biological macromolecule.

Enzyme Endogenous biocatalyst; converts one or several substrate/s into one or several product/s.

Inhibitor Ligand that prevents the binding of a substrate to its enzyme, either in a direct (competitive) or indirect (allosteric) manner, reversibly or irreversibly.

Receptor A membrane-bound or soluble protein or protein complex, which exerts a physiological effect (intrinsic effect), after binding of an agonist, via several steps.

Agonist A receptor ligand that mediates a receptor response (intrinsic effect).

Antagonist A receptor ligand, which prevents the action of an agonist, in a direct (competitive) or indirect (allosteric) manner.

Partial Agonist A (high-affinity) antagonist, which itself has more or less pronounced intrinsic activity.

Ion channel A pore, formed by proteins, that allows the diffusion of certain ions through the cell membrane along a concentration gradient; the channel opening is either ligand- or voltage-controlled.

Transporter A protein, which transports molecules or ions through the cell membrane, against a concentration gradient, under energy consumption.

Pharmacophore A pharmacophorepharmacophore is the ensemble of steric and electronic features that is necessary to ensure the optimal supramolecular interactions with a specific biological target structure to trigger (or to block) its biological response.

ADMET Absorption, distribution, metabolism, excretion and toxicology.

SAHA Suberoylanilide hydroxamic acid.

QSAR Quantitative structure-activity relationship.

CADD Computer-aided drug design.

References

Accelrys Inc. http://www.accelrys.com

Adams CP, Brantner VV. Estimating the cost of new drug development: Is it really $802 million? Health Aff. 2006;25:420–8.

Barillari C, Marcou G, Rognan D. Hot-spots-guided receptor-based pharmacophores (HSP harm): a knowledge-based approach to identify ligand-anchoring atoms in protein cavities and prioritize structure-based pharmacophores. J Chem Inf Model. 2008;48:1396–410.

Barnum D, Greene J, Smellie A, Sprague P. Identification of common functional configurations among molecules. J Chem Inf Comput Sci. 1996;36:563–71.

Blum KA, Advani A, Fernandez L, Van Der Jagt R, Brandwein J, Kambhampati S, Kassis J, Davis M, Bonfils C, Dubay M, Dumouchel J, Drouin M, Lucas DM, Martell RE, Byrd JC. Phase II study of the histone deacetylase inhibitor MGCD0103 in patients with previously treated chronic lymphocytic leukaemia. Br J Haematol. 2009;147(4):507–14.

Bodor N. In: Keverling Bruissman JA, editor. Strategies in drug research. Amsterdam: Elsevier; 1982. p. 137.

Bodor N. In: Mutschler E, Winterfeldt E, editors. Trends in medicinal chemistry. Weinheim: VCH; 1987. p. 195.

Bohm HJ. The computer program LUDI: a new method for the de novo design of enzyme inhibitors. J Comput Aid Mol Des. 1992;6:61–78.

Borlak J. Handbook of toxicogenomics -strategies and applications. Weinhem: Verlag Gmbh& Co., Wiley-VCH; 2005.

Bradley D, Rathkopf D, Dunn R, Stadler WM, Liu G, Smith DC, Pili R, Zwiebel J, Scher H, Hussain M. Vorinostat in advanced prostate cancer patients progressing on prior chemotherapy. Cancer. 2009;115(23):5541–9.

Chemical Computing Group. http://www.chemcomp.com

Chen J, Lai LH. Pocket v.2: further developments on receptor-based pharmacophore modelling. J Chem Inf Model. 2006;46:2684–91.

Cho SJ, Zheng W, Tropsha A. Focus-2d: a new approach to the design of targeted combinatorial chemical libraries. In Pacific symposium on biocomputing; 1998. p. 305–16.

Cramer RD, Patterson DE, Bunce JD. Comparative molecular field analysis (comfa). 1. effect of shape on binding of steroids to carrier proteins. J Am Chem Soc. 1988;110(18):5959–67.

Crum-Brown A, Fraser TR. On the connection between chemical constitution and physiological action. part i. - on the physiological action of the salts of the ammonium bases, derived from strychnia, brucia, thebaia, codeia, morphia, and nicotia. Trans Royal Soc Edinb. 1968;25:151–203.

Dixon SL, Samondyrev AM, Knoll EH, Rao SN, Shaw DE, Friesner RA. PHASE: a new engine for pharmacophore perception, 3D QSAR model development, and 3D database screening. 1. Methodology and preliminary results. J Comput Aid Mol Des. 2006;20:647–71.
Drews J. Quest of tomorrow's medicines. New York: Springer; 1999.
Dunn WJ, Rogers D. Genetic partial least squares in QSAR. London: Academic Press; 1996. p. 109–30.
Franke R. Theoretical drug design methods. In: Nauta T, Rekker RF, editors. Pharmacochemistry library, vol. 7. Amsterdam: Elsevier; 1984.
Free SM, Wilson J. Mathematical contribution to structure-activity studies. J Med Chem. 1964;7:395–9.
Gao H, Williams C, Labute P, Bajorath J. Binary quantitative structure-activity relationship (QSAR) analysis of estrogen receptor ligands. J Chem Inf Model. 1999;39:164–8.
Goodford PJ. A computational procedure for determining energetically favorable binding sites on biologically important macromolecules. J Med Chem. 1985;28:849–57.
Hansch C. In: Ramsden CA, editor. Comprehensive medicinal chemistry, vol. 4. New York: Pergamon Press; 1990. p. 5–8.
Hughes B. Nat Rev Drug Discov. 2009;8:93–6.
Jones G, Willet P. GASP: genetic algorithm superimposition program. In: Gu¨ner OF, editor. Pharmacophore perception, development, and use in drug design. LaJolla: International University Line; 2000. p. 85–106.
Kandakatla N, Ramakrishnan G. Molecular Docking of designed benzamide derivatives as HDAC inhibitors. Int J Pharm Pharm Sci. 2014a;6(4):324–8.
Kandakatla N, Ramakrishnan G. Ligand based pharmacophore modelling and virtual screening studies for identification of novel inhibitors for HDAC2. Adv Bioinf. 2014b;2014:11, Article ID 812148. doi:10.1155/2014/812148.
Khan N, Jeffers M, Kumar S, Hackett C, Boldog F, Khramtsov N, Qian X, Mills E, Berghs SC, Carey N, Finn PW, Collins LS, Tumber A, Ritchie JW, Jensen PB, Lichenstein HS, Sehested M. Determination of the class and isoform selectivity of small molecule HDAC inhibitors. Biochem J. 2007;409(2):581–9.
Kitchen DB, Decornez H, Furr JR, Bajorath J. Docking and scoring in virtual screening for drug discovery: methods and applications. Nat Rev Drug Discov. 2004;3:935–49.
Koh JT. Making virtual screening a reality. Proc Natl Acad Sci U S A. 2003;100:6902–3.
Kubinyi H. Chance favors the prepared mind. From serendipity to rational drug design. J Recept Signal Transduct Res. 1999;19:15–39.
Leach AR, Gillet VJ. An introduction to chemoinformatics. The Netherlands: Kluwer Academic Publishers/Springer; 2003.
Li H, Sutter J, Hoffmann R. HypoGen: an automated system for generating 3D predictive pharmacophore models. In: Gu¨ner OF, editor. Pharmacophore perception, development, and use in drug design. LaJolla: International University Line; 2000. p. 171–89.
Lipinski CA, Lombardo F, Dominy BW, Feeny PJ. Adv Drug Deliv Rev. 1997;23:4–25.
Livingstone DJ. Predicting chemical toxicity and fate. Boca Raton: CRC Press LLC; 2004. p. 151–70.
Lowis D. R., (1997) Hqsar: a new, highly predictive QSAR technique. Technical report.
Martin YC. DISCO: what we did right and what we missed. In: Gu¨ner OF, editor. Pharmacophore perception, development, and use in drug design. LaJolla: International University Line; 2000. p. 49–68.
Moitessier N, Englebienne P, Lee D, Lawandi J, and Corbeil C.R, (2008) Towards the development of universal, fast and highly accurate docking/scoring methods: a long way to go. Br J Pharmacol, vol. 153 Suppl 1, pp. S7–26.
Ortuso F, Langer T, Alcaro S. GBPM: GRID based pharmacophore model. Concept and application studies to protein–protein recognition. Bioinformatics. 2006;22:1449–55.
Perola E, Walters WP, Charifson PS. Proteins: Struct. Funct Bioinf. 2004;56:235–49.

Poptodorov K, Liuu T, Hoffmann RD, Hoffmann H, et al. Pharmacophore model generation software tools. In: Langer T, Hoffmann RD, editors. Pharmacophores and pharmacophore searches. Weinheim: Wiley–VCH; 2006. p. 17–47.

Price waterhouse coopers, price waterhouse coopers pharma. An industrial revolution in r&d. 2005. http://www.pwc.com/gx/eng/about/ind/pharma/industrial_revolution.pdf

Rester U. From virtuality to reality – virtual screening in lead discovery and lead optimization: a medicinal chemistry perspective. Curr Opin Drug Discov Devel. 2008;11:559–68.

Reynolds CA, King PM, Richards WG. Free energy calculations in molecular biophysics. Mol Phys. 1992;76(2):251–75.

Ryan J, Newman A, Jacobs M, editors. The pharmaceutical century. Ten decades of drug discovery. Washington: Supplement to ACS Publications, American Chemical Society; 2000.

Schrodinger Inc. http://www.schrodinger.com

Selassie CD. History of quantitative structure-activity relationships, vol. 6. New York: Wiley; 2003.

Silverman R. The organic Chemistry of drug design and drug action. 2nd ed. Burlington: Elsevier; 2004.

Stahl M, Rarey M. J Med Chem. 2001;44:1035–42.

Teramoto R, Fukunishi H. J Chem Inf Model. 2007;47:526–34.

Tintori C. Targets looking for drugs: a multistep computational protocol for the development of structure-based pharmacophores and their applications for hit discovery. J Chem Inf Model. 2008;48:2166–79.

Tripos Inc. http://www.tripos.com

VLife MDS. http://www.vlifesciences.com/

Wolber G, Langer T. LigandScout: 3-D pharmacophores derived from protein bound ligands and their use as virtual screening filters. J Chem Inf Model. 2005;45:160–9.

Walkinshaw DR, Yang XJ. Histone deacetylace inhibitors as novel anticancer therapeutics. Curr Oncol. 2008;15(5):237–43.

Zhang L, Gallicchio E, Friesner R, Levy R. Solvent models for protein–ligand binding: Comparison of implicit solvent poisson and surface generalized born models with explicit solvent simulations. J Comput Chem. 2001;22(p):591–607.

Wolber G, Seidel T, Bendix F, Langer T. Molecule-pharmacophore super positioning and pattern matching in computational drug design. Drug Discov Today. 2008;13:23–9.

Wagner JM, Hackanson B, Lubbert M, Jung M. Histone deacetylase inhibitors in recent clinical trials for cancer therapy. Clin Epigenetics. 2010;1(3–4):117–36.

Chapter 2
Translational Bioinformatics and Drug Discovery

Pramodkumar Pyarelal Gupta

Abstract With drug pipelines running dry and a slew of blockbuster medicines about to lose patent protection, the voices arguing that the traditional drug development process is too expensive and inefficient to survive are getting louder. To overcome the cost and accelerate the discovery of novel drug, in silico methodologies have made an enormous contribution. This chapter discusses the paradigm of bioinformatics and its translational approaches in drug discovery. Public domain database and efficient data mining approaches are the most optimum criteria for identification and selection of data, whereas genomic technologies such as microarray and next-generation sequencing (NGS) stand for its target identification and validation process. The use of molecular docking and QSAR techniques under the structure- and ligand-based discovery helps in screening the chemical data from nonfunctional to functional ones in terms of activity and toxicity. However, pharmacokinetic and pharmacodynamic (PKPD) simulation can help produce desired concentrations and least side effects with an approximately computed dose regimen.

Keywords Chemical database • Drug discovery • NGS • QSAR • Translational bioinformatics

2.1 Introduction

2.1.1 Translational Bioinformatics

Translational bioinformatics is the evolution of conventional in silico science that deals with storage, analysis, and knowledge extraction from voluminous genomic, proteomic, sequence, and structural data. Translational bioinformatics takes account of research in the development of novel techniques for the integration of

P.P. Gupta (✉)
School of Biotechnology and Bioinformatics, D Y Patil University, Plot 50, Sector 15, CBD Belapur, Navi Mumbai, Maharashtra, India
e-mail: pramodkumar785@gmail.com; pramod.gupta@dypatil.edu

© Shanghai Jiao Tong University Press, Shanghai and Springer Science+Business Media Dordrecht 2017

D.-Q. Wei et al. (eds.), *Translational Bioinformatics and Its Application*, Translational Medicine Research, DOI 10.1007/978-94-024-1045-7_2

clinical and biological data that serves as a source input to designed algorithms and includes the methodology to transform the biological observations into desired knowledge that benefits the scientists, clinicians, and patients that we will see in this chapter. Complicated biological network mechanisms of disease and structure of molecules involved pose several experimental challenges in the drug discovery processes. These complications arise from independent operation of the different parts involved in drug development process with little interaction between clinical practitioners, academic institutions, and pharmaceutical industries (Portela and Soares-da-Silva 2015). Specially, the research in drug development is purpose specific and performed by highly specialized scientists and researchers in their respective fields considering few inputs from clinicians and medical practitioners in strategy design for future therapies (Portela and Soares-da-Silva 2015). Translational research is a road map in which novel therapies will link the experimental discoveries with computational techniques in delivering the clinical needs to the market. Theoretical/computational techniques offer valuable visions in experimental discoveries with pharmacological and pathophysiological mechanisms and virtual development of new prospects in designing and synthesis of novel and better molecular entities with time and cost-effectiveness (Raza 2006).

2.2 Supporting Resources

2.2.1 Online Database

Sequence database such as NCBI, EMBL, or UniProt imparts a mammoth contribution to disease, diagnosis, and drug development industry. Structure database such as Protein Databank incorporates structures evaluated by the 3D crystallography, NMR, and hybrid technology and plays a key role in the structural bioinformatics (Berman 2008). SCOP (Hubbard et al. 1999) and CATH (Oreng et al. 1997) classify the structure on the basis of structural and domain features, whereas PDBsum describes the graphical overview of the deposited 3D structure in a more precise form (Laskowski et al. 1997).

Database that handles reaction and kinetics between the genes, proteins, enzymes, and chemical components with their signal activity is known as metabolic pathway database. MetaCyc (http://metacyc.org) holds experimentally identified biochemical pathways which can be used as a reference data set for the metabolism design and analysis (Zhang et al. 2005). KEGG (http://www.genome.jp/kegg/) is a database for understanding complex functions of the biological system such as cell, organism, and ecosystem by combining the knowledge from genomic and molecular information. KEGG executes a computational representation of the biological system in a wired network diagram (system information) consisting of molecular building blocks of genes and proteins (genomic information) and chemical substances (chemical information) (Kanehisa et al. 2002). The BioCyc database data sets contain a group of organism-specific pathway/genome databases (PGDBs). They provide reference to genome and metabolic pathways of a few thousand organisms (Caspi et al. 2011). BRENDA (BRaunschweig ENzyme DAtabase) is an enzyme database established in 1987 at the Helmholtz Centre for Infection Research, formerly known as German National Research Centre for Biotechnology, and is currently maintained by the Department of Bioinformatics and Biochemistry at the TU Braunschweig. BRENDA is supplemented by enzyme-specific data classified by their biochemical reaction (Scheer et al. 2011). Other databases are also available such as Panther (Thomas et al. 2003), Reactome (Croft et al. 2010), HumanCyc (Miles et al. 2010), Mint (Licata et al. 2012), etc.

2.2.2 *Small Chemical Structure Database*

The online free access chemical databases assist the scientific community in identifying the previous experimental and nonexperimental chemical entities which can be an auxiliary/further tested for similar or different therapeutic applications. Online publically available small chemical structure databases such as PubChem (Bolton et al. 2008), DrugBank (Wishart et al. 2006), ZINC database (Irwin and Shoichet 2005), eMolecules (https://www.emolecules.com/), etc., listed in Table 2.1 regularly share their information on the basis of knowledge exposure. More than thousands of structures are deposited annually in these public databases with millions of compounds tested for known or unknown activities (http://depth-first.com/articles/2011/10/12/sixty-four-free-chemistry-databases/).

2.3 Chemical Data Mining Strategies

The exhaustive and fast designed algorithms compete in the identification of structurally similar compounds. Methodology including structural similarity searching and clustering of small molecules plays an important role in screening of compounds with identical or common scaffold in drug discovery pipelines. To search, analyze, and assemble the diverse compounds from a public database is critical to enable the full utilization of existing resources. However, most of the software in this area is only commercially available, and open source is at high demand with optimum accuracy and precision. The long-term goal of the *ChemmineR* project is to narrow this resource gap by providing free access to a flexible and expandable open-source framework for the analysis of small molecule data from chemical genomics, agrochemical, and drug discovery screens (Cao et al. 2008). Based on screening data from PubChem BioAssay database, Pouliot et al. used reported adverse event data with experimental molecular data and generated a logistic regression model to correlate and predict post-marketing ADRs (Shah and Tenenbaum 2012; Pouliot et al. 2011). In a similar way, an existing data mining

Table 2.1 List of chemical structure database

Sr no	Database	Link
1	ChEMBL	https://www.ebi.ac.uk/chembl/
2	ChemDB/Chemical Search	http://cdb.ics.uci.edu/cgibin/ChemicalSearchWeb.py
3	ChemSpider	http://www.chemspider.com/
4	ChemIDplus	http://chem.sis.nlm.nih.gov/chemidplus/
5	CoCoCo	http://cococo.unibo.it/
6	Comparative Toxicogenomics Database (CTD)	http://ctdbase.org/
7	DNP (Dictionary of Natural Products)	http://dnp.chemnetbase.com/intro/index.jsp
8	DrugBank	http://www.drugbank.ca/
9	e-Drug3D	http://chemoinfo.ipmc.cnrs.fr/MOLDB/index.html
10	GLL (GPCR Ligand Library)	http://cavasotto-lab.net/Databases/GDD/
11	GLIDA (GPCR-Ligand Database)	http://pharminfo.pharm.kyoto-u.ac.jp/services/glida/
12	Glide Fragment Library	http://www.schrodinger.com/Glide/Fragment-Library
13	Glide Ligand Decoys Set	http://www.schrodinger.com/Glide/Ligand-Decoys-Set
14	KEGG DRUG	http://www.genome.jp/kegg/drug/
15	KKB (Kinase Knowledgebase)	http://www.eidogen.com/kinasekb.php
16	Ligand Expo	http://ligand-expo.rutgers.edu/
17	MMsINC	http://mms.dsfarm.unipd.it/MMsINC/search/
18	Mcule database	https://mcule.com/pricing/
19	PubChem	https://pubchem.ncbi.nlm.nih.gov/
20	PubChem Mobile	https://play.google.com/store/apps/details?id=com.bim.pubchem
21	SPRESIweb	http://www.spresi.com/
22	The Cambridge Structural Database (CSD)	https://www.ccdc.cam.ac.uk/solutions/csd-system/components/csd/
23	SuperDrug database	http://bioinf.charite.de/superdrug/
24	TCM	http://tcm.cmu.edu.tw/
25	Virtual Library Repository	http://nbcr.ucsd.edu/wordpress2/
26	ZINClick	http://www.symech.it/index.asp?catID=31&lang=en
27	Zinc database	http://zinc.docking.org/

algorithm was enhanced by using molecular fingerprints with chemical information that codifies the structural features or functional groups to augment the ADE signals generated from adverse event reports (Shah and Tenenbaum 2012; Vilar et al. 2011).

National Cancer Institute (NCI) database is one of the first amalgamated public efforts in distributing the large data sets according to their bioactivity information in a searchable database format for the cancer and HIV research community (Voigt et al. 2001; Ihlenfeldt et al. 2002; Couzin 2003). ChemBank, PubChem, ZINC, and other public databases have also joined the race in screening the database on the basis of structure similarity and biological activity. Online and open-sources software are useful resources in cheminformatics software development (Girke et al. 2005).

Liu et al. (2012) demonstrated the ability to predict adverse drug reactions (ADRs) by integrating chemical, biological, and phenotypic properties of drugs. They showed that data fusion approaches are promising for large-scale ADR predictions in both preclinical and post-marketing phases (Shah 2012).

2.4 Genomic Technologies

The completion of human (*Homo sapiens*) and mouse (*Mus musculus*) genome sequence projects has increased the number of gene annotations and made it possible for bioinformaticians to develop new approaches that help experimental researchers tackle biological problems (Jin et al. 2004).

Microarray technique also known as chip-based technique was launched in the early 1990s which helped the scientists to monitor the expression of many genes concurrently, and this technology became a powerful and gold standard tool for analyzing, studying, and understanding the expression and regulation of a number of genes in parallel (Tavera-Mendoza et al. 2006). Analyzing multiple genes at the same time revealed detailed genomic and proteomic information which may lay the foundation for identification of novel target or receptor. The outputs from the microarray analysis strengthen the translational research in drug discovery and development method by generating the results from chip-based technology. Microarrays have been used to slice up nuclear receptor functions both in normal and disease states, in tissues, and in cell models. Numerous studies on nuclear receptor gene regulation for identification of downstream signaling pathways have been carried out in an experiment (Tavera-Mendoza et al. 2006). In a similar experiment, activation of PPAR is studied in a high cholesterol context trailed by microarray studies and results in a potential target gene of triglyceride-lowering drugs (Tavera-Mendoza et al. 2006; Frederiksen et al. 2004).

2.4.1 Next-Generation Sequencing (NGS)

The main application of sequencing technology is to sequence out biological data from an organism, including molecular cloning, gene identification comparative studies, and evolutionary studies. The first-generation sequencing method such as "Sanger sequencing" has been estimated to cost US$2.7 billion for the Human

Genome Project (HGP), whereas the identical procedure costs only US$1.5 million with the next-generation sequencing (NGS) method (Morini et al. 2015).

In the past few years, the NGS-based procedure has expanded its growth and application by attracting the attention from the most cutting-edge technologies. Technological advancement and increased automation, in the field of benchtop sequencing and high-throughput sequencing, have also decreased the cost and facilitated the use of sequencing technology by laboratories of all sizes involved in studies ranging from plants to human diseases (Benjamin 2015). NGS refers to those DNA sequencing methods that came after capillary-based Sanger sequencing (first generation) back in 2005. Current next-generation DNA and RNA sequencing companies include Illumina (TruSeq, HiSeq), Life Technologies (Ion Torrent, SOLiD), Complete Genomics (DNA nanoball sequencing), 454 Sequencing (pyrosequencing), and Oxford Nanopore Technologies (GridION) (Carlson 2012).

2.4.2 NGS and Personalized Medicine

Sudden cardiac death (SCD) is commonly defined as a natural death from unexplained cardiac causes. Young athlete's community is the most affected group by SCD. The most common factor identified is the adrenergic stress during the competitive sports activity for arrhythmias and SCD in the presence of inherited cardiac disease such as cardiomyopathy, primary arrhythmia syndrome, or vascular diseases. Hence, study and molecular analysis of cardiac channelopathies and cardiomyopathies would allow early diagnosis and prevention of SCD in a significant percentage of young individuals. To gain a fruitful result, one should design an appropriate and well-defined NGS diagnostic protocol and must verify in a validation phase that all the details such as mutation identified in a previous group of individuals by Sanger sequencing method must also be detectable by new advanced sequencing techniques. By contrast, novel variants identified by NGS must also be confirmed by Sanger sequencing to evaluate the reproducibility of the NGS approach (Fig. 2.1) (Morini et al. 2015).

Research published in *Nature Medicine* reports that NGS sequencing has revealed genomic alterations directly associated with clinically available therapeutics or a relevant clinical trial of a targeted therapy in 72% of 24 non-small cell lung cancer (NSCLC) tumors and in 52.5% of 40 colorectal cancer (CRC) tumors. Two novel gene fusions, KIF5B-RET in NSCLC and C2orf44-ALK in CRC, were among the alterations that might be treated by drugs. The fusion of C2orf44 and ALK produces an overexpression of anaplastic lymphoma kinase (ALK), the target of crizotinib (Xalkori), approved for the treatment of ALK-positive NSCLC, which suggests the possibility that ALK-positive CRC patients may respond to ALK-inhibitor treatment (Fig. 2.2) (Carlson 2012).

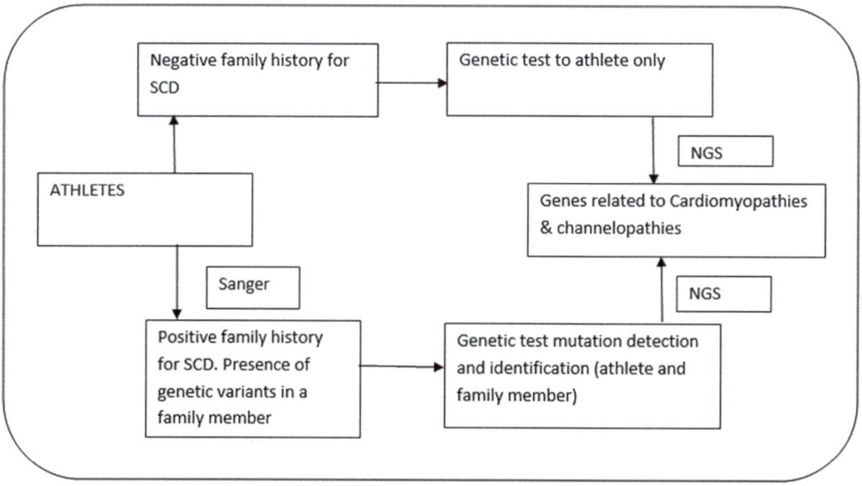

Fig. 2.1 NGS protocol for sudden cardiac death conditions

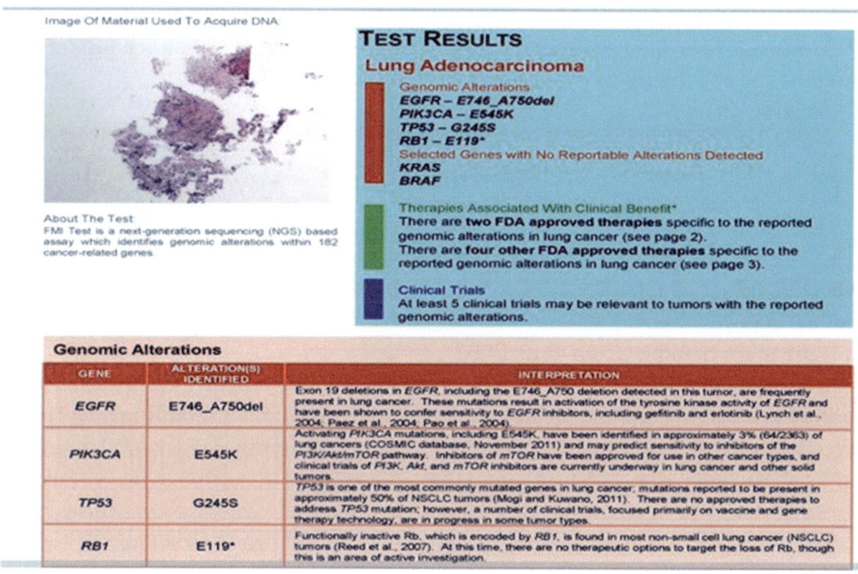
Image Of Material Used To Acquire DNA

About The Test
FMI Test is a next-generation sequencing (NGS) based assay which identifies genomic alterations within 182 cancer-related genes

TEST RESULTS

Lung Adenocarcinoma

Genomic Alterations
EGFR – E746_A750del
PIK3CA – E545K
TP53 – G245S
RB1 – E119*
Selected Genes with No Reportable Alterations Detected
KRAS
BRAF

Therapies Associated With Clinical Benefit*
There are **two FDA approved therapies** specific to the reported genomic alterations in lung cancer (see page 2).
There are **four other FDA approved therapies** specific to the reported genomic alterations in lung cancer (see page 3).

Clinical Trials
At least 5 clinical trials may be relevant to tumors with the reported genomic alterations.

Genomic Alterations

GENE	ALTERATION(S) IDENTIFIED	INTERPRETATION
EGFR	E746_A750del	Exon 19 deletions in *EGFR*, including the E746_A750 deletion detected in this tumor, are frequently present in lung cancer. These mutations result in activation of the tyrosine kinase activity of *EGFR* and have been shown to confer sensitivity to *EGFR* inhibitors, including gefitinib and erlotinib (Lynch et al., 2004; Paez et al., 2004; Pao et al., 2004).
PIK3CA	E545K	Activating *PIK3CA* mutations, including E545K, have been identified in approximately 3% (64/2363) of lung cancers (COSMIC database, November 2011) and may predict sensitivity to inhibitors of the *PI3K/Akt/mTOR* pathway. Inhibitors of *mTOR* have been approved for use in other cancer types, and clinical trials of *PI3K*, *Akt*, and *mTOR* inhibitors are currently underway in lung cancer and other solid tumors.
TP53	G245S	*TP53* is one of the most commonly mutated genes in lung cancer; mutations reported to be present in approximately 50% of NSCLC tumors (Mogi and Kuwano, 2011). There are no approved therapies to address *TP53* mutation; however, a number of clinical trials, focused primarily on vaccine and gene therapy technology, are in progress in some tumor types.
RB1	E119*	Functionally inactive Rb, which is encoded by *RB1*, is found in most non-small cell lung cancer (NSCLC) tumors (Reed et al., 2007). At this time, there are no therapeutic options to target the loss of Rb, though this is an area of active investigation.

Fig. 2.2 Test result for genomic alterations (Carlson 2012)

2.5 Structure-Based Drug Discovery

In recent years, structure-based drug discovery (SBDD) is a rapidly rising methodology in overall drug discovery and development industry. The boom of genomic, proteomic, and related structural data has delivered a number of novel targets and future prospects in lead discovery. In early 1980s the capability of rational drug design with protein structure was an unidentified object to structural biologists. The first success stories of SBDD were published in the early 1990s, and it now becomes an integral and major subject of inquiry in many research and academic organizations (Amy 2003; Roberts et al. 1990; Erickson et al. 1990; Dorsey et al. 1994).

The iterative process of SBDD principally initiates with identification, cloning, purification, and 3D structure determination of the target protein or nucleic acid by any of the following methods: X-ray crystallography, NMR, homology modeling, or various hybrid technologies. Known or calculated active sites are positioned by the computer-based algorithms and targeted by known and unknown 3D chemical compounds, ligands, or drugs identified by specific industry, organization, academic, and research groups from private and public databases. The generated complexes are ranked on the basis of binding energy, pharmacophoric interaction points, and types of interaction such as hydrogen bonding, electrostatic interaction, van der Waals interaction, etc., given in Eq. 2.1. The optimum-generated complexes are then tested with the suitable biochemical assay and knowledge is generated for further evaluation. One with the least micromolar inhibition in in vitro conditions reveals a path to scientists that the compound can be optimized to increase its potency. A repeated cycle of design, synthesis, testing, and evaluation process to a lead compound may produce a patentable market product in binding and specificity to the target (Fig. 2.3) (Amy 2003).

Binding energy:

$$\begin{aligned}\Delta G = & \left(V^{L-L}\text{ bound} - V^{L-L}\text{ Unbound}\right) + \left(V^{P-P}\text{ bound} - V^{P-P}\text{ Unbound}\right) \\ & + \left(V^{P-L}\text{ bound} - V^{P-L}\text{ Unbound} + \Delta S_{\text{conf.}}\ .\ .\ .\right)\end{aligned} \tag{2.1}$$

where P refers to the protein, L refers to the ligand, V represents the pair-wise evaluations mentioned above, and ΔS_{conf} denotes the loss of conformational entropy upon binding (Ruth et al. 2007).

In comparative docking analysis between known and unknown compounds with respect to a common target, ideally, the generated ligand poses (conformations) that are closest to the experimental or known structure conformation should be ranked highest. In order, the analysis could be achieved by quantifying the similarity between a native ligand and a generated ligand pose, where root-mean-square deviation (RMSD) can be calculated between both the ligand structures (Raschka 2014):

$$\text{RMSD}\,(a,b) = \sqrt{\frac{1}{n}\sum\nolimits_{i=1}^{n}\,(a_{ix}-b_{ix})^2 + \left(a_{iy}-b_{iy}\right)^2 + (a_{iz}-b_{iz})^2} \tag{2.2}$$

where a_i refers to the atoms of molecule 1 and b_i to the atoms of molecule 2. The subscripts x, y, and z denote the x-y-z coordinates for every atom.

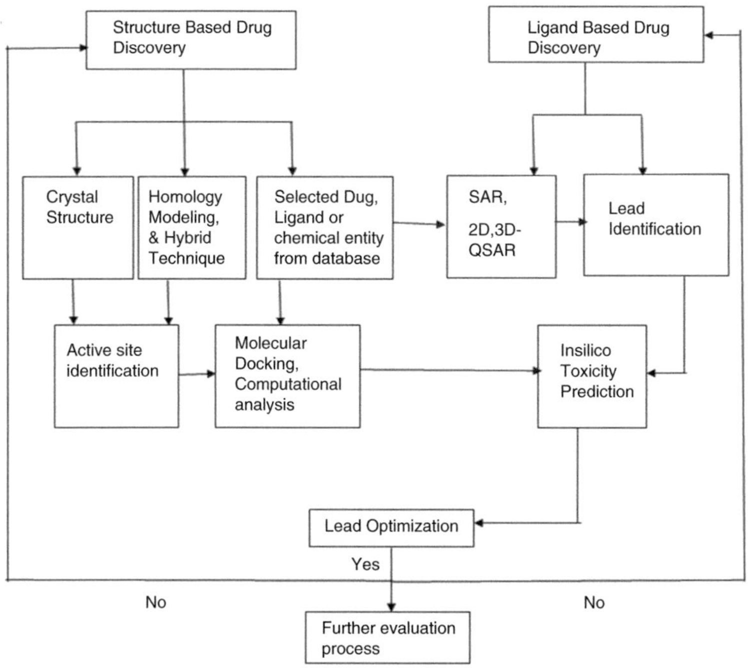

Fig. 2.3 Diagrammatic representation of a structure- and ligand-based drug discovery pipeline

2.5.1 Molecular Docking

The molecular docking is a computational technique to model the interaction between a protein macromolecule known as receptor or target and a small chemical entity/ligand/drug molecule/a protein macromolecule depending on the type of study a scientist carries out. It elucidates the behavior of a ligand molecule with the active site of a receptor protein and its fundamental biochemical process. The docking process involves two basic steps: prediction of ligand conformation within the active site of receptor protein and finally the assessment of binding energies (Meng et al. 2011; McConkey et al. 2002).

Fischer originally proposed a docking mechanism for ligand-receptor binding studies, which is the lock-and-key model, where a ligand fits into a receptor as a key and the receptor behaves as a lock. The primary early docking studies were based on this theory and receptor and ligand were considered as rigid bodies. Koshland put forward an "induced-fit" theory that takes the lock-and-key model a step further and suggests that there is a continuous change in the receptor protein conformation because of the interaction between the ligand and the protein. The theory proposes to treat both ligand and receptor as a flexible entity during docking that could describe the binding events more accurately than under rigid conditions (Fischer 1894; Kuntz et al. 1982; Koshland 1963; Hammes 2002).

Site-specific docking strategies significantly increase the docking efficiency. In many conditions the binding site is unknown. One can predict the putative binding site using commercial software such as SYBYL-X Suite (SYBYL-X-SuiteS: YBYL 8.0), SiteMap – Schrodinger (Halgren 2007), BioPredicta – VLife Molecular Design Suite (MDS) (www.vlifesciences.com), Discovery Studio (Dassault Systèmes BIOVIA 2015), FLEXX (Rarey et al. 1996), Molegro Virtual Docker System (Thomsen and Christensen 2006), ICM-Pro – Molsoft (An et al. 2005), etc. This can also be performed using online servers, e.g., Cast P (Dundas et al. 2006), GRID (Goodford 1985; Kastenholz et al. 2000), POCKET (Levitt and Banaszak 1992), SurfNet (Laskowski 1995; Glaser et al. 2006), PASS (Brady and Jr Stouten 2000), and MMC (Mezei 2003). Docking without any assumption about the binding site is called blind docking.

The main application of molecular docking lies in the structure-based virtual screening for identification of new active compounds for a particular target protein. Molecular docking technique takes a path of translational science and combines the computational output and experimental data in analyzing various biochemical reactions and interactions to study the biological system (Kubinyi 2006; Kroemer 2007; Venhorst et al. 2003; Williams et al. 2003; Meng et al. 2009).

High-throughput screening (HTS) has low rates of success to identify the optimum novel inhibitors of DNA gyrase. Boehm et al. applied de novo design methodology and successfully obtained several new inhibitors (Boehm et al. 2000). Initially, 3D complex structure of DNA gyrase with known inhibitors, ciprofloxacin and novobiocin, was analyzed and patterns of common residual interactions were calculated. Both inhibitors donate one hydrogen bond to Asp 73 and accept one hydrogen bond from a conserved water molecule. In addition, lipophilic fragments are required in the molecule to have lipophilic interaction with the receptor protein. Based on the existing knowledge, LUDI and CATALYST were employed to search and identify similar chemical structure in the Available Chemical Directory (ACD) and Roche Compound Inventory (RIC), resulting in 600 compounds. Close structural analogs of these compounds were considered and 3000 compounds were tested using biased screening. One hundred fifty compounds were selected and clustered into 14 classes of which 7 classes were proved to be the novel and true inhibitor. Succeeding hit optimization was strongly dependent on 3D structures of the binding site and generated a potent DNA gyrase inhibitor (Boehm et al. 2000).

Retinoblastoma (RB), a cancer of the eye, occurs in young children. Researchers have reported their lab findings that fatty acid synthase (FASN) is a promising diagnostic/prognostic and therapeutic target for retinoblastoma. Three inhibitors that target various domains of FASN and are potential anticancer drugs (i.e., cerulenin, triclosan, and orlistat) were considered in the previous studies (Vandhana et al. 2011; Kuhajda et al. 1994; Steven et al. 2004). The experimental data for cerulenin, triclosan, and orlistat gave an IC50 of 3.54 μg/ml, 7.29 μg/ml, and 145.25 μM, respectively, with a dose-dependent decrease in the viability of retinoblastoma cancer cells (Deepa et al. 2010). The crystal structure KS-MAT didomain of human FASN [PDB ID: 3HHD] was also used for docking with cerulenin (Pappenberger et al. 2010) and revealed the binding energy of −5.82 kcal/mol.

As there are no data available for enoyl reductase from human FASN in public database, the crystallized structure of ER domain [PDB ID: 2VZ8] was considered as a template for human ER domain. Furthermore, this model was subjected for docking with triclosan and exhibited a binding energy of −5.73 kcal/mol (Deepa et al. 2010). Pemble et al. considered crystallized 3D complex structure of the human TE domain with orlistat (PDB-ID: 2PX6) in his experiment. Based on the crystal structure, data re-docking was performed using auto dock and binding energy was found to be −2.97 kcal/mol. All these findings have indicated the predictive accuracy of the in silico methods adopted (Pemble et al. 2007).

2.6 Ligand-Based Drug Discovery

The identification of new lead molecule from millions of compound via traditional approach is time consuming and very costly. Since the 1960s, scientists from diverse life science background have put enormous efforts to identify the quantitative parameters that determine the biological activity, in what is known as QSAR/QSPR studies (Nantasenamat et al. 2009). The origin of QSAR was long back in 1863 by Cros in the field of toxicology, where he proposed the relationship between toxicity of primary aliphatic alcohol with their water solubility (Nantasenamat et al. 2009). Crum-Brown and Fraser hypothesized the relationship between chemical constitution and physiological action in 1968 (Crum-Brown and Fraser 1868). A separate discovery was led by Richet (1893), Meyer (1899), and Overton (1901) and showed a linear correlation between lipophilicity (e.g., oil-water partition coefficients) and biological effects (e.g., narcotic effects and toxicity) (Nantasenamat et al. 2009). Hammett (1935, 1937) presented a method to account for substituent effects on reaction mechanisms through the use of an equation which took two parameters into consideration, namely, (i) the substituent constant and (ii) the reaction constant (Nantasenamat et al. 2009; Crum-Brown and Fraser 1868).

Hammett quantified the effect of substituents on any reaction by defining an empirical electronic substituent parameter (σ), which is derived from the acidity constants, Ka's of substituted benzoic acids (Fig. 2.4) (https://web.viu.ca/krogh/chem331/LFER%20Hammett%202012.pdf).

$$\log\left(\frac{KX}{KH}\right) = \rho\sigma \text{ or } pKH - pKX = \rho\sigma \tag{2.3}$$

For the ionization of benzoic acid in pure water at 25 °C (the reference reaction), the constant ρ is defined as 1.00. Thus, the electronic substituent parameter (σ) is defined as

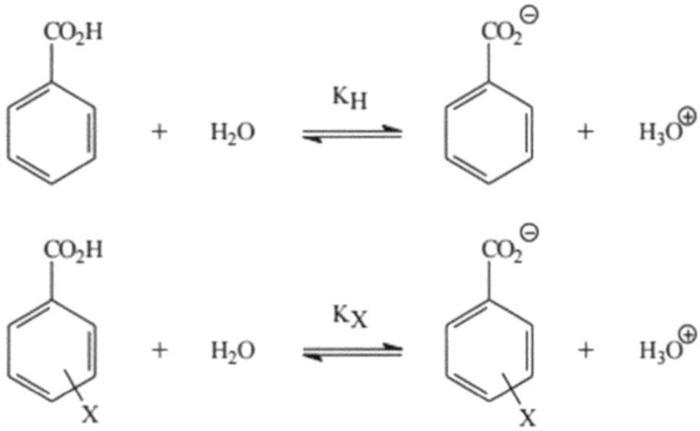

Fig. 2.4 The Hammett equation relates the relative magnitude of the equilibrium constants to a reaction constant ρ and a substituent constant σ Eq. 2.3

$$\sigma = \log\left(\frac{KX}{KH}\right) \tag{2.4}$$

The reaction constant is a measure of how sensitive a particular reaction is to changes in electronic effects of substituent groups (1–5). The reaction constant depends on the nature of the chemical reaction as well as the reaction conditions (solvent, temperature, etc.). Both the sign and magnitude of the reaction constant are indicative of the extent of charge buildup during the reaction progress. Reactions with $\rho > 0$ are favored by electron-withdrawing groups (i.e., the stabilization of negative charge). Those with $\rho < 0$ are favored by electron-donating groups (i.e., the stabilization of positive charge). The greater the magnitude of ρ, the more sensitive the reaction is to electronic substituent effects (Nantasenamat et al. 2009).

In 1956 Taft proposed an approach for separating polar, steric, and resonance effects of substituents in aliphatic compounds (Nantasenamat et al. 2009). In 1964 Hansch and Fujita put forward their linear Hansch equation using the contributions of Hammett and Taft that stood as a mechanistic basis of QSAR/QSPR development. Hansch et al. in late 1960s identified the nonlinear (parabolic) dependence of biological activity with log *P* and gave the following equation:

$$\log(1/C) = a\log P - b\left(\log P^2\right) + c \tag{2.5}$$

where $1/C$ = measure of biological activity, $\log P$ = log of octanol-water partition coefficient, and *a*, *b*, and *c* = regression coefficients (Nantasenamat et al. 2009; Corwin and Toshio 1964).

2.6.1 *Quantitative Structural Activity Relationship (QSAR)*

The discovery of clinically germane inhibitors is a challenging assignment, and the quantitative structural activity relationship (QSAR) methodology has become a very expedient and principally widespread technique for ligand-based drug design and discovery. More than 1000 2D and 3D molecular descriptors are discovered and identified by the scientific community; a few are listed here such as Individual (Mol. Wt, Volume, H-AcceptorCount, H-DonorCount, RotatableBondCount, XlogP, slogp, smr, polarizabilityAHC, and polarizabilityAHP), Retention Index (chi), Atomic valence connectivity index (chiv), Path Count, Chi Chain, Chiv Chain, Chain Path Count, Cluster, Path Cluster, Kappa, Element Count, Dipole Moment, Electrostatic, Distance Based Topological, Estate Numbers, Estate Contributions, Information Theory Index, Semi Empirical, Hydrophobicity XlogpA, Hydrophobicity XlogpK, Hydrophobicity SlogpA, Hydrophobicity SlogpK, and Polar Surface Area (http://www.vlifesciences.com/support/QSAR_Descriptor_Definations_faqs_Answer.php).

2.6.1.1 Model Development

QSAR is among the most extensively used computational technique for ligand-based design, and Bohari et al. have recently reviewed the application of a variety of molecular descriptors like quantum chemical, molecular mechanics, conceptual density functional theory (DFT), and molecular docking-based descriptors for predicting biological activity (Bohari et al. 2011). A summary of relevant data analysis method, regression analysis, and model validation process is provided below along with some examples.

2.6.1.2 Data Analysis Method

Principal components analysis (PCA) and cluster analysis are two widely used methods in 2D and 3D QSAR data analysis. PCA was first invented by Karl Pearson in 1901 and is one of the most popular and primary data reduction techniques. PCA aims at data transformation from large multidimensions to low-dimensional representation, known as data reduction (Pearson 1901; http://www.doc.ic.ac.uk/~dfg/ProbabilisticInference/IDAPILecture15.pdf). Cluster analysis technique is used to partition the data set (with typical molecular properties) into class and categories.

2.6.1.3 Regression Method

Regression analysis is a statistical process for estimating the relationships among dependent and independent variables by the use of modeling techniques implementing on several variables.

Partial least square (PLS) regression technique is used when the number of descriptors (independent variables) is greater than the number of compounds (data points) and/or there are any factors leading to correlations between variables (Martens and Naes 1989; Höskuldsson 1988; Eriksson et al. 2001).

Multiple linear regression (MLR) is an easily interpretable mathematical expression and primary method to construct QSAR/QSPR models, but it often fails in modeling highly correlated data sets. A few new methods have been developed using MLR such as best multiple linear regression (BMLR), heuristic method (HM), genetic algorithm-based multiple linear regression (GA-MLR), stepwise MLR, factor analysis MLR, and so on. Other methods such as self-learning and machine learning algorithms have also been developed to fit the data into the equations such as neural network (NN), support vector machine (SVM), and its variants: least square support vector machine (LS-SVM), grid search support vector machine (GS-SVM), potential support vector machine (P-SVM), and genetic algorithms support vector machine (GASVM) (Liu and Long 2009).

2.6.1.4 2D QSAR (Girgis et al. 2015)

Girgis and his team synthesized a total of 19 dispiro [3H-indole-3,2′-pyrrolidine-3′,3″-piperidines] (Fig. 2.5) of which 11–19 analogs were screened against HeLa (cervical). Compounds 13, 14, and 16 reveal higher potency (IC50 = 4.87, 5.75, and 7.25 μM, respectively) against HeLa (cervical) cell line than the standard reference cisplatin (IC50 = 7.71 μM) (clinically used against cervical carcinoma). See Table 2.2.

Structure–activity relationships (SAR) based on the experimental antitumor activity against HeLa (cervical carcinoma) reveal that the nature of the substituent attached to the phenyl group at C-4′ and consequently the exocyclic olefinic linkage seem to be a controlling factor governing the antitumor properties. Substitution of this phenyl group by fluorine atom enhances the observed antitumor properties more than two chlorine atoms, as exhibited in pairs 11, 13 (IC50 = 16.69, 4.87 μM, respectively) and 12, 14 (IC50 = 12.71, 5.75 μM, respectively) (Tables 2.3 and 2.4).

The basic idea behind QSAR is to generate a relationship between the chemical structure of an organic compound and its physiochemical properties. In the partial pharmacologically active data set mentioned in the present study, external data points were also considered. Spiro-alkaloids with similar scaffold are considered as an external data point and their biological activities were determined, but the same standard technique is earlier followed in the present study.

For the QSAR model development, compounds 11, 13, 15–17, and 19 were considered from Table 2.2 in addition to compounds 20–44 from Table 2.3. Thirty-one derivatives of spiro-alkaloids were used as a training set. The test set (external data set for validation) from synthesized analogs was considered representing high and low potent antitumor active agents 12, 14, and 18 (Table 2.2). Selected compounds geometry is optimized using molecular mechanics force field (MM^+), followed by a semiempirical AM1 method implemented in the Hyperchem. A total of 728 two-dimensional molecular descriptors were calculated using CODESSA-Pro software including constitutional, topological, geometrical, charge-related, semiempirical, molecular-type, atomic-type, and bond-type descriptors for the training set (Table 2.3) and test set (Table 2.4) data. Log property (1/log) and

Fig. 2.5 Synthesized dispiro [3H-indole-3,20-pyrrolidine-30,300-piperidines] derivatives (Girgis et al. 2015)

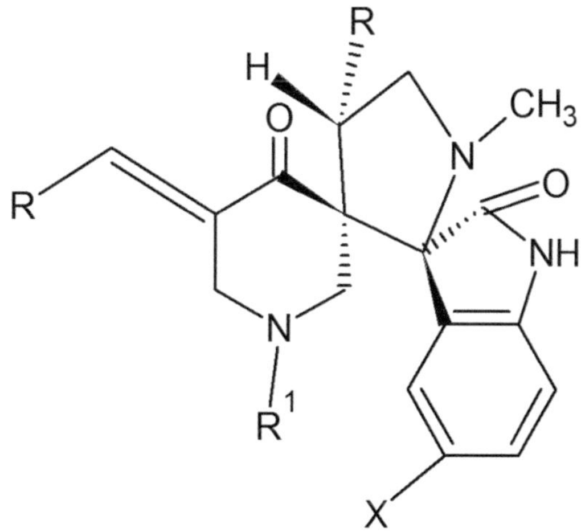

Table 2.2 Antitumor properties of the synthesized compounds 11–19 (tested against HeLa)

No	*R*	*R*[1]	*X*	IC50[a] at, μg/ml (μM) HeLa
11	2,4-$Cl_2C_6H_3$	Et	H	10.27 (16.69)
12	2,4-$Cl_2C_6H_3$	Et	Cl	8.26 (12.71)
13	4-FC_6H_4	Et	H	2.50 (4.87)
14	4-FC_6H_4	Et	Cl	3.15 (5.75)
15	2-Thienyl	Et	H	5.33 (10.89)
16	2-Thienyl	Et	Cl	3.80 (7.25)
17	3-Pyridinyl	Me	H	9.35 (20.08)
18	3-Pyridinyl	Et	H	5.16 (10.76)
19	3-Pyridinyl	Et	Cl	11.58 (22.53)
*	Doxorubicin hydrochloride	–	–	4.19 (7.22)
**	Cisplatin	–	–	4.19 (7.71)

[a]IC50 = concentration required to produce 50% inhibition of cell growth compared to control experimental data
Girgis et al. (2015)
* and ** stands for standard drug

biological activity/IC 50 value were considered for all the training and test sets against HeLa (cervical) cell lines of the training set compounds for the present QSAR modeling.

Best multi-linear regression (BMLR) was utilized which performs a stepwise search for the best *n*-parameter regression equations (where *n* stands for the number of descriptors used), based on the highest R^2 (squared correlation coefficient), $R_{cv}{}^2$OO (squared cross-validation "leave-one-out (LOO)" coefficient), $R_{cv}{}^2$MO (squared cross-validation "leave-many-out (LMO)" coefficient), Fisher statistical significance criteria (*F*) values, and standard deviation (S^2). Statistical characteristics of the QSAR models are presented in Table 2.5.

Table 2.3 Observed and predicated values of training set compounds 11, 13, 15–17, and 19–44 according to the multi-linear QSAR models

					HeLa cervical cell line		
Entry	Comp	*R*	R^1	*X*	Observed IC50 (μM)	Estimated IC50 (μM)	Error
1	11	2,4-Cl2C6H3	Et	H	16.69	12.26	4.43
2	13	4-FC6H4	Et	H	4.87	5.94	1.07
3	15	2-Thienyl	Et	H	10.89	10.48	0.41
4	16	2-Thienyl	Et	Cl	7.25	7.86	0.61
5	17	3-Pyridinyl	Me	H	20.08	26.07	5.99
6	19	3-Pyridinyl	Et	Cl	22.53	20.89	1.64
7	20	Ph	Me	H	6.21	5.92	0.29
8	21	Ph	Me	Cl	5.92	5.41	0.51
9	22	4-ClC6H4	Me	H	6.74	6.3	0.44
10	23	4-ClC6H4	Me	Cl	5.08	5.72	0.64
11	24	4-ClC6H4	Et	Cl	4.96	5.28	0.32
12	25	4-ClC6H4	Me	OMe	5.78	5.9	0.12
13	26	4-ClC6H4	Et	OMe	5.2	5.43	0.23
14	27	4-FC6H4	Me	H	6.51	5.95	0.56
15	28	4-FC6H4	Me	Cl	5.15	5.71	0.56
16	29	4-FC6H4	Me	OMe	5.44	6.21	0.77
17	30	4-H3CC6H4	Me	H	8.64	7.09	1.55
18	31	4-H3CC6H4	Me	Cl	6.65	6.71	0.06
19	32	4-H3CC6H4	Et	Cl	5.55	7.78	2.23
20	33	4-H3CC6H4	Me	OMe	6.96	7.68	0.72
21	34	4-H3COC6H4	Me	H	6.45	7.17	0.72
22	35	4-H3COC6H4	Et	H	7.22	6.54	0.68
23	36	4-H3COC6H4	Me	Cl	11.2	6.53	4.67
24	37	4-H3COC6H4	Et	Cl	8.74	6.27	2.47
25	38	4-H3COC6H4	Me	OMe	6.1	6.94	0.84
26	39	4-H3COC6H4	Et	OMe	5.51	7.84	2.33
27	40	4-Me2NC6H4	Me	Cl	24.36	20.24	4.12
28	41	2-Thienyl	Me	H	8.94	8.18	0.76
29	42	2-Thienyl	Me	Cl	6.86	7.98	1.12
30	43	2-Thienyl	Me	OMe	9.65	10.77	1.12
31	44	5-Methyl-2-furanyl	Me	Cl	9.88	8.46	1.42

Girgis et al. (2015)

Descriptors enlisted in the table are the chief contributors in the model development. Above all Min # HA and # HD molecular-type descriptor explaining the bioactive agent as hydrogen acceptor/donor is important in governing the QSAR model with its t-criterion (9.200) and minimum coefficient with (0.247). The second largest contributing molecular descriptor is FNSA-2 fractional PNSA (PNSA-2/TMSA), which is a charge-related descriptor with t-criterion (5.546)

Table 2.4 Observed and predicated values of external test set compounds 12, 14, and 18 according to the multi-linear QSAR models

					HeLa cervical cell line		
Entry	Comp	*R*	R^1	*X*	Observed IC50 (μM)	Estimated IC50 (μM)	Error
1	12	2,4-Cl2C6H3	Et	Cl	12.71	8.99	3.72
2	14	4-FC6H4	Et	Cl	5.75	5.64	0.11
3	18	3-Pyridinyl	Et	H	10.76	23.7	12.94

Girgis et al. (2015)

Table 2.5 Descriptor of the best multi-linear QSAR model for the HeLa (cervical) tumor cell line active agents

$N = 31$, $n = 3$, $R^2 = 0.815$, $R_{cv}{}^2OO = 0.738$, $R_{cv}{}^2MO = 0.776$, $F = 39.615$, $s^2 = 0.008$					
Entry	ID	Coefficient	s	T	Descriptor
1	0	0.141	0.185	0.763	Intercept
2	D1	0.247	0.027	9.2	Min.(#HA, #HD) (MOPAC PC)
3	D2	0.596	0.107	5.546	FNSA-2 fractional PNSA (PNSA-2/TMSA) (MOPAC PC)
4	D3	0.426	0.096	4.424	HASA-2/SQRT(TMSA) (Zefirov PC) (all)

Girgis et al. (2015)

and has the highest coefficient value of 0.596 controlling the QSAR model that is given by

$$\mathrm{FNSA2} = \frac{\mathrm{PNSA2}}{\mathrm{TMSA}} \tag{2.6}$$

The third and last molecular descriptor of HeLa QSAR is depicted with t-criterion (4.424), and the second most effective parameter controlling the QSAR model based on its coefficient (0.426) is HASA-2/SQRT(TMSA), which is also a charge-related descriptor. The area-weighted surface charge of hydrogen-bonding acceptor atoms (HASA2) is determined by

$$\mathrm{HASA2} = \sum_A \frac{q_A \sqrt{S_A}}{\sqrt{S_{tot}}} \quad A \epsilon\, X_{H-\text{acceptor}} \tag{2.7}$$

2.6.1.5 QSAR Model Validation

The reliability and statistical validity of QSAR model solely depend on the internal and external validation procedures. In the present QSAR model, the internal validation is assessed by CODESSA-Pro technique employing both leave one out (LOO) and leave many out (LMO). The observed correlations from the internal validation are $R_{cv}^2OO = 0.738$ and $R_{cv}^2MO = 0.776$. The squared correlation coefficient of the designed QSAR model is $R^2 = 0.815$, the standard deviation of the regression is $S^2 = 0.008$, and the Fischer test value is $F = 39.615$ that reflects the ratio of the variance explained by the model and the variance due to their errors. The most potent synthesized analog 13, from the training set, exhibited an IC50 of 5.94 μM on the HeLa QSAR model with an experimental value of 4.87 μM and an error of 1.07. The other compounds from the training data set 16, 20–29, 31, 33–35, 38, and 42 relative to cisplatin standard reference clinically used against cervical carcinoma (IC50 = 7.71 μM) showed predicted experimental values with an error range of 0.06–1.12. Compounds 32 and 39 were considered potent analogs against cervical carcinoma (IC50 = 5.55, 5.51) and had predicted values (IC50 = 7.78, 7.84) with a greater error range of 2.23 and 2.33, respectively. Among the mild antitumor active agents against HeLa cell line, compounds 15, 30, 37, 41, 43, and 44 (IC50 range = 8.64–10.89 μM) revealed predicted potency (IC50 range = 6.27–10.77 μM) with a relatively larger error range (0.41–2.47) than the high potent analogs. Among the low potent analogs against HeLa cell lines, compounds 11, 17, 19, 36, and 40 (IC50 range = 11.20–24.36 μM) revealed large deviation in the predicted potency (IC50 range = 6.53–26.07 μM) with an error range of 1.64–5.99 (Table 2.5). From all the above statistical observations, the attained HeLa QSAR model can be considered a good predicative model to produce high potent HeLa antitumor hits compared to those of mild or low potency.

Compounds 12, 14, and 18 were selected for the purpose of validating and examining the predictive ability. The selected test set exhibited experimentally high or low potency against the tested cell line. Table 2.4 reveals the experimental and predicted IC50 values of the test set. Compound 14, considered as high potent against the HeLa cell line relative to the standard reference (cisplatin), had an experimental value of IC50 = 5.75 μM and a predicted value of IC50 = 5.64 μM with a minimum error of 0.11. However, compounds 12 and 18, considered low potent activity against HeLa cell line, had experimental values of IC50 = 12.71 and 10.76 μM and predicted IC50 values of 8.99 and 23.70 μM along with much greater error values of 3.72 and 12.94, respectively.

2.7 Pharmacokinetic and Pharmacodynamic (PKPD) Simulation (Nielsen and Friberg 2013)

Rowland and Tozer state in 2011 that pharmacokinetic (PK) has been defined as "how the body handles the drug" and pharmacodynamic (PD) has been defined as "how the drug affects the body." PK and PD are the vital mechanisms of the modern drug development process. Characterization of PKPD effectively suggests that the concentration that leads to desired effects and least side effects, with an appropriate dose regimen, can be computed.

2.7.1 Pharmacokinetics

Being a central part of clinical pharmacology, PK designates the link between drug dosing and drug concentration-time profile in the body. The determination of drug concentration (C) in plasma and its change from an initial concentration (C_0) with respect to time (t) is given by an exponential function:

$$C(t) = C_0 * e^{-k_e * t} \tag{2.8}$$

Equation 2.8 describes the single PK model with decline in concentration by single distribution phase. Considering the elimination rate for a given system, the change over the time points is directly proportional to the concentration or amount remaining in the system and elimination rate constant (k_e), which is of the first order and has a unit of per time (h^{-1}):

$$\frac{dc}{dt} = -k_e * C \tag{2.9}$$

where k_e is the parameter to be estimated based on the data and is inversely related to half-life ($t_{1/2}$) of the drug. From Eqs. 2.8 and 2.9, it follows that once k_e is known, the drug concentration can be predicted at any time point for a given C_0.

k_e is determined by the apparent volume of distribution (V_d) as well as clearance (CL) that describe the elimination capacity, which is typically governed by liver and kidney function. For a drug with immediate distribution and a CL value independent of concentration, k_e can be described as

$$k_e = \frac{CL}{V_d} \tag{2.10}$$

Often the nature of a drug is more complex because the distribution of the drug inside the body is not immediate due to the effect of its surrounding environment. Hence, the concentration-time course of drug distribution can be better explained by two or more compartments. The differential equations for a two-compartment model can be written as

$$\frac{dA_c}{dt} = -\frac{CL}{V_c} * A_c - \frac{Q}{V_c} * A_c + \frac{Q}{V_p} * A_p \tag{2.11}$$

$$\frac{dA_p}{dt} = -\frac{Q*}{V_p} A_p + \frac{Q}{V_c} * A_C \tag{2.12}$$

where A_c and A_p are the amounts in the central and peripheral compartments and V_c and V_p are the corresponding volumes of distribution. Q represents intercompartmental clearance. An intravenously administered dose would be given into the central compartment.

The total exposure is often described as the area under the concentration-time curve (AUC). AUC is obtained by integrating the drug concentration-time profile and can also be computed as the systemically available dose over *CL*. The bioavailability, *F*, determines the fraction of an extravascular dose that reaches the systemic circulation and is thereby a measure of the extent of absorption. The rate of absorption is often characterized by a first-order rate constant, k_a.

2.7.2 Pharmacodynamics

Pharmacodynamics/PD designates the association among concentration and both the desired and undesirable effects by the given drug. The mathematical function describing the PKPD relationship is a sigmoidal. $E_{\max}$ model given by

$$E(t) = E_0 + \frac{E_{max} * C\ (t)^{\gamma}}{EC_{50}^{y} + C\ (t)^{\gamma}} \tag{2.13}$$

where $E_{\max}$ is the maximum effect that can be achieved by the drug in the investigated system and EC50 is the drug concentration that results in half of the maximum effect. EC50 is inversely related to the potency. γ is the Hill or sigmoidicity factor that determines the steepness of the relationship but is in many cases not statistically significant from 1.

However, there are often situations where sufficiently high concentrations cannot be achieved to estimate $E_{\max}$, and simplifications can be made to estimate fewer parameters. When C « EC50, the $E_{\max}$ model collapses to a linear model ($\gamma = 1$) or a power function ($\gamma \neq 1$) with coefficient slope as shown below:

$$E\ (t) = E_0 + \text{Slope} * C(t)^{\gamma} \tag{2.14}$$

The underlying E_0 is not always constant over the study period. For example, the effect variable may vary because of an underlying disease, such as fluctuations in glucose in the event of diabetes or a diurnal rhythm in blood pressure.

2.8 Conclusion

Translational science in bioinformatics and drug discovery provides a powerful method especially when used as a tool within an armamentarium for discovering new target, drug leads, and novel approach in diagnostic and treatment for the betterment of society. Genomic technologies and NGS methods have proven to be the keystone of advanced research. The identification of genes' role in disease and disorder makes it possible to design personalized medicine approach, where a single or a few genes can be targeted or may act as a biomarker in the diagnosis and treatment of disease and disorder. Data from public domain chemical libraries selected for appropriate target with structure-based and ligand-based discovery can create a very promising lead which may continue to clinical trials. Simulation study of pharmacokinetic and pharmacodynamic behavior of a chemical compound helps us estimate the concentration and dose value in computed form that can significantly reduce the overconcentration and dosing effects. As bioinformatics develops further, it is expected that genomics, proteomics, drug discovery, and computational power will continuously explode with new advances in therapeutic applications; new targets and leads may be brought to marketplace more rapidly each year.

References

Amy CA. The Process of structure-based drug design. Chem Biol. 2003;10:787–97. doi:10.1016/j.chembiol.2003.09.002.

An J, Totrov M, Abagyan R. Pocketome via comprehensive identification and classification of ligand binding envelopes. Mol Cell Proteomics. 2005;4:752–61. doi:10.1074/mcp.M400159-MCP200.

Benjamin B. Next generation sequencing and translational research: from bench to bedside. 2015. http://www2.mlo-online.com/features/201208/lab-management/next-generation-sequencing-and-translational-research-from-bench-to-bedside.aspx. Accessed on 10 Sept 2015.

Berman HM. The protein data bank: a historical perspective. Acta Crystallogr Sect A Found Crystallogr. 2008;A64(1):88–95. doi:10.1107/S0108767307035623.

Boehm HJ, Boehringer M, Bur D, Gmuender H, Huber W, Klaus W, et al. Novel inhibitors of DNA gyrase: 3D structure based biased needle screening, hit validation by biophysical methods, and 3D guided optimization. A promising alternative to random screening. J Med Chem. 2000;43:2664–74. doi:10.1021/jm000017s.

Bohari MH, Srivastava HK, Sastry GN. Analogue-based approaches in anti-cancer compound modelling: the relevance of QSAR Models. Org Med Chem Lett. 2011;1:3. doi:10.1186/2191-2858-1-3.

Bolton E, Wang Y, Thiessen PA, Bryant SH. PubChem: integrated platform of small molecules and biological activities. Chapter 12. In: Wheeler RA, Spellmeyer DC, editors. Annual reports in computational chemistry. Oxford: Elsevier; 2008. p. 217–41. doi:10.1016/S1574-1400(08)00012-1.

Brady GP, Jr Stouten PF. Fast prediction and visualization of protein binding pockets with PASS. J Comput Aided Mol Des. 2000;14:383–401. doi:10.1023/A:1008124202956.

Cao Y, Charisi A, Cheng L-C, Jiang T, Girke T. ChemmineR: a compound mining framework for R. Bioinformatics. 2008;24:1733–4. doi:10.1093/bioinformatics/btn307.

Carlson B. Next generation sequencing: the next iteration of personalized medicine: next generation sequencing, along with expanding databases like the cancer genome atlas, has the potential to aid rational drug discovery and streamline clinical trials. Biotechnol Healthc. 2012;9(2):21–5.

Caspi R, Altman T, Dreher K, Fulcher CA, Subhraveti P, Keseler IM, et al. The Meta Cyc database of metabolic pathways and enzymes and the Bio Cyc collection of pathway/genome databases. Nucleic Acids Res. 2011;40:742–53. doi:10.1093/nar/gkr1014.

Corwin H, Toshio F. p-σ-π analysis. a method for the correlation of biological activity and chemical structure. J Am Chem Soc. 1964;86:1616–26. doi:10.1021/ja01062a035.

Couzin J. NIH dives into drug discovery. Science. 2003;302:218–21. doi:10.1126/science.302.5643.218.

Croft D, O'Kelly G, Wu G, Haw R, Gillespie M, Matthews L, et al. Reactome: a database of reactions, pathways and biological processes. Nucleic Acids Res. 2010;39:691–7. doi:10.1093/nar/gkq1018.

Crum-Brown A, Fraser TR. On the connection between chemical constitution and physiological action. Pt 1. On the physiological action of the salts of the ammonium bases, derived from Strychnia, Brucia, Thebia, Codeia, Morphia, and Nicotia. T Roy Soc Edin. 1868;25:151–203.

Dassault Systèmes BIOVIA. Discovery studio modeling environment, Release 4.5. San Diego: Dassault Systèmes; 2015.

Deepa PR, Vandhana S, Muthukumaran S, Umashankar V, Jayanthi U, Krishnakumar S. Chemical inhibition of fatty acid synthase: molecular docking analysis and biochemical validation in ocular cancer cells. J Ocul Biol Dis Infor. 2010;3:117–28. doi:10.1007/s12177-011-9065-7.

Dorsey BD, Levin RB, McDaniel SL, Vacca JP, Guare JP, Darke PL, et al. L-735,524: the design of a potent and orally available HIV protease inhibitor. J Med Chem. 1994;37:3443–51. doi:10.1021/jm00047a001.

Dundas J, Ouyang Z, Tseng J, Binkowski A, Turpaz Y, Liang J. CASTp: computed atlas of surface topography of proteins with structural and topographical mapping of functionally annotated residues. Nucleic Acids Res. 2006;34(Web Server issue):W116–8. doi:10.1093/nar/gkl282.

Erickson J, Neidhart D, VanDrie J, Kempf D, Wang X, Norbeck D, et al. Design, activity and 2.8 A° crystal structure of a C2 symmetric inhibitor complexed to HIV-1 protease. Science. 1990;249:527–33. doi:10.1126/science.2200122.

Eriksson L, Johansson E, Kettaneh-Wold N, Wold S. Chapter 18, process analytical technology (PAT) and quality by design (QbD) multi- and megavariate data analysis: principles and applications. Umetrics: Umeå; 2001.

Fischer E. Einfluss der configuration auf die wirkung der enzyme. Ber Dt Chem Ges. 1894;27:2985–93.

Frederiksen KS, Wulff EM, Sauerberg P, Mogensen JP, Jeppesen L, Fleckner J. Prediction of PPAR-α ligand-mediated physiological changes using gene expression profiles. J Lipid Res. 2004;45:592–601. doi:10.1194/jlr.M300239-JLR200.

Girgis AS, Panda SS, Aziz MN, Steel PJ, Dennis Hall C, Katritzky AR. Rational design, synthesis, and 2D-QSAR study of anti-oncological alkaloids against hepatoma and cervical carcinoma. RSC Adv. 2015;5:28554–69. doi:10.1039/C4RA16663A.

Girke T, Cheng L-C, Raikhel N. ChemMine. A compound mining database for chemical genomics1. Plant Physiol. 2005;138:573–77. doi: http://dx.doi.org/10.1104/pp.105.062687

Glaser F, Morris RJ, Najmanovich RJ, Laskowski RA, Thornton JM. A method for localizing ligand binding pockets in protein structures. Proteins. 2006;62:479–88. doi:10.1002/prot.20769.

Goodford PJ. A computational procedure for determining energetically favorable binding sites on biologically important macromolecules. J Med Chem. 1985;28:849–57. doi:10.1021/jm00145a002.

Halgren T. New method for fast and accurate binding-site identification and analysis. Chem Biol Drug Des. 2007;69:146–8. doi:10.1111/j.1747-0285.2007.00483.x.

Hammes GG. Multiple conformational changes in enzyme catalysis. Biochemistry. 2002;41(26):8221–8. doi:10.1021/bi0260839.

Hammett LP. Some relations between reaction rates and equilibrium constants. Chem Rev. 1935;17:125–36.

Hammett LP. The effect of structure upon the reactions of organic compounds. Benzene derivatives. J Am Chem Soc. 1937;59:96–103.

Höskuldsson A. PLS regression methods. J Chemomet. 1988;2:211–28. doi:10.1002/cem.1180020306.

http://depth-first.com/articles/2011/10/12/sixty-four-free-chemistry-databases/. Accessed on 25 Aug 2015.

http://www.vlifesciences.com/support/QSAR_Descriptor_Definations_faqs_Answer.php. Accessed on 22 Sept 2015.

https://web.viu.ca/krogh/chem331/LFER%20Hammett%202012.pdf.
https://www.emolecules.com/. Accessed on 14 Sept 2015.
Hubbard TJ, Ailey B, Brenner SE, Murzin AG, Chothia C. SCOP: a structural classification of proteins database. Nucleic Acids Res. 1999;27(1):254–6. doi:10.1093/nar/27.1.254.
Ihlenfeldt WD, Voigt JH, Bienfait B, Oellien F, Nicklaus MC. Enhanced CACTVS browser of the open NCI database. J Chem Inf Comput Sci. 2002;42:46–57. doi:10.1021/ci010056s.
Irwin JJ, Shoichet BK. J Chem Inf Model. 2005;45(1):177–82. doi:10.1021/ci049714+.
Jin VX, Leu Y-W, Liyanarachchi S, Sun H, Fan M, Nephew KP, et al. Identifying estrogen receptor a target genes using integrated computational genomics and chromatin immunoprecipitation microarray. Nucleic Acids Res. 2004;32:6627–35. doi:10.1093/nar/gkh1005.
Kanehisa M, Goto S, Kawashima S, Nakaya A. The KEGG databases at GenomeNet. Nucleic Acids Res. 2002;30(1):42–6. doi:10.1093/nar/30.1.42.
Kastenholz MA, Pastor M, Cruciani G, Haaksma EE, Fox T. GRID/CPCA: a new computational tool to design selective ligands. J Med Chem. 2000;43:3033–44. doi:10.1021/jm000934y.
Koshland Jr DE. Correlation of structure and function in enzyme action. Science. 1963;142:1533–41. doi:10.1126/science.142.3599.1533.
Kroemer RT. Structure-based drug design: docking and scoring. Curr Protein Pept Sci. 2007;8:312–28.
Kubinyi H. Success stories of computer-aided design. In: Ekins S, Wang B, editors. Computer applications in pharmaceutical research and development, Wiley series in drug discovery and development. New York: Wiley-Interscience; 2006. p. 377–424.
Kuhajda FP, Jenner K, Wood FD, Hennigar RA, Jacobs LB, Dick JD, et al. Fatty acid synthesis: a potential selective target for antineoplastic therapy. Proc Natl Acad Sci. 1994;91:6379–83. doi:10.1073/pnas.91.14.6379.
Kuntz ID, Blaney JM, Oatley SJ, Langridge R, Ferrin TE. A geometric approach to macromolecule-ligand interactions. J Mol Biol. 1982;161:269–88. doi:10.1016/0022-2836(82)90153-X.
Laskowski RA. SURFNET: a program for visualizing molecular surfaces, cavities, and intermolecular interactions. J Mol Graph. 1995;13:323–30. doi:10.1016/0263-7855(95)00073-9.
Laskowski RA, Hutchinson EG, Michie AD, Wallace AC, Jones ML, Thornton JM. PDBsum: a web-based database of summaries and analyses of all PDB structures. Trends Biochem Sci. 1997;22(12):488–90. doi:10.1016/S0968-0004(97)01140-7.
Lecture 15: Principal component analysis. DOC493: intelligent data analysis and probabilistic inference lecture. http://www.doc.ic.ac.uk/~dfg/ProbabilisticInference/IDAPILecture15.pdf. Accessed on 18 Aug 2015.
Levitt DG, Banaszak LJ. POCKET: a computer graphics method for identifying and displaying protein cavities and their surrounding amino acids. J Mol Graph. 1992;10:229–34. doi:10.1016/0263-7855(92)80074-N.
Licata L, Briganti L, Peluso D, Perfetto L, Iannuccelli M, Galeota E, et al. MINT, the molecular interaction database: 2012 update. Nucleic Acids Res. 2012;40(Database issue):D857–61. doi:10.1093/nar/gkr930.
Liu P, Long W. Current mathematical methods used in QSAR/QSPR studies. Int J Mol Sci. 2009;10:1978–98. doi:10.3390/ijms10051978.
Liu M, et al. Large-scale prediction of adverse drug reactions by integrating chemical, biological, and phenotypic properties of drugs. J Am Med Inform Assoc. 2012;19(e1):e28–35. doi:http://dx.doi.org/10.1136/amiajnl-2011-000699.
Martens H, Naes T. Multivariate calibration. Chichester: Wiley; 1989.
McConkey BJ, Sobolev V, Edelman M. The performance of current methods in ligand-protein docking. Curr Sci. 2002;83:845–85.
Meng XY, Zheng QC, Zhang HX. A comparative analysis of binding sites between mouse CYP2C38 and CYP2C39 based on homology modeling, molecular dynamics simulation and

docking studies. Biochim Biophys Acta. 2009;1794:1066–72. doi:10.1016/j.bbapap.2009.03.021.

Meng XY, Zhang HX, Mezei M, Cui M. Molecular docking: a powerful approach for structure-based drug discovery. Curr Comput Aided Drug Des. 2011;7:146–57. doi:10.2174/157340911795677602.

Meyer H. Zur Theorie der Alkoholnarkose. Arch Exp Pathol Pharm. 1899;42:109–18.

Mezei M. A new method for mapping macromolecular topography. J Mol Graph Model. 2003;21:463–72. doi:10.1016/S1093-3263(02)00203-6.

Miles T, Tomer A, Carol AF, Ron C, Markus K, Suzanne P, et al. Beyond the genome (BTG) is a (PGDB) pathway genome database: HumanCyc. Genome Biol. 2010;11(Suppl 1):O12. doi:10.1186/gb-2010-11-s1-o12.

Molecular docking, estimating free energies of binding, and AutoDock's semi-empirical force field – written by Sebastian Raschka July 26, 2014. http://sebastianraschka.com/Articles/2014_autodock_energycomps.html. Accessed on 28 Sept 2015.

Morini E, Sangiuolo F, Caporossi D, Novelli G, Amati F. Application of next generation sequencing for personalized medicine for sudden cardiac death. Front Genet. 2015;6:55. doi:10.3389/fgene.2015.00055.

Nantasenamat C, Isarankura-Na-Ayudhya C, Thanakorn Naenna T, Prachayasittikul VA. Practical overview of quantitative structure-activity relationship. EXCLI J. 2009;8:74–88.

Nielsen EI, Friberg LE. Pharmacokinetic-pharmacodynamic modeling of antibacterial drugs. Pharmacol Rev. 2013;65:1053–90. doi:10.1124/pr.111.005769.

Oreng CA, Michie AD, Jones S, Jones DT, Swindells MB, Thornton JM. CATH--a hierarchic classification of protein domain structures. Structure. 1997;5(8):1093–108. doi:10.1016/S0969-2126(97)00260-8.

Overton CE. Studien über die Narkose. Jena: Fischer; 1901.

Pappenberger G, Benz J, Gsell B, Hennig M, Ruf A, Stihle M, et al. Structure of the human fatty acid synthase KS-MAT didomain as a framework for inhibitor design. J Mol Biol. 2010;397:508–19. doi:10.1016/j.jmb.2010.01.066.

Pearson K. On lines and planes of closest fit to systems of points in space. Philos Mag. 1901;2:559–72.

Pemble CW, Johnson LC, Kridel SJ, Lowther WT. Crystal structure of the thioesterase domain of human fatty acid synthase inhibited by orlistat. Nat Struct Mol Biol. 2007;14:704–9. doi:10.1038/nsmb1265.

Portela C, Soares-da-Silva P. The translational approach between computational chemistry and clinical expertise in drug development. 2015. http://sigarra.up.pt/fmup/pt/publs_pesquisa.show_publ_file?pct_gdoc_id=42752. Accessed on 15 Aug 2015.

Pouliot Y, Chiang AP, Butte AJ. Predicting adverse drug reactions using publicly available PubChem BioAssay data. Clin Pharmacol Ther. 2011;90:90–9. doi:10.1038/clpt.2011.81.

Rarey M, Kramer B, Lengauer T, Klebe G. A fast flexible docking method using an incremental construction algorithm. J Mol Biol. 1996;261:470–89. doi:10.1006/jmbi.1996.0477.

Raza M. A role for physicians in ethnopharmacology and drug discovery. J Ethnopharm. 2006;104:297–301. doi:10.1016/j.jep.2006.01.007.

Richet MC. Note sur le rapport entre la toxicité et les propriétes physiques des corps. Compt Rend Soc Biol (Paris). 1893;45:775–6.

Roberts N, Martin J, Kinchington D, Broadhurst A, Craig J, Duncan I, et al. Rational design of peptide-based HIV proteinase inhibitors. Science. 1990;248:358–61. doi:10.1126/science.2183354.

Ruth H, Morris GM, Olson AJ, Goodsell DS. A semiempirical free energy force field with charge-based desolvation. J Comput Chem. 2007;28:1145–52. doi:10.1002/jcc.20634.

Scheer M, Grote A, Chang A, Schomburg I, Munaretto C, Rother M, et al. BRENDA, the enzyme information system in 2011. Nucleic Acids Res. 2011;39:D670–6. doi:10.1093/nar/gkq1089.

Shah NH. Survey: translational bioinformatics embraces big data. Yearb Med Inform. 2012;7:130–4.

Shah NH, Tenenbaum JD. The coming age of data-driven medicine: translational bioinformatics' next frontier. J Am Med Inform Assoc. 2012;19:e2–4. doi:10.1136/amiajnl-2012-000969.

Steven JK, Fumiko A, Natasha R, Jeffrey WS. Orlistat is a novel inhibitor of fatty acid synthase with antitumor activity. Cancer Res. 2004;64:2070–5. doi:10.1158/0008-5472.CAN-03-3645.

SYBYL-X-SuiteS: YBYL 8.0. Tripos International, 1699 South Hanley Rd., St. Louis, Missouri, 63144, USA.

Tavera-Mendoza LE, Mader S, White JH. Genome-wide approaches for identification of nuclear receptor target genes. Nucl Recept Signal. 2006;4:e018. doi:10.1621/nrs.04018.

Thomas PD, Kejariwal A, Campbell MJ, Mi H, Diemer K, Guo N, et al. PANTHER: a browsable database of gene products organized by biological function, using curated protein family and subfamily classification. Nucleic Acids Res. 2003;31:334–41. doi:10.1093/nar/gkg115.

Thomsen R, Christensen MH. MolDock: a new technique for high-accuracy molecular docking. J Med Chem. 2006;49:3315–21. doi:10.1021/jm051197e.

Vandhana S, Deepa PR, Jayanthi U, Biswas J, Krishnakumar S. Clinico-pathological correlations of fatty acid synthase expression in retinoblastoma: an Indian cohort study. Exp Mol Pathol. 2011;90:29–37. doi:10.1016/j.yexmp.2010.11.007.

Venhorst J, ter Laak AM, Commandeur JN, Funae Y, Hiroi T, Vermeulen NP. Homology modeling of rat and human cytochrome P450 2D (CYP2D) isoforms and computational rationalization of experimental ligand-binding specificities. J Med Chem. 2003;46:74–86. doi:10.1021/jm0209578.

Vilar S, Harpaz R, Chase HS, Costanzi S, Rabadan R, Friedman C. Facilitating adverse drug event detection in pharmacovigilance databases using molecular structure similarity: application to rhabdomyolysis. J Am Med Inform Assoc. 2011;18 (Suppl 1):i73–80. doi:http://dx.doi.org/10.1136/amiajnl-2011-000417 i73-i80.

VLifeMDS: Molecular Design Suite, VLife Sciences Technologies Pvt. Ltd., Pune, India, 2010 (www.vlifesciences.com)

Voigt JH, Bienfait B, Wang S, Nicklaus MC. Comparison of the NCI open database with seven large chemical structural databases. J Chem Inf Comput Sci. 2001;41:702–12. doi:10.1021/ci000150t.

Williams PA, Cosme J, Ward A, Angove HC, Matak Vinkovic D, Jhoti H. Crystal structure of human cytochrome P450 2C9 with bound warfarin. Nature. 2003;424:464–8. doi:10.1038/nature01862.

Wishart DS, Knox C, Guo AC, Shrivastava S, Hassanali M, Stothard P, et al. DrugBank: a comprehensive resource for in silico drug discovery and exploration. Nucleic Acids Res. 2006;34(suppl 1):D668–72. doi:10.1093/nar/gkj067.

Zhang P, Foerster H, Tissier CP, Mueller L, Paley S, Karp PD, Rhee SY. MetaCyc and AraCyc. Metabolic pathway databases for plant research. Plant Physiol. 2005;138(1):27–37. doi:http://dx.doi.org/10.1104/pp.105.060376.

Chapter 3
Translational Research in Drug Discovery and Development

Neha Arora, Pawan Kumar Maurya, and Puneet Kacker

Abstract Translational research facilitates the application of basic scientific discoveries in clinical and community settings to prevent and treat human diseases. The translation of knowledge and innovations from basic laboratory experiments to point-of-care patient applications; production of new drugs, devices, and healthcare products; and promising treatments for patients is referred to as benchside to bedside transition. Numerous opportunities encompass translational research. However, there are several obstacles involved in the process that make the translational journey quite challenging. The major challenges that hamper the growth of translational research include insufficient resources, inadequate funding and infrastructure, shortage of qualified researchers, and lack of sufficient experience in essential techniques. Translational drug discovery and development is an exceedingly difficult, expensive, time-consuming, and risky process. Despite thousands of pharmaceutical companies working to develop and get new drugs to market, and billions of dollars spent every year, only a few new molecular entities (NMEs) receive marketing approval from the FDA per year. Translational drug discovery demands both the need for cooperation between clinical and pharmacological research and the significance of the role of academia in target identification and drug discovery, design, and development. This chapter highlights an overview of translational research in a drug discovery and development perspective. We further discussed associated opportunities and challenges, as well as possible strategies that could be used to overcome the challenges. Certain strategies like prioritizing research area,

N. Arora
Department of Botany, Banaras Hindu University, Varanasi, Uttar Pradesh 221005, India

P.K. Maurya
Amity Institute of Biotechnology, Amity University, Sector-125, Noida, Uttar Pradesh 201313, India

Interdisciplinary Laboratory of Clinical Neurosciences (LiNC), Department of Psychiatry, Universidade Federal de São Paulo, São Paulo, Brazil

P. Kacker (✉)
Pharma Analytics Division, Excelra Knowledge Solutions Pvt. Ltd., 6th floor, Wing B, NSL SEZ Arena, Plot No. 6, Survey No. 1, IDA Uppal, Hyderabad, Telangana 500039, India
e-mail: puneet.kacker@gmail.com

© Shanghai Jiao Tong University Press, Shanghai and Springer Science+Business Media Dordrecht 2017

D.-Q. Wei et al. (eds.), *Translational Bioinformatics and Its Application*, Translational Medicine Research, DOI 10.1007/978-94-024-1045-7_3

clearer vision on the project, committed team of researchers, established infrastructure, sufficient funding, and meaningful collaborations could be highly beneficial in accelerating the hunt to discover new drugs and for the establishment of successful translational drug discovery process.

Keywords Translational research • Drug discovery and development • Opportunities • Challenges • Drug repurposing

List of Abbreviations

BrIDGs	Bridging Interventional Development Gaps
FASEB	Federation of American Societies for Experimental Biology
FDA	Food and Drug Administration
GWAS	Genome-wide association study
ITHS	Institute of Translational Health Sciences
MHRA	Medicines and Healthcare Products Regulatory Agency
NCATS	National Center for Advancing Translational Science
NCI	National Cancer Institute
NIH	National Institutes of Health
TRWG	Translational Research Working Group
CRC	Colorectal cancer
ALL	Acute lymphoblastic leukemia

3.1 Translational Research

Translational research is basically translating knowledge obtained from laboratory science into clinical practice in order to improve human health. It involves the process of applying ideas, insights, and discoveries unveiled through basic scientific researches for the welfare of mankind. The knowledge acquired, mechanisms devised, and techniques developed using basic science researches are effectively translated into new approaches for prevention, diagnosis, and treatment of diseases (Fang and Casadevall 2010). The applications of such basic scientific discoveries in clinical and community settings thereby are instrumental in bridging the gap between biomedical science and medical practice (Zerhouni 2005).

3.1.1 How Is It Different from Traditional Research?

Basic and translational research can be considered as complementary areas of human endeavor that differ primarily in integration and practicality, respectively. Whereas basic science contributes in deriving deeper knowledge in the desired

Table 3.1 Difference between basic and translational research

Terms of comparison	Basic research	Translational research
Research orientation	Also referred to as traditional research wherein the motivation lies in acquiring knowledge. This type of research is mostly exploratory and often leads to great discoveries	Also referred to as advanced research wherein the motivation is to get results. This research is more of practical approach that refines the discoveries into useful products
Scientific approach	It is the style of scientific inquiry which is bottom-up	The scientific inquiry is top-down
Organization	It is generally performed by academia constituting scientists and biologists involved in benchwork	Generally performed largely by engineers employed by industries and/or by government organizations
Type of invention	Basic research is revolutionary	Translational research is evolutionary
Research application	The results and findings of basic research are sometimes shelved without an obvious immediate use	As this type of research is goal oriented, its results are of immediate use to be used outside of academia

field, the significance of translational science lies entirely in its practicality (Koshland 1993; Selep 2013). Some of the differences in approach and goals of the two kinds of researches are summarized in Table 3.1.

3.1.2 Translation Continuum from Benchside to Bedside

Since basic and translational researches are complementary to each other, they assist each other in their further development. While basic research takes up the task of unveiling promising novel ideas for their use in translational research, the translational research on the other hand raises new questions for the researchers of basic sciences to address. Thus, basic research generally plays the part of generating new ideas, and applied research conveys these ideas in the more refined and applicable form to the market so as to be implemented for the betterment of the population (Drolet and Lorenzi 2011). This translation of the nurtured research ideas going long way till their application in more elaborate, productive, valuable, profitable, and promising manner to enhance human health and well-being is often referred to as benchside to bedside transition (Keramaris et al. 2008). Translational research helps turn early-stage innovations starting from basic laboratory experiments progressing through the several rounds of clinical trials to point-of-care patient applications (Tufts 2015); production of new drugs, devices, and healthcare products; and promising treatments for patients, thereby advancing the innovation to make it attractive for further development and commercialization by the medical industry or healthcare sectors (Woolf 2008).

3.1.3 Translational Research Phases

Translational research has often been described in phases of translation, also known as "T-phases" that revolve around the development of evidence-based guidelines. The Institute of Translational Health Sciences (ITHS) has adopted a model of five phases (T_0–T_4), which is adapted from the Khoury et al.'s (2007) description of four phases. T_0 phase is characterized by the identification of opportunities and approaches to health problems that need to be addressed, whereas T_1 phase attempts to translate basic discovery into a candidate health application. T_2 phase assesses the value of application for health practice leading to the development of evidence-based guidelines which are moved into health practice through dissemination and diffusion research in T_3 phase. The final evaluation of the health outcomes of the health practice is then performed in T_4 phase. An outline of the transformation of basic research from benchside to translational research till bedside progressing through different research phases is illustrated in Fig. 3.1.

3.1.4 Translational and Clinical Science

Biomedical research community in today's world has taken up the task of translating the remarkable scientific innovations into health benefits. For realizing this objective and developing the ideas and strategies, the US National Institutes of Health (NIH) initiated a series of consultations with the research community in order to define major scientific trends collectively, with the goal of identifying thematic areas that the whole of the NIH needed to address. This initiative led to the development of the NIH road map for medical research, which is based on three fundamental themes (Zerhouni 2003):

(i) *New Pathways to Discovery*: This theme is aimed at the identification of the need to stimulate the development of novel approaches to unravel the complexity of biologic systems and their regulation. Implementation groups in this area are Molecular Libraries and Imaging; Building Blocks, Biological Pathways, and Networks; Structural Biology; Bioinformatics and Computational Biology; and Nanomedicine.

(ii) *Research Teams of the Future*: Under this theme, the main objective is to explore out ways to reduce the cultural and administrative barriers that often hamper the research which is done at the interface of preexisting disciplines and to invoke an era in which scientists can cooperate in new and different ways. NIH also developed an innovative program called as the Pioneer Award, wherein unprecedented intellectual freedom is provided to highly creative thinkers who are engaged in investigating problems of biomedical and behavioral importance. Implementation groups in this area are Interdisciplinary Research; Interdisciplinary Health Research Training: Behavior, Environment

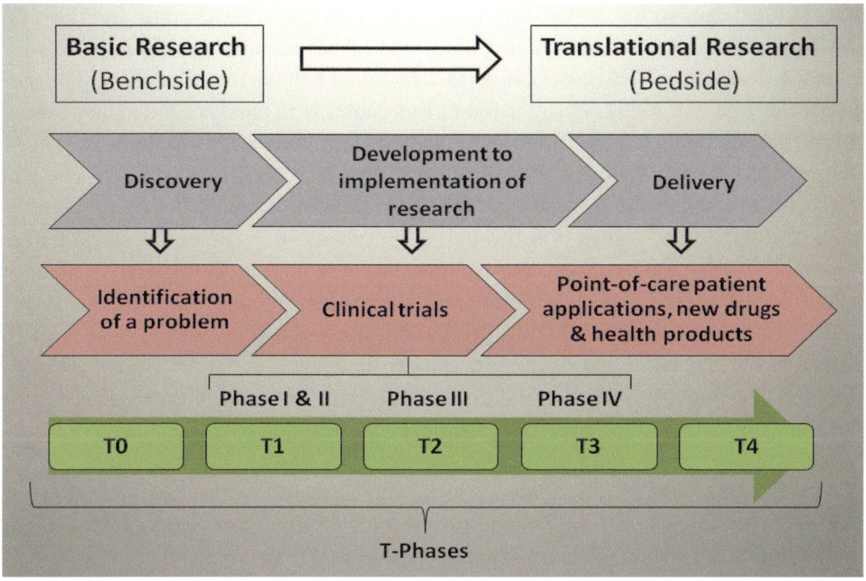

Fig. 3.1 Translation of basic science from benchside to bedside

and Biology; High-Risk Research; NIH Director's Pioneer Award (NDPA); and Public/Private Partnerships.

(iii) *Reengineering the Clinical Research Enterprise*: There have been concerns to bring together the basic, translational, and clinical researchers for better and fruitful interactions. Moreover, the new investigators are lesser interested in clinical research which is preventing the scientists to go on for patient-oriented research. This has called for an immediate need for instigating renovations in translational and clinical science by the NIH, which is the main objective of this theme. Implementation groups in this area are Harmonization of Clinical Research Regulatory Processes, Integration of Clinical Research Networks Clinical Research Informatics: National Electronic Clinical Trials and Research System (NECTRS), Regional Translational Research Centers, Enabling Technologies for Improved Assessment of Clinical Outcomes, and Dynamic Assessment of Patient-Reported Chronic Disease Outcomes.

3.1.5 Reengineering Translational Science

Due to immense economic stresses and patent expirations, pharmaceutical companies are turning down their investments in research (Wilson 2011). Furthermore, biotechnology companies are finding it very difficult to obtain venture capital for projects that need many years of support for achieving long-term profitability (Ernst and Young 2010). Realizing the need to pursue opportunities for disruptive translational innovation and reengineering the process of developing diagnostics, devices, and therapeutics across a wide range of human diseases through translational research, the NIH has established a National Center for Advancing Translational Science (NCATS). The mission of this center is to catalyze the generation of innovative methods and technologies for the development and implementation of diagnostics and therapeutics (Ferrell 2009). The long timelines, steep costs, and high failure rates in the translational pathway compel the initiation of revolutionizing the science of translation through comprehensive, systematic, and creative approach. NIH aims to shape and sharpen this new vision through a transparent scientific environment, via NIH-based online resources, thereby ensuring the proper and wider dispersal of complete information about the successes and failures in research swiftly to all the stakeholders (Zerhouni 2003). NCATS aims to offer unparalleled opportunities to researchers for intense focus on the reengineering of the translational process, beginning from the initial target identification to first-in-human application of small molecules, biologics, diagnostics, and devices (Collins 2011). Besides NCATS, there are various other research institutes and centers all over the world that are dedicated toward performing translational research in different fields of scientific advancement (Table 3.2).

3.1.6 Opportunities in Translational Research

Translational research is a new area of investigation that involves the integrated application of advanced technologies that include multiple disciplines of science like physiology, pathophysiology, natural history of disease, genetics, and proof-of-concept studies of drugs and devices (Zerhouni 2005). Recent research breakthroughs, most importantly, completion of the Human Genome Project, offer a pool of nonending opportunities for basic investigators to work and make further advancements in these areas. Other accomplishments like advances in information technology; biocomputing; high-throughput technologies for screening, identifying, and studying compounds of interest; and novel imaging capabilities also tend to provide direct and immediate rewards for individual investigators and the institutions that support their work (Hobin et al. 2012).

3.1.6.1 Opportunities for Researchers

For basic researchers, engaging in translational research benefits them in contributing to the understanding and treatment of human diseases and participating in the development of solutions to medical and public health problems that serves as a source of intellectual inspiration and stimulation. On personal front, translational research provides opportunities to researchers to develop their own science and learn new methods that paves way for initiating new projects, gives directions for existing projects, and increases publication rates. Furthermore, it helps in

Table 3.2 Partial list of dedicated translational research institutes and centers across the globe

Institute	Weblink	Country	Comments
Translational Research Institute Australia	www.tri.edu.au/	Australia	Aims at comprehensive medical research and biopharmaceutical facility. The institute currently hosts four flagship programs: (1) immunotherapy, (2) diagnostic imaging, (3) microbiome, and (4) gynecological cancer
The Centre for Drug Research and Development	www.cdrd.ca/	Canada	Alliance of Translational Research Centres established to accelerate global drug development. Their project portfolio includes 13 technologies commercialized till date in the wider area of immunotherapy, neuroscience, anti-infective, oncology, fibrosis, inflammation, and regenerative medicines
National Research Center for Translational Medicine	www.natureindex.com/institution-outputs/china/national-research-center-for-translational-medicine/556d6532140ba05c398b4570	China	First of five institutions meant to bridge the gap between basic research and clinical application by putting researchers, doctors, and patients under one roof
Translational Health Science and Technology Institute	www.thsti.res.in/	India	The emphasis is on fast-tracking healthcare solutions that would meet the needs of a rapidly developing economy in need of healthcare intensity for its large population
Translational Research Informatics Center	www.tri-kobe.org/	Japan	The center is supported by the Ministry of Education, Culture, Sports, Science and Technology. The center's aim is to develop methods for improved prognosis in important disease areas

(continued)

Table 3.2 (continued)

Institute	Weblink	Country	Comments
European Infrastructure for Translational Research (representative organization)	www.eatris.eu/	Multiple European countries	One-stop access to over 70 academic research centers in Europe. Their research services are focused around the following technologies: (1) ATMP and biologics, (2) biomarkers, (3) imaging and tracing, (4) small molecules, and (5) vaccines
Centre for Translational Research and Diagnostics	https://www.csi.nus.edu.sg	Singapore	The center is equipped with three major facilities: (1) the NUHS Tissue Repository, (2) the Translational Interface molecular pathology facility, and (3) the Diagnostic Molecular Oncology Centre with excellence in clinical sample and data management, translational research and clinical trial support, and the development and deployment of novel diagnostics into the clinic
National Center for Advancing Transla tional Sciences	www.ncats.nih.gov/	USA	Centers at the NIH; established to transform the translational process so that new treatments and cures for disease can be delivered to patients faster
Center for Comparative Medicine and Translational Research	https://cvm.ncsu.edu/research/centers/ccmtr/	USA	Promotes scientific discovery and facilitates its clinical application to achieve the goal of improving the health of animals and humans
Translational Research Institute for Metabolism and Diabetes	http://www.tri-md.org/	USA	Joint venture between Florida Hospital and Sanford-Burnham Medical Research Institute; dedicated to the study of obesity, metabolism, diabetes, and the

(continued)

Table 3.2 (continued)

Institute	Weblink	Country	Comments
			metabolic origins of cardiovascular disease
Center for Translational Injury Research	http://cetir-tmc.org/	USA	The goal of the center is to lead research and development of next-generation medical technologies in the areas of hemostasis, resuscitation, and computerized decision support for trauma patients
Translational Geno mics Research Institute	https://www.tgen.org/	USA	Primary focus is to discover the genetic cause of disease. The institute's thrust areas include disorders in the areas of oncology, neurogenomics, and metabolic diseases
Center for Translational Medicine, the University of Texas Southwestern Medical Center	http://www.utsouthwestern.edu/research/translational-medicine/index.html	USA	CTM is a member of the national Clinical and Translational Science Award (CTSA) consortium, a group of 62 medical research institutions, funded by the National Institutes of Health (NIH), that work together to improve the way biomedical research is conducted across the country, to reduce the time it takes for laboratory discoveries to become treatments for patients, to engage communities in clinical research efforts, and to train a new generation of clinical and translational researchers
The Institute for Translational Medicine and Therapeutics	http://www.itmat.upenn.edu/	USA	ITMAT includes faculty, basic research space, and the Clinical and Translational Research Center (CTRC), which now includes the former

(continued)

Table 3.2 (continued)

Institute	Weblink	Country	Comments
			General Clinical Research Center (GCRC) of both Penn and the Children's Hospital of Philadelphia (CHOP). It supports research at the interface of basic and clinical research focusing on developing new and safer medicines
Duke Translational Medicine Institute	https://www.dtmi.duke.edu/what-we-do/translational-medicine-at-duke/	USA	DTMI strives to overcome the obstacles to developing discoveries into devices, drugs, or therapies to improve health. The major areas of research are as diverse as ophthalmology, cancer screening, and a device for screening blood for transfusions
Center for Translational Medicine, the University of Minnesota	http://www.researchservices.umn.edu/services-name/center-translational-medicine/	USA	The center solicits and evaluates promising research leads; identifies necessary resources; provides expertise for the preclinical evaluation and testing of novel reagents, GMP manufacture of clinical products, and IND/IDE development and submission; and supports phase I clinical trial design and implementation

promoting interdisciplinary collaborations between clinicians, clinical researchers, and basic investigators that can provide exposure to new areas of science and also can generate new ideas. The basic researchers in this way also can mentor clinical colleagues in basic science methods.

3.1.6.2 Opportunities for Institutions

Besides researchers, the institutions facilitating and encouraging translational research also benefit from these programs. They have ample opportunities to provide unique training experiences to undergraduates, graduate students, and postdoctoral fellows, thereby motivating them and encouraging new talent to enter biomedical research. Moreover, due to the promotion of the development of new drugs, devices, and other medical interventions, they are able to accomplish their biomedical research missions, attract more and more patients, and enhance their status and repute. Translational research opportunities help institutions and organizations to facilitate their investigators with easy and affordable access to resources, collaborators, and expensive shared equipment and facilities. They are also able to attract public-private partnerships, leverage federal and nonfederal resources, and attain support from funding agencies in new and lucrative projects.

3.1.7 Challenges in Translational Research

Although there are numerous opportunities encompassing translational research, there are several obstacles as well involved in the process that makes the translational journey more and more challenging. The major challenges that limit professional interest and hamper the translational enterprise are insufficient resources, inadequate funding and infrastructure for developing research programs, shortage of qualified investigators, and lack of sufficient experience with essential methods and techniques as well as with complex regulatory requirements (Hait 2005). Other issues that have been repeatedly debated include academic cultural differences between basic scientists and clinicians that hinder collaboration. These differences arise due to communication gap, differences in education and training, and different goals and targets. The culture of valuing clinical care over research sidelines the basic researchers who tend to show little interest in the research.

Moreover, lack of incentives and rewards for the researchers discourage them to take initiative in novel research. Regulatory and ethical issues that are involved in human research, tissue banking, intellectual property rights, and toxicology and manufacturing regulations have become more gruesome with expanding work in the fields of cell and gene therapies and tissue engineering. Getting approvals from regulatory agencies like the Food and Drug Administration (FDA) and Medicines and Healthcare Products Regulatory Agency (MHRA) has become much more difficult and complicated. All of these issues contribute to several checkpoints in translational research phases including the "valley of death" that exists between preclinical research and clinical trials (Butler 2008). The Federation of American Societies for Experimental Biology (FASEB) has made recommendations to deal with these challenges by emphasizing upon the roles and responsibilities of institutions, professional societies, funding organizations, and individual scientists (http://www.faseb.org/Portals/0/PDFs/opa/TranslationalReportFINAL.pdf).

3.1.8 Controversies in Translational Research

Since the establishment of the NCATS, there have been a lot of controversies within the research community on the purpose, structure, and funding of the center. The mission of this center is to experiment with innovative approaches to reduce, remove, or bypass the bottlenecks often associated with the translational pipeline. Although efficient implementation of the translational research process is an important step in the emergence of an advanced research scenario, however many basic research scientists raised the concerns on the development of NCATS as it would take off the focus away from basic research entirely to the translational research (McClure 2012). This debate by the basic researchers is apt and reasonable as advances in the treatments that are being made at present are the result of enormous efforts made by the basic researchers over the decades that laid the foundation for further discoveries. Therefore, this calls for an understanding of the significance of the basic research and considering the investments made in basic research, for all intents and purposes, as an investment in translational research.

3.2 Translational Drug Discovery

The past decade has witnessed an increased emphasis on laboratory-based translational research which has been instrumental in enabling clearer understanding of the disease mechanisms and in the development of novel approaches to varied scientific areas like in gene therapy, RNA interference, and stem cells (Littman et al. 2007). The increasing adoption of translational research is leading to novel integrated discovery nexuses that may change the landscape of drug discovery. Drug discovery is the first step in the creation of new drugs that takes place in academic institutions, biotech companies, and large pharmaceutical organizations. These sectors, though, used to operate independently with minimal collaboration between those at the forefront of discovery research and those with experience in developing drugs, but with the emergence of translational research, have come closer for seeking collaboration to pool the expertise required to generate new therapies by linking laboratory discoveries directly to unmet clinical needs (Fishburn 2013). However, despite the huge investments made in drug discovery process in the past decade, there still remains a shortage of new drugs. The reasons behind this could be attributed to the continued existence of a standard drug development model that has not attuned to changes in science and public perception of drug companies. Furthermore, the pace of drug development process lags behind in the USA due to a high profit margin that prevents reform in the absence of economic pressures (Fitzgerald 2005). The World Health Organization report on "Priority Medicines for Europe and the World" (WHO Geneva 2004) specifies several "high-burden" diseases for which no active treatment is currently available, including infectious diseases, a range of chronic diseases of the central nervous system and the cardiovascular system, autoimmune disorders, and cancer. The number of patients with these chronic diseases is continually growing in the aging population.

Translational drug discovery covers the entire spectrum from target identification to the evaluation of the efficacy and safety of novel medicines in clinical practice. It requires data generated from molecular investigations, healthcare, and clinical research. A vast and diverse amount of data thus produced pose various challenges like integrating fragmented databases, facilitating secondary usage of patient data in clinical research, and generating information systems for easy and immediate use to clinicians and biomedical researchers. Hence, for the management and integration of clinical and molecular data, a large number of web-based user-friendly databases and tools are being designed that are available online (Table 3.3).

Translational drug discovery demands both the need for cooperation between clinical and pharmacological research and the significance of the role of academia in target identification and drug discovery, design, and development. In the past decade, an important trend has been observed wherein an increasing proportion of innovative new drugs emerge from small biotech companies typically working in close collaboration with academia. An example of one such drug is the peptide vaccine for HPV-induced cervical cancer.

Discovery and development of safe and effective new drugs is an exceedingly difficult, expensive, and time-consuming process. Despite thousands of pharmaceutical companies working to develop and get new drugs to market, and approximately $50 billion spent every year, only 23 new molecular entities (NMEs) per year (on average) have received marketing approval from the FDA during the last 10 years. The most concerning fact is that, while expenditures have increased steadily since the mid-1990s, the number of drugs reaching the market has declined to a relatively low plateau (Fishburn 2011).

Although the developmental process of new drugs has been slow, the overall research and development investment has yielded important breakthroughs in basic cellular and molecular biology and in producing novel technologies to advance drug development. These advancements include the identification of all the genes in the human genome (the Human Genome Project), the use of microchip-based robotics for rapidly testing large numbers of potential new drug compounds, and the creation of cell-based systems for large-scale synthesis of protein and antibody therapeutics. The mismatch between scientific progress and poor productivity of drugs has led many scientists to reexamine the existing strategies for creating new

Table 3.3 Partial list of databases and tools dedicated for accelerating translational research in drug discovery

Databases		
Name	**Contents**	**References**
Neuroblastoma patients (NeuPAT)	An intranet-based database integration for neuroblastoma patients	Villamon et al. (2013)
Diet, Genomics, and Immunology Laboratory (DGIL) Porcine	Contains functional information on genes commonly studied in humans, pigs, and rodents	http://www.ars.usda.gov/Main/docs.htm?docid=6065
Stanford Translational Research Integrated Database Environment (STRIDE)	A Stanford project which consists of three components: clinical data warehouse, research data management applications, and biospecimen data management system	Lowe et al. (2009)
Cancer Survivors Against Radon (canSAR)	Database that integrates data from biology, chemistry, pharmacology, structural biology, cellular networks, and clinical annotations. A tool is also built that applies machine learning methods for several useful drug discovery predictions	Bulusu et al. (2014)
Repository of Molecular Brain Neoplasia Data (Rembrandt)	A cancer clinical genomic database and a web-based data mining and analysis platform aimed at facilitating discovery by connecting the dots between clinical information and genomic characterization data	Madhavan et al. (2009)

(continued)

Table 3.3 (continued)

Research Electronic Data Capture (REDCap)	It is a free web-based application designed to support data capture for research studies. It is a metadata-driven EDC software solution and workflow methodology for designing clinical and transla tional research databases	Harris et al. (2009)
Tools		
Name	**Purpose**	**Reference**
PRISYM CLINTRIAL	Clinical trial management and patient stratification system	http://www.prisymid.com/solutions/clinical-trials/prisym-clintrial/
GenetRx	Patient stratification based on gene expression biomarker	http://www.genebiomarkers.com/applications/patient-stratification.php/
HLA Twin	It is a dual-algorithm genotyping software	http://www.omixon.com/hla-twin/
RANDI2	It is a web-based application that supports many randomization algorithms, free configurable patient properties, stratification, and definition of inclusion criteria for easy management of multicenter clinical trials	Schrimpf et al. (2010)
HERMES	It is a free simulation software used for taking the key decisions of minimization or stratification using various modeling parameters	Fron Chabouis et al. (2014)

drugs. Presently, it takes an average of 12–15 years to bring a new drug to the market because the process involves sequential stages of discovery, preclinical development, clinical trials (phases I, II, III), and FDA review (Fishburn 2011). Moreover, the successful development of a single drug often starts with the synthesis and testing of thousands of different candidate drug molecules.

3.2.1 Translational Drug Development for Diseases

Drug development is the process of bringing a new pharmaceutical drug to the market once a lead compound has been identified during drug discovery process. Across all diseases, translational drug discovery and development are lengthy, costly, and risky processes. The average cost for the development of new drug has been estimated to be greater than $1 billion. There are several challenges in neuroscience drug development such as target identification and validation, significant disillusionment with the use of animal model to evaluate efficacy, lacking biomarkers, and stratification of populations for clinical trials (Lally and MacCabe 2015). Various challenges in drug discovery and development for diseases are summarized in Fig. 3.2 and Table 3.4.

3.2.1.1 Nervous System Disorders

The prevalence and burden of neurological disorders impel the leadership within industry, academia, and government to take initiative for curing these disorders. Despite intensive research over many years, the treatment of brain disorders remains a major health issue (Morinet 2014). In spite of high prevalence, enormous contributions to disability worldwide, and substantial economic burden, there are no disease-altering therapies for neurodegenerative disorders (Karnati et al. 2015). Compared with other disease areas, failure rates in late-stage clinical trials are disproportionately high for neurodegenerative disorders. Many drug companies

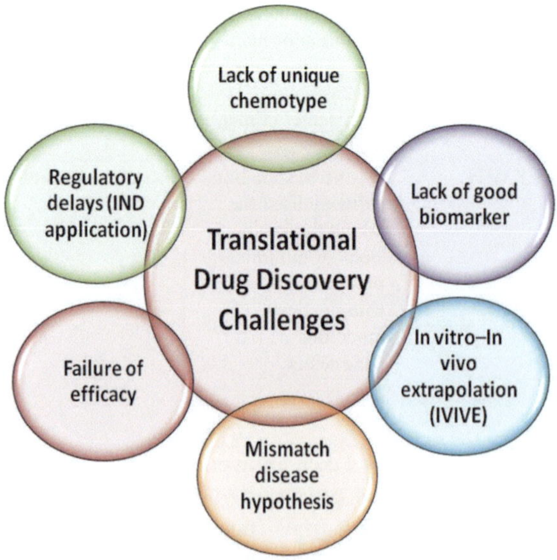

Fig. 3.2 Some of the major challenges in drug discovery and development

Table 3.4 List of important diseases and challenges associated with their drug discovery and development projects

Disease	Major challenges	References
Alzheimer's disease (AD)	Exact cause for AD onset is still unknown	Gu et al. (2015)
	Limited effectiveness of the cognitive tests	Wesnes and Edgar (2014)
	Problems associated with blood-brain barrier (BBB) penetration of drugs and its pharmacokinetic properties	Butini et al. (2013)
Schizophrenia (SZ)	Limitation in identifying, validating, developing, and clinically deploying new treatments	Millan et al. (2015)
	First- and second-generation antipsychotic drugs based upon the dopamine hypothesis are limited	Winchester et al. (2014)
Bipolar disease (BD)	Limitation in clinically deploying new treatments	Millan et al. (2015)
	Challenges in improving methods and tools to generate, integrate, and analyze high-dimensional data	Hoertel et al. (2013)
	Technical challenges related to the identification and validation of candidate genes and peripheral biomarkers	Le-Niculescu et al. (2011)
Major depressive disorder (MDD)	Limited efficacy and a pronounced delay to onset of action and provoke distressing side effects	Millan (2006)
Cancer	The sequencing of increasingly larger numbers of cancer genomes	McDermott et al. (2011) and Sellers (2011)
	Identification of key driver mutations and matching drug therapies	Greaves and Maley (2012)
	Heterogeneous populations in cancers are likely to include drug-resistant stem cells and a range of host cells that are involved in tumor progression	Jordan et al. (2006) and De Palma and Hanahan (2012)
	Selecting and validating the best targets	Benson et al. (2006)
	Druggability gap	Verdine and Walensky (2007) and Paul et al. (2010)
Diabetes	Safety concerns of GLP1 analogues	Drucker et al. (2010)
	Failure of antidiabetic medications like troglitazone, rosiglitazone, and pioglitazone	Henney (2000, Nissen and Wolski (2007) and Hillaire-Buys et al. (2011)
Cardiovascular disease (CVD)	Failure of translating good preclinical data into a safe and effective medicine (e.g., CETP inhibitor torcetrapib and vasopeptidase inhibitor omapatrilat)	Tall et al. (2007) and Ferdinand et al. (2001)
	Lower efficacy of the thrombolytic drugs (e.g., streptokinase and accelerated tissue plasminogen activator [tPA])	White and Van de Werf (1998)
	Lesser opportunities for young biotech companies	Katherine (2007)

have divested themselves almost entirely from neuroscience research program to other therapeutic areas.

Since pathophysiology of brain disorders is poorly understood, it is difficult to identify promising molecular targets and its validation (Feustel et al. 2012). It is also difficult to choose an animal model, because of poor understanding of disease mechanism. Brain disorders are associated with defect in synaptic communication and functional connectivity. Target identification is a critical factor in the drug discovery and development. Target validation is an iterative process of increasing confidence in a target, which can be conceptualized as continuing through phase III clinical trials. Along with target validation, it is critical to establish the therapeutic levels of a drug that can be reliably delivered to the brain and those levels at which the drug binds its target, thereby modifying the disease pathway in the desired direction. In the absence of this information, clinical trials are not eligible for testing the target validation hypothesis. Furthermore, failure of animal models to predict accurately the efficacy of drugs with new mechanism for neurological disorders has been a central problem in drug development. Owing to differences between animals and humans in cell types, transmitter function, and anatomy, the usage of animal models is not reliable.

Developing and integrating new approaches that utilize combination of animal and nonanimal models of disease mechanisms, along with new tools, technologies, and techniques, might illuminate the underlying biological mechanism of disease and improve target identification, validation, and therapeutic development. Current research paradigm might need to change, particularly for clinical studies. For the development of drugs, better understanding of disease mechanisms and improved ability to translate such discoveries into biomarkers and therapeutics is required. Several opportunities in translational drug discovery are illustrated in Fig. 3.3.

3.2.1.2 Psychiatric Disorders

Innovation is important for the identification and novel pharmacological therapies to meet the treatment needs of patients with psychiatric disorders including schizophrenia (SZ), bipolar disorder (BD), and major depressive disorder (MDD) (Maurya et al. 2016). The process of drug discovery encompasses a period of intense research and development efforts that typically take 13–15 years. It involves search for target, optimization to allow candidate drug selection, and human testing to achieve proof of mechanism, principle, and concept, followed by regulatory approval. In the last decade, large-scale candidate gene and genome-wide association studies have generated a growing list of "risk" genes for psychiatric illness (Hess et al. 2016). Multiple risk genes were identified, each making a small contribution to such disorders. These human genome-based approaches of understanding the location and function of specific gene products and their relevance to disease pathophysiology have rewarded the field of biological psychiatry with some novel target ideas. New drug candidates can act in psychiatric disorders to show their effects by modifying certain pathways in the biological system, including

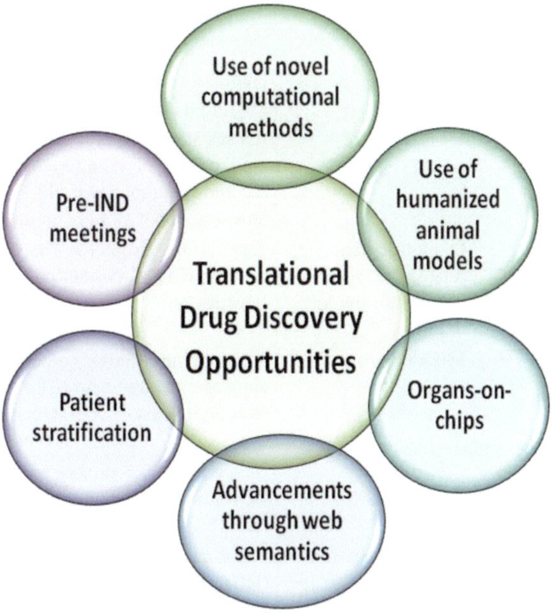

Fig. 3.3 Figure showing open opportunities with translational research in drug discovery

inflammatory pathway, cell-mediated immune (CMI) pathway, oxidative and nitrosative stress (O&NS) processes, antioxidant system (enzymatic and nonenzymatic), mitochondria, and neuroprogression (Table 3.5).

3.2.1.3 Cancer

Despite advances in diagnosis and therapies, cancer is still the leading cause of death worldwide. Genome-wide studies have been extensively used over the past decades as a powerful tool in defining the signature of different cancers and in predicting outcome and response to therapies.

MicroRNAs (miRNAs) are a contemporary class of tiny noncoding endogenous RNA molecules, only 18–25 nucleotides long. In human genome, the expression of each gene is tightly regulated to control the function and environment of each cell. In the nucleus, the template for genetic information is encoded in DNA segments, which are transcribed into RNA molecules. These molecules are transported from the nucleus to the cytoplasm, where they are translated into proteins. The activity of genes is controlled at the level of DNA, RNA, and protein. Different RNAs have different degrees of stability due to their unique interaction with cellular degradation machinery, which is regulated by cellular signals. The recent discovery of miRNAs, which alter RNA stability, ignited a growing interest in gaining further knowledge of gene regulation at the RNA level. About 5300 human genes have been implicated as targets for miRNAs, making them one of the most abundant classes of regulatory genes in humans. miRNAs recognize their target mRNAs

Table 3.5 Various drug candidates that can be used for psychiatric disorders in different modified pathways

S. No.	Pathways	Name of drugs (antidepressants)	References
1.	Inflammatory	Celecoxib, eicosapentaenoic acid (EPA), statins, acetylsalicylic acid, minocycline, interleukin-1 receptor antagonist (IL-1RA), etanercept, ketamine, curcumin	Maes et al. (2012), Najjar et al. (2013) and Lotrich et al. (2014)
2.	Oxidative and nitrosative processes, antioxidant defense	Zinc, N-acetylcysteine (NAC), coenzyme Q10, curcumin, liquiritin	Maes et al. (2011) and Doboszewska et al. (2016)
3.	Mitochondria	Minocycline, statins, celecoxib, eicosapentaenoic acid (EPA), N-acetylcysteine (NAC), coenzyme Q10, curcumin, resveratrol	Morel and Singer (2014), Pandya et al. (2013) and Ungvari et al. (2011)
4.	Cell-mediated immune (CMI) pathway	Indoleamine-2,3-dioxygenase (IDO) blockade, minocycline	Munn and Mellor (2013) and Dean et al. (2014)
5.	Neuroprogression	Neuronal cell adhesion molecule (NCAM), vascular endothelial growth factor (VEGF), vascular growth factor (VGF), fibroblast growth factor receptor (FGFR), minocycline, ginseng, N-acetylcysteine (NAC), zinc, coenzyme Q10, curcumin	Nowacka and Obuchowicz (2012), Wędzony et al. (2013), Elsayed et al. (2012) and Berk et al. (2012)

based on sequence complementarity and act on +region of miRNA, important for mRNA target recognition, which is located at the end of the mature miRNA sequence from bases 2 to 8. This is often referred to as the "seed sequence" (Bartel 2004). Given the importance of miRNAs in regulating cellular differentiation and proliferation, it is not surprising that their misregulation is linked to cancer. In cancer, miRNAs function as regulatory molecules, acting as oncogenes or tumor suppressors. Amplification or overexpression of miRNAs can downregulate tumor suppressors or other genes involved in cell differentiation, thereby contributing to tumor formation by stimulating proliferation, angiogenesis, and invasion. Similarly, miRNAs can downregulate different proteins with oncogenic activity; i.e., they act as tumor suppressors. MicroRNA genes are evolutionarily conserved and are located within the introns or exons of protein-coding genes, as well as in intergenic areas. miRNA genes are transcribed by RNA polymerase II or III into pri-miRNAs. Pri-miRNAs are next cleaved into approx. 70 nucleotide-long precursor miRNAs (pre-miRNAs) by the nuclear microprocessor complex formed by the RNase III Drosha and DiGeorge syndrome critical region gene 8 (DGCR8). The average human pre-miRNA contains a 33-base-pair hairpin stem, a terminal loop, and two single-stranded flanking regions upstream and downstream of the hairpin. Pre-miRNAs are next transported by the exportin-5/Ran GTPase complex into the cytoplasm, where miRNAs undergo maturation. In the cytoplasm, pre-miRNAs are cleaved by RNase III Dicer into B22 nucleotide-long miRNA duplex and are unwound by helicase. The passenger strand is degraded, and the selected guide strand together with Ago protein activates RISC (RNA-induced silencing complex), resulting in mRNA degradation or translational inhibition, depending on the percentage of sequence complementarity between the miRNA 50-seed and mRNA 30-UTR element.

Over the past few years, cancer death rates have shown an overall decrease compared with previous years. This trend is largely due to the development and implementation of improved cancer screening methods and treatment strategies (Gilliland et al. 2016; Matter 2015). Drug discovery is a risky, costly, and time-consuming process depending on multidisciplinary methods to create safe and effective medicines. Although considerable progress has been made by high-throughput screening methods in drug design, the cost of developing contemporary approved drugs did not match that in the past decade (Zhou et al. 2016). Despite these steps toward improving survival and reducing mortality rates, breast cancer still remains the leading cause of death among women younger than 85 years. As with many cancers, progress in early breast cancer detection has been inadequate, and methods for determining diagnosis and prognosis of breast cancer are still limited to invasive procedures, such as tissue biopsies for histological examination. Advances in understanding the cancer cell at the molecular level have enabled development of several targeted therapies that have advanced the treatment of relevant patient subgroups. In colorectal cancer (CRC), miRNAs have evolved in the regulation of chemoresistance to various CRC treatments and the stemness of CRC stem cells (CRSCs), sequentially modulating the sensitivity of CRC cells to anticancer treatments. Targeting miRNAs thus may be a novel plan for eradicating CRSCs, resensitizing drug-resistant cells to anticancer agents, improving drug competence, and developing novel biological agents for CRC treatments (Liu et al. 2016). Genomics-based predictors of drug response have the potential to improve outcomes associated with cancer therapy. Oncoproteomics is an important innovation in the early diagnosis, management, and development of personalized treatment of acute lymphoblastic leukemia (ALL). As inherent factors are not completely known, radiation exposure, benzene chemical exposure, certain viral exposures such as infection with the human T-cell lymphoma/leukemia virus-1, as well as some inherited syndromes may raise the risk of ALL – each ALL patient may modify the susceptibility of therapy. Shotgun proteomic strategies to unravel ALL aberrant signaling networks nowadays are very promising (López Villar et al. 2015). Osteosarcoma (OS), the most common primary bone cancer in dogs, is commonly treated with adjuvant doxorubicin or carboplatin following amputation of the affected limb. Literature shows that intra- and interspecies gene expression models can successfully predict response in canine OS, which may improve outcome in dogs and serve as preclinical validation for similar methods in human cancer research (Fowles et al. 2016). Solid tumors account for approximately 30% of all childhood cancers. An increased efflux rate of the antineoplastic drugs from cancer cells by action of members of the ATP-binding cassette (ABC) transporters is one of the most important mechanisms of multidrug resistance (Fruci et al. 2016).

Epidemiological studies indicate that natural products are also used as anticancer agents. Agents targeting the genetic and/or epigenetic machinery offer potential for the development of anticancer drugs (Cho 2010). Accumulating evidence has demonstrated that some common natural products [such as epigallocatechin-3-gallate (EGCG), curcumin, genistein, sulforaphane (SFN), and resveratrol] have anticancer properties through the mechanisms of altering epigenetic processes [including DNA methylation, histone modification, chromatin remodeling, microRNA (miRNA) regulation] and targeting cancer stem cells (CSCs). These bioactive compounds are able to revert epigenetic alterations in a variety of cancers in vitro and in vivo. They exert anticancer effects by targeting various signaling pathways related to the initiation, progression, and metastasis of cancer (Wang et al. 2013).

The National Cancer Institute (NCI) has initiated the prioritization of cancer antigens so as to pave a way to a well-vetted, priority-ranked list of cancer vaccine target antigens based on predefined and pre-weighted objective criteria. By doing this, NCI also aims to test the new approach for prioritizing translational research opportunities based on an analytic hierarchy process. The elucidation and weighting of criteria for assessing cancer antigens would enable the immunologists to determine the characteristics and provide them with the experimental data needed to select the most promising antigens for further development and testing in clinical trials (Cheever et al. 2009). The NCI embarked on this new approach of identification, prioritization, and funding of translational cancer research due to the recommendations of the Translational Research Working Group (TRWG) (Old 2008). The focus is on evaluation of a method to select cancer antigens for subsequent development through the immune response modifier pathway, which is one of the six TRWG pathways leading from basic laboratory discoveries to final testing in clinical trials (Hawk et al. 2008; Cheever et al. 2008).

3.3 Strategies to Accelerate Translational Research in Drug Development

With the ever-increasing number of unmet medical needs, researchers all around the world are striving for the cure. Not only the diseases that are coming to the scientific knowledge are new but also are complex. Many a times, these new diseases come as outbreaks and spread epidemically in no time (Stadler et al. 2003; Gostin et al. 2014). In such vulnerable situations, well-strategized translational research efforts can provide immediate hope for the cure. A prioritized research area, clear vision on the project, well-established infrastructure, strong team of committed researchers, sufficient funding, meaningful collaboration that can address the demand for extended project activity, and use of new scientific

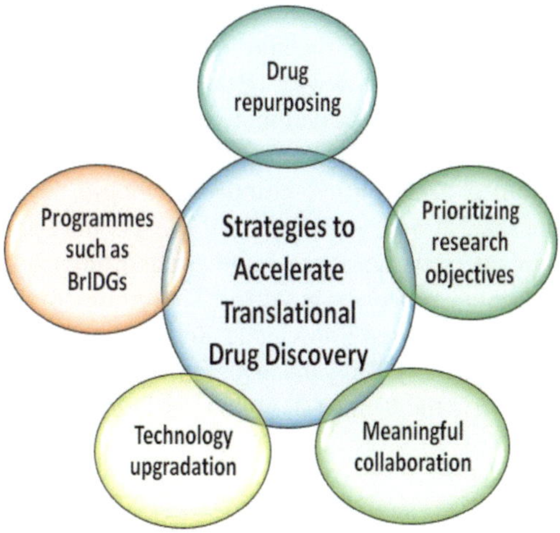

Fig. 3.4 Figure showing some of the strategies that could be employed in translational drug discovery

methods can characterize successful translational research strategies (Khanna 2012). Some of the components of a good strategy helpful in accelerating the translational research in drug discovery and development are highlighted in Fig. 3.4.

3.3.1 Prioritizing Area of Research and Objectives

It is nearly impossible for any single organization to find a cure for every human disease known at any moment of time. Therefore, it is extremely important to prioritize the thrust area and the objectives. A loosely defined objective stating the definition of translational research will certainly not be helpful to lead an organization to a set destination (Sugarman and McKenna 2003). Objectives must be concise and focused to research area and must clearly address the research need with tentative road map and millstones. Understanding the importance of focused research, many centers around the world have been dedicated for specific diseases of the kidney, metabolic disorders, cancer, etc. Such dedicated centers have not only become the models for other scientific organizations but also have enhanced the scientific knowledge in terms of their contribution (Andrews 2013).

3.3.2 Meaningful Collaboration

Collaboration is the integral part of any drug discovery project. Many of the collaborations are actually seen based on their names or brand value (Butler 2008). However, in translational research, collaboration is the most important thing. Collaboration should be made to further progress the discovery from one phase to another (Ioannidis 2004). If an organization is good in basic research (identification of new drug targets), then a fruitful collaboration with an organization competent in developing the new chemical moieties, which can synthesize and test the compounds against the target, would be beneficial (Watson et al. 2008). Small pharmaceutical companies should decide their limit of expanding research because at many times, with one or more lead discoveries, they invest a lot in clinical trials and such huge expenditures in a new assignment lead their initial discovery processes to suffer. Therefore, collaboration not only improves scientific research through knowledge sharing but also helps in maintaining the risk associated with any drug discovery project. A right collaboration in a drug discovery project is the best strategy in translational research (Pober et al. 2001).

3.3.3 Technology Upgradation

For successful translation of drug discovery project, it has to be ensured that quality and speed must go hand in hand. It is important to include the latest technology to produce high-quality results with speed and accuracy. One of the major bottlenecks in translational drug discovery is the limited extrapolation of results generated from preclinical studies as predicted clinical outcome. Therefore, in such instances, it is difficult to get confidence with the data produced with such preclinical experiments (Huh et al. 2010). Drug discovery phases, where the use of the latest technology such as tissue-on-a-chip and organ-on-a-chip could be beneficial to a greater extent, should be encouraged to produce reliable results speedily in a cost-effective manner (Ioannidis 2004).

3.3.4 Bridging Interventional Development Gaps (BrIDGs) Scheme

This is a special scheme under NIH-NCATS program (https://commonfund.nih.gov/raidoverview). The idea behind this unique concept is that certain critical resources are needed for the development of new therapeutic agents and it is difficult for an initial discovery to attract private sector partner to advance project with significant commercial potential. This is due to high risk associated with not-so-common disorders. Under BrIDGs, the investigator gets the chance to receive access to NIH experts and contractors who conduct preclinical studies free of cost for the investigator. At present, four services – synthesis, formulation, pharmacokinetic, and toxicology services – are available under this scheme. The success of this scheme is very encouraging as can be seen from data of 2014 published on the

BrIDGs website. Out of 15 supported projects, five BrIDGs-supported candidates have gone as far as phase II human clinical trials.

3.3.5 Drug Repurposing

Finding a novel indication for an existing drug is called drug repurposing (Issa et al. 2013; Law et al. 2013; Oprea et al. 2011; Buchan et al. 2011; Padhy and Gupta 2011; Ashburn and Thor 2004). This approach is gaining greater interest among the scientist around the world due to its direct market applicability and comparatively low financial risk. In this approach, since the starting point is mostly a molecule passed in clinical phase I, the risk associated with the toxicity becomes negligible. Drug repurposing approaches can be highly useful for orphan diseases and diseases for developing countries where pharmaceutical companies show lesser interest due to low financial returns. There are many instances where drug repurposing approaches have been successfully applied. A number of methods have been used to find alternative indication based on the same target or an alternative target (Sardana et al. 2011).

3.3.5.1 Computational Chemistry

Wide combinations of computational approaches can be utilized for identification of nonobvious indications for a given compound (Ekins et al. 2011; Hurle et al. 2013; Achenbach et al. 2011; Sanseau and Koehler 2011). Broadly, all these approaches can be divided into two methods: (1) ligand based (Gregori-Puigjane and Mestres 2008; Gong et al. 2013) and (2) structure based (Kharkar et al. 2014; Blondeau et al. 2010; Issa et al. 2015). Ligand-based methods are based on the notion that if compound C1 of target T1 significantly matches with a known compound C2 of target T2, then compound C1 could also work for target T2. A variety of sub-methods such as 2D fingerprint-based similarity, 3D shape similarity, scaffold matching, and comparison of ligand pharmacophore or atomic property fields can be utilized. The structure-based methods require the structural information of the target and its compound. Sub-methods such as high-throughput virtual target screening (Gfeller et al. 2014; Santiago et al. 2012), interacting pharmacophore (Liu et al. 2010), and active site similarity have been explored under this category (Haupt et al. 2013). A graphical overview of two computational chemistry methods is given in Fig. 3.5.

3.3.5.2 Literature Mining

A large amount of scientific findings are recorded in the form of publications. There are more than 25 million articles indexed alone in PubMed. It is practically

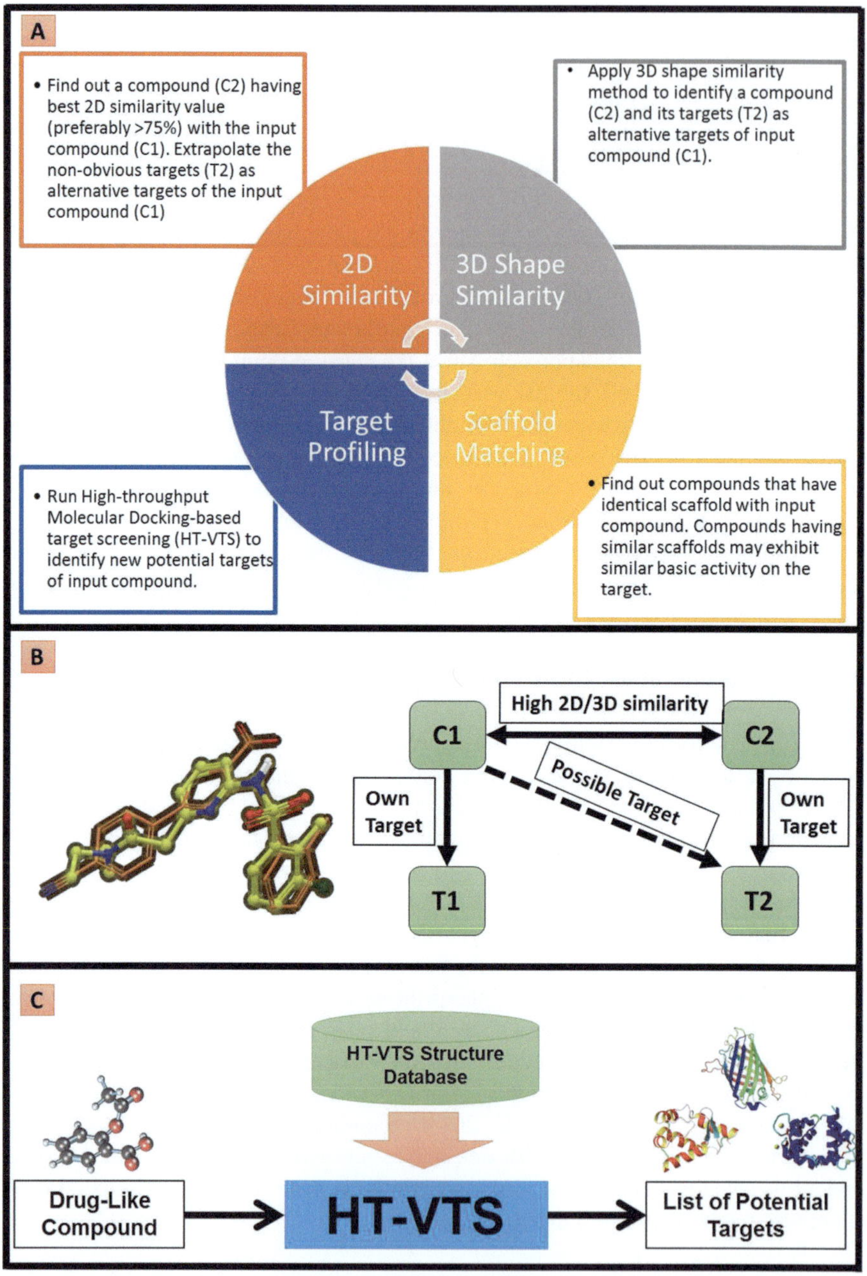

Fig. 3.5 Computational chemistry methods for drug repurposing. (**a**) An overview of four approaches. (**b**) 2D or 3D shape-based similarity method for drug repurposing. (**c**) High-throughput virtual target screening (HT-VTS) method for the identification of potential alternative target

impossible to manually scan all the articles related to a drug, disease, concept, or their associated combination. Therefore, a large number of literature mining techniques aiming to effectively extract the relation between keywords and present the biomedical interrelation have evolved (Frijters et al. 2010). In particular, ontologies have been extensively utilized in the biomedical domain either as controlled vocabularies or to provide the skeleton for mapping relations between concepts in biology and medicine (Andronis et al. 2011).

3.3.5.3 Genome-Wide Association Study (GWAS)

GWAS studies are not only limited to curate the biology of diseases but also provide translational opportunities for drug discovery and development through drug repurposing. In this approach, disease association for a given target gene is looked by means of single-nucleotide polymorphism (SNP) (Sanseau et al. 2012). Curated databases are available where information related to SNP and their association to pathologies is indexed. One of the extensively used databases is provided by the National Human Genome Research Institute (http://www.genome.gov/gwastudies/). Using this resource, Sanseau et al. (2012) performed an analysis to unveil potential new indications for protein targets through GWAS. The underlying concept behind the approach is that the association between a SNP and a trait from a GWAS can be extrapolated as a relation between a gene and a disease.

Other than the abovementioned methods, adverse events (Yang and Agarwal 2011), electronic health records (Xu et al. 2015), and web semantics (Chen et al. 2012) are also reported as useful methods for finding alternative indications.

3.4 Conclusion

Translational research, in recent years, has gained wide attention among the scientists across the globe. Traditionally, the industries were mostly considered as "product-driven" and academics as "knowledge-driven" centers. However, with the increasing burden of unmet medical needs, it becomes difficult and unfair for industries to be held responsible to sought solutions for all such medical urgencies. Therefore, understanding the need, governments across the globe have initiated to open nonprofit centers to conduct translational research. These centers have brought opportunities for professionals from multiple disciplines to come together, collaborate, exchange ideas, and focus their efforts to achieve the goal of having a disease-free world. However, like any other discipline, the field of translational research also comes with several challenges. Some of these challenges include wide coverage of chemical space around active ingredient to protect intellectual property, poor knowledge of disease pathophysiology, unknown differences in disease progression in animal vs. human, and regulatory delays, to name a few. The exceptionally large amount of scientific data are being generated on day-to-day basis and being stored in central repositories such as databases and publications, which, in turn, again are accessed from all corners of the globe and have opened several avenues to retrieve data, develop tools, build analytics, and generate meaningful hypothesis to facilitate translational research. However, one has to understand that the tools and techniques being utilized conventionally have now become obsolete to process such data. High-end technology – driven by intelligent algorithms such as machine learning and natural language processing – is now being utilized not only to mine deeper into the information stack but also to establish scientifically meaningful relationship in a complex physiological network. Thanks for the enhancement of technology, several opportunities can also be seen to attain high success rate in translational research. Organ-on-chip is one of such technologies being used to reduce attrition at later stages of drug discovery campaign. Accurate knowledge of patient stratification would be a key in clustering the patient population and to ensure that investigational new therapy should be delivered to the right patient and produce optimum benefits. Pre-Investigational New Drug (IND) meeting can ease the understanding of the technicalities associated with lengthy documentation and therefore reduce the time for later revisions. Not limited to this, the strategies such as drug repurposing, more strategic collaborations could be highly beneficial in accelerating the hunt to discover new drugs. The field of translational research is still an evolving discipline which not only needs dynamic workforce but also requires significant amount of monetary investment to put the right and advanced technology in place. The challenges in finding new and better therapies would never be less, but courage of facing such challenges has certainly strengthened with translational research. The continuous rise of diseases and their outbreaks and complexity have made the translational research as a necessity and not mere a scientific concept. With the hope that translational research will come as a ray of hope by bringing better and more effective therapies to millions of patients in the form of novel and highly efficient drugs, we all wish to live in a better future in the years to come.

References

Achenbach J, Tiikkainen P, Franke L, Proschak E. Computational tools for polypharmacology and repurposing. Future Med Chem. 2011;3:961–8.

Andrews J. Prioritization criteria methodology for future research needs proposals within the effective health care program: PiCMe-prioritization criteria methods. Methods future res needs reports. Rockville: Agency for Healthcare Research and Quality (US); 2013.

Andronis C, Sharma A, Virvilis V, Deftereos S, Persidis A. Literature mining, ontologies and information visualization for drug repurposing. Brief Bioinform. 2011;12:357–68.

Ashburn TT, Thor KB. Drug repositioning: identifying and developing new uses for existing drugs. Nat Rev Drug Discov. 2004;3:673–83.

Bartel DP. MicroRNAs: genomics, biogenesis, mechanism, and function. Cell. 2004;116 (2):281–97.

Benson JD, Chen YN, Cornell-Kennon SA, Dorsch M, Kim S, Leszczyniecka M, Sellers WR, Lengauer C. Validating cancer drug targets. Nature. 2006;441:451–6.

Berk M, Dean OM, Cotton SM, Gama CS, Kapczinski F, Fernandes B, Kohlmann K, Jeavons S, Hewitt K, Moss K, Allwang C, Schapkaitz I, Cobb H, Bush AI, Dodd S, Malhi GS. Maintenance N-acetyl cysteine treatment for bipolar disorder: a double-blind randomized placebo controlled trial. BMC Med. 2012;10:91.

Blondeau S, Do QT, Scior T, Bernard P, Morin-Allory L. Reverse pharmacognosy: another way to harness the generosity of nature. Curr Pharm Des. 2010;16:1682–96.

Buchan NS, Rajpal DK, Webster Y, Alatorre C, Gudivada RC, Zheng C, Sanseau P, Koehler J. The role of translational bioinformatics in drug discovery. Drug Discov Today. 2011;16:426–34.

Bulusu KC, Tym JE, Coker EA, Schierz AC, Al-Lazikani B. canSAR: updated cancer research and drug discovery knowledgebase. Nucleic Acids Res. 2014;42:D1040–7.

Butini S, Brogi S, Novellino E, Campiani G, Ghosh AK, Brindisi M, Gemma S. The structural evolution of β-secretase inhibitors: a focus on the development of small-molecule inhibitors. Curr Top Med Chem. 2013;13(15):1787–807.

Butler D. Translational research: crossing the valley of death. Nature. 2008;453:840–2.

Cheever MA, Schlom J, Weiner LM, et al. Translational Research Working Group developmental pathway for immune response modifiers. Clin Cancer Res. 2008;14:5692–9.

Cheever MA, Allison JP, Ferris AS, et al. The prioritization of cancer antigens: a national cancer institute pilot project for the acceleration of translational research. Clin Cancer Res. 2009;15 (17):5323–37.

Chen B, Ding Y, Wild DJ. Assessing drug target association using semantic linked data. PLoS Comput Biol. 2012;8:e1002574.

Cho WC. Conquering cancer through discovery research. IUBMB Life. 2010;62(9):655–9.

Collins FS. Reengineering translational science: the time is right. Sci Transl Med. 2011;3 (90):90cm17.

Dean OM, Maes M, Ashton M, et al. Protocol and rationale-the efficacy of minocycline as an adjunctive treatment for major depressive disorder: a double blind, randomised, placebo controlled trial. Clin Psychopharmacol Neurosci. 2014;12(3):180–8.

Doboszewska U, Szewczyk B, Sowa-Kućma M, Noworyta-Sokołowska K, Misztak P, Gołębiowska J, Młyniec K, Ostachowicz B, Krośniak M, Wojtanowska-Krośniak A, Gołembiowska K, Lankosz M, Piekoszewski W, Nowak G. Alterations of bio-elements, oxidative, and inflammatory status in the zinc deficiency model in rats. Neurotox Res. 2016;29(1):143–54.

Drolet BC, Lorenzi NM. Translational research: understanding the continuum from bench to bedside. Transl Res. 2011;157(1):1–5.

Drucker DJ, Sherman SI, Gorelick FS, Bergenstal RM, Sherwin RS, Buse JB. Incretin-based therapies for the treatment of type 2 diabetes: evaluation of the risks and benefits. Diabetes Care. 2010;33:428–33.

Ekins S, Williams AJ, Krasowski MD, Freundlich JS. In silico repositioning of approved drugs for rare and neglected diseases. Drug Discov Today. 2011;16:298–310.

Elsayed M, Banasr M, Duric V, Fournier NM, Licznerski P, Duman RS. Antidepressant effects of fibroblast growth factor-2 in behavioral and cellular models of depression. Biol Psychiatry. 2012;72(4):258–65.

Ernst & Young. Beyond borders: global biotechnology report. New York: Ernst & Young; 2010. http://www.ey.com/publication/vwluassets/beyond_borders/$file/beyond_borders_2010.pdf.

Fang FC, Casadevall A. Lost in translation – basic science in the era of translational research. Infect Immun. 2010;78(2):563–6.

Ferdinand K, Saini R, Lewin A, Yellen L, Barbosa JA, Kushnir E. Efficacy and safety of omapatrilat with hydrochlorothiazide for the treatment of hypertension in subjects nonresponsive to hydrochlorothiazide alone. Am J Hypertens. 2001;14(8 pt 1):788–93.

Ferrell CB. Reengineering clinical research science: a focus on translational research. Behav Modif. 2009;33(1):7–23.

Feustel SM, Meissner M, Liesenfeld O. Toxoplasma gondii and the blood-brain barrier. Virulence. 2012;3:182–92.

Fishburn CS. Translational research: improving the efficiency of drug development from bench to bedside and back again. Health New. 2011;9:1–5.
Fishburn CS. Translational research: the changing landscape of drug discovery. Drug Discov Today. 2013;18(9–10):487–94.
Fitzgerald GA. Opinion: anticipating change in drug development: the emerging era of translational medicine and therapeutics. Nat Rev Drug Discov. 2005;4(10):815–8.
Fowles JS, Dailey DD, Gustafson DL, Thamm DH, Duval DL. The Flint Animal Cancer Center (FACC) canine tumour cell line panel: a resource for veterinary drug discovery, comparative oncology and translational medicine. Vet Comp Oncol. 2016 May 19. doi: 10.1111/vco.12192. [Epub ahead of print]
Frijters R, van Vugt M, Smeets R, van Schaik R, de Vlieg J, Alkema W. Literature mining for the discovery of hidden connections between drugs, genes and diseases. PLoS Comput Biol. 2010;6:e1000943.
Fron Chabouis H, Chabouis F, Gillaizeau F, Durieux P, Chatellier G, Ruse ND, Attal JP. Randomization in clinical trials: stratification or minimization? The HERMES free simulation software. Clin Oral Invest. 2014;18:25–34.
Fruci D, Cho WC, Nobili V, et al. Drug transporters and multiple drug resistance in pediatric solid tumors. Curr Drug Metab. 2016;17(4):308–16.
Gfeller D, Grosdidier A, Wirth M, Daina A, Michielin O, Zoete V. SwissTargetPrediction: a web server for target prediction of bioactive small molecules. Nucleic Acids Res. 2014;42:W32–8.
Gilliland CT, Zuk D, Kocis P, et al. Putting translational science on to a global stage. Nat Rev Drug Discov. 2016;15(4):217–8.
Gong J, Cai C, Liu X, Ku X, Jiang H, Gao D, Li H. ChemMapper: a versatile web server for exploring pharmacology and chemical structure association based on molecular 3D similarity method. Bioinformatics. 2013;29:1827–9.
Gostin LO, Lucey D, Phelan A. The Ebola epidemic: a global health emergency. JAMA. 2014;312:1095–6.
Greaves M, Maley CC. Clonal evolution in cancer. Nature. 2012;481(7381):306–13.
Gregori-Puigjane E, Mestres J. A ligand-based approach to mining the chemogenomic space of drugs. Comb Chem High Throughput Screen. 2008;11:669–76.
Gu X, Chen H, Gao X. Nanotherapeutic strategies for the treatment of Alzheimer's disease. Ther Deliv. 2015;6(2):177–95.
Hait WN. Translating research into clinical practice: deliberations from the American Association for Cancer Research. Clin Cancer Res. 2005;11(12):4275–7.
Harris PA, Taylor R, Thielke R, Payne J, Gonzalez N, Conde JG. Research electronic data capture (REDCap) – a metadata-driven methodology and workflow process for providing translational research informatics support. J Biomed Inform. 2009;42:377–81.
Haupt VJ, Daminelli S, Schroeder M. Drug promiscuity in PDB: protein binding site similarity is key. PLoS One. 2013;8:e65894.
Hawk ET, Matrisian LM, Nelson WG, et al. The translational research working group developmental pathways: introduction and overview. Clin Cancer Res. 2008;14:5664–71.
Henney JE. Withdrawal of troglitazone and cisapride. J Am Med Assoc. 2000;283:2228.
Hess JL, Kawaguchi DM, Wagner KE, Faraone SV, Glatt SJ. The influence of genes on "positive valence systems" constructs: a systematic review. Am J Med Genet B Neuropsychiatr Genet. 2016;171:92–110.
Hillaire-Buys D, Faillie JL, Montastruc JL. Pioglitazone and bladder cancer. Lancet. 2011;378:1543–4.
Hobin JA, Deschamps AM, Bockman R, Cohen S, Dechow P, Eng C, Galey W, Morris M, Prabhakar S, Raj U, Rubenstein P, Smith JA, Stover P, Sung N, Talman W, Galbraith R. Engaging basic scientists in translational research: identifying opportunities, overcoming obstacles. J Transl Med. 2012;13(10):72.
Hoertel N, de Maricourt P, Gorwood P. Novel routes to bipolar disorder drug discovery. Expert Opin Drug Discovery. 2013;8(8):907–18.

Huh D, Matthews BD, Mammoto A, Montoya-Zavala M, Hsin HY, Ingber DE. Reconstituting organ-level lung functions on a chip. Science. 2010;328:1662–8.

Hurle MR, Yang L, Xie Q, Rajpal DK, Sanseau P, Agarwal P. Computational drug repositioning: from data to therapeutics. Clin Pharmacol Ther. 2013;93:335–41.

Ioannidis JP. Materializing research promises: opportunities, priorities and conflicts in translational medicine. J Transl Med. 2004;2:5.

Issa NT, Byers SW, Dakshanamurthy S. Drug repurposing: translational pharmacology, chemistry, computers and the clinic. Curr Top Med Chem. 2013;13:2328–36.

Issa NT, Peters OJ, Byers SW, Dakshanamurthy S. RepurposeVS: a drug repurposing-focused computational method for accurate drug-target signature predictions. Comb Chem High Throughput Screen. 2015;18(8):784–94.

Jordan CT, Guzman ML, Noble M. Cancer stem cells. N Engl J Med. 2006;355:1253–61.

Karnati HK, Panigrahi MK, Gutti RK, Greig NH, Tamargo IA. miRNAs: key players in neurodegenerative disorders and epilepsy. J Alzheimers Dis. 2015;48:563–80.

Katherine TA. Will biotechnology keep the heart healthy? Biotechnol Healthc. 2007;4(4):43–8.

Keramaris NC, Kanakaris NK, Tzioupis C, Kontakis G, Giannoudis PV. Translational research: from benchside to bedside. Injury. 2008;39(6):643–50.

Khanna I. Drug discovery in pharmaceutical industry: productivity challenges and trends. Drug Discov Today. 2012;17:1088–102.

Kharkar PS, Warrier S, Gaud RS. Reverse docking: a powerful tool for drug repositioning and drug rescue. Future Med Chem. 2014;6:333–42.

Khoury MJ, Gwinn M, Yoon PW, Dowling N, Moore CA, Bradley L. The continuum of translation research in genomic medicine: how can we accelerate the appropriate integration of human genome discoveries into health care and disease prevention? Genitourin Med. 2007;9 (10):665–74.

Koshland DE. Basic research (I). Science. 1993;259:291.

Lally J, MacCabe JH. Antipsychotic medication in schizophrenia: a review. Br Med Bull. 2015;114(1):169–79.

Law GL, Tisoncik-Go J, Korth MJ, Katze MG. Drug repurposing: a better approach for infectious disease drug discovery? Curr Opin Immunol. 2013;25:588–92.

Le-Niculescu H, Balaraman Y, Patel SD, Ayalew M, Gupta J, Kuczenski R, Shekhar A, Schork N, Geyer MA, Niculescu AB. Convergent functional genomics of anxiety disorders: translational identification of genes, biomarkers, pathways and mechanisms. Transl Psychiatry. 2011;1:e9.

Littman BH, Di Mario L, Plebani M, Marincola FM. What's next in translational medicine? Clin Sci (London, England). 2007;112(4):217–27.

Liu X, Ouyang S, Yu B, Liu Y, Huang K, Gong J, Zheng S, Li Z, Li H, Jiang H. PharmMapper server: a web server for potential drug target identification using pharmacophore mapping approach. Nucleic Acids Res. 2010;38:W609–14.

Liu X, Fu Q, Du Y, et al. MicroRNA as regulators of cancer stem cells and chemoresistance in colorectal cancer. Curr Cancer Drug Targets. 2016;16:738–54.

Lotrich FE, Butters MA, Aizenstein H, Marron MM, Reynolds CF, Gildengers AG. The relationship between interleukin-1 receptor antagonist and cognitive function in older adults with bipolar disorder. Int J Geriatr Psychopharmacol. 2014;29(6):635–44.

Lowe HJ, Ferris TA, Hernandez PM, Weber SC. STRIDE – an integrated standards-based translational research informatics platform. AMIA Annu Symp Proc. 2009;2009:391–5.

Madhavan S, Zenklusen JC, Kotliarov Y, Sahni H, Fine HA, Buetow K. Rembrandt: helping personalized medicine become a reality through integrative translational research. Mol Cancer Res. 2009;7:157–67.

Maes M, Galecki P, Chang YS, Berk M. A review on the oxidative and nitrosative stress (O&NS) pathways in major depression and their possible contribution to the (neuro)degenerative processes in that illness. Prog Neuro-Psychopharmacol Biol Psychiatry. 2011;35(3):676–92.

Maes M, Fišar Z, Medina M, Scapagnini G, Nowak G, Berk M. New drug targets in depression: inflammatory, cell-mediated immune, oxidative and nitrosative stress, mitochondrial,

antioxidant, and neuroprogressive pathways. And new drug candidates – Nrf2 activators and GSK-3 inhibitors. Inflammopharmacology. 2012;20(3):127–50.
Matter A. Bridging academic science and clinical research in the search for novel targeted anti-cancer agents. Cancer Biol Med. 2015;12(4):316–27.
Maurya PK, Noto C, Rizzo LB, Rios AC, Nunes SO, Barbosa DS, Sethi S, Zeni M, Mansur RB, Maes M, Brietzke E. The role of oxidative and nitrosative stress in accelerated aging and major depressive disorder. Prog Neuropsychopharmacol Biol Psychiatry. 2016;65:134–44.
McClure J. The value of basic research shouldn't be lost in translation. ASBMB; 2012.
McDermott U, Downing JR, Stratton MR. Genomics and the continuum of cancer care. N Engl J Med. 2011;364:340–50.
Millan MJ. Multi-target strategies for the improved treatment of depressive states: conceptual foundations and neuronal substrates, drug discovery and therapeutic application. Pharmacol Ther. 2006;110(2):135–370.
Millan MJ, Goodwin GM, Meyer-Lindenberg A, Ögren SO. 60 years of advances in neuropsychopharmacology for improving brain health, renewed hope for progress. Eur Neurol. 2015;5(5):591–8.
Morel J, Singer M. Statins, fibrates, thiazolidinediones and resveratrol as adjunctive therapies in sepsis: could mitochondria be a common target? Intensive Care Med Exp. 2014;2:9.
Morinet F. Aging of the brain, dementias, role of infectious proteins: facts and theories. Interdiscip Top Gerontol. 2014;39:177–86.
Munn DH, Mellor AL. Indoleamine 2,3 dioxygenase and metabolic control of immune responses. Trends Immunol. 2013;34(3):137–43.
Najjar S, Pearlman DM, Alper K, Najjar A, Devinsky O. Neuroinflammation and psychiatric illness. J Neuroinflammation. 2013;10:43.
Nissen SE, Wolski K. Effect of rosiglitazone on the risk of myocardial infarction and death from cardiovascular causes. N Engl J Med. 2007;356:2457–71.
Nowacka MM, Obuchowicz E. Vascular endothelial growth factor (VEGF) and its role in the central nervous system: a new element in the neurotrophic hypothesis of antidepressant drug action. Neuropeptides. 2012;46(1):1–10.
Old LJ. Cancer vaccines: an overview. Cancer Immun. 2008;8(1):1.
Oprea TI, Bauman JE, Bologa CG, et al. Drug repurposing from an academic perspective. Drug Discov Today Ther Strateg. 2011;8:61–9.
Padhy BM, Gupta YK. Drug repositioning: re-investigating existing drugs for new therapeutic indications. J Postgrad Med. 2011;57:153–60.
Palma M, Hanahan D. The biology of personalized cancer medicine: facing individual complexities underlying hallmark capabilities. Mol Oncol. 2012;6:111–27.
Pandya CD, Howell KR, Pillai A. Antioxidants as potential therapeutics for neuropsychiatric disorders. Prog Neuro-Psychopharmacol Biol Psychiatry. 2013;46:214–23.
Paul SM, Mytelka DS, Dunwiddie CT, Persinger CC, Munos BH, Lidborg SR, Schacht AL. How to improve R&D productivity: the pharmaceutical industry's grand challenge. Nat Rev Drug Discov. 2010;9:203–14.
Pober JS, Neuhauser CS, Pober JM. Obstacles facing translational research in academic medical centers. FASEB J. 2001;15:2303–13.
Sanseau P, Koehler J. Editorial: computational methods for drug repurposing. Brief Bioinform. 2011;12:301–2.
Sanseau P, Agarwal P, Barnes MR, Pastinen T, Richards JB, Cardon LR, Mooser V. Use of genome-wide association studies for drug repositioning. Nat Biotechnol. 2012;30:317–20.
Santiago DN, Pevzner Y, Durand AA, Tran M, Scheerer RR, Daniel K, Sung SS, Woodcock HL, Guida WC, Brooks WH. Virtual target screening: validation using kinase inhibitors. J Chem Inf Model. 2012;52:2192–203.
Sardana D, Zhu C, Zhang M, Gudivada RC, Yang L, Jegga AG. Drug repositioning for orphan diseases. Brief Bioinform. 2011;12:346–56.

Schrimpf D, Plotnicki L, Pilz LR. Web-based open source application for the randomization process in clinical trials: RANDI2. Int J Clin Pharmacol Ther. 2010;48:465–7.
Selep M. Translational research vs. basic science: comparing apples to upside-down apples. PLOS Blogs. 2013.
Sellers WR. A blueprint for advancing genetics-based cancer therapy. Cell. 2011;147:26–31.
Stadler K, Masignani V, Eickmann M, Becker S, Abrignani S, Klenk HD, Rappuoli R. SARS – beginning to understand a new virus. Nat Rev Microbiol. 2003;1:209–18.
Sugarman J, McKenna WG. Ethical hurdles for translational research. Radiat Res. 2003;160:1–4.
Tall AR, Yvan-Charvet L, Wang N. The failure of torcetrapib. Was it the molecule or the mechanism? Arterioscler Thromb Vasc Biol. 2007;27:257–60.
Tufts. What is translational science. http://tuftsctsi.org/. Tufts Clinical and Translational Science Institute; 2015.
Ungvari Z, Sonntag WE, de Cabo R, Baur JA, Csiszar A. Mitochondrial protection by resveratrol. Exerc Sport Sci Rev. 2011;39(3):128–32.
Verdine GL, Walensky LD. The challenge of drugging undruggable targets in cancer: lessons learned from targeting BCL-2 family members. Clin Cancer Res. 2007;13:7264–70.
Villamon E, Piqueras M, Meseguer J, Blanquer I, Berbegall AP, Tadeo I, Hernandez V, Navarro S, Noguera R. NeuPAT: an intranet database supporting translational research in neuroblastic tumors. Comput Biol Med. 2013;43:219–28.
Villar EL, Wang X, Madero L, Cho WC. Application of oncoproteomics to aberrant signalling networks in changing the treatment paradigm in acute lymphoblastic leukaemia. J Cell Mol Med. 2015;19(1):46–52.
Wang Y, Li Y, Liu X, et al. Genetic and epigenetic studies for determining molecular targets of natural product anticancer agents. Curr Cancer Drug Targets. 2013;13(5):506–18.
Watson MS, Epstein C, Howell RR, Jones MC, Korf BR, McCabe ER, Simpson JL. Developing a national collaborative study system for rare genetic diseases. Genitourin Med. 2008;10:325–9.
Wędzony K, Chocyk A, Maćkowiak M. Potential roles of NCAM/PSA-NCAM proteins in depression and the mechanism of action of antidepressant drugs. Pharmacol Rep. 2013;65 (6):1471–8.
Wesnes KA, Edgar CJ. The role of human cognitive neuroscience in drug discovery for the dementias. Curr Opin Pharmacol. 2014;14:62–73.
White HD, Van de Werf FJJ. Clinical cardiology: new frontiers thrombolysis for acute myocardial infarction harvey. Circulation. 1998;97:1632–46.
WHO Geneva. 2004. http://www.who.int/whr/2004/en/
Wilson D. Drug firms face billions in losses in '11 as patents end. The New York Times; 2011. http://www.nytimes.com/2011/03/07/business/07drug.html.
Winchester CL, Pratt JA, Morris BJ. Risk genes for schizophrenia: translational opportunities for drug discovery. Pharmacol Ther. 2014;143(1):34–50.
Woolf SH. The meaning of translational research and why it matters. JAMA. 2008;299(2):211–3.
Xu H, Aldrich MC, Chen Q, et al. Validating drug repurposing signals using electronic health records: a case study of metformin associated with reduced cancer mortality. J Am Med Inform Assoc. 2015;22:179–91.
Yang L, Agarwal P. Systematic drug repositioning based on clinical side-effects. PLoS One. 2011;6:e28025.
Zerhouni E. The NIH roadmap. Science. 2003;302:63–72.
Zerhouni EA. Translational and clinical science – time for a new vision. N Engl J Med. 2005;353:1621–3.
Zhou W, Wang Y, Lu A, et al. Systems pharmacology in small molecular drug discovery. Int J Mol Sci. 2016;17(2):246.

Chapter 4
Exploring the Potential of Herbal Ligands Toward Multidrug-Resistant Bacterial Pathogens by Computational Drug Discovery

Sinosh Skariyachan

Abstract The emergence of multidrug resistance (MDR), extensive drug resistance (XDR), and pan-drug resistance (PDR) has become a critical issue worldwide. The available drugs are no longer effective therapeutic remedy against such bacterial pathogens. This necessitates alternative therapy remedies. Computational drug discovery plays a central role in designing novel phytotherapeutics against drug-resistant bacterial pathogens. This chapter initially describes the recent issues and concerns associated with bacterial extreme resistance. Further, it demonstrates the utility of herbal-based compounds as probable lead molecules against various drug targets of multidrug-resistant bacteria by molecular docking approaches.

Keywords Multidrug resistance • Probable lead molecules • Therapeutic remedy • Molecular docking • Phytotherapeutics • Probable lead molecules • Extreme resistance • Drug targets

4.1 Introduction

The development of multidrug resistance is a major healthcare burden in the treatment of pathogenic bacteria by distinct antimicrobial agents. Moreover, it is not just an issue confined only to bacteria but all microorganisms that have the efficiency to mutate and deliver the new drugs unsuccessful (Carlet 2014). Most of the pathogenic strains have become drug resistant, and some have become resistant to multiple conventionally used antibiotics and chemotherapeutic agents; they emerged as multidrug-resistant (MDR) strains or superbugs (Nikaido 2009; Carlet 2014). Recent studies revealed that antibiotics have lost their status as the "miracle

S. Skariyachan (✉)
Department of Biotechnology Engineering, Dayananda Sagar Institutions, Bangalore 560 078, India

Visvesvaraya Technological University, Belagavi, Karnataka, India
e-mail: sinosh-bt@dayanandasagar.edu

© Shanghai Jiao Tong University Press, Shanghai and Springer Science+Business Media Dordrecht 2017

D.-Q. Wei et al. (eds.), *Translational Bioinformatics and Its Application*,
Translational Medicine Research, DOI 10.1007/978-94-024-1045-7_4

drug," and "treatment failure" is a new and often observed situation (Schjørring and Krogfelt 2011; Gowrishankar et al. 2013). Since the bacteria became resistant to many conventional antibiotics, there is a necessity to identify probable drug targets and screen for alternative therapeutic substances. One promising method is to prevent such drug-resistant pathogens by novel therapeutic compounds that are not based on existing synthetic antimicrobial agents. There is also a need for a deeper understanding of the mechanisms by which bacteria gain resistance to antibiotics which will aid in identifying novel targets for drugs or treatment. There are reports suggesting that several herbs produce bioactive compounds which are effective therapeutic agents (Nair et al. 2005). These medicinal plants are well studied and their bioactive compounds have been separated (Briskin 2000). Moreover, the bioactivity assay, modes of action, and inhibitory properties against various drug targets for many herbal-derived compounds are studied. Molecular docking-based studies pave new insight to screen natural herbal ligands which have ideal drug likeliness and pharmacokinetic properties (Bharath et al. 2011).

Computational drug discovery is the fundamental concept of structure-based drug design that uses a variety of computational methods to screen novel lead molecules with selectivity, efficacy, and safety (Lionta et al. 2014). The study of receptor-ligand interaction is the main focus of rational drug design, and the prediction of such interactions by computational approaches has profound scope and applications. Molecular docking is the prime component in computer-assisted molecular design. Molecular docking plays a vital role to understand the binding mechanism of herbal ligands toward various drug targets and inhibition of the pathways or any other means. Both rigid-body docking and flexible-body docking are playing vital roles in this dimension. The utility of best docking program, simulations and scoring, ranking, and docked conformations helps to hypothesize the probable mechanism. This provides profound scope and insight to further experimental analysis and screening of novel natural therapeutic substances (Lionta et al. 2014).

This chapter focuses the recent concerns and issues associated with multidrug resistance of bacterial pathogens and scope of molecular docking-based approaches for the discovery of novel herbal therapeutics against multidrug-resistant strains. The main strategy to achieve application for phytomedicine toward MDR is molecular docking-based studies and further in vitro and in vivo evaluation for the proposed approach.

4.2 Recent Issues Associated with MDR Bacteria

The increase in multidrug resistance poses a foremost healthcare threat. In the context of an almost complete absence of new chemotherapeutic drugs in progress, antibiotic resistance (ABR) has become one of the main healthcare implications (Boucher et al. 2009). According to Margret Chan, director general of World Health Organization, *Post antibiotic era is almost upon us.* Similarly, David Cameron, prime minister, UK, recently called for *a global action to tackle the growing threat of resistance to antibiotics*. Antibiotics are a unique class of therapeutic remedy because of their major impact in society. The application of an antibiotic in a person can select for ABR that can spread across human populations, animals, and the environment, making an antibacterial used in one person unproductive for many others. As bacteria acquire resistance mechanisms, the altered bacterial genetic material coding for resistance can be transferred between bacterial populations, expanding the reach and coverage of bacterial resistance. Treatment failures because of multidrug-resistant (MDR) bacteria arise very commonly in hospitals, in particular in the intensive care unit, and increasingly spreading in the other areas such as food, water, and air. *Methicillin-resistant Staphylococcus aureus* (MRSA) infections, especially due to community-acquired MRSA (DeLeo et al. 2010), are tremendously widespread in many European countries (European Center for Diseases Control and Prevention, EARSS-Net database. http://www.ecdc.europa.eu), the USA, South America, and Asia (Morcillo et al. 2015). MRSA infection accounts for 44% of all hospital-associated infections in the USA, and as many as 92% of persons hospitalized for MRSA have community-acquired MRSA (CA-MRSA) (Gould et al. 2008). There are newly developed agents that are active against vancomycin-resistant MRSA, such as linezolid and quinupristin/dalfopristin known as vancomycin-resistant enterococci (VRE). These bacteria are also very common, with large variations between countries ranging from 1 to >50% (Mutters et al. 2015). The predominance of *Escherichia coli* and *Klebsiella pneumoniae* harboring extended-spectrum β-lactamases is widespread across the world reaching 50–70% for *E. coli* in some European or Asian countries (Lowe et al. 2012). One of the study revealed that prevalence of *K. pneumoniae* with carbapenemases was going from 1 to >50% (Nordmann et al. 2009). Furthermore, a serious threat may be the emergence of Gram-negative bacteria that are resistant to all classes of the available chemotherapeutic agents referred to as pan resistance (Enani 2015). The emergence of "pan-resistant strains," mainly belonging to *Pseudomonas aeruginosa* and *Acinetobacter baumannii*, occurred in the recent past, after most of the major pharmaceutical industries stopped the development of new chemotherapeutic agents against bacterial infections (Nikaido 2009). One of the main global health concerns is the emergence and spread of drug-resistant tubercle bacilli across the world. The high burden of multidrug-resistant tuberculosis (MDR-TB) and the emergence and rise of advanced forms of drug resistance such as extensively drug-resistant TB (XDR-TB) and extremely drug-resistant TB (XXDR-TB) are some of the major concerns in the global healthcare sectors (Dalal et al. 2015).

In addition to clinical and hospital-associated cases, the multidrug resistance is spread across the environmental sectors. The lake, river, water storage tanks, etc. have become a cesspool of antibiotic-resistant bacteria (Thevenon et al. 2012). Due to massive accumulation of organic and industrial effluents especially sewage from hospitals and pharmaceutical industries, the natural status of the water bodies changed in terms of nutritional contents, dissolved oxygen, temperature, pH, and other physiochemical parameters. These create an ideal environment for the growth, survival, adaptation, and rapid proliferation of many pathogenic microorganisms especially bacterial coliforms. Along with the rapid multiplication, bacteria acquire many additional features due to the sudden changes in their chromosomes; an important concern is the acquisition of drug-resistant genes. These ingested coliforms are able to transfer drug resistance to other sensitive coliforms or enteric pathogens (Truman et al. 2014). The prevalence of carbapenem-resistant *E. coli* that harbored *NDM-1* gene in drinking water and sewage samples in New Delhi, India, was recently reported (Walsh et al. 2012). The superbugs carried various drug resistance genes in tap and springwaters in coastal region of Turkey (Ozgumus et al. 2007), and drinking water biofilms in Mainz, Germany, were also reported (Schwartz et al. 2003). Further, the prevalence of many pathogenic bacteria and their genes responsible for multidrug resistance toward β-lactam, amoxicillin/ampicillin (bla_{TEM}), streptomycin/spectinomycin (*aadA*), tetracycline (*tet*), chloramphenicol (*cmlA*), methicillin (*mec*), and vancomycin (*van*) in various aquatic ecosystems was also reported (Thevenon et al. 2012). Similarly, the prevalence of sulfonamide resistance genes in many aquatic environments in Tianjin, China (Gao et al. 2012), and cefotaxime and ciprofloxacin resistance genes in hospital-associated wastewater samples in Madhya Pradesh, India, were also reported (Diwan et al. 2012). A multidrug-resistant strain of *Salmonella* serovar *typhimurium* definitive type 104 (DT104) (resistant to sulfamethoxazole, tetracycline, streptomycin, chloramphenicol, and ampicillin) emerged across the USA during the 1990s (Glynn et al. 1998). In 2000, the Center for Disease Control and Prevention and several state health departments have identified a surge in the incidence of *Salmonella* serovar *Newport* (known as Newport-MDRAmpC), particularly multiple drug-resistant strains. These strains were also resistant to sulfamethoxazole, tetracycline, streptomycin, chloramphenicol, and ampicillin. Moreover, Newport-MDRAmpC isolates were resistant to cefoxitin, amoxicillin/clavulanic acid, ceftiofur, and cephalothin and showed decreased sensitivity to ceftriaxone (Gupta et al. 2003).

The infections due to MDR pathogens require very complex associations of high doses of old and new antibiotics, and mortality rate is very high. It is expected that at a minimum 25,000 patients in Europe and 23,000 in the USA die each year from infections caused by resistant bacteria (CDC, ECDC). The cost of ABR is incredible, whether measured as the personal and societal burden of illness, death rates, or healthcare costs. The WHO theme for the year 2011 was antimicrobial resistance (AMR), prioritizing the enhanced threat of a return to the pre-antibiotic era, when millions of lives were lost annually due to the MDR pathogens. In the European Union (EU), drug-resistant infections are estimated to generate healthcare costs of 1.5 billion euros per annum. In 2009, the EU has declared November 18 as "European Antibiotic Awareness Day," on each year to promote the cautious use of antimicrobial drugs (Gyles 2011).

4.3 Mechanism of Antibiotic Resistance: Recent Perspective

The expansion of bacterial resistance to antibiotics that had been available in nature prior to antibiotics was considered in chemotherapy. It has been reported that most pathogenic bacteria acquire resistance genes from the natural environments especially soils and water. The entire molecular and genetics cascade responsible for multidrug resistance (antibiotic resistome) has been superior to provide the basic framework for understanding the ecology of resistance. The antibiotic resistome comprises a set of all antibiotic resistance genes including those distributing in pathogenic bacteria, antibiotic producers, and benign nonpathogenic organisms found either free living or commensals of other organisms (Tavares et al. 2013). Most of the antibiotic producers live in soils and water, and as an ecological consequence, most of the susceptible bacteria in their locality, including human and animal pathogens, vanish, but some build up resistance to these natural habitats thought to manage the microbial population (Cox and Wright 2013).

The bacteria have become multidrug resistant by natural means or by acquired resistance. The natural resistance (intrinsic resistance) is due to some genes responsible for resistance to its own antibiotics. Acquired resistance is due to the mutation in bacterial chromosomes or the acquisition of mobile genetic elements (plasmid or transposons) which harbor the drug resistance genes (Martinez 2008). The resistance can be transferred between bacteria by horizontal gene transfer via transformation, transduction, or conjugation. Many drug resistance genes present in plasmids, facilitating their transfer, and develop multidrug-resistant bacteria. Thus, antibiotic resistance genes may be shared among different bacteria. Common biochemical and genetic aspects of antibiotic resistance mechanism are illustrated in Fig. 4.1. Further in detail, the probable mechanisms of antibiotic resistance that are reviewed by Nikaido (2009) are explained below.

4.3.1 *Alteration of the Target Protein by Mutation*

The bacteria can become resistant through mutations that make the target protein less susceptible to antibiotics. In the case of fluoroquinolone, the resistance is probably due to mutations in DNA topoisomerases, one of the target enzymes (Hooper 2000). The resistance of this antibiotic that is easily transferred to other cells on plasmids depends on the mode of action of the drug. The transfer of the drug-resistant enzyme gene is unable to make the bacteria completely resistant, and the mutated target gene will be transferred. This will be more prevalent in the presence of selective pressure by clonal selection. Similarly, the resistance acquired from target modification is conferred by the *erm* gene, which is responsible for the resistance toward macrolide (such as erythromycin), lincosamide, and streptogramin B. The *erm* gene is a plasmid-encoded gene which produces the

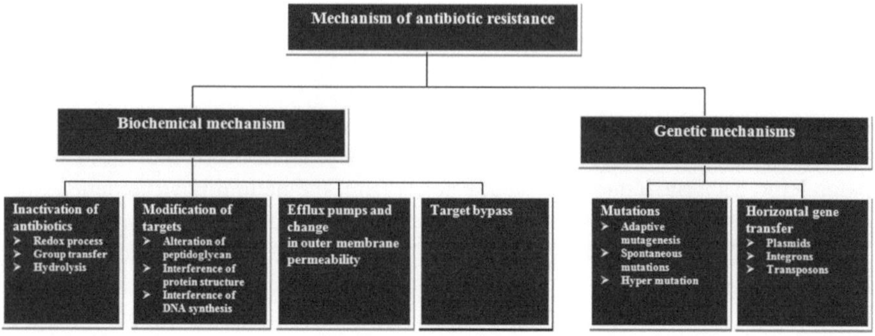

Fig. 4.1 Biochemical and genetic mechanism behind the evolution of drug resistance in bacteria

methylation of adenine at position 2058 of the 50S rRNA (Weisblum 1995). Furthermore, the sulfa drugs select drug-resistant mutants of the respective enzymes. The production of drug-resistant target enzymes from plasmids can make the bacteria resistant, and the resistant genes is widespread on plasmids in the case of sulfa drugs (Huovinen et al. 1995).

4.3.2 *Inactivation of the Drug by Various Enzymes*

This is the most common mechanism for natural resistance by bacteria. The antibiotic groups such as β-lactams (penicillins, cephalosporins, and carbapenems such as imipenem) inactivated via enzymatic hydrolysis by β-lactamases and aminoglycosides (amikacin, kanamycin, tobramycin, etc.) by enzymatic phosphorylation by aminoglycoside phosphoryltransferase (APH), adenylation by aminoglycoside adenyltransferase or nucleotidyltransferase, and acetylation by aminoglycoside acetyltransferase (AAC). The encoded genes for these inactivating enzymes can easily produce resistance as additional genetic components on plasmids.

4.3.3 *Gene Acquisition for Less Susceptible Target Proteins from Other Species*

This concept is based on the sequence data of penicillin-binding proteins (PBPs) or DD-transpeptidase, major penicillin target, which revealed that penicillin resistance observed in *Streptococcus pneumoniae* and *Neisseria meningitidis* was due to the production of mosaic proteins, parts of which came from other bacterial species (Spratt 1994, pp. 388–393). Similarly, methicillin-resistant *Staphylococcus aureus* contains a methicillin-resistant penicillin-binding protein (PBP), called PBP-2A or

2′, whose expression is induced by methicillin and other semisynthetic penicillin. The gene for this new PBP is located in a 30–60-kb large segment of DNA, which apparently came from other bacterial species and also contains other antibiotic resistance genes (de Lencastre et al. 2007).

4.3.4 Target Bypassing

The antibiotic vancomycin has an unusual mode of action. Instead of inhibiting an enzyme, vancomycin binds to a substrate, the lipid-linked disaccharide pentapeptide, a precursor of cell wall peptidoglycan. Studies revealed that the end of the pentapeptide, D-Ala-D-Ala, where vancomycin binds, was replaced in the resistant strain by an ester structure, D-Ala-D-lactic acid, which is not bound by vancomycin (Courvalin 2006). Production of this altered structure requires the involvement of many imported genes. Vancomycin resistance is common among enterococci. Since the enterococci are naturally resistant to aminoglycosides, β-lactams, tetracycline, and macrolides, these vancomycin-resistant strains of enterococci become predominant in a hospital environment, colonize the patients, and cause infections that are difficult to treat.

4.3.5 Declining Drug Access to Targets

The drug entrances can be reduced by an active efflux process especially by decreasing the influx across the outer membrane barrier. The main mechanisms are (i) local inhibition of drug access, (ii) drug-specific efflux pumps, and (iii) nonspecific inhibition of drug access.

4.3.6 Local Inhibition of Drug Access

Tet(S) or Tet(M) proteins, produced by Gram-positive bacteria, bind to ribosomes with high affinity and change the conformation of ribosomes, thereby preventing the association of tetracyclines to ribosomes (Connell et al. 2003). Similarly, Qnr proteins are thought to protect DNA topoisomerases from fluoroquinolones (Robicsek et al. 2006).

4.3.6.1 Drug-Specific Efflux Pumps

Drug resistance due to active efflux was discovered with TetA, the tetracycline resistance protein in Gram-negative bacteria. This protein catalyzes a protonmotive-force-dependent outward pumping of Mg-tetracycline complex (Tamura et al. 2003).

4.3.6.2 Nonspecific Inhibition of Drug Access

Reports suggested that porin, a membrane protein, mutants are found in some of the bacteria as a means of last-line resistance to the recent version of β-lactams that withstand inactivation by β-lactamases. Mutations within the coding sequences of the porin probably reduce the permeation rates of β-lactams without disturbing those of smaller molecules in the nutrient medium (Achouak et al. 2001).

4.4 Need for an Alternative Therapy

The antibiotic resistance became sustainable in the environment as already resistant bacteria emerged as new dominant population and evolved as superbugs (Schjørring and Krogfelt 2011; Gowrishankar et al. 2013). Since the bacteria became resistant to many conventional antibiotics, there is a necessity to identify probable drug targets and screen for alternative therapeutic substances. One promising method is to prevent such drug-resistant pathogens by novel therapeutic compounds that are not based on existing synthetic antimicrobial agents (Chah et al. 2006). The new approaches which have to be implemented include identification of novel molecular markers, screening of novel lead molecules for drug development, identification of novel treatment methods, and identification of a sample bacteria and its susceptibility to antibiotic treatment. There is also a need for a deeper understanding of the mechanisms by which bacteria gain resistance to antibiotics which will aid in identifying novel targets for drugs or treatment (Daniels 2011). Studying the genetic variation among plasmids from different bacterial species or strains is a key step toward understanding the mechanism of virulence and their evolution. Understanding their virulence helps in designing more effective drugs against the antibiotic-resistant microorganisms. The recent availability of new sequencing technologies provides the capability for rapid and cost-effective sequencing of small genomes (Siegel et al. 2006). Drug discovery and development are complex, laborious, and interdisciplinary approaches. For the pharmaceutical industry, the time span required to introduce a new drug to market is approximately 12–14 years and costing up to \$1.2–\$1.4 billion. For every 10,000 compounds that are tested in animal models, around 10 will qualify for clinical trials in order to get one drug on the market (Pandey et al. 2010).

By considering all the socio-environmental issues, there is a pressing need for screening novel lead molecules. The new approaches which have to be implemented include identification of new molecular markers, identification of novel lead molecules for drug development, identification of novel treatment methods, and identification of a sample bacteria and its susceptibility to antibiotic treatment. There is also a need for a deeper understanding of the mechanisms by which bacteria gain resistance to antibiotics which will aid in identifying novel targets for drugs or treatment (Daniels 2011). Advanced drug discovery process has been revolutionized with the advent of computational biology, genomics, proteomics, combinatorial chemistry, high-throughput screening, and structure-based design. The important aspects of computation in drug developments are virtual screening, de novo design, in silico ADMET prediction, and determination of receptor-ligand interactions. In silico ADMET prediction, screening is performed alongside of the in vitro data generated, for analyzing the target structures for possible binding conformation, generating bioactive conformation, checking the drug likeliness of ligands, docking these molecules with the target, ranking them according to their binding affinities, and further optimizing the molecules to improve the binding characteristics. Computational biology tools provide the advantage of delivering new therapeutic agents with ideal drug likeliness and pharmacophoric properties. High-performance computing and data management tools are enabling the access of large amount of complex biological data into executable knowledge in advanced drug discovery process.

4.5 Scope of Computer-Assisted or Structure-Based Drug Discovery

The screening and characterization of lead molecule showing therapeutic property against a biological target and standardization of the druggish properties and efficiency of these molecules are the initial stages of drug screening. For this purpose, many pharmaceutical industries have adopted the experimental screening of large chemical libraries against a therapeutically appropriate target (high-throughput screening or HTS) to discover new lead compounds. Through HTS, bioactive compounds, drug-resistant genes, or toxins, which amend a particular metabolic pathway, can be identified; these provide an initial insight for drug discovery and for knowing the role of a particular biochemical process in biological sciences. Even though HTS remains as the main attraction for drug discovery in the pharmaceutical industry, the various demerits of this approach that include the high capital cost, time, and the ambiguity of the mode of action of the bioactive lead molecules have turned to the increasing service of rational, structure-based drug design (SBDD) with the use of computational biology approaches. The important stages of structure-based drug discovery are illustrated in Fig. 4.2 (Lionta et al. 2014).

Computer-assisted drug discovery (CADD) is now being used for the identification of active drug candidates and selection and optimization of lead molecules which transform biologically active compounds into suitable drugs by improving their pharmacokinetic and drug likeliness properties. Computer-aided virtual screening is used to screen novel lead molecules from various chemical scaffolds

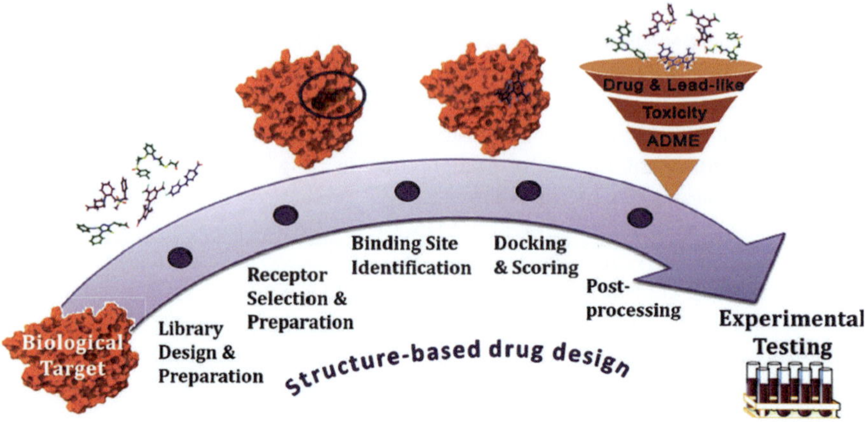

Fig. 4.2 Various stages of structure-based virtual screening ranging from receptor and library preprocessing to docking, scoring, and post-processing of top-scoring hits (Lionta et al. 2014)

by searching chemical structure databases and other resources (Kapetanovic 2008). CADD is the fundamental concept of structure-based drug design that uses a variety of computational methods to screen novel lead molecules with selectivity, efficacy, and safety. The study of receptor-ligand interaction is the main focus of rational drug design, and the prediction of such interactions by computational approaches has profound scope and applications (Lyskov and Gray 2008).

At present, structure-based drug discovery (SBDD) is the vital approach to the resourceful development of various therapeutic leads and to the understanding of metabolic processes especially the molecular-level mechanisms. SBDD is a well-established approach than the traditional way of drug discovery process to demonstrate the molecular mechanisms of a disease and utilizes the understanding of the three-dimensional (3D) structure of the biological target in the process. By the application of various bioinformatics approaches and the 3D structural information of the target protein, it is possible to explore the molecular interactions concerned with the protein-ligand binding and thus deduce the experimental results in molecular level. The utility of computer science and information technology in drug discovery provides the additional benefit of delivering novel drug candidates cost-effectively and quickly (Lionta et al. 2014).

The main concepts behind structure-based drug design methods are virtual screening (VS) and de novo drug design; these approaches serve as an alternative efficient approach to HTS. The main concept of virtual screening includes large libraries of drug-like molecules that are commercially obtainable and are screened computationally against probable targets of known structure, and those that are predicted to have better binding potential are validated experimentally (Lavecchia and Di Giovanni 2013). However, virtual screening does not offer molecules that are structurally "novel" as these molecules have been previously synthesized by various medicinal chemists. In the de novo drug design process, the information obtained from the 3D cavity of the receptor is used to design structurally relevant molecules that have not been synthesized previously by chemical intuition or any other methods (Jorgensen 2004).

Computer-assisted drug screening has recently had an important accomplishment: novel biologically active molecules have been predicted along with their receptor-bound conformation, and in quite a few cases, the success rates have been greater than with conventional high-throughput screening (Lavecchia and Di Giovanni 2013; Benod et al. 2013). Furthermore, though it is unusual to deliver lead molecules in the nanomolar (nM) concentration through virtual screening, several recent studies have demonstrated that the identification of nM leads from virtual screening approaches (Heifetz et al. 2013). Hence, computational biology methods play a vital role in the drug discovery and development process in the pharmaceutical sectors.

Computational biology became increasingly important in various areas such as gene and protein prediction, comparative or homology modeling, functional site location, characterization of active site for binding, docking of lead molecules into receptor-binding sites, protein-protein interactions, and molecular simulations. The outcome of computational studies yields information that is sometimes beyond current experimental possibilities and can be used to guide and improve a vast array of experiments (Gago 2004). Studies emphasize that the recognition of remote protein homologies is a major aspect of the structural and functional annotation of newly determined antibiotic resistance genes. PSI-BLAST is used for genome annotation using the widely used homology-searching program (Muller et al. 1999).

The primary necessity of computer-aided drug design is the three-dimensional structure of the resistant gene products or other drug targets such as toxins. However, the three-dimensional structures of most of the targets are not available in native forms. Hence, there is a need for an accurate three-dimensional model. This can be achieved by comparative modeling or homology modeling. Comparative modeling of proteins is a predictive technique to build high-resolution atomic model for a given amino acid sequence based on the structures of templates that have been experimentally determined. The ultimate goal of this modeling is to predict a structure from its sequence with an accuracy that is comparable to the best results achieved experimentally (Marti-Renom et al. 2000).

4.5.1 Scope of Molecular Docking Studies

A lead molecule is usually a small organic molecule, also known as ligand that binds to the target protein or receptors and changes the physiological function of the receptor, thus, leading to a therapeutic impact. Molecular docking or computer-assisted docking is an exceptionally useful means to achieve the understanding of receptor-ligand interactions which is a fundamental concept behind structure-based drug discovery. Computational docking is the method of computationally predicting the interaction and binding affinity of the lead molecule or inhibitor in the binding cavity of the protein. Molecular docking methods depend on search algorithms which determine the interaction of ligand in the binding cavity and a scoring function which calculates the binding efficiency, how perfectly the ligand interacts with the receptor (Dhanik and Kavraki 2012). The main forces that stabilized the receptor-ligand interactions are weak interactions such as hydrogen bonds, hydrophobic interactions, van der Waals forces, and electrostatic interaction. Hence, the main parameters required to evaluate a stable docked complexes are number of hydrogen bonding, extent of electrostatic interactions, and negative binding energy (kcal/mol). There are various methods that have been developed to explain the principles and concepts behind computational docking problems; some of the main concepts of molecular docking are:

- Molecular docking methods play a vital role in the drug discovery and development process.
- The docking methods identify the interaction of a ligand molecule in the binding cavity of receptor and determine the binding efficiency.
- There are two main important approaches for docking studies: (i) rigid-body docking and (ii) flexible body docking. The rigid-body docking approaches consider both the receptor and ligand as rigid bodies. However, flexible-body docking approaches consider the ligand as a flexible molecule, and flexible receptor approaches consider both the ligand and the protein as flexible molecules. In most of the cases, the docking programs consider the ligand as a flexible molecule and protein as a rigid molecule.
- The fundamental concepts involved in the docking studies are conformation search (by algorithm) and a scoring function that evaluate the binding capacity and efficiency.
- The flexibility of the protein is an essential component to determine the accuracy of various docking programs.
- There are various efforts that have been made to demonstrate the flexibility of protein in molecular docking studies; however, more studies need to be carried out.

Molecular docking has been an ideal option for the modeling of three-dimensional structure of the protein-ligand complex and evaluating the stability that estimates the specific biological recognition. However, there are few issues associated with these approaches: primarily, investigating the conformational space of ligands that interact with the receptor, and, secondly, ranking the conformations according to their estimated binding affinities (scoring) (Koehler and Villar 2000). More clearly, with the help of scoring function, the conformation of ligand is generated and compared to the previous conformations. The present conformation is further considered or discarded on the basis of the total score for that conformation. Furthermore, a new conformation is generated, and the search process continues until it covers all possible conformations. Hence, searching conformation and scoring can be coupled in docking process (Shoichet et al. 2002). Hence, it is very essential to identify better scoring functions so that the maximum rank ordered conformation would have higher experimental binding affinity with the receptor.

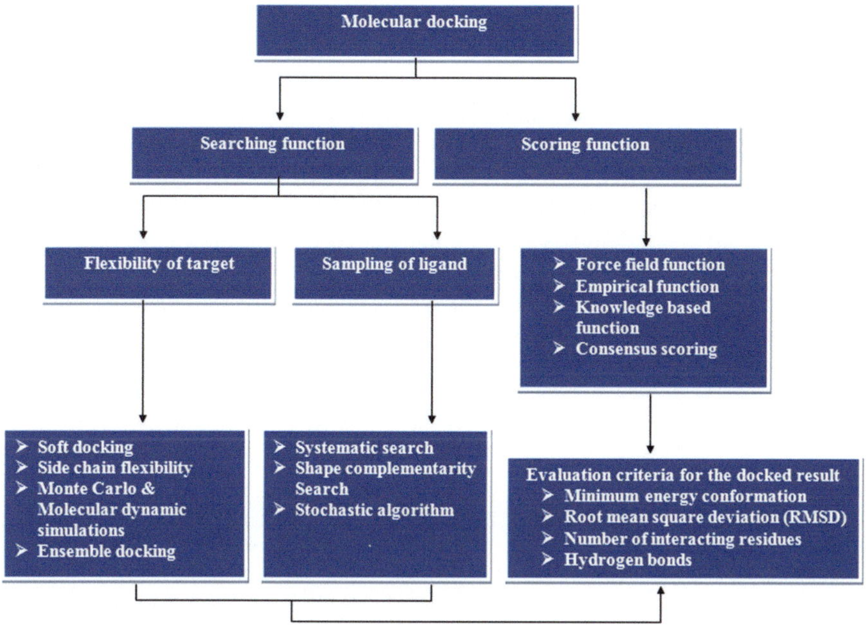

Fig. 4.3 Overview of molecular docking shows the steps involved in searching function and scoring function

The overview of molecular docking is illustrated in Fig. 4.3. The docking algorithm utilizes various approaches for conformational search in order to search conformational space of the ligand. The main approaches are:

(a) Systematic torsion exploration, which places the small molecules in the predicted binding pocket after considering the possible degrees of freedom
(b) Stochastic or random torsion exploration about rotatable bonds, such as genetic algorithms or Monte Carlo method to "achieve" new minimum energy conformers
(c) Molecular dynamics simulation approach and energy minimization for exploring the stable energy landscape of a compound (Lionta et al. 2014)

Scoring function is another critical step in docking process. The estimation of binding affinity between the receptor and ligands is the main logic of scoring function. The scoring functions have two main responsibilities. First, these functions serve as an objective function to distinguish between various poses of a single ligand in the receptor-binding pocket. Second, the scoring functions are essential to determine binding capabilities of various receptor-ligand complexes and to rank them as per the binding energies. The main factors that influence the receptor-ligand interactions are hydrogen bonding, van der Waals and dispersion interactions, hydrophobic effects, steric and electrostatic interactions, and solvation effects which are directed by various kinetic and thermodynamic principles (Reddy et al. 2007). The various approaches of scoring functions include shape and chemical complementary scoring, force field scoring, empirical scoring functions, and knowledge-based scoring functions. These methods are more or less combinations of ensemble-averaged terms and comprise a compromise between real and computational effort. The most effective search algorithm stops functioning in the absence of an ideal scoring function. The popular scoring functions currently available are grouped as (a) force field-based, (b) empirical-based, (c) knowledge-based, and (d) consensus-based scoring functions (Perola et al. 2004). A comprehensive list of various docking software available for public domains for the effective protein-ligand docking studies is reviewed in Table 4.1.

Table 4.1 List of the most popular protein-ligand docking programs available as of the middle of 2015

Docking software/ program	Year of establishment	Country of origin	References
DOCK	1988	USA	Ewing et al. (2001)
AutoDock	1990	USA	Morris et al. (1998)
SOFTDocking	1991	USA	Jiang and Kim (1991)
DockVision	1992	Canada	Hart and Read (1992)
LUDI	1992	Germany	Bohm (1992, pp. 61–78)
ADAM	1994	Japan	Mizutani et al. (1994)
FLOG	1994	USA	Miller et al. (1994)
SYSDOC	1994	USA	Luty et al. (1995)
DIVALI	1995	USA	Clark (1995, pp. 1210–1226)
GOLD	1995	UK	Jones et al. (1997)
FlexX	1996	Germany	Kramer et al. (1999)
Hammerhead	1996	USA	Welch et al. (1996)
LIGIN	1996	Israel/Germany	Sobolev et al. (1996)
FTDOCK	1997	UK	Gabb et al. (1997)
ICM-Dock	1997	USA	Totrov and Abagyan (1997)
QXP	1997	USA	McMartin and Bohacek (1997)
PRO LEADS	1998	UK	Baxter et al. (1998)
SANDOCK	1998	UK	Burkhard et al. (1998)
MCDOCK	1999	USA	Liu and Wang (1999)
PRODOCK	1999	USA	Trosset and Scheraga (1999)
SFDOCK	1999	China	Rodinger and Pomes (2000)
DARWIN	2000	USA	Taylor and Burnett (2000)
EUDOC	2001	USA	Pang et al. (2001)
FLEXE	2001	Germany	Claussen et al. (2001)
FDS	2003	UK	Taylor et al. (2003)
FRED	2003	USA/UK	McGann et al. (2003)
LigandFit	2003	USA	Venkatachalam et al. (2003)

(continued)

Table 4.1 (continued)

Docking software/ program	Year of establishment	Country of origin	References
PhDOCK	2003	USA	Joseph-McCarthy et al. (2003)
Surflex	2003	USA	Jain (2003, pp. 499–511)
iGEMDOCK	2004	Taiwan	Yang and Chen (2004)
Glide	2004	USA	Halgren et al. (2004)
ProPose	2004	Germany	Seifert et al. (2004)
YUCCA	2005	USA	Choi (2005, pp. 1517–1524)
eHiTS	2006	Canada/UK	Zsoldos et al. (2007)
MolDock	2006	Denmark	Thomsen and Christensen (2006)
PLANTS	2006	Belgium/ Germany	Korb et al. (2006)
PSI-DOCK	2006	China	Pei et al. (2006)
EADock	2007	Switzerland	Grosdidier et al. (2007)
FLIPDock	2007	USA	Zhao and Sanner (2007)
MDock	2007	USA	Huang and Zou (2007)
ParDOCK	2007	India	Gupta et al. (2007)
PSO@AUTODOCK	2007	Germany	Namasivayam and Gunther (2007)
SODOCK	2007	Taiwan	Chen et al. (2008)
Lead finder	2008	Russia/Canada	Stroganov et al. (2008)
MS-DOCK	2008	France	Sauton et al. (2008)
Q-Dock	2008	USA	Brylinski and Skolnick (2008)
MADAMM	2009	Portugal	Cerqueira et al. (2009)
AutoDock Vina	2010	USA	Trott and Olson (2010)
AADS	2011	India	Singh et al. (2011)
BetaDock	2011	South Korea	Kim et al. (2011)
LigDockCSA	2011	South Korea	Shin et al. (2011)
PythDock	2011	South Korea	Chung et al. (2011)
VoteDock	2011	Poland	Plewczynski et al. (2011)
idTarget	2012	Taiwan	Wang et al. (2012)
EpiDOCK	2013		Atanasova et al. (2013)
rDock	2013	UK	Ruiz-Carmona et al. (2014)
FIPSDock	2013	China	Liu et al. (2013)
DINC	2013	USA	Dhanik et al. (2013)
iStar	2014	UK	Li et al. (2014)
PharmDock	2014	USA	Hu and Lill (2014)
MoDock	2015	China	Gu et al. (2015)

4.6 Herbal Bioactive Compounds as Novel Therapeutics Against MDR Bacteria

There are reports suggesting that several herbs produce bioactive compounds which are effective therapeutic agents (Nair et al. 2005). These medicinal plants are well studied and their bioactive compounds have been separated. Their structural and functional mechanisms have also been established. Moreover, the bioactivity assay, modes of action, and inhibitory properties against various drug targets for many herbal-derived compounds are well studied (Briskin 2000). Computer-aided drug design (CADD) is an effective platform to screen several herbal lead molecules with better pharmacokinetic features and bioavailability (Bharath et al. 2011).

There are many databases which host the complete information of various lead molecules. The three-dimensional structures of most of the ligands are elucidated experimentally and can be retrieved from various databases. The most popular small molecule databases are ZINC (Irwin and Shoichet 2005), NCBI PubChem (Wang et al. 2012), Chemspider (Little et al. 2012), Drug Bank (Wishart et al. 2008), KEGG (Kanehisa 2002), etc.

There are many reports revealing the utility of computer-aided virtual screening toward the screening of novel therapeutic agents with better pharmacokinetic properties. Recent reports revealed the inhibitory properties of bioactive compounds screened from essential oils toward various drug targets of *Streptococcus mutans* (Galvão et al. 2012) by computational virtual screening. Similar reports showed that phytochemical compounds screened from few medicinal plants have significant inhibitory properties against various drug targets of multidrug-resistant clinical isolates (Dahiya and Purkayastha 2012). Similarly, the inhibitory activity of kurarinone, a bioactive flavonoid isolated from *Sophora flavescens*, against drug targets of methicillin-resistant *Staphylococcus aureus*, vancomycin-resistant *Streptococcus* sps., and *Streptococcus mutans* was also reported (Chen et al. 2005). Recently, it has been suggested that novel herbal inhibitors screened by computational virtual screening demonstrated good inhibitory properties against streptolysin-O of MDR *Streptococcus pyogenes* (Skariyachan et al. 2014). Similarly, a study also suggested that the herbal leads screened by in silico approach were found to have better inhibitory activities against the MDR gene products of *Vibrio cholerae*, *Salmonella typhi*, and *Staphylococcus aureus* (Skariyachan et al. 2013). Furthermore, previous studies have identified many novel lead molecules against virulent toxins of many superbugs (Skariyachan et al. 2012).

The computational redesign of bacterial biotin carboxylase inhibitors using structure-based virtual screening was recently reported (Brylinski and Waldrop 2014). Further, the identification of novel inhibitors of the glyoxylate shunt in MDR Gram-negative pathogens was also reported (Fahnoe et al. 2012). In silico discovery and virtual screening of multi-target inhibitors for various drug targets in *Mycobacterium tuberculosis* were recently reported (Chung et al. 2013). Similarly, another report revealed the utility of polyphosphate kinase (PPK) as a novel antimicrobial drug target and its high-throughput virtual screening toward MDR

E. coli isolates (Saha and Verma 2013). The inhibitory properties of novel lead candidates toward bacterial serine protease from various MDR isolates by computational virtual screening were also recently reported (Mandal et al. 2014). Furthermore, recent study demonstrated that herbal-based compounds such as nimbolide and isomargololone showed an appreciable IC50 value and significant binding properties toward New Delhi metallo-beta-lactamase 1 (bla_{NDM}) in comparison with 14 β-lactam antibiotics. The docking result of the antibacterial herbal compounds demonstrated that nimbolide (1.34 μM), isomargolonone (1.25 μM), margolone (5.25 μM), margolonone (5.34 μM), acetyl aleuritolic acid (0.2772 μM), and harmane (4.32 μM) had IC50 value lower than β-lactam antibiotics; this implies the therapeutic potential of herbal-based ligands over conventional drugs (Thakur et al. 2013). Similarly, lanatoside C and daidzein, two natural herbal leads, were identified as natural compound inhibitors against multidrug efflux pumps of *Escherichia coli* and *Pseudomonas aeruginosa* using computer-assisted virtual screening and in vitro validation (Aparna et al. 2014).

4.6.1 Relevance of Computational Discovery of Novel Herbal Therapeutics Toward MDR Bacterial Targets

The study of receptor-ligand interactions plays a vital role in understanding the screen novel therapeutic agents against multidrug-resistant pathogens. Molecular docking is the fundamental approach to study such kind of interactions for structure-based drug discovery. The following sections will explain how the docking studies are useful screen novel herbal therapeutic agents against various MDR pathogens.

The binding properties of various phytoligands toward the probable drug targets of multidrug-resistant *Salmonella typhi* and *Vibrio cholerae* and methicillin- and vancomycin-resistant *Staphylococcus aureus* were explored by molecular docking studies (Skariyachan et al. 2013). The genes responsible for multidrug properties of these organisms were screened. The selection of the genes was based on literature studies (Chen et al. 2010; Martínez 2012; Reimer et al. 2011; Hiramatsu et al. 1992; Weigel et al. 2003). Aminoglycoside phosphotransferase (aph; Uniprot ID: E2D0Y8), virulent protein for kanamycin resistance (Chen et al. 2010), and dihydrofolate reductase (dhfr; Uniprot ID: A7DY50), responsible factor for trimethoprim resistance of *Salmonella typhi* (Martínez 2012), were selected. Similarly, dihydrofolate reductase type I (dfrA1; Uniprot ID: G7TU76) and virulent factor for trimethoprim resistance from *Vibrio cholerae* (Reimer et al. 2011) were selected. Methicillin-resistant gene (Mec1; Uniprot ID: P68261) (Hiramatsu et al. 1992) and vancomycin-resistant gene (VanH; Uniprot ID: Q7BWD8) from *Staphylococcus aureus* (Weigel et al. 2003) were also selected.

Molecular docking studies suggested that baicalein, a type of flavonoid, commonly present in the root of *Scutellaria baicalensis*, and luteolin, another flavonoid

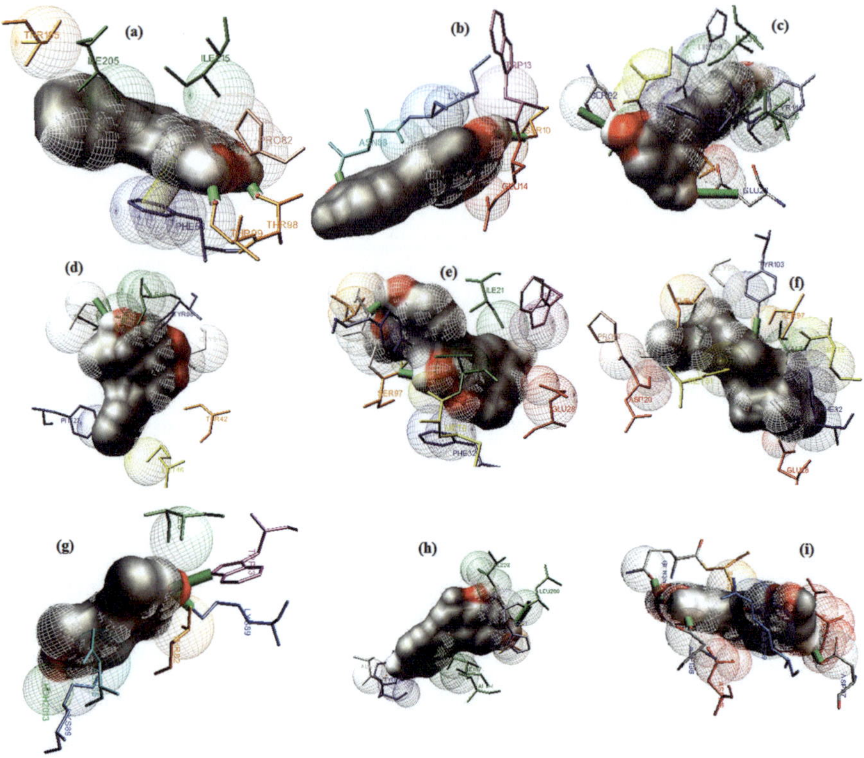

Fig. 4.4 Binding efficiency of phytoligand toward various drug targets of MDR pathogenic bacteria. Interaction between *aph* of *Salmonella typhi* and (**a**) baicalein and (**b**) luteolin. Interaction between *dhfr* of *Salmonella typhi* and (**c**) resveratrol and (**d**) wogonin. Interaction between dfrA1 of *Vibrio cholerae* and (**e**) herniarin and (**f**) pyrocide. Interaction between *mecI* of methicillin *Staphylococcus aureus* and taraxacin (**g**). Interaction between *vanH* of vancomycin-resistant *Staphylococcus aureus* and (**h**) apigenin and (**i**) luteolin

present in *Terminalia chebula*, showed the best interactions with aminoglycoside phosphotransferase (*aph*, drug-resistant gene for kanamycin resistance) of *Salmonella typhi*. The binding energy of *aph*-baicalein docked complex was estimated to be −6.39 kcal/mol, and the interactions were stabilized by two hydrogen bonds (Thr 98, Thr 99) (Fig. 4.4a). Similarly, the binding energy of *aph*-luteolin was estimated to be −6.42 kcal/mol, and the interactions were stabilized by two hydrogen bonds (Asn 88 and Ser 10) (Fig. 4.4b). From this study, it is clear that these phytoligands have significant binding and inhibitory properties toward kanamycin-resistant protein. Resveratrol, a natural phytoalexin commonly present in *Vitis vinifera* (grape), and wogonin, an O-methylated flavone found in *Scutellaria baicalensis* (baikal skullcap), showed significant inhibitory activities against dihydrofolate reductase (*dhfr*, gene product responsible for trimethoprim resistance) of *Salmonella typhi*. From the docking studies, it is clear that the binding energy of resveratrol toward *dhfr* was identified as −7.58 kcal/mol, and the interaction was stabilized by two hydrogen bonds (Glu 23, Ser 92) (Fig. 4.4c). The binding energy of wogonin against *dhfr* was estimated to be −7.28 kcal/mol (Fig. 4.4d). The interactions were stabilized by two hydrogen bonds (Ala 3, Gly 93). The antimicrobial effects of resveratrol and wogonin against various bacterial pathogens and their toxins have been studied (Schrader 2010, pp. 1676–1689; Chan 2002, pp. 99–104), which revealed that these phytochemicals have significant inhibitory properties toward virulent factors of many MDR pathogens. Herniarin, a natural chemical compound found in *Herniaria glabra* (smooth rupturewort), and pyrocide, a common flavone present in *Daucus carota* (carrot), showed the best binding activity toward dihydrofolate reductase (*dfrA1*, trimethoprim-resistant protein) of *Vibrio cholerae*. The docking studies suggested that the docked complex of *dfrA1* herniarin was stabilized by two hydrogen bonds (Ser 97, Gly 98) with binding energy of −8.06 kcal/mol (Fig. 4.4e). Studies on the antifungal and antibacterial activities of various herniarin derivatives revealed that these phytoligands showed good inhibitory activities against various enteric bacterial pathogens (Céspedes et al. 2006). Similarly, the interaction between pyrocide and *dfrA1* was stabilized by a hydrogen bond (Tyr 103) with binding energy of −8.93 kcal/mol (Fig. 4.4f). Though pyrocide exhibits better binding energies (−8.93 kcal/mol), the number of interactions with receptor is only Tyr 103 residue. Hence, better simulation studies are essential to screen this ligand, and present data pave significant insight for such studies. Luteolin and taraxacin, a sesquiterpene guaianolide present in *Taraxacum officinale* (weber), showed better binding properties toward *mecI* protein (gene code for methicillin resistance) of *Staphylococcus aureus*. The molecular docking studies revealed that the docked complex of *mecI* and luteolin were stabilized by two hydrogen bonds (Ala 101, Tyr 102) with binding energy of −7.58 kcal/mol. Similarly, taraxacin binds with *mecI* by the formation of two hydrogen bonds (Trp 13 and Lys 89) with the binding energy of −7.28 kcal/mol (Fig. 4.4g). This study depicts that Gly, Lys, His, and Thr are the main conserved residues which play a major functional role in the receptor (Kahlon et al. 2012). From this study, it is clear that these phytochemicals have significant inhibitory properties toward probable drug targets of MDR pathogens. Many studies revealed the inhibitory properties of taraxacin and luteolin (Ahmad et al. 2000) against various pathogenic microorganisms by different mechanisms. Apigenin, a flavone found in *Coffea arabica* (coffee), and luteolin were found to interact against *vanH* (gene responsible for vancomycin resistance) protein. The docked complex of *vanH*–apigenin was stabilized by two hydrogen bonds (Tyr 102, Leu 200; binding energy −6.07 kcal/mol) (Fig. 4.4h). Similarly, luteolin interacted with *vanH* by three hydrogen bonds (Gln 35, Asp 198, and Asp 64; binding energy −6.32 kcal/mol), and the main residues present in the active sites are Gln 35, Ser 36, Asp 64, Asp 67, Asp 198, Asp 216, and Arg 219 (Fig. 4.4i) (Skariyachan et al. 2013). The antimicrobial activities of all these lead molecules are well studied. A significant inhibitory property of apigenin toward drug-resistant *Enterobacter cloacae* was recently reported in comparison with the known chemotherapeutic agent, ceftazidime (Eumkeb and Chukrathok 2013). The current study identified various phytoligands which showed effective binding and conformational changes in drug targets. The binding efficiency of these phytoligands toward various drug-resistant proteins paves better understanding of the inhibitory mechanism of herbal leads, and such studies have high relevance in clinical and preclinical trials.

In another study, the author suggested that afzelin and gallocatechin, two important herbal ligands, demonstrated good binding affinities toward bla_{TEM} (gene products responsible for β-lactam resistance) of multidrug-resistant bacteria. Afzelin is a flavonol glycoside commonly present in the medicinal herb *Nymphaea odorata* (fragrant water lily). The molecular docking studies suggested that afzelin binds to bla_{TEM} with an energy of −7.44 kcal/mol, and the interaction is stabilized with three hydrogen bonds (Fig. 4.5a). Similarly, gallocatechol or gallocatechin is a flavanol commonly present in *Camellia sinensis* (green tea). It was found to be a noncarcinogenic compound to both rat and mouse models. The molecular docking suggested that the phytoligand binds bla_{TEM} with three hydrogen bonds by the binding energy of −6.36 kcal/mol (Fig. 4.5b). The antibacterial potential of Azelin (azelaic acid) against various clinical pathogens is reported (Fluhr and Degitz 2010). Similarly, the inhibitory potential of gallocatechin against drug targets of various multidrug-resistant isolates was also reported (Radji et al. 2013).

The in silico data provides significant insights for further experimental validation of novel inhibitors against the drug targets of MDR pathogens. A recent study reported by Wang et al. (2015) revealed the discovery of novel New Delhi metallo-β-lactamase-1 inhibitors by multistep virtual screening and docking studies (Wang et al. 2015). The *NDM-1* enzyme provides bacterial resistance against the β-lactam ring of antibiotics by its hydrolytic activity. Inhibition of *NDM-1* may stop the hydrolysis of β-lactam ring and plays a vital role against antibacterial resistance. The study focused the screening of potential *NDM-1* inhibitors by multistep virtual screening and molecular docking simulations. The study demonstrated that they have screened 2,800,000 lead-like molecules from the ZINC database and generated 298 compounds, and the binding efficiency was studied by molecular docking

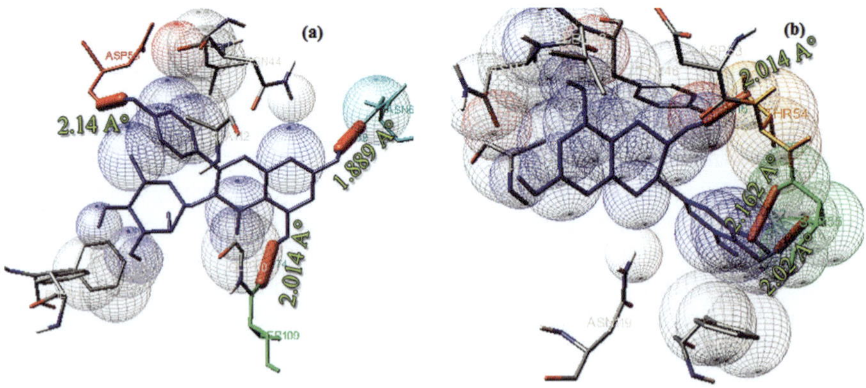

Fig. 4.5 Receptor-ligand interaction between (**a**) afzelin and bla_{TEM} and (**b**) gallocatechin and bla_{TEM} studied by molecular docking

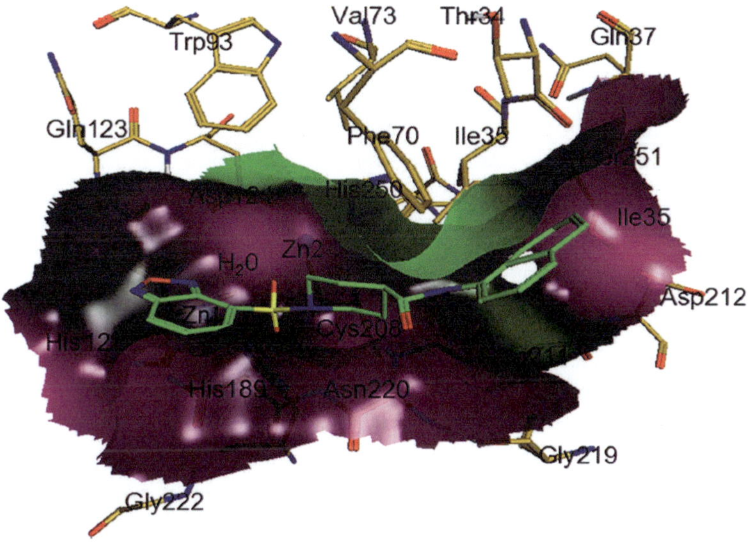

Fig. 4.6 The docked conformation showing the interaction between the active site of *NDM-1* and VNI-41. VNI-41 and adjacent *NDM-1* residues shown in stick representation, and the binding cavity is shown as molecular surface (Wang et al. 2015)

with *NDM-1*. The best lead molecules obtained from virtual screening and docking analysis were experimentally validated. Three novel *NDM-1* inhibitors with IC50 μM values were validated. Among the tested molecules, VNI-41 showed better inhibitory properties against *NDM-1* with an IC50 of 29.6 ± 1.3 μM. Molecular dynamic simulation based on the docking studies revealed that VNI-41 interacted with the active site (Fig. 4.6) (Wang et al. 2015). This study clearly demonstrated the possibility of applying virtual screening especially molecular docking studies in discovering novel and potential inhibitors against *NDM-1*, a metallo-β-lactamase of various multidrug-resistant bacterial pathogens.

Similar studies conducted by Thakur et al. (2013) suggested that molecular docking studies pave significant insight to design novel natural compounds against *NDM-1* gene products of various MDR pathogens. They have used molecular docking approach to design novel natural inhibitors against *NDM-1* receptor of MDR pathogens. The study suggested that lead molecules from botanicals such as nimbolide and isomargololone, bioactive compounds derived from *Azadirachta indica* (Neem tree), demonstrated good IC50 value as well as significant binding potential toward *NDM-1*. The study further suggested that the natural compounds expressed better binding affinity to *NDM-1* in comparison with conventional *β*-lactam antibiotics (Thakur et al. 2013).

4.7 Conclusion

As many bacteria emerged as extreme drug-resistant strains, designing of alternative remedies has prime scope and therapeutic relevance. Thus, there is a priority to screen new leads. The exploration of phytomedicine through molecular docking-based approaches serves as ultimate platforms to discover novel inhibitors against these drug targets, and present concepts offer outstanding landmarks for further in vitro and in vivo studies.

4.8 Future Perspectives

Molecular docking approaches are an effective platform to discover novel lead molecules against various drug targets of multidrug-resistant bacteria when conventional therapies seem to have failed. Molecular docking provides a comprehensive profile of the receptor-ligand interaction which is the fundamental concept of structure-based drug discovery. However, further experimental analysis is required to appreciate the hypothesis. Hence, the herbal bioactive compounds hypothesized needed to be isolated and characterized. The purified lead molecules need to be tested in vitro and in vivo to validate the proposed hypothesis-based molecular docking studies. The current approach has profound scope and applications in the development of future therapies against multidrug-resistant pathogens.

References

Achouak W, Heulin T, Pagès JM. Multiple facets of bacterial porins. FEMS Microbiol Lett. 2001;199:1–7.

Ahmad VU, Yasmeen S, Ali Z, Khan MA, Choudhary MI, Akhtar F, et al. Taraxacin, a new guaianolide from *Taraxacum wallichii*. J Nat Prod. 2000;63:1010–1.

Aparna V, Dineshkumar K, Mohanalakshmi N, Velmurugan D, Hopper W. Identification of natural compound inhibitors for multidrug efflux pumps of *Escherichia coli* and *Pseudomonas aeruginosa* using *in silico* high-throughput virtual screening and in vitro validation. PLoS One. 2014;9:e101840.

Atanasova M, Patronov A, Dimitrov I, Flower DR, Doytchinova I. EpiDOCK: a molecular docking-based tool for MHC class II binding prediction. Protein Eng Des Sel. 2013;26:631–4.

Baxter CA, Murray CW, Clark DE, Westhead DR, Eldridge MD. Flexible docking using Tabu search and an empirical estimate of binding affinity. Proteins. 1998;33:367–82.

Benod C, Carlsson J, Uthayaruban R, Hwang P, Irwin JJ, Doak AK, et al. Structure based discovery of antagonists of nuclear receptor LRH-1. J Biol Chem. 2013;288:19830–44.

Bharath EN, Manjula SN, Vijaychand A. *In silico* drug design tool for overcoming the innovation deficit in the drug discovery process. Int J Pharm Pharm Sci. 2011;3:8–12.

Bohm HJ. The computer program LUDI: a new method for the de novo design of enzyme inhibitors. J Comput Aided Mol Des. 1992;6:61–78.

Boucher HW, Talbot GH, Bradley JS, Edwards JE, Gilbert D, Rice LB, et al. Bad bugs, no drugs: no ESKAPE! An update from the Infectious Diseases Society of America. Clin Infect Dis. 2009;48:1–12.

Briskin DP. Medicinal plants and phytomedicines. Linking plant biochemistry and physiology to human health. Plant Physiol. 2000;124:507–14.

Brylinski M, Skolnick J. Q-Dock: low-resolution flexible ligand docking with pocket-specific threading restraints. J Biol Chem. 2008;29:1574–88.

Brylinski M, Waldrop GL. Computational redesign of bacterial biotin carboxylase inhibitors using structure-based virtual screening of combinatorial libraries. Molecules. 2014;19:4021–45.

Burkhard P, Taylor P, Walkinshaw MD. An example of a protein ligand found by database mining: description of the docking method and its verification by a 2.3 A X-ray structure of a thrombin-ligand complex. J Mol Biol. 1998;277:449–66.

Carlet J. Antibiotic resistance: protecting antibiotics – the declaration of the world alliance against antibiotic resistance. Indian J Crit Care Med. 2014;18:643–5.

Cerqueira NMFS, Bras NF, Fernandes PA, Ramos MJ. MADAMM: a multistaged docking with an automated molecular modeling protocol. Proteins. 2009;74:192–206.

Céspedes CL, Avila JG, Martínez A, Serrato B, Calderón-Mugica JC, Salgado-Garciglia R. Antifungal and antibacterial activities of Mexican tarragon (*Tagetes lucida*). J Agric Food Chem. 2006;54:3521–7.

Chah KF, Eze CA, Emuelosi CE, Esimone CO. Antibacterial and wound healing properties of methanolic extracts of some Nigerian medicinal plants. J Ethnopharmacol. 2006;104:164–7.

Chan MM. Antimicrobial effect of resveratrol on dermatophytes and bacterial pathogens of the skin. Biochem Pharmacol. 2002;63:99–104.

Chen L, Cheng X, Shi W, Lu Q, Go VL, Heber D, Ma L. Inhibition of growth of Streptococcus mutans, methicillin-resistant *Staphylococcus aureus*, and vancomycin-resistant enterococci by kurarinone, a bioactive flavonoid isolated from *Sophora flavescens*. J Clin Microbiol. 2005;43:3574–5.

Chen HM, Liu BF, Huang HL, Hwang SF, Ho SY. SODOCK: swarm optimization for highly flexible protein-ligand docking. J Biol Chem. 2008;28:612–23.

Chen CY, Lindsey RL, Strobaugh Jr TP, Frye JG, Meinersmann RJ, Chen CY, Meinersmann RJ. Prevalence of ColE1-like plasmids and kanamycin resistance genes in *Salmonella enterica serovars*. Appl Environ Microbiol. 2010;76:6707–14.

Choi V. YUCCA: an efficient algorithm for small-molecule docking. Chem Biodivers. 2005;22:1517–24.

Chung JY, Cho SJ, Hah JM. A python-based docking program utilizing a receptor bound ligand shape: PythDock. Arch Pharm Res. 2011;34:1451–8.

Chung BK, Dick T, Lee DY. *In silico* analyses for the discovery of tuberculosis drug targets. J Antimicrob Chemother. 2013;68:2701–9.

Clark KP. Flexible ligand docking without parameter adjustment across four ligand-receptor complexes. J Comput Chem. 1995;16:1210–26.

Claussen H, Buning C, Rarey M, Lengauer T. FlexE: efficient molecular docking considering protein structure variations. J Mol Biol. 2001;308:377–95.

Connell SR, Tracz DM, Nierhaus KH, Taylor DE. Ribosomal protection proteins and their mechanism of tetracycline resistance. Antimicrob Agents Chemother. 2003;47:3675–81.

Courvalin P. Vancomycin resistance in gram-positive cocci. Clin Infect Dis. 2006;42(Suppl 1): S25–34.

Cox G, Wright GD. Intrinsic antibiotic resistance: mechanisms, origins, challenges and solutions. Int J Med Microbiol. 2013;303:287–92.

Dahiya P, Purkayastha S. Phytochemical screening and antimicrobial activity of some medicinal plants against multi-drug resistant bacteria from clinical isolates. Indian J Pharm Sci. 2012;74:443–50.

Dalal A, Pawaskar A, Das M, Desai R, Prabhudesai P, Chhajed P, et al. Resistance patterns among multidrug-resistant tuberculosis patients in greater metropolitan Mumbai: trends over time. PLoS One. 2015;10:e0116798.

Daniels R. Surviving the first hours in sepsis: getting the basics right (an intensivist's perspective). J Antimicrob Chemother. 2011;66:11–23.

de Lencastre H, Oliveira D, Tomasz A. Antibiotic resistant *Staphylococcus aureus*: a paradigm of adaptive power. Curr Opin Microbiol. 2007;10:428–35.

DeLeo FR, Otto M, Kreiswirth BN, Chambers HF. Community-associated methicillin-resistant *Staphylococcus aureus*. Lancet. 2010;375:1557–68.

Dhanik A, Kavraki LE. Protein–ligand interactions: computational docking. eLS. Wiley;2012.

Dhanik A, McMurray JS, Kavraki LE. DINC: a new AutoDock-based protocol for docking large ligands. BMC Struct Biol. 2013;13(Suppl 1):S11. doi:10.1186/1472-6807-13-S1-S11.

Diwan V, Chandran SP, Tamhankar AJ, Lundborg CS, Macaden R. Identification of extended-spectrum β-lactamase and quinolone resistance genes in *Escherichia coli* isolated from hospital wastewater from central India. J Antimicrob Chemother. 2012;67:857–9.

Enani MA. Antimicrobial resistance. Insights from the declaration of World Alliance Against Antibiotic Resistance. Saudi Med J. 2015;36:11–2.

Eumkeb G, Chukrathok S. Synergistic activity and mechanism of action of ceftazidime and apigenin combination against ceftazidime-resistant *Enterobacter cloacae*. Phytomedicine. 2013;20:262–9.

Ewing TJA, Makino S, Skillman AG, Kuntz ID. DOCK 4.0: search strategies for automated molecular docking of flexible molecule databases. J Comput Aided Mol Des. 2001;15:411–28.

Fahnoe KC, Flanaganm ME, Gibson G, Shanmugasundaram V, Che Y, Tomaras AP. -Non-traditional antibacterial screening approaches for the identification of novel inhibitors of the glyoxylate shunt in gram-negative pathogens. PLoS One. 2012;7:e51732.

Fluhr JW, Degitz K. Antibiotics, azelaic acid and benzoyl peroxide in topical acne therapy. J Dtsch Dermatol Ges. 2010;8:S24–30.

Gabb HA, Jackson RM, Sternberg MJ. Modelling protein docking using shape complementarity, electrostatics and biochemical information. J Mol Biol. 1997;272:106–20.

Gago F. Modelling and simulation: a computational perspective in anticancer drug discovery. Curr Med Chem Anti-Cancer Agents. 2004;4:401–3.

Galvão LC, Furletti VF, Bersan SM, da Cunha MG, Ruiz AL, de Carvalho JE, et al. Antimicrobial activity of essential oils against *Streptococcus mutans* and their antiproliferative effects. Evid Based Complement Alternat Med. 2012;2012:751435.

Gao P, Mao D, Luo Y, Wanga L, Xu B, Xu L. Occurrence of sulfonamide and tetracycline-resistant bacteria and resistance genes in aquaculture environment. Water Res. 2012;46:2355–64.

Glynn MK, Bopp C, Dewitt W, Dabney P, Mokhtar M, Angulo FJ. Emergence of multidrug-resistant *Salmonella enterica serotype typhimurium DT104* infections in the United States. N Engl J Med. 1998;338:1333–8.

Gould DJ, Moralejo D, Drey N, Chudleigh J. Interventions to improve hand hygiene compliance in patient care. Cochrane Database Syst Rev. 2008;9:CD005186. doi:10.1002/14651858.CD005186.pub3.

Gowrishankar S, Thenmozhi R, Balaji K, Pandian SK. Emergence of methicillin-resistant, vancomycin-intermediate *Staphylococcus aureus* among patients associated with group A Streptococcal pharyngitis infection in southern India. Infect Genet Evol. 2013;14:83–389.

Grosdidier A, Zoete V, Michielin O. EADock: docking of small molecules into protein active sites with a multiobjective evolutionary optimization. Proteins. 2007;67:1010–25.

Gu J, Yang X, Kang L, Wu J, Wang X. MoDock: a multi-objective strategy improves the accuracy for molecular docking. Algorithms Mol Biol. 2015;10:8. doi:10.1186/s13015-015-0034-8.

Gupta A, Fontana J, Crowe C, Bolstorff B, Stout A, Duyne SV, Angulo FJ. Emergence of multidrug-resistant *Salmonella enterica Serotype Newport* infections resistant to expanded-spectrum cephalosporins in the United States. J Infect Dis. 2003;188:1707–16.

Gupta A, Gandhimathi A, Sharma P, Jayaram B. ParDOCK: an all atom energy based Monte Carlo docking protocol for protein-ligand complexes. Protein Pept Lett. 2007;14:632–46.
Gyles C. The growing problem of antimicrobial resistance. Can Vet J. 2011;52:817–20.
Halgren TA, Murphy RB, Friesner RA, Beard HS, Frye LL, Pollard WT, Banks JL. Glide: a new approach for rapid, accurate docking and scoring. 2. Enrichment factors in database screening. J Med Chem. 2004;47:1750–9.
Hart TN, Read RJ. A multiple-start Monte Carlo docking method. Proteins. 1992;1992:206–22.
Heifetz A, Barker O, Verquin G, Wimmer N, Meutermans W, Pal S, et al. Fighting obesity with a sugar based library: discovery of novel MCH-1R antagonists by a new computational-VAST approach for exploration of GPCR binding sites. J Chem Inf Model. 2013;53:1084–99.
Hiramatsu K, Asada K, Suzuki E, Okonogi K, Yokota T. Molecular cloning and nucleotide sequence determination of the regulator region of mecA gene in methicillin-resistant *Staphylococcus aureus* (MRSA). FEBS Lett. 1992;298:133–6.
Hooper DC. Mechanisms of action and resistance of older and newer fluoroquinolones. Clin Infect Dis. 2000;31(Suppl 2):S24–8.
Hu B, Lill MA. PharmDock: a pharmacophore-based docking program. J Cheminformatics. 2014;6:14. doi:10.1186/1758-2946-6-14.
Huang SY, Zou XQ. Ensemble docking of multiple protein structures: considering protein structural variations in molecular docking. Proteins-Struct Funct Bioinf. 2007;66:399–421.
Huovinen P, Sundström L, Swedberg G, Sköld O. Trimethoprim and sulfonamide resistance. Antimicrob Agents Chemother. 1995;39:279–89.
Irwin JJ, Shoichet BK. ZINC--a free database of commercially available compounds for virtual screening. J Chem Inf Model. 2005;45:177–82.
Jain AN. Surflex: fully automated flexible molecular docking using a molecular similarity-based search engine. J Med Chem. 2003;46:499–511.
Jiang F, Kim SH. "Soft docking": matching of molecular surface cubes. J Mol Biol. 1991;219:79–102.
Jones G, Willett P, Glen RC, Leach AR, Taylor R. Development and validation of a genetic algorithm for flexible docking. J Mol Biol. 1997;267:727–48.
Jorgensen WL. The many roles of computation in drug discovery. Science. 2004;303:1813–8.
Joseph-McCarthy D, Thomas BE, Belmarsh M, Moustakas D, Alvarez JC. Pharmacophore-based molecular docking to account for ligand flexibility. Proteins. 2003;51:172–88.
Kahlon AK, Darokar MP, Sharma A. Probing the evolutionary conserved regions within functional site of drug-resistant target proteins of *Staphylococcus aureus*: *In silico* phylogenetic motif profiling approach. Indian J Biochem Biophys. 2012;49:442–50.
Kanehisa M. The KEGG database. Novartis Found Symp. 2002;247:91–101.
Kapetanovic IM. Computer-aided drug discovery and development (CADDD): *in silico*-chemico-biological approach. Chem Biol Interact. 2008;171:165–76.
Kim DS, Kim CM, Won CI, Kim JK, Ryu J, Cho Y, Bhak J. BetaDock: shape-priority docking method based on beta-complex. J Biomol Struct Dyn. 2011;29:219–42.
Koehler TH, Villar OH. Design of screening libraries biased for pharmaceutical discovery. J Comput Chem. 2000;21:1145–52.
Korb O, Stutzle T, Exner TE. PLANTS: application of ant colony optimization to structure-based drug design. Berlin: Springer; 2006. p. 2006.
Kramer B, Rarey M, Lengauer T. Evaluation of the FlexX incremental construction algorithm for protein-ligand docking. Proteins. 1999;37:228–41.
Lavecchia A, Di Giovanni C. Virtual screening strategies in drug discovery: a critical review. Curr Med Chem. 2013;20:2839–60.
Li H, Leung KS, Ballester PJ, Wong MH. istar: a web platform for large-scale protein-ligand docking. PLoS One. 2014;9:e85678.
Lionta E, Spyrou G, Vassilatis DK, Cournia Z. Structure-based virtual screening for drug discovery: principles, applications and recent advances. Curr Top Med Chem. 2014;14:1923–38.

Little JL, Williams AJ, Pshenichnov A, Tkachenko V. Identification of "known unknowns" utilizing accurate mass data and ChemSpider. J Am Soc Mass Spectrom. 2012;23:179–85.

Liu M, Wang S. MCDOCK: a Monte Carlo simulation approach to the molecular docking problem. J Comput Aided Mol Des. 1999;13:435–51.

Liu Y, Zhao L, Li W, Zhao D, Song M, Yang Y. FIPSDock: a new molecular docking technique driven by fully informed swarm optimization algorithm. J Comput Chem. 2013;34:67–75.

Lowe CF, McGeer A, Muller MP, Katz K. For the Toronto ESBL working group. Decreased susceptibility to non-carbapenem antimicrobials in extended-spectrum-B-lactamase-producing *Escherichia coli* and *Klebsiella pneumoniae* isolates in Toronto, Canada. Antimicrob Agents Chemother. 2012;56:3977–80.

Luty BA, Wasserman ZR, Stouten PFW, Hodge CN, Zacharias M, McCammon JAA. Molecular mechanics grid method for evaluation of ligand-receptor interactions. J Comput Chem. 1995;16:454–64.

Lyskov S, Gray JJ. The Rosetta Dock server for local protein-protein docking. Nucleic Acids Res. 2008;36:W233–8.

Mandal SM, Porto WF, De D, Phule A, Korpole S, Ghosh AK, et al. Screening of serine protease inhibitors with antimicrobial activity using iron oxide nanoparticles functionalized with dextran conjugated trypsin and in silico analyses of bacterial serine protease inhibition. Analyst. 2014;139:464–72.

Martinez JL. Antibiotics and antibiotic resistance genes in natural environments. Science. 2008;321:365–7.

Martínez JL. Natural antibiotic resistance and contamination by antibiotic resistance determinants: the two ages in the evolution of resistance to antimicrobials. Front Microbiol. 2012; doi:10.3389/fmicb.2012.00001.

Marti-Renom MA, Stuart A, Fiser A, Sánchez R, Melo F, Sali A. Comparative protein structure modeling of genes and genomes. Annu Rev Biophys Biomol Struct. 2000;29:291–325.

McGann MR, Almond HR, Nicholls A, Grant JA, Brown FK. Gaussian docking functions. Biopolymers. 2003;68:76–90.

McMartin C, Bohacek RS. QXP: powerful, rapid computer algorithms for structure-based drug design. J Comput Aided Mol Des. 1997;11:333–44.

Miller MD, Kearsley SK, Underwood DJ, Sheridan RP. FLOG: a system to select 'quasi-flexible' ligands complementary to a receptor of known three-dimensional structure. J Comput Aided Mol Des. 1994;8:153–74.

Mizutani MY, Tomioka N, Itai A. Rational automatic search method for stable docking models of protein and ligand. J Mol Biol. 1994;243:310–26.

Morcillo A, Castro B, Rodríguez-Alvarez C, Abreu R, Aguirre-Jaime A, Arias A. Descriptive analysis of antibiotic-resistant patterns of methicillin-resistant *Staphylococcus aureus* (MRSA) st398 isolated from healthy swine. Int J Environ Res Public Health. 2015;12:611–22.

Morris GM, Goodsell DS, Halliday RS, Huey R, Hart WE, Belew RK, Olson AJ. Automated docking using a Lamarckian genetic algorithm and an empirical binding free energy function. J Comput Chem. 1998;19:1639–62.

Muller A, MacCallum RM, Sternberg MJ. Benchmarking PSI-BLAST in genome annotation. J Mol Biol. 1999;293:1257–71.

Mutters NT, Werner G, Tacconelli E, Mischnik A. Treatment options for serious infections caused by vancomycin-resistant enterococci. Dtsch Med Wochenschr. 2015;140:42–5.

Nair R, Kalariya T, Chanda S. Antibacterial activity of some selected Indian medicinal flora. Turk J Biol. 2005;29:41–7.

Namasivayam V, Gunther R. PSO@AUTODOCK: a fast flexible molecular docking program based on swarm intelligence. Chem Biol Drug Des. 2007;70:475–84.

Nikaido H. Multidrug resistance in bacteria. Annu Rev Biochem. 2009;78:19–46.

Nordmann P, Cuzon G, Naas T. The real threat of *Klebsiella pneumoniae* carbapenemase-producing bacteria. Lancet Infect Dis. 2009;9:228–36.

Ozgumus OB, Celik-Sevim E, Alpay-Karaoglu S, Sandalli C, Sevim A. Molecular characterization of antibiotic resistant *Escherichia coli* strains isolated from tap and spring waters in a coastal region in Turkey. J Microbiol. 2007;45:379–87.

Pandey S, Pandey P, Tiwari G, Tiwari R. Bioanalysis in drug discovery and development. Pharm Methods. 2010;1:14–24.

Pang YP, Perola E, Xu K, Prendergast FG. EUDOC: a computer program for identification of drug interaction sites in macromolecules and drug leads from chemical databases. J Comput Chem. 2001;22:1750–71.

Pei JF, Wang Q, Liu ZM, Li QL, Yang K, Lai LH. PSIDOCK: towards highly efficient and accurate flexible ligand docking. Proteins. 2006;62:934–46.

Perola E, Walters WP, Charifson PS. A detailed comparison of current docking and scoring methods on systems of pharmaceutical relevance. Proteins-Struct Funct Bioinf. 2004;56:235–49.

Plewczynski D, Lazniewski M, Von Grotthuss M, Rychlewski L, Ginalski K. Vote Dock: consensus docking method for prediction of protein-ligand interactions. J Comput Chem. 2011;32:568–81.

Radji M, Agustama RA, Elya B, Tjampakasari CR. Antimicrobial activity of green tea extract against isolates of methicillin-resistant *Staphylococcus aureus* and multi-drug resistant *Pseudomonas aeruginosa*. Asian Pacific J Trop Biomed. 2013;3:663–7.

Reddy AS, Pati SP, Kumar PP, Pradeep HN, Sastry GN. Virtual screening in drug discovery – a computational perspective. Curr Protein Pept Sci. 2007;8:329–51.

Reimer AR, Van Domselaar G, Stroika S, Walker M, Kent H, Tarr C, V. cholerae Outbreak Genomics Task Force. Comparative genomics of *Vibrio cholerae* from Haiti, Asia, and Africa. Emerg Infect Dis. 2011;17:2113–21.

Robicsek A, Jacoby GA, Hooper DC. The worldwide emergence of plasmid-mediated quinolone resistance. Lancet Infect Dis. 2006;6:629–40.

Rodinger T, Pomes R. Enhancing the accuracy, the efficiency and the scope of free energy simulation. Curr Opin Struct Biol. 2000;15:164–70.

Ruiz-Carmona S, Alvarez-Garcia D, Foloppe N, Garmendia-Doval AB, Juhos S, Schmidtke P, et al. rDock: a fast, versatile and open source program for docking ligands to proteins and nucleic acids. PLoS Comput Biol. 2014;2014(10):e1003571.

Saha SB, Verma V. *In silico* analysis of *Escherichia coli* polyphosphate kinase (PPK) as a novel antimicrobial drug target and its high throughput virtual screening against PubChem library. Bioinformation. 2013;9:518–23.

Sauton N, Lagorce D, Villoutreix BO, Miteva MA. MSDOCK: accurate multiple conformation generator and rigid docking protocol for multi-step virtual ligand screening. BMC Bioinforma. 2008;2008:9.

Schjørring S, Krogfelt KA. Assessment of bacterial antibiotic resistance transfer in the gut. Int J Microbiol. 2011;2011:312956. doi:10.1155/2011/312956.

Schrader KK. Plant natural compounds with antibacterial activity towards common pathogens of pond-cultured channel catfish (*Ictalurus punctatus*). Toxins (Basel). 2010;2:1676–89.

Schwartz T, Kohnen W, Jansen B, Obst U. Detection of antibiotic-resistant bacteria and their resistance genes in wastewater, surface water, and drinking water biofilms. FEMS Microbiol Ecol. 2003;43:325–35.

Seifert MHJ, Schmitt F, Herz T, Kramer B. ProPose: a docking engine based on a fully configurable protein-ligand interaction model. J Mol Model. 2004;10:342–57.

Shin WH, Heo L, Lee J, Ko J, Seok C, Lee J. LigDock-CSA: protein-ligand docking using conformational space annealing. J Comput Chem. 2011;32:3226–32.

Shoichet BK, McGovern SL, Wei B, Irwin JJ. Lead discovery using molecular docking. Curr Opin Chem Biol. 2002;6:439–46.

Siegel JD, Rhinehart E, Jackson M, Chiarello L. Management of multidrug-resistant organisms in healthcare settings. CDC – MDRO guidelines – HICPAC. 2006. Accessed 2 Mar 2015.

Singh T, Biswas D, Jayaram B. AADS – an automated active site identification, docking, and scoring protocol for protein targets based on physicochemical descriptors. J Chem Inf Model. 2011;51:2515–27.

Skariyachan S, Mahajanakatti AB, Sharma N, Karanth S, Rao S, Rajeswari N. Structure based virtual screening of novel inhibitors against multidrug resistant superbugs. Bioinformation. 2012;8:420–5.

Skariyachan S, Jayaprakash N, Bharadwaj N, Narayanappa R. Exploring insights for virulent gene inhibition of multidrug resistant *Salmonella typhi*, *Vibrio cholerae*, and *Staphylococcus aureus* by potential phytoligands via *in silico* screening. J Biomol Struct Dyn. 2013;32:1379–95.

Skariyachan S, Narayan NS, Aggimath TS, Nagaraj S, Reddy MS, Narayanappa R. Molecular modeling on streptolysin-O of multidrug resistant Streptococcus pyogenes and computer aided screening and in vitro assay for novel herbal inhibitors. Curr Comput-Aided Drug Des. 2014;10:59–74.

Sobolev V, Wade RC, Vriend G, Edelman M. Molecular docking using surface complementarity. Proteins. 1996;25:120–9.

Spratt BG. Resistance to antibiotics mediated by target alterations. Science. 1994;264:388–93.

Stroganov OV, Novikov FN, Stroylov VS, Kulkov V, Chilov GG. Lead finder: an approach to improve accuracy of protein-ligand docking, binding energy estimation, and virtual screening. J Chem Inf Model. 2008;48:2371–85.

Tamura N, Konishi S, Yamaguchi A. Mechanisms of drug/H+ antiport: complete cysteine-scanning mutagenesis and the protein engineering approach. Curr Opin Chem Biol. 2003;7:570–9.

Tavares LS, Silva CS, de Souza VC, da Silva VL, Diniz CG, Santos MO. Strategies and molecular tools to fight antimicrobial resistance: resistome, transcriptome, and antimicrobial peptides. Front Microbiol. 2013;4:412. doi:10.3389/fmicb.2013.00412.

Taylor JS, Burnett RM. DARWIN: a program for docking flexible molecules. Proteins. 2000;41:173–91.

Taylor RD, Jewsbury PJ, Essex JW. FDS: flexible ligand and receptor docking with a continuum solvent model and soft-core energy function. J Comput Chem. 2003;24:1637–56.

Thakur PK, Kumar J, Ray D, Anjum F, Hassan MI. Search of potential inhibitor against New Delhi metallo-beta-lactamase 1 from a series of antibacterial natural compounds. J Nat Sci Biol Med. 2013;4:51–6.

Thevenon F, Adatte T, Wildi W, Poté J. Antibiotic resistant bacteria/genes dissemination in lacustrine sediments highly increased following cultural eutrophication of Lake Geneva (Switzerland). Chemosphere. 2012;86:468–76.

Thomsen R, Christensen MH. MolDock: a new technique for high-accuracy molecular docking. J Med Chem. 2006;49:3315–21.

Totrov M, Abagyan R. Flexible protein-ligand docking by global energy optimization in internal coordinates. Proteins. 1997;1:215–20.

Trosset JY, Scheraga HA. Prodock: software package for protein modeling and docking. J Comput Chem. 1999;20:412–27.

Trott O, Olson AJ. AutoDock Vina: improving the speed and accuracy of docking with a new scoring function, efficient optimization, and multithreading. J Comput Chem. 2010;31:455–61.

Truman AW, Kwun MJ, Cheng J, Yang SH, Suh JW, Hong HJ. Antibiotic resistance mechanisms inform discovery: identification and characterization of a novel amycolatopsis strain producing ristocetin. Antimicrob Agents Chemother. 2014;58:5687–95.

Venkatachalam CM, Jiang X, Oldfield T, Waldman M. LigandFit: a novel method for the shape directed rapid docking of ligands to protein active-sites. J Mol Graph Model. 2003;2003:289–307.

Walsh TR, Weeks J, Livermore DM, Toleman MA. Dissemination of NDM-1 positive bacteria in the New Delhi environment and its implications for human health: an environmental point prevalence study. Lancet Infect Dis. 2012;1:355–62.

Wang JC, Chu PY, Chen CM, Lin JH. idTarget: a web server for identifying protein targets of small chemical molecules with robust scoring functions and a divide-and-conquer docking approach. Nucleic Acids Res. 2012;40(Web Server issue):W393–9.

Wang X, Lu M, Shi Y, Ou Y, Cheng X. Discovery of novel Delhi metallo-β-lactamases-1 inhibitors by multistep virtual screening. PLoS One. 2015;10:e0118290.

Weigel LM, Clewell DB, Gill SR, Clark NC, McDougal LK, Flannagan SE, Tenover FC. Genetic analysis of a high-level vancomycin-resistant isolate of *Staphylococcus aureus*. Science. 2003;302:1569–71.

Weisblum B. Erythromycin resistance by ribosome modification. Antimicrob Agents Chemother. 1995;39:577–85.

Welch W, Ruppert J, Jain AN. Hammerhead: fast, fully automated docking of flexible ligands to protein binding sites. Chem Biol. 1996;3:449–62.

Wishart DS, Knox C, Guo AC, Cheng D, Shrivastava S, Tzur D, et al. DrugBank: a knowledgebase for drugs, drug actions and drug targets. Nucleic Acids Res. 2008;36(Database issue):D901–6.

Yang JM, Chen CC. GEMDOCK: a generic evolutionary method for molecular docking. Proteins. 2004;55:288–304.

Zhao Y, Sanner MF. FLIPDock: docking flexible ligands into flexible receptors. Proteins. 2007;68:726–37.

Zsoldos Z, Reid D, Simon A, Sadjad SB, Johnson AP. eHiTS: a new fast, exhaustive flexible ligand docking system. J Mol Graph Model. 2007;26:198–212.

Part II
Protein-Ligand Interactions and Drug Development

Chapter 5
The Progress of New Targets of Anti-HIV and Its Inhibitors

Ke Z. Wu and Ai X. Li

Abstract HIV-1 virus is the largest genetic variation in human pathogens, with a high reproduction, high mutation, and high reorganization. At present, commonly prescribed drugs of anti-AIDS mainly contain nucleoside analogue reverse transcriptase inhibitor, non-nucleoside reverse transcriptase inhibitor, protease inhibitor, and integrase inhibitor. With rapid development in biotechnology during the latest decades, it has gradually uncovered not only the details of fusion and endocytosis between HIV and the host cells but also the necessary enzymes of HIV-1 during the whole life cycle, which brings about great progress in the field of anti-AIDS drugs development. In this article, we focus on some crucial proteins and cofactors correlated with the virus or the human defense function. The cofactor CCR5 and the viral envelope protein gp120 are significant in the initial process of fusion between HIV-1 and the host cells. Both of them become important targets of anti-HIV, and numerous inhibitors have been developed in which some have entered various stages of clinical trials or even been approved for marketing. Besides, the target of virus infectivity factor (Vif) and TRIM5-α protein is correlating with the host defense system. The inhibition of the former and the expression of the latter will increase the ability of response to the viral invasion. Both of them are still at the experimental stage. New targets and some corresponding inhibitors have been referred in this review; it is hoped that it can provide some clues for the drug development of anti-HIV.

Keywords Anti-HIV • Targets • Inhibitors

K.Z. Wu • A.X. Li (✉)
The Pharmacy of General Unit of Armed Police Hospital in Jiangxi, Nanchang 330001, China

Drug Design Laboratory of the Basic Science Department, Logistics University of Chinese People's Armed Police Force, Tianjin 300309, China
e-mail: liaixiu2006@126.com

© Shanghai Jiao Tong University Press, Shanghai and Springer Science+Business Media Dordrecht 2017

D.-Q. Wei et al. (eds.), *Translational Bioinformatics and Its Application*, Translational Medicine Research, DOI 10.1007/978-94-024-1045-7_5

5.1 Introduction

Acquired immune deficiency syndrome (AIDS), caused by human immunodeficiency virus (HIV), is a worldwide serious infectious disease. Up to the end of 2013, there have been 78,000,000 HIV patients over the world, and more than 3,900,000 people have died from AIDS. Heretofore, 26 anti-HIV drugs have been approved by FDA mainly distributing in six targets with different mechanisms including those well-known zidovudine and indinavir (Rower et al. 2012; Geng et al. 2010). Belonging to a reverse transcription virus, the viral nucleic acid is RNA and is enclosed by protein capsid in cubical symmetry which then recognizes the membrane of the host cell. This virus has no necessary genetic materials from the human body for its reproduction (Hollox and Hoh 2014). The biotechnology, rapidly developing during the latest decades, has gradually uncovered not only the details of fusion and endocytosis between HIV and the host cells but also the necessary enzymes of HIV during the whole life cycle that brings about great progress in the field of anti-AIDS drugs development then. Normally, according to the mechanism of infection process, the targets of anti-HIV can be divided into protease, reverse transcriptase, integrase, and so on. Unfortunately, during the drug therapy (highly active antiretroviral therapy, HAART) for AIDS, more and more drug-resistant virus emerge, and many HIV sufferers have to face the embarrassment of no cure for certain conditions. Therefore, searching for new targets and developing effective inhibitors of anti-AIDS have practical significance now.

Through overall documents consulting, this article mainly concerned about the new targets and corresponding inhibitors in recent years which aim to raise rational drug therapy strategies of anti-AIDS.

5.2 Research Situation

5.2.1 The Target CCR5 and Its Inhibitors

CC chemokine receptor 5 (CCR5) (Fig. 5.1), one of the coreceptors of HIV-1, is a new target of anti-HIV therapy (Lucia 2010). Experiments have shown that HIV-1 infects the human body by combining with the CD4 cell which is one of the most important immune cells in the human immune system. But subsequent studies have also found that the invasion of HIV-1 will not be successful only mediated by the CD4 cells, and one or more coreceptors are necessary during the initial process. Thus, a series of coreceptors of HIV-1 including CXCR4, CCR5, and integrin $\alpha 4\beta 7$ have been discovered in laboratory (Aiamkitsumrit et al. 2014). As members of G protein-coupled receptors (GPCR) superfamily, CCR5 is analogous in the structural features to most other chemokine receptors which are essentially composed of seven transmembrane regions and three extracellular loops (Tan et al. 2013). From the view of pathology, the interaction between gp120 and CCR5 can be

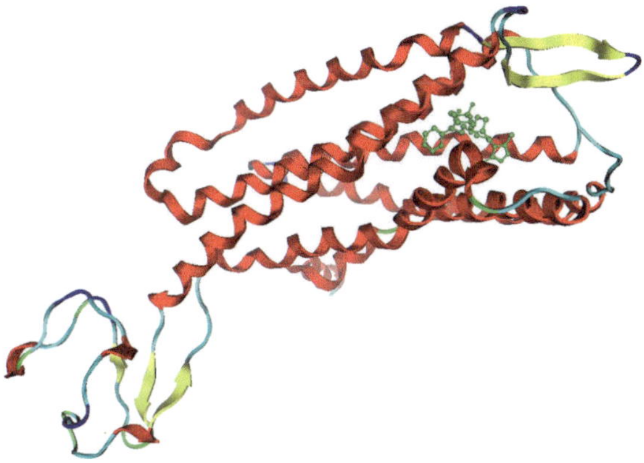

Fig. 5.1 CC chemokine receptor 5 (slab *ribbon*) and its inhibitor maraviroc (*green*)

divided into two steps. Firstly, the N-terminal region of CCR5 recognizes and combines with gp120 in a rational conformation. Secondly, as result of the conformation change of the two molecules, the interaction between the extracellular loops of CCR5 and the V3 region of gp120 finally causes the membrane fusion and the genetic materials' inner flow (Berro et al. 2012).

As an ideal target of anti-HIV, the inhibitors of CCR5 block the combination between HIV-1 and the cell membrane receptor by changing the conformation of CCR5 that disturbs the recognition of gp120 or the internalization of it (Kaqiampakis et al. 2011). The virus invading to the host cells may be interrupted with the combination of CCR5 and result in the declines in HIV infection rates. Nowadays, there are several kinds of CCR5 inhibitors including CCR5 derivatives, synthetic compounds, and peptide compounds.

5.2.1.1 Derivatives of CCR5

As the natural ligands of CCR5, β-RANTES such as MIP-1α (Mikawa et al. 2005) and MIP-1β (Kim et al. 2001) (Fig. 5.2a) certainly is the antagonist of HIV-1 receptors which protects the host cells by inducing the endocytosis of CCR5 to some extent. However, it is not suitable for the β-RANTES to become real drugs on account of their short half-life period (less than 10 min) and potential inflammatory response of these natural compounds. Researches show that CC-RANTES in a high concentration can help slow disease progression, while some different studies proposed that the high concentration may also activate cells to worsen the HIV-1 infection (Trkola et al. 1999). Thus, the Chemotactic Factor RANTES becomes more aggressive than the natural ligands because the former ones hinder the internalization of the receptor while not inducing the signaling pathways.

CC-chemokine RANTES (3–68) (Schols et al. 1998) (Fig. 5.2b), missing two NH_2-terminal residues, has been isolated from leukocytes and tumor cells. It has been proved to be an effective CCR5 receptor antagonist. RANTES (9–68) (Polo et al. 2000) is another CCR5 inhibitor which missed eight amino acids of the NH_2-

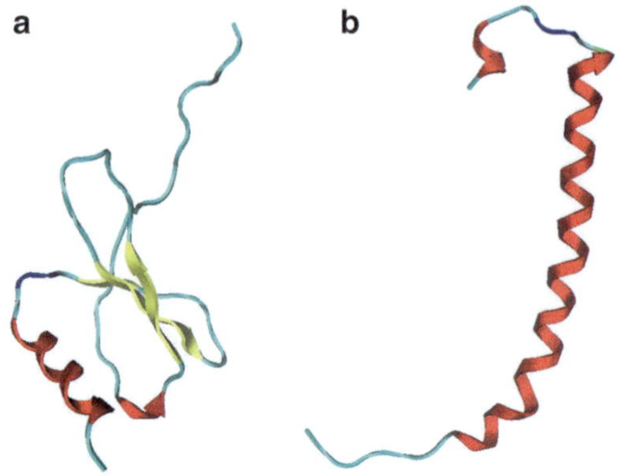

Fig. 5.2 The derivative inhibitors of CCR5. (**a**) MIP-1β. (**b**) CC-chemokine RANTES (3–68)

terminal, and the experimental data shows a lower inhibitory activity than the RANTES (3–68). Besides, other RANTES-based modified derivatives, such as AOP-RANTES, MET-RANTES, and PSC-RANTES (Lobritz et al. 2013), are also effective CCR5 antagonists. The derivatives are able to reduce the CCR5 expression level on the cell surface to realize the antiviral purpose.

5.2.1.2 Peptide Compounds

Peptide compounds specially recognize the particular extracellular region of CCR5 that has less poisonous side effect but will not be easily digested and degraded by the host (Wu et al. 2012). Appeared on the market in 2003, T20 is a peptide CCR5 inhibitor belonging to the HIV-1 fusion inhibitors which are derived from sequences 643–678 of transmembrane protein gp140. Peptide T (Maria et al. 2005) is another derivative of gp120 that is derived from the 185–192 amino acid sequences, and it has been proved to be nontoxic with inhibitory activities against HIV-1. There are also some polypeptides like peptide S, cDDR-MAP derived from CCR5 structure itself which possesses the ability of blocking the binding between gp120 and CCR5. Besides, a dodecapeptide (sequence: AFDWTFVPSLIL) screened from the peptide database by phage display has shown specific binding to CCR5 whose binding domain may belong to the ECL2 of it.

5.2.1.3 Non-peptide Compounds

Now, the non-peptide compounds are predominant in the developing of CCR5 inhibitors. This kind of antagonists, without potential inflammatory response effect, has advantages of low-cost production contrasting to the peptide compounds which also can be injected intravenously. There are several kinds of non-peptide inhibitors including TAK-779 (Ni et al. 2009), SCH-351125, maraviroc, etc. (Fig. 5.3). TAK-779, a small molecule antagonist of CCR5, has been developed by Takeda

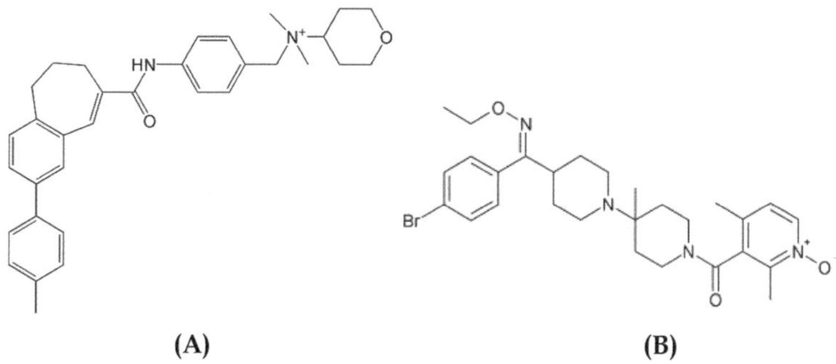

Fig. 5.3 The non-peptide inhibitors of CCR5. (**a**) Structure of TAK-779. (**b**) Structure of SCH-351125

Company in Japan and mainly used for the CCR5 receptor. As the first CCR5 drug, maraviroc has been approved for marketing in 2007, and it has been advised in the drug-combined therapy with the antiretroviral drugs. SCH-351125 (Marjan et al. 2007) is another high specific CCR5 antagonist with oxime piperidine structure which changes the conformation of extracellular domain by combining with the 1, 2, 3, and 7 transmembrane domains of CCR5. It is the first non-peptide CCR5 antagonist in the clinic that the side effect of prolonging the heart QT interval at high concentration prevents it from going further in clinical practice. But the structure reformation based on SCH-351125 has been proven to be successful. A series of derivatives have been received, in which SCH-417690 was testified to be enhanced by ten times in activity as its predecessor. Besides, without the cardiovascular side effect, SCH-417690 processed the advantage in pharmacokinetic parameters which has entered phase III clinical trial stage now.

5.2.2 The Target gp120 and Its Inhibitor

During the process of HIV-1 infecting the host cell, the viral envelope protein named gp120 primarily mixes together with the target cell membrane (Zhou et al. 2007). For this reason, the virus infection can be inhibited in the initial stage if the fusion process is blocked, which is also believed to be a promising drug therapy strategy.

HIV gp120, a member of glycoprotein, is composed of five variable regions (V1–V5) and another five relatively conservative regions (C1–C5) (Kwon et al. 2012). Researches show that there are four heparin sulfate binding sites on the surface of gp120 including V2, V3, C-terminal, and CD4 binding sites. After combining with CD4 at the last site, the other binding sites are exposed to the coreceptors like CX, CR4, or CCR5 and are easily recognizable (Schnur et al. 2011). The crystal structure of natural gp120 has been resolved in 2005, although that was from the simian immunodeficiency virus (SIV) rather than the HIV-1.

Compared with the structure binding with CD4, the natural gp120 is never found to have the bridge piece layer structure between the inner and the outer domains. The conformation of gp120 will significantly change while it combines with CD4 molecule, and a bridge piece layer structure forms between the V1/V2 and β20/β21 domains after then which commonly construct the binding site for the coreceptor in the next step (Shrivastava et al. 2012). Researches have shown that a stable α2-helix structure domain (Tan and Rader 2009) (Resides 335–352) locating on the outer region of gp120 binding with CD4 would be a target for developing new anti-HIV drugs.

5.2.2.1 Polypeptide Inhibitors

Gp120 exists in tri-polymer commonly between which the distances of any two CD4 binding pockets are 3–6 nm (Fig. 5.4). Based on this, a series of polypeptides simulating the tri-polymer CD4 have been designed in which a polypeptide G1 (ARQPSFDLQCGF) (Choi et al. 2001) simulates the Phe43 and β-folding of CD4 with good gp120 inhibiting activity. On this basis, several similar polypeptides have been synthesized after structure modification to G1, one of which even has an IC_{50} 1 μmol/L.

Another polypeptide (12p1) composed of 12 amino acids (RINNIPWSEAMM) has anti-HIV activity in vitro. It is able to combine with gp120 to prevent the binding between gp120 and CD4. Studies have demonstrated that 12pl inhibits CD4 binding with gp120 through allosteric effect rather than competitively inhibiting the binding of CD4 (Biorn et al. 2004). Recently, some researchers made a dimeric compound linked by 12pl and *Nostoc ellipsosporum* of *Cyanobacteria* (Zappe et al. 2008) which embodied stronger antiviral activity than the 12pl monomer.

5.2.2.2 Macromolecular Inhibitors

Researches indicated that solvable CD4 significantly reduced the virus load of HIV-1 in laboratory, but when it applied in clinical trials, it was weak to inhibit the viral strain. Further studies have found that some derivatives of solvable CD4 inhibit various HIV strains of which the half-time period in vivo remarkably extended. A recombinant fusion protein CD4-IgG2 designed by Progenics Pharmaceuticals Company could effectively reduce the viral load after clinical experiments (Alexandre et al. 2011). Furthermore, the induced allergic reaction and other side effects were not found that may have a good clinical application value.

The gp120 targeting broad-spectrum neutralizing antibody (Solanki et al. 2014) is another breakthrough in the research field of anti-HIV recently. These antibodies usually combine with the HIV envelope protein to prevent the virus from entering into the target cell; besides, they also clear the HIV particles and infected host cells through complement activation and other methods. Burton et al. isolated two broad-spectrum neutralizing antibodies PG9 and PG16 (Doores and Burton 2010) from

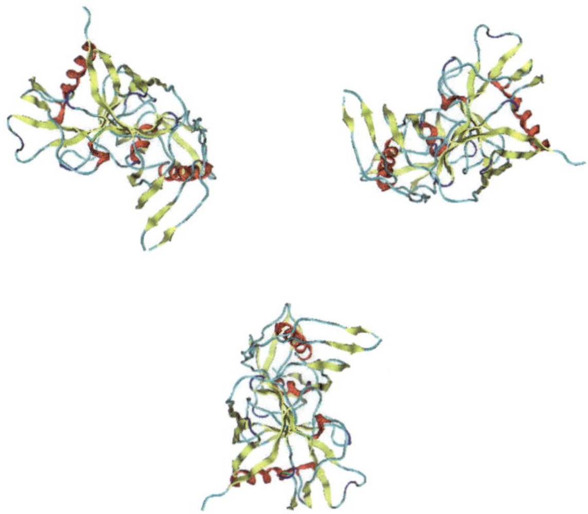

Fig. 5.4 HIV-1 gp120 forming timers on the surface of the viral membrane

the infected B-cells which neutralize two HIV subtype strains. Both of them are able to neutralize 127 and 119 plants, respectively, from the total 162 viral strains at a low concentration less than 1 mg/L which are confirmed to bind with the variable loops of conservative V1/V2 and V3 of gp120 in the next researches. Corti et al. separated another neutralizing antibody HJ16 that recognizes the binding domain of nearby CD4 (Corti et al. 2010). Experimental data showed that this antibody neutralized various subtypes of 92 HIV virus strains at a ratio of 36%. And the binding site recognized by HJ16 located in the conservative region of gp120 between the joint part of internal and external domains which may prove to be an important target for drug and vaccine design.

5.2.2.3 Small Molecular Inhibitors

The polypeptide inhibitors are the primary infusion inhibitors so far which have some shortcomings of low bioavailability and expensive cost restricting for use in clinical practice. Thus, the developing of small molecular inhibitors is necessary to increase the choices for patients. And the current developed inhibitors mainly block the binding between gp120 and CD4 or targeting to the phe43 binding pocket whose chemical structures include BMS-378806, NBD-556 [(Liu et al. 2014), (Zhao et al. 2005)], and their analogues (Fig. 5.5).

5.2.3 The Target Vif and Its Inhibitors

Innate immunity plays an important part in defending the HIV infection of human bodies, one of which called APOBEC3G (A3G) belonging to the apolipoprotein B

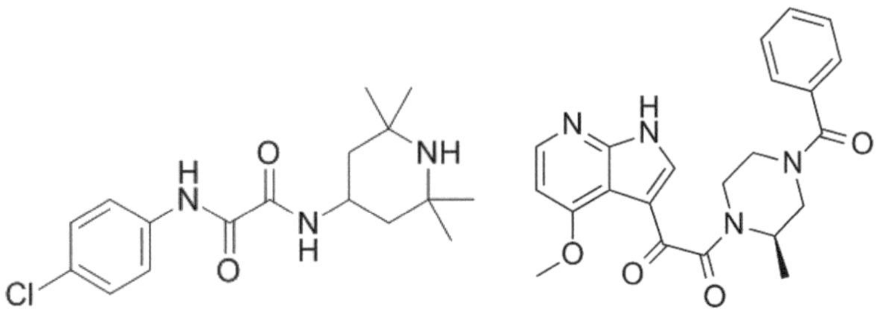

(A) (B)

Fig. 5.5 Small molecular inhibitors of gp120. (**a**) Structure of NBD-556. (**b**) Structure of BMS-378806

mRNA-editing enzyme catalytic polypeptide protein (APOBEC) inhibits the replication of HIV-1 by cytosine deaminase mechanism (Zhou et al. 2014). Meanwhile, the virus infectivity factor (Vif) combines with A3G to induce the degradation of it that increases the risk of infection for 100 times. It is no doubt that the interaction between A3G and Vif becomes the hotspot of anti-HIV.

The structure of Vif, composed of 192 amino acid residues, contains four domains, that is, N-terminal, HCCH domain, SLQYLA, and PPLP sequence motif. The N-terminal region is a critical section of Vif in which amino acid residues 40–44 and 85–99 are the binding sites of A3G, and residues 1–21, which are highly conservative, are the tryptophan-rich domains which play an important part in recognizing and inhibiting A3G. HCCH domain contains residues H108, C114, C113, and H139 and a chelating Zn^{2+} forming a zinc finger that mediates the combining with Cullin5 (Cul5) (Wang et al. 2014). Moreover, ^{144}SLQYLA^{149} motif induces the interaction between Vif and ElonginC.

5.2.3.1 The Pharmaceutical Research Strategy Based on the Target Vif-A3G

Vif collects Cul5, ElonginC, and Rbxl to form a Skpl-cullin-F-box complex to inhibit the virus replication through A3G degradation by ubiquitin-protease way (Wang et al. 2011). There are two methods at present to reach this goal: to prevent the degradation of A3G directly or to increase the package quantity of A3G.

Based on knowledge in the region of interaction between Vif and A3G, Straska et al. screened a compound RN-18 (Nathans et al. 2008) that degraded Vif to improve the A3G level, and Shan et al. also screened IMB-26/35 (Cen et al. 2010) (Fig. 5.6) from an 8634 compound database which competitively inhibited the combining of Vif and A3G. Another approach to prevent the degradation by Vif-mediated protease way lies in the structure modification of A3G whose tiny change will significantly reduce the sensibility of Vif. Researches showed that a

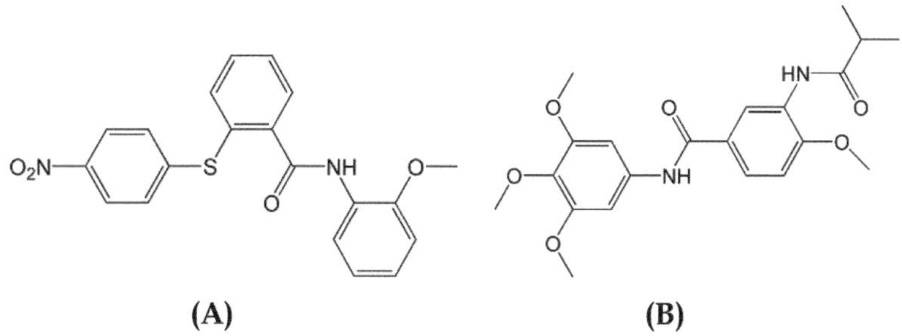

Fig. 5.6 Structure of Vif inhibitors RN-18 (**a**) and IMB-26 (**b**)

site-directed mutation D128K of A3G hindered the binding with Vif that the A3G was protected finally. Li et al. (2008) merged A3G and UBA2 together to form a fusion protein which would not be degraded by the ubiquitin. That is, the pathway of ubiquitin protease degradation has been blocked up.

Enhancing the ability of entering the progeny virus of A3G is another method to express its activity as possible. Green et al. built a Nef7-A3G fusion protein in which the Nef mutant called Nef7 increased the targeting transport efficiency of A3G while did not destroy the antiviral ability of it.

5.2.4 *The Target TRIM5-α*

Although HIV-1 is capable to infect numerous kinds of mammalian cells including human beings, it cannot infect some primates such as *Macaca rhesus*. The subsequent studies have found that there is a new protein in these primates' body named TRIM5-α which can effectively inhibit the replication of HIV-1 in vivo. As a critical inhibiting factor of the primate immune system to inhibit the retrovirus infection, it provides a new way for the treatment of AIDS.

5.2.4.1 Mechanism of TRIM5-αProtein Preventing the HIV-1 Infection

Researchers have found that TRIM5-α protein may package the HIV-1 capsid to hold back the genetic materials released that finally control the replication of HIV-1 in vivo (Zhang et al. 2013). After HIV-1 invading the host cells, the capsid and particle of the virus separate. Then, TRIM5-α protein seeks the released viral capsid for binding. And with the function of E3 ubiquitin-protein ligase in the RING domain (Lienlaf et al. 2011; Roa et al. 2012), it accelerates the degradation of the capsid to inhibit the replication of HIV-1 effectively. However, the existence of similar TRIM5-α protein was also confirmed in the cells of mankind whose activity of inhibiting HIV-1 was much sparser compared with the one in the monkey cells. It may attribute to the reason of genetic variation between different species that the powers of TRIM5-α protein in diverse individuals are not the same.

5.2.4.2 The Prospect of TRIM5-αProtein for Treating AIDS

Gene therapy (Anderson 2013) opens up board prospect for HIV-1 infection by which the TRIM5-α gene can be put into the target cells and then integrate with the genetic materials of the host. With the expression products of it, the uninfected host cells will be protected. Scientists are trying to crack the gene code of TRIM5-α, after then it could be transferred into the microorganism or the in vitro cells to produce abundant TRIM5-α proteins. Based on the new protein, other ways are to develop new medicines and vaccines by genetic engineering techniques or the new animal infection models prepared by TRIM5-α proteins.

5.3 Conclusion

At present, specific drugs and targets of anti-HIV have been developed, however, human beings are still having a hard time to eradicate the risk of this virus. The reason lies in its high reproduction, high mutation, and high reorganization. Thus, we urgently need to find new targets and more effective drugs of anti-HIV for the sake of human health. Based on the recent researches, several new targets including CCR5, gp120, Vif, and TRIM5-α have been found, and some corresponding inhibitors have been gradually used in the clinic.

With more binding sites on CCR5 receptor to combine, polypeptide inhibitors may also overcome the resistance produced by the small molecular inhibitors. For this reason, the polypeptide inhibitors may be the main research directions that will cover the shortages of the later ones. On the contrary, the polypeptide inhibitors of gp120 is low in bioavailability with expensive price, so the development of some small molecular entry inhibitors will be necessary. As the crystal structures of gp120 antigen/antibody complex have been resolved, the virtual screening may be effective technological means for searching new small molecular inhibitors of gp120. While the inhibition mechanisms have been elucidated, both Vif and TRIM5-α are hot new anti-HIV targets in recent years toward which some potential active compounds have been designed or screened. It is believed that the main tasks of anti-HIV in the future would be to find effective special targets so as to develop more powerful, high sensitive, and low side effect inhibitors.

Acknowledgements Supported by the National Natural Science Foundation of China (No. 30472166, No. 81241114).

References

Aiamkitsumrit B, Dampier W, Martin-Garcia J, Nonnemacher MR, Pirrone V, Ivanova T, Zhong W, Kilareski E, Aldigun H, Frantz B, Rimbey M, Wojno A, Passic S, Williams JW, Shah S, Blakey B, Parikh N, Jacobson JM, Moldover B, Wigdahl B. Defining Differential Genetic Signatures in CXCR4- and the CCR5-Utilizing HIV-1 Co-Linear Sequences PloS One 2014; 9: 1–22 doi: 10.1371/journal.pone.0107389 . PMID:25265194

Alexandre KB, Gray ES, Pantophlet R, Moore PL, McMahon JB, Chakauya E, O'Keefe BR, Chikwamba R, Morris L. Binding of the mannose-specific lectin, griffithsin, to HIV-1 gp120 exposes the CD4-binding site. J Virol. 2011;85:9039–50. doi:10.1128/JVI.02675-10. PMID:21697467

Anderson JS. Using TRIM5α as an HIV therapeutic: the alpha gene? Expert Opin Biol Ther. 2013;13:1029–38. doi:10.1517/14712598.2013.779251. PMID:23480791

Berro R, Klasse PJ, Jakobsen MR, Gorry PR, Moore JP, Sanders RW. V3 determinants of HIV-1 escape from the CCR5 inhibitors Maraviroc and Vicriviroc. Virology 2012; 427: 158–165 doi:10.1016/j.virol.2012.02.006. PMID:22424737

Biorn AC, Cocklin S, Madani N, Si Z, Ivanovic T, Samanen J, Van Ryk DI, Pantophlet R, Burton DR, Freire E, Sodroski J, Chaiken IM. Mode of action for linear peptide inhibitors of HIV-1 gp120 interactions. Biochemistry. 2004;43:1928–38. doi:10.1021/bi035088i. PMID:14967033

Cen S, Peng ZG, Li XY, Li ZR, Ma J, Wang YM, Fan B, You XF, Wang YP, Liu F, Shao RG, Zhao LX, Yu L, Jiang JD. Small molecular compounds inhibit HIV-1 replication through specifically stabilizing APOBEC3G. J Biol Chem. 2010;285:16546–52. doi:10.1074/jbc.M109.085308. PMID:20363737

Choi YH, Rho WS, Kim ND, Park SJ, Shin DH, Kim JW, Im SH, Won HS, Lee CW, Chae CB, Sung YC. Short peptides with induced beta-turn inhibit the interaction between HIV-1 gp120 and CD4. J Med Chem. 2001;44:1356–63. doi:10.1021/jm000403+. PMID:11311058

Corti D, Langedijk JP, Hinz A, Seaman MS, Vanzetta F, Fernandez-Rodriguez BM, Silacci C, Pinna D, Jarrossay D, Balla-Jhagjhoorsingh S, Willems B, Zekveld MJ, Dreja H, O'Sullivan E, Pade C, Orkin C, Jeffs SA, Montefiori DC, Davis D, Weissenhorn W, McKnight A, Heeney JL, Sallusto F, Sattentau QJ, Weiss RA, Lanzavecchia A. Analysis of memory B cell responses and isolation of novel monoclonal antibodies with neutralizing breadth from HIV-1-infected individuals. PLoS One. 2010;5:e8805. doi:10.1371/journal.pone.0008805. PMID: 20098712

Doores K, Burton DR. Variable loop glycan dependency of the broad and potent HIV-1-neutralizing antibodies PG9 and PG16. J Virol. 2010;84:10510–21. doi:10.1128/JVI.00552-10. PMID:20686044

Geng QM, Li HP, Bao ZY, Liu YJ, Zhuang DM, Li L, Liu SY, Li JY. Indinavir resistance evolution in one human immunodeficiency virus type 1 infected patient revealed by single-genome amplification. Virologica Sinica 2010; 25: 316–328 doi:10.1007/sI2250-010-3122-4. PMID:20960178

Hollox EJ, Hoh BP. Human gene copy number variation and infectious disease. Hum Genet. 2014;133:1217–33. doi:10.1007/S00439-014-1457-x. PMID:25110110

Kaqiampakis I, Gharibi A, Mankowski MK, Snyder BA, Ptak RG, Alatas K, LiWang PJ. Potent strategy to inhibit HIV-1 by binding both gp120 and gp41. Antimicrob Agents Chemother. 2011;55:264–75. doi:10.1128/AAC.00376-10. PMID:20956603

Kim S, Jao S, Laurence JS, LiWang PJ. Structural comparison of monomeric variants of the chemokine MIP-1β having differing ability to bind the receptor CCR5. Biochemistry. 2001;40:10782–91. doi:10.1021/bi011065x. PMID:11535053

Kwon YD, Finzi A, Wu X, Dogo-Isonagie C, Lee LK, Moore LR, Schmidt SD, Stuckey J, Yang Y, Zhou T, Zhu J, Vicic DA, Debnath AK, Shapiro L, Bewley CA, Mascola JR, Sodroski JG, Kwong PD. Unliganded HIV-1 gp120 core structures assume the CD4-bound conformation with regulation by quaternary interactions and variable loops. Proc Natl Acad Sci USA. 2012;109:5663–8. doi:10.1073/pnas.1112391109. PMID:22451932

Li L, Liang D, Li JY, Zhao RY. APOBEC3G-UBA2 fusion as a potential strategy for stable expression of APOBEC3G and inhibition of HIV-1 replication. Retrovirology. 2008;5:72–85. doi:10.1186/1742-4690-5-72. PMID:18680593

Lienlaf M, Hayashi F, Di Nunzio F, Tochio N, Kigawa T, Yokoyama S, Diaz-Griffero F. Contribution of E3-ubiquitin ligase activity to HIV-1 restriction by TRIM5alpha(rh): structure of the RING domain of TRIM5alpha. J Virol. 2011;85:8725–37. doi:10.1128/JVI.00497-11. PMID:21734049

Liu T, Huang B, Zhan P, De CE, Liu X. Discovery of small molecular inhibitors targeting HIV-1 gp120-CD4 interaction derived from BMS-378806. Eur J Med Chem. 2014;86:481–90. doi:10.1016/j.ejmech.2014.09.012. PMID:25203778

Lobritz MA, Ratcliff AN, Marozsan AJ, Dudley DM, Tilton JC, Arts EJ. Multifaceted mechanism of HIV inhibition and resistance to CCR5 inhibitors PSC-RANTES and maraviroc. Antimicrob Agents Chemother. 2013;57:2640–50. doi:10.1128/AAC.02511-12. PMID:23529732

Lucia L. CCR5: from natural resistance to a new anti-HIV strategy. Viruses. 2010;2:574–600. doi:10.3390/v2020574. PMID:21994649

Maria TP, Francis WR, Candace BP, Michael RR. Chemokine receptor-5 (CCR5) is a receptor for the HIV entry inhibitor peptide T (DAPTA). Anriviral. 2005;67:83–92. doi:10.1016/j.antiviral.2005.03.007. PMID:16002156

Marjan DG, Marcel BMT, Jean PO, Julien RL, Jean MN, Daisy IP, Arreaza MG, Jason SS, Maarten K, Jan DB, Menno AR. Expression of the chemokine receptor CCR5 in psoriasis and results of a randomized placebo controlled trial with a CCR5 inhibitor. Arch Dermatol Res. 2007;299:305–13. doi:10.1007/s00403-007-0764-7. PMID:17647003

Mikawa AY, Malavazil TSA, Abrao EP, Da CP. The beta-chemokines MIP-1alpha and RANTES and lipoprotein metabolism in HIV-infected Brazilian patients. Braz J Infect Dis. 2005;9:315–23. PMID:16270124

Nathans R, Cao H, Sharova N, Ali A, Sharkey M, Stranska R, Stevenson M, Rana TM. Small-molecule inhibition of HIV-1 Vif. Nat Biotechnol. 2008;26:1187–92. doi:10.1038/nbt.1496. PMID:18806783

Ni J, Zhu YN, Zhong XG, Ding Y, Hou LF, Tong XK, Tang W, Ono S, Yang YF, Zuo JP. The chemokine receptor antagonist, TAK-779, decreased experimental autoimmune encephalomyelitis by reducing inflammatory cell migration into the central nervous system, without affecting T cell function. British journal of Pharmacology. 2009;158:2046–56. doi:10.1111/j.1476-5381.2009.00528.x. PMID:20050195

Polo S, Nardese V, De SC, Arcelloni C, Paroni R, Sironi F, Verani A, Rizzi M, Boloqnesi M, Lusso P. Enhancement of the HIV-1 inhibitory activity of ranted by modification of the N-terminal region: dissociation from CCR5 activation. Eur J Immunol. 2000;30:3190–8

Roa A, Hayashi F, Yang Y, Lienlaf M, Zhou J, Shi J, Watanabe S, Kigawa T, Yokoyama S, Aiken C, Diaz-Griffero F. RING domain mutations uncouple TRIM5α restriction of HIV-1 from inhibition of reverse transcription and acceleration of uncoating. J Virol. 2012;(86):1717–27. doi:10.1128/JVI.05811-11. PMID:22114335

Rower JE, Meditz A, Gardner EM, Lichtenstein K, Predhomme J, Bushman LR, Klein B, Zheng JH, Mawhinney S, Anderson PL. Effect of HIV-1 infection and sex on the cellular pharmacology of the antiretroviral drugs zidovudine and lamivudine. Antimicrob Agents Chemother 2012; 55: 3011–3019 doi:10.1128/AAC.06337-11. PMID:22391541

Schnur E, Noah E, Ayzenshtat I, Sargsyan H, Inui T, Ding FX, Arshava B, Sagi Y, Kessler N, Levy R, Scherf T, Naider F, Anglister J. The conformation and orientation of a 27-residue CCR5 peptide in a ternary complex with HIV-1 gp120 and a CD4-mimic peptide. J Mol Biol. 2011;410:778–97. doi:10.1016/j.jmb.2011.04.023. PMID:21763489

Schols D, Proost P, Struyf S, Wuyts A, De MI, Scharpe S, Van DJ, De CE. CD26-processed RANTES(3-68), but not intact RANTES, has potent anti-HIV-1 activity. Antiviral Res. 1998;39:175–87. PMID:9833958

Shrivastava IH, Wendel K, Lalonde JM. Spontaneous rearrangement of the β20/β21 strands in simulations of unliganded HIV-1 glycoprotein, gp120. Biochemistry. 2012;51:7783–93. doi:10.1021/bi300878d. PMID:22963284

Solanki AK, Rathore YS, Basmalia MD, Dhoke RR, Nath SK, Nihalani D, Ashish. Global shape and ligand binding efficiency of the HIV-1 neutralizing antibodies differs from the ones which cannot neutralize. J Biol Chem 2014; 289: 1–38 doi:10.1074/jbc.M114.563486. PMID:25331945

Tan H, Rader AJ. Identification of putative, stable binding regions through flexibility analysis of HIV-1 gp120. Proteins. 2009;74:881–94. doi:10.1002/prot.22196. PMID:18704932

Tan Q, Zhu Y, Li L, Chen, Z., Han, GW, Kufareva, ILT, Ma L, Fenalti G, Li J, Zhang, WX, Yang H, Jiang H, Cherezov V, Liu H, Stevens RC, Zha Q, Wu B. Structure of the CCR5 chemokine receptor-HIV entry inhibitor maraviroc complex. Science 2013; 341: 1387–1390 doi:10.1126/science.1241475. PMID:24030490

Trkola A, Gordon C, Matthews J, Maxwell E, Ketas T, Czaplewski L, Proudfoot AE, Moore JP. The CC-chemokine RANTES increases the attachment of human immunodeficiency virus type 1 to target cells via glycosaminoglycans and also activates a signal transduction pathway that enhances viral infectivity. J Virol. 1999;73:6370–9. PMID:10400729

Wang J, Zhang W, Lv M, Zuo T, Kong W, Yu X. Identification of a Cullin5-ElonginB-ElonginC E3 complex in degradation of feline immunodeficiency virus Vif-mediated feline APOBEC3 proteins. J Virol. 2011;85:12482–91. doi:10.1128/JVI.05218-11. PMID:21957297

Wang Y, Kinlock BL, Shao Q, Turner TM, Liu B. HIV-1 Vif inhibits G to A hypermutations catalyzed by virus-encapsidated APOBEC3G to maintain HIV-1 infectivity. Retrovirology. 2014;11:89–99. doi:10.1186/s12977-014-0089-5. PMID:25304135

Wu Y, Deng R, Wu W. Study on CCR5 analogs and affinity peptides. Protein Eng Des Sel. 2012;25:97–105. doi:10.1093/protein/gzr062. PMID:22238429

Zappe H, Snell ME, Bossard MJ. PEGylation of cyanovirin-N, an entry inhibitor of HIV. Adv Drug Deliv Rev. 2008;60:79–87. doi:10.1016/j.addr.2007.05.016. PMID:17884238

Zhang G, Qiu W, Xiang R, Ling F, Zhuo M, Du H, Wang J, Wang X. TRIM5α polymorphism identification in cynomolgus macaques of Vietnamese origin and Chinese rhesus macaques. Am J Primatol. 2013;75:938–46. doi:10.1002/ajp.22158. PMID:23775958

Zhao Q, Ma L, Jiang S, Lu H, Liu S, He Y, Strick N, Neamati N, Debnath AK. Identification of N-phenyl-N′-(2,2,6,6-tetramethyl-piperidin-4-yl)-oxalamides as a new class of HIV-1 entry inhibitors that prevent gp120 binding to CD4. Virology. 2005;339:213–25. doi:10.1016/j.virol.2005.06.008. PMID:15996703

Zhou T, Xu L, Dey B, Hessell AJ, Van RD, Xiang SH, Yang X, Zhang MY, Zwick MB, Arthos J, Burton DR, Dimitrov DS, Sodroski J, Wyatt R, Nabel GJ, Kwong PD. Structural definition of a conserved neutralization epitope on HIV-1 gp120. Nature. 2007;445:732–7. doi:10.1038/nature05580. PMID:17301785

Zhou D, Wang Y, Tokunaqa K, Huang F, Sun B, Yang R. The HIV-1 accessory protein Vpr induces the degradation of the anti-HIV-1 agent APOBEC3G through a VprBP-mediated proteasomal pathway. Virus Res. 2014;195:25–34. doi:10.1016/j.virusres.2014.08.021. PMID:25200749

Chapter 6
Recent Studies on Mechanisms of New Drug Candidates for Alzheimer's Disease Interacting with Amyloid-β Protofibrils Using Molecular Dynamics Simulations

Huai-Meng Fan, Qin Xu, and Dong-Qing Wei

Abstract Alzheimer's disease (AD) is the most common form of dementia. The aggregation and deposition of amyloid-β (Aβ) peptide in the brain is one of its characteristic hallmarks. In order to inhibit or even destabilize Aβ fibrils, a number of candidate molecules have been proposed in recent years. Although experiments have suggested their interactions with Aβ peptides, the molecular details are generally unclear. The determination of a three-dimensional model of Aβ protofibril boosted the exploration of the details at the atomic level, such as the binding sites, the critical residues, and the key interactions for such compounds to bind or to degrade Aβ protofibril using molecular dynamics (MD) simulations. Focused on this emerging strategy, this review looks through a bunch of Aβ-interacting compounds. In particular, the MD simulations on one of the novel drug candidates, wgx-50, identified by our group recently are described in more details so as to show a typical work flow of these studies. The structural features of these compounds revealed by MD simulations may provide new macroscopic translational information for the structure-based drug design for Alzheimer's disease.

Keywords Alzheimer's disease • Amyloid-β peptide • Anti-aggregation • Molecular dynamics simulations • Structure-based drug design

6.1 Introduction

The aggregation of amyloidogenic proteins is associated with many severe neurodegenerative diseases, including Alzheimer's disease (AD), Parkinson's disease (PD), Huntington's disease (HD), and type II diabetes (Temussi et al. 2003;

H.-M. Fan • Q. Xu (✉) • D.-Q. Wei (✉)
State Key Laboratory of Microbial Metabolism and School of Life Sciences and Biotechnology, Shanghai Jiao Tong University, Shanghai 200240, China
e-mail: xuqin523@sjtu.edu.cn; dqwei@sjtu.edu.cn

© Shanghai Jiao Tong University Press, Shanghai and Springer Science+Business Media Dordrecht 2017

D.-Q. Wei et al. (eds.), *Translational Bioinformatics and Its Application*, Translational Medicine Research, DOI 10.1007/978-94-024-1045-7_6

Cohen and Kelly 2003; Dobson 2003; Chiti and Dobson 2006; DeToma et al. 2012; Eisenberg and Jucker 2012). Among these diseases, AD is the most common one, which is associated with aggregation of amyloid-β (Aβ) peptide into amyloid plaques (Monsonego et al. 2003). Identifying candidates including short peptides or small molecules that targeting Aβ is an attractive strategy for treating AD.

Aβ is a 39- to 43-residue-long peptide, generated by sequential cleavage of β- and γ-secretases from the amyloid precursor protein (APP). The predominant components of the fibrillar deposits in the brains of AD patients are 40- and 42-residue-long Aβ peptides (Aβ40 and Aβ42), with Aβ42 considered to be more neurotoxic than Aβ40 (Haass and Selkoe 2007; Mucke et al. 2000; Jarrett and Lansbury 1993; Selkoe 1999). However, both of them may have a common structural motif of strand-loop-strand, which is aligned into two stacked β-sheets and further assembled into cross-β structures in several hypothetical ways. One of the major challenges to identify binding sites on Aβ is the lack of high-resolution crystal structures of the Aβ aggregates, which make the molecular details of the interactions between the inhibitors and the Aβ peptide unknown. This situation has gradually been improved by progress in determinations of the structural models of Aβ fibrils, with the help of the newly developed NMR spectroscopy, especially after a three-dimensional structure of a pentameric protofibril was determined in 2005 (Luhrs et al. 2005) (Fig. 6.1). In this structure, the disordered N-terminal residues 1–16 of each peptide monomer were missing; only residues 17–42 were included, since they form a strand-loop-strand motif, which is expected to contribute to the stability of Aβ fibril most. Each U-shape peptide consisted of an N-terminal β-strand (β1) including residues V18-S26, a C-terminal β-strand (β2) including residues I31-A42, and a loop (residues N27-A30) linking them. Five identical peptides of residues 17–42 were aligned into two antiparallelly stacked β-sheets, where the directions of backbone hydrogen bonds were parallel to the fibril axis and the β-strands were perpendicular to it.

Recent in vitro studies have suggested that some polyphenolic compounds from red wine and green tea may bind to Aβ, inhibit Aβ aggregation, and destabilize preformed fibrils (Ono et al. 2003; Hamaguchi et al. 2006). In vivo experiments on the Alzheimer's mouse model found lowered level of amyloid plaque and improved memory and cognitive ability after feeding of red wine (Ho et al. 2009; Wang et al. 2008). Inspired by these results, Riviere et al. proposed a hypothesis that the interactions between resveratrol derivative and Aβ could shift the equilibration of Aβ polymorphism from β-sheets into disordered monomers (Riviere et al. 2009). Following this hypothesis, many compounds were found to be capable of interacting with Aβ, as listed in Table 6.1. and described with more details in the second section below.

Because of the polymorphism of Aβ oligomers or their higher-order polymers, many obstacles were encountered with traditional experimental methods in the studies on details of binding mechanism, interaction dynamics, and structural alterations of Aβ peptides by the potential drug candidates. Complementary to experimental studies, computational methods like molecular docking and all-atom molecular dynamics (MD) simulations can provide atomic-level information on the

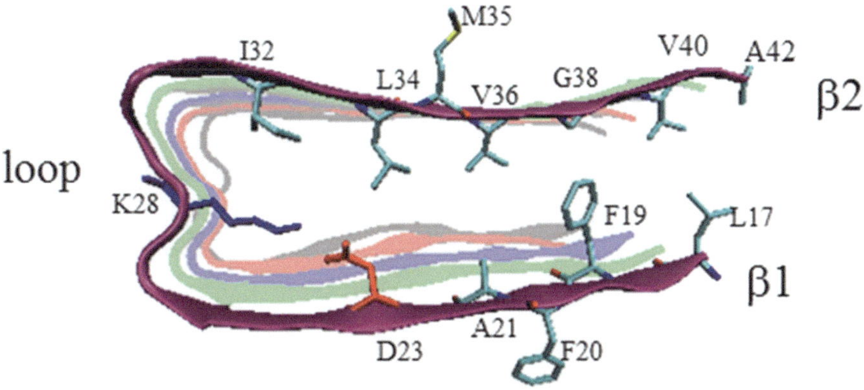

Fig. 6.1 A typical structure of Aβ pentamer as a unit for higher-order aggregation. The residues important to the stabilization of the fibril are shown in sticks, with the hydrophobic residues in *cyan*, the negative D23 in *red*, and the positive K28 in *blue*. The motif of the two layers of β strands and the loop are also labeled

interactions between amyloid peptides and small compounds. These novel methods have been successfully employed in exploring the mechanisms of many Aβ aggregation inhibitors in recent years. In the third section of this article, more details of the MD simulations and analyses on novel drug candidate wgx-50 are provided as an example.

6.2 Recent Progresses on Compounds Binding to Amyloid-β Protofibrils

In recent years, many compounds have been found to bind to the Aβ protofibrils (Table 6.1.). Unraveling the mechanisms of these compounds to interact with Aβ would be quite useful in development of drugs degrading Aβ aggregation and treat Alzheimer's disease. More details about these compounds are summarized below.

6.2.1 EGCG

(-)-Epigallocatechin-3-gallate (EGCG), the major polyphenolic component of green tea (Bieschke et al. 2010; Guo et al. 2010), has been found to directly bind to the Aβ42 and to redirect the aggregation of the peptide to a disordered off-pathway reaction resulting in unstructured, nontoxic aggregates (Ehrnhoefer et al. 2008). However, it seems unlikely that experimental approaches can provide valuable information on the molecular interactions between Aβ and EGCG. Liu et al. (Liu et al. 2011) investigated the molecular mechanism of the inhibition effect

Table 6.1 Possible compounds interacting with Aβ protofibril with the key references exploring atomic-scale mechanisms by simulations

Compounds	Type	Main binding sites	Key MD references
EGCG	Polyphenolic component of green tea	Surface(F19_F20,K28, L34–37,I41)	Liu, JPCB 2011
Ibuprofen	Nonsteroidal anti-inflammatory drug	Edge	Raman, Biophy J 2009; Takeda, JPCB 2010; Chang, Biophys J, 2010
$[Ru(bpy)_2dppz]^{2+}$	Fluorescent dye	Surface(V18_F20); Edge	Cook, JACS, 2013
O4	Orcein-related small organic molecule	Surface(V18_A21, V24_N27,I31_V39)	Sun, JPCB, 2015
ThCT and ThNT	β-sheet breakers	Edge	Autiero, Molecular bioSystems, 2013a, b
DMF	Water-soluble fullerene derivative	Surface(L17_A21, N27_I31,I31_I41)	Zhou, JPCB, 2014
ThT and PIB	Fluorescent dye	Surface(V18_F20, I31_M35,V24_S26); Inside(D23,K28,I32,L34); Edge	Wu, JMB 2008, Biophys. J 2011, 2012
Morin	Polyphenolic compounds from food products	Inside(D23,K28,I32,L34); Edge	Lemkul, Biochemistry, 2010
Wgx-50	Compounds from natural flavoring vegetation, Sichuan pepper	Surface(V18_F20, I31_M35); Inside(D23, K28,I32,L34)	Fan, JPCB, 2015

Abbreviation or alternative name: *ThT* thioflavin-T, *PIB* Pittsburgh compound B, *$[Ru(bpy)_2dppz]^{2+}$* $[Ru(2,2'\text{-bipyridine})_2\text{dipyrido}[3,2\text{-a}:2',3'\text{-c}]\text{-phenazine}]^{2+}$, *ThCT, Ac-LPFFD-Th; ThNT* Th-succinyl-LPFFD-NH_2, *DMF* 1,2-(dimethoxymethano)fullerene, *O4* 8-bis(2,4-dihydroxyphenyl)-7-hydroxy-phenoxazin-3-one, *EGCG* (−)-epigallocatechin-3-gallate, *Wgx-50* N-[2-(3,4-dimethoxyphenyl) ethyl]-3-phenyl-acrylamide. The two-dimensional structures of some of these compounds are shown in Fig. 6.2

of EGCG on the conformational transition of Aβ42 using a series of MD simulations and the MM-PBSA method. The conformational transition of the peptide at different EGCG concentrations, the favorite interactions between the peptide and EGCG, and the driving force were all explored. In addition, the key residues of the peptide were classified according to the energy contribution of each residue using the MM-PBSA method. The results of free energy decomposition calculated by this method indicate that the nonpolar term contributes major binding free energy of the EGCG-Aβ42 complex, while polar interactions play a minor role. It has been recognized that the nonpolar interactions are mainly provided by the hydrophobic residues, while polar interactions are mainly formed by the main chain of Aβ42. These observations are helpful in understanding the inhibition mechanism of EGCG on the conformational transition of Aβ42 and useful for exploring more effective agents for the inhibition of Aβ42 aggregation.

6.2.2 Ibuprofen

The nonsteroidal anti-inflammatory drug ibuprofen is considered one of the promising candidates to reduce Aβ aggregation (Xia 2003). Biomedical studies suggest that treatment with ibuprofen reduces the amount of Aβ deposits and improves memory in mice models, decreases the amount of Aβ oligomers in mice brain (McKee et al. 2008), and lowers the risk of AD in humans (Vlad et al. 2008), but the mechanism of this drug against AD is unclear. Experimental in vitro studies have shown that ibuprofen reduces accumulation of Aβ fibrils by interrupting fibril elongation and at least partially dissociate preformed Aβ fibrils (Hirohata et al. 2005). Despite the experimental progress, the interaction between Aβ and ibuprofen is still not well understood at the molecular level. The Klimov group studied ibuprofen by implicit-solvent molecular dynamics and replica exchange simulations, suggesting that ibuprofen may bind to the ends of amyloid fibrils to prevent fibril growth by a competitive mechanism. They found that concave (CV) fibril edge has significantly higher binding affinity for ibuprofen than the convex edge (Raman et al. 2009; Takeda et al. 2010; Chang et al. 2010).

6.2.3 $[Ru(bpy)_2dppz]^2$

Recent studies have revealed that molecules capable of binding on the surface of Aβ fibrils might be able to restrict Aβ monomers to its surface, inhibiting the formation of Aβ (Cohen et al. 2013). It was found that ruthenium dipyridophenazine complexes can bind to Aβ fibrils, increasing the photoluminescence intensity remarkably (Cook et al. 2011). Different from most dyes for Aβ detection, these ruthenium dyes are not planar and are easy to modify, making them potential parent complexes capable of inhibiting Aβ aggregation or reducing the production of toxic species induced by Aβ fibrils. Even so, this still requires understanding of the detailed interactions between ruthenium dipyridophenazine complexes and Aβ. Recently, Cook has combined biophysical and computational studies to elucidate the binding modes of $[Ru(bpy)_2dppz]^{2+}$ (bpy=2,2′-bipyridine; dppz=dipyrido[3,2-a:2′,3-′-c]-phenazine) to Aβ40 fibrils. Ruthenium dipyridophenazine metal complexes have been widely used in a variety of applications including cell viability studies (Jimenez-Hernandez et al. 2000), DNA detection (Erkkila et al. 1999), solubilization of carbon nanotubes (Jain et al. 2011), and cell imaging (Puckett and Barton 2008) but have hardly been used for peptide research. The computations combining molecular docking (both rigid and flexible) and all-atom molecular dynamics (MD) simulations predicted the plausible binding site in the hydrophobic cleft formed on the surface of Aβ fibrils between Val18 and Phe20, which could also clarify the enhancement in photoluminescence upon binding. In contrast to the binding site of these complexes in DNA, this binding site was parallel to the fibril axis. The result was confirmed by binding studies in an Aβ fragment (Aβ25–35) that lacked the necessary residues for the binding site which showed low photoluminescence response. The agreement between the experimental and computational studies suggests a valuable method for studying the interaction of amyloid-binding molecules to Aβ (Cook et al. 2013).

6.2.4 *O4*

Recently, Bieschke et al. have reported that the orcein-related small molecule 8-bis (2,4-dihydroxyphenyl)-7-hydroxyphenoxazin-3-one (termed O4) is capable of reducing the concentration of small toxic Aβ oligomers by promoting Aβ fibrillization. In the presence of O4, the inhibition of long-term potentiation caused by Aβ oligomers in hippocampal rat brain slices was greatly controlled (Bieschke et al. 2012). This suggested that the small molecule O4 stabilizes the Aβ protofibrils by binding to two hydrophobic regions of Aβ, accelerating Aβ fibrillization. Sun et al. studied the structural stability of the fibril-like Aβ (17–42) trimer by performing atomistic molecular dynamics simulations. They found that the Aβ (17–42) trimer is unstable without O4, whereas the stability of its structure is greatly enhanced with O4. Four promising binding sites were found around residues of F20, S26, and M35: the CHC site, the turn site, and two hydrophobic-groove sites. The binding of O4 at the CHC site is mostly stabilized by hydrophobic interactions. Hydrogen-bonding interaction between O4 and S26 is important in the turn site. The two hydrophobic grooves near M35 also assist binding of O4 by hydrophobic interaction (Sun et al. 2015).

6.2.5 *ThCT and ThNT*

The β-sheet breakers (BSB) are promising therapeutic strategies, aimed at preventing the deposition of insoluble protein fibrils in Alzheimer's disease (Bieler and Soto 2004). Although various experimental studies have supported the hypothesis of such ligands preventing the formation of soluble Aβ oligomers or inhibiting the accumulation of Aβ peptides, their interaction mechanism has not yet been precisely elucidated. Many efforts have been made in the development of BSB based on a peptide scaffold homologous to the hydrophobic core region (HCR) of Aβ1–42, residues 17–20, which is the region mainly targeted by BSB ligands, containing critical elements for Aβ self-assembly, as experimentally demonstrated by inhibition of aggregation induced by mutations of V18, F19, and F20 (Hilbich et al. 1992; Esler et al. 1996). This drives the binding through a self-recognition mechanism, involving additional binding layers at the end of the oligomer (Lowe et al. 2001; Zhang et al. 2003). Appending polar groups to the peptide scaffold significantly improves the BSB affinity for Aβ oligomers and enhances the binding (Cairo et al. 2002; Reinke and Gestwicki 2011). The trehalose-conjugated peptides

Ac-LPFFD-Th (ThCT) and Th-succinyl-LPFFD-NH_2 (ThNT)(Bona et al. 2009) were designed to combine a polar group with the peptide portion LPFFD, the well-known BSB Soto's peptide (Soto et al. 1998), which were proposed as effective Aβ inhibitors by binding to the ends of the growing fibril. The all-atom molecular dynamics (MD) simulations suggested that the binding on the two protofibril ends occurs through different binding modes. Particularly, binding on the odd edge (chain A) guided by a hydrophobic cleft entailed a significant structure destabilization, deducing (inducing?) a partial β structure loss. The energetically favored hydrophobic cleft perceived on the odd edge may signify a new window for designing new molecules with improved anti-aggregating features (Autiero et al. 2013a, b).

6.2.6 DMF

The process of Aβ aggregation is associated with many factors such as metal ions, membranes, or nanoparticles (NPs) (Miller et al. 2010; DeToma et al. 2012; Cabaleiro-Lago et al. 2010; Cabaleiro-Lago et al. 2008). The carbon-based NPs, including fullerenes (C60, 58) and carbon nanotubes (CNTs) (Linse et al. 2007), have attracted increasingly more attention in AD. However, the poor solubility of carbon NPs has been a major obstacle in the potential biomedical applications. Thus, the water-soluble derivatives of C60 fullerenes have been synthesized and are shown to be effective in the treatment of neurodegenerative diseases (Dugan et al. 1997, 2001). In vitro studies reported that 1,2-(dimethoxymethano)fullerene (DMF), a fullerene derivative, strongly inhibits the Aβ peptide aggregation at the early stage. Other water-soluble fullerene C60 derivatives were also reported to be capable of inhibiting amyloid fibrillation and reducing the cytotoxicity of Aβ peptides (Podolski et al. 2007). Zhou et al. have recently investigated the detailed interaction of DMF molecule with full-length Aβ42 using multiple all-atom explicit-solvent molecular dynamics (MD) simulations. Starting from different initial states of DMF molecule, the simulations showed that the DMF binds to the Aβ protofibril in three dominant binding sites: the central hydrophobic core (CHC) site (17LVFFA21), the turn site (27NKGAI31), and the C-terminal β-sheet site consisting of hydrophobic residues (31IIGLMVGGVVI41). The importance of π-stacking interactions and hydrophobic interactions was revealed by binding energy analyses. Particularly, the binding of DMF to the turn region can disrupt the D23–K28 salt bridge which is important for the Aβ fibrillation. These results provide better understanding of the binding mechanism and fibril inhibition effect of fullerene derivatives and may offer help in the therapeutic drug design using fullerene derivatives against AD (Zhou et al. 2014).

6.2.7 ThT and PIB

Thioflavin-T (ThT) is one of the most commonly used fluorescent dyes to detect the presence of amyloid fibrils. ThT exhibits a red shift in its excitation and an emission enhancement when bound to amyloid fibrils (Furumoto et al. 2007). Although ThT offers a reliable method of visualizing amyloid aggregates in vitro, it is weakly hydrophobic and does not readily enter the brain, and its binding affinity to fibrils is low. Thus many derivatives of ThT have been developed over the past few years (Raji et al. 2008). Among them, the Pittsburgh compound B (PIB), developed by Mathis et al. (Mathis et al. 2003), was the most promising candidate. This dye is currently being assessed clinically in many positron emission tomography (PET) centers all around the world for direct visualization of amyloid plaques in the brains of AD patients. The resulting PIB molecule shows increased binding affinity, improved brain clearance abilities, and increased lipophilicity over ThT. It is interesting that the modifications made in PIB led to slight changes in the fluorescence properties, and the uncharged derivative does not demonstrate the red shift in excitation and the emission enhancement upon binding to amyloid fibrils (Klunk et al. 2001). The detail of the binding sites of PIB on Aβ amyloid fibrils is not well known at atomic resolution. Using molecular dynamics simulations, Wu and coworkers characterized the binding sites of ThT and PIB on protofibrils of both $A\beta_{9-40}$ and $A\beta_{17-42}$. Simulations showed that they both bind to the grooves formed by hydrophobic or aromatic residues on the β-sheet surface along the fibril axis. The lack of the charge and two methyl groups in PIB not only improves its hydrophobicity but also allows it to insert more deeply into aromatic/hydrophobic grooves. This significantly increases the steric, aromatic, and hydrophobic interactions, leading to stronger binding (Wu et al. 2008, 2011, 2012).

6.2.8 Morin

Recent in vitro experiments have suggested that polyphenolic compounds (flavonoids) from food products such as green tea and red wine may be effective in targeting Aβ (Ono et al. 2003). Among these flavonoids, morin was found to be remarkably effective in inhibiting Aβ aggregation suggesting a direct physical interaction between these molecules and Aβ. Additionally, in vivo oral administration of red wine or polyphenolic extracts reduced amyloid plaque burden and concomitantly improved memory and cognitive ability in Alzheimer's mouse model (Ho et al. 2009; Wang et al. 2008). Riviere et al. (Riviere et al. 2009) demonstrated that the resveratrol derivative piceid was able to destabilize Aβ oligomers and fibrils. Although the therapeutic potential of polyphenols has been shown both in vitro and in vivo, the exact mechanism of these compounds remains obscure. Lemkul and Bevan identified the mechanism of Aβ destabilization by morin, an effective anti-aggregation flavonoid, employing atomistic explicit-solvent molecular dynamics (MD) simulations. They found that morin bound to the ends of $A\beta_{17-42}$ to block further combination of incoming peptides and also entered into the hydrophobic core to disrupt interior interactions like D23-K28 salt bridges and backbone H-bonds (Lemkul and Bevan 2010).

6.2.9 Wgx-50

N-[2-(3,4-Dimethoxyphenyl)ethyl]-3-phenyl-acrylamide or named as wgx-50 (earlier as gx-50) was designed by Dongqing Wei et al. (Gu et al. 2009) using structure-based drug design method and was found to be an ingredient of a natural flavoring vegetation, Sichuan pepper (*Zanthoxylum bungeanum*) (Fig. 6.1). Based on a series of biological experiments, this novel drug candidate was suggested to be an effective therapeutic agent for Alzheimer's disease (AD). In vitro experiments demonstrated that wgx-50 could reduce neuronal calcium toxicity and inhibit Aβ-induced neuronal apoptosis; in vivo experiments revealed that wgx-50 could pass through the blood-brain barrier, decrease the accumulation of Aβ in the cerebral cortex, and improve the cognitive abilities of mice (Tang et al. 2013). Found in natural food products, nontoxic in clinically relevant doses, able to cross the blood-brain barrier (Ho et al. 2009; Wang et al. 2008), and effective in inhibiting Aβ aggregation, all these unique properties give wgx-50 many advantages to be an attractive therapeutic candidate. In a recent MD simulation, wgx-50 was found to be inserted into the hydrophobic interior of the Aβ protofibril spontaneously, which may provide more hints to design drug candidates to interact with Aβ. More details of this study are described below in the third section as an example of MD simulations to reveal molecular details of the interactions between the compounds and Aβ protofibrils.

6.2.10 Structure Features of the Compounds

The design of drug candidate compounds to interact with the Aβ fibril and to relieve Alzheimer's disease is still in very early stage, and the results from experiments and simulations are still too diverse and hard to be developed into some common rules. However, comparisons between the structures of these compounds could possibly provide some hints for structure-based drug design for Aβ anti-aggregation: (1) Hydrophobic aromatic group is a common character in many of the Aβ-binding compounds, and hydrophobic interaction is a key reason for these compounds to bind with the Aβ fibril, no matter on the β-sheet surface, on the fibril edge, or inside the double sheet. (2) The size of the compounds is another key feature to determine the binding site of these compounds. Only the smaller molecules in Fig. 6.2b can penetrate into the interior of the cross-β subunit and deform the protofibril significantly. Those huge compounds in Fig. 6.2a, as well as the peptide derivatives

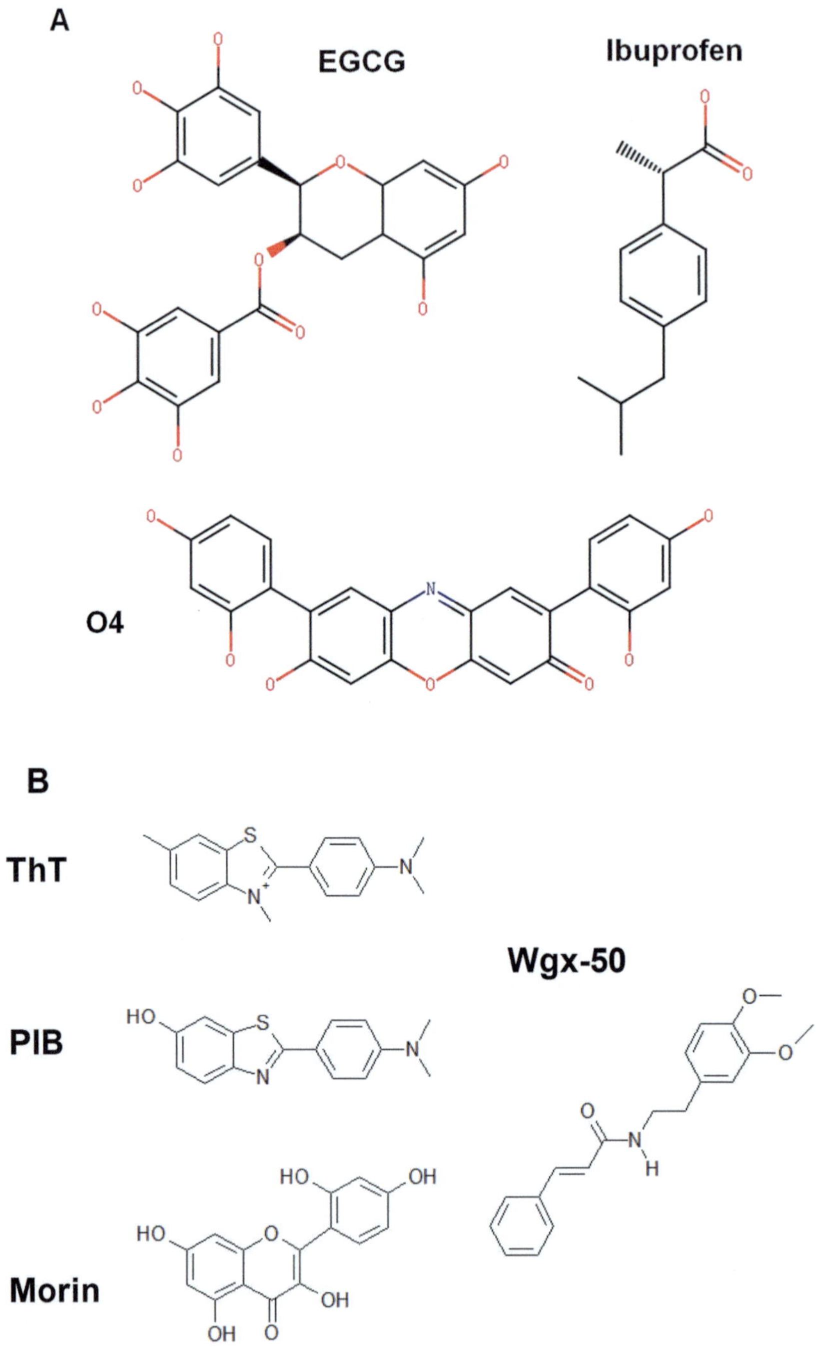

Fig. 6.2 Diagrams of several compounds possibly binding to Aβ protofibril on surface or edges (**a**) or in the interior (**b**)

(ThCT and ThNT) and the polymer DMF, are only possible to bind to the surface or edge of Aβ protofibril to compete with the incoming peptides and inhibit growth of Aβ aggregation. (3) Polar groups in the compounds might be helpful to destabilize the D23-K28 salt bridge when bound, but their effect might be a double-edged sword. In some cases they could be an advantage. For example, the hydroxyl of PIB results in stronger binding than ThT, and in MD simulations of morin, the disruption of the salt bridges was attributed to the abundant hydroxyls. However, in the case of wgx-50, without the hydroxyls, the hydrophobic groups of wgx-50 are enough to make it spontaneously insert into the double sheets and deform the cross-β subunit mainly by steric clashes, although partial disruption of the salt bridges was also observed as a consequence of the deformation. On the other hand, the small molecule ibuprofen was not reported to be inserted into the hydrophobic interior of the protofibril, possibly because its strong polar carboxyl group would keep it out in the solution. In addition, the too hydrophilic surface of the compounds might be a disadvantage for them to get through the blood-brain barrier. (4) Derived from natural products, the compounds EGCG, morin, and wgx-50 might be easier to be developed into clinical drugs in the future.

6.3 Reveal the Molecular Mechanisms Using Molecular Dynamics Simulations: An Example from Wgx-50

Using all-atom molecular dynamics simulations to explore the detailed mechanism at atomic level of the drug candidates to bind or to destabilize the Aβ protofibril is realized ever since the determination of the solid-state NMR structure of Aβ protofibril by Luhrs et al. (PDB entry 2BEG) (Luhrs et al. 2005). The work flow of these simulations is similar, as shown in Fig. 6.3.

Here is an example of the simulations to destabilize the Aβ protofibril with a wgx-50 molecule. The models of Aβ protofibrils for simulations were constructed based on the resolved NMR structure mentioned above, with the disordered N-terminal residues 1–16 missing. This protofibril models were solvated in a water box composed by ~11,000 TIP3P (Jorgensen et al. 1983) water molecules so that the sides of box be more than 11 Å away from the model. Several simulation systems were then set up with or without the compound wgx-50 added, in addition to positive sodium ions that neutralize the system. Under periodic boundary conditions on the simulation box, each system was first minimized by the steepest descent method, then equilibrated with positional restraints on peptide heavy atoms, and at last simulated for 150 ns with all the positional restraints released. The molecular dynamics (MD) simulations were under NPT conditions at 300 K (as well as 320 K) and 1 atm pressure, using the GROMACS package (Spoel et al. 2005) and the AMBER ff03 force field (Duan et al. 2003). Binding free energies between the compounds and the protofibril, as well as the interchain interactions, were calculated by molecular mechanics-generalized born surface area (MM-GBSA) in the

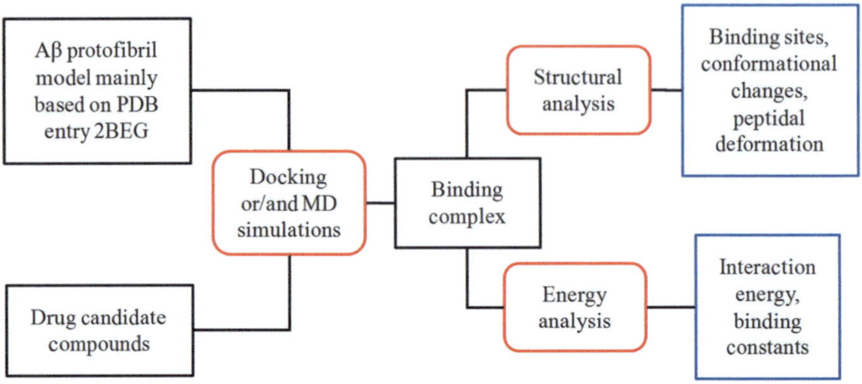

Fig. 6.3 A common work flow to explore atomic-level mechanisms of compounds to bind or to degrade Aβ protofibril using molecular dynamics simulations. The key techniques are in *red rounded squares*, with the working objects in *black squares*. The important translational information obtained is illustrated in the *blue* frames

AMBER package (Kollman et al. 2000). In addition, there were a series of structural analyses to help describe the deformation of the protofibril, such as the traditional RMSD of backbone, RMSF of Cα atoms of the peptides, fluctuations of the D23-K28 salt bridge, average intra-chain and interchain distances, as well as hydrophobic interactions between the compound and the protofibril.

Three possible binding sites were found in four 150 ns simulations at 300 K. Two exterior binding sites A and B are in the V18-F20 groove on the β1 sheet layer and the I31-M35 groove on the β2 sheet layer, respectively. These two grooves are composed of side chains of hydrophobic/aromatic residues and favor the hydrophobic wgx-50 to bind. Earlier simulations also supported that hydrophobic/aromatic and steric interactions are stabilizing forces for binding of several ligands (Wu et al. 2011). Site C is in the interior of the pentamer, against the side chains of I32 and L34 and the salt bridge between D23 and K28. All the three binding sites were also detected in simulations on other Aβ-ligand complexes, although not at the same time. Cook predicted the hydrophobic cleft between V18 and F20 as a promising binding site for $[Ru(bpy)_2dppz]^{2+}$ (Cook et al. 2013). Both surface sites were characterized in the study of binding ThT and its derivatives to Aβ fibril (Wu et al. 2011). Interior site similar to the partial insertion of morin into the hydrophobic core was also founded in Lemkul's simulations (Autiero et al. 2013a). The major hydrophobic interaction is between the aromatic rings of wgx-50 and the side chains of I32 and L34 on the β2 portion, which may be a crucial stabilization element in aggregation and elongation of Aβ-aggregates (Masman et al. 2009; Buchete and Hummer 2007).

However, only insertion of wgx-50 into the interior site caused significant destabilization of the protofibril. The aromatic ring of wgx-50 was packed against the side chains of I32 and L34 on β2, partially disrupted the salt bridges of D23-K28 which are crucial to the stabilization of the loop region (Fig. 6.4). The insertion also

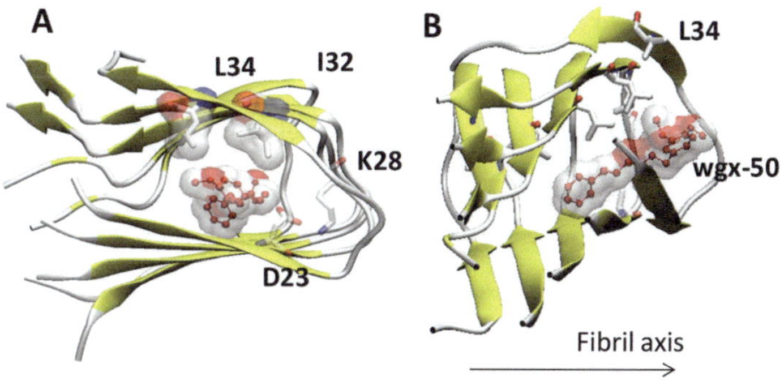

Fig. 6.4 The insertion of the wgx-50 into the interior of a pentameric Aβ protofibril after 150 ns molecular dynamics simulations at 300 K, shown from the top view (**a**) and the alternate view (**b**). The compound wgx-50 is shown in ball and stick. The side chains of the important residues D23, K28, I32, and L34 are shown in *stick*

increased the distances of two β-sheet layers which are characteristics of the stacked β-strand-loop-β-strand motifs. The disruption of the compact cross-β subunit weakened the interactions between the peptide chains: the interchain binding energies were lowered, the number of interchain backbone hydrogen bonds was decreased, and the average distances between the peptide chains were increased, resulting in destabilization of the protofibril aggregates. Merge up: The results above were confirmed by simulations repeated at 320 K, where deeper insertion of wgx-50 molecule into the pentamer was detected.

6.4 Conclusion and Perspective

In this review, we summarized and compared several drug candidate compounds for Alzheimer's disease binding to amyloid-β peptide fibrils, especially those studied recently by molecular dynamics simulations. Different with the compounds ($[Ru(bpy)2dppz]^{2+}$, ibuprofen, ThCT and ThNT, DMF, O4, EGCG), which only bind to the surface or edge of the Aβ to inhibit its further growth, ThT, PIB, morin, and wgx-50 are capable of binding to the interior of Aβ and may destabilize the Aβ aggregation. The molecular size might be an important reason for this difference. And, the composition of hydrophobic and hydrophilic groups in these compounds may also affect the pattern of binding. The molecular dynamics simulations of these compounds provide more details of the mechanism at atomic level for these novel drug candidates to interact with the Aβ protofibril and to inhibit Aβ aggregations, such as the binding sites, the critical residues, and the key interactions, which might be useful to future structure-based drug design for Alzheimer's disease. Although the current results of structural analyses are quite limited and too complicated to be summarized into accurate rules, some hints are still quite inspiring, such as the indispensability of hydrophobic aromatic rings, effect of composition of polar groups, and the influence of molecular size. With the accumulation of more results, these structural features might be summarized into some quantitative structure-activity relationship (QSAR) models or be utilized to train statistically meaningful prediction program, which may be powerful in high-throughput drug screening and lead optimizations. This computer-aided prescreening would help greatly reduce costs in experiments and times even long before the drug candidates reach the stage of clinic trials.

Acknowledgment Dong-Qing Wei is supported by grants from the National High-Tech R&D Program (863 Program Contract No. 2012AA020307), the National Basic Research Program of China (973 Program Contract No. 2012CB721000), the Key Project of Shanghai Science and Technology Commission (Contract No. 11JC1406400), and the PhD Programs Foundation of Ministry of Education of China (Contract No., 20120073110057). Qin Xu is supported by grants from the National Natural Science Foundation of China for Young Scholars (Grant No. 31400704). We thank Dr. Yelin Chen from Interdisciplinary Research Center on Biology and Chemistry, Shanghai Institute of Organic Chemistry, Chinese Academy of Sciences, China, for his helpful discussions and revisions on this manuscript.

References

Autiero I, Langella E, Saviano M. Insights into the mechanism of interaction between trehalose-conjugated beta-sheet breaker peptides and Aβ(1-42) fibrils by molecular dynamics simulations. Mol BioSyst. 2013a;9(11):2835–41.

Autiero I, Saviano M, Langella E. In silico investigation and targeting of amyloid beta oligomers of different size. Mol BioSyst. 2013b;9(8):2118–24.

Bieler S, Soto C. beta-sheet breakers for Alzheimer's disease therapy. Curr Drug Targets. 2004;5 (6):553–8.

Bieschke J, et al. EGCG remodels mature alpha-synuclein and amyloid-beta fibrils and reduces cellular toxicity. Proc Natl Acad Sci U S A. 2010;107(17):7710–5.

Bieschke J, et al. Small-molecule conversion of toxic oligomers to nontoxic beta-sheet-rich amyloid fibrils. Nat Chem Biol. 2012;8(1):93–101.

Bona P, et al. Design and synthesis of new trehalose-conjugated pentapeptides as inhibitors of A beta(1-42) fibrillogenesis and toxicity. J Pept Sci. 2009;15(3):220–8.

Buchete N-V, Hummer G. Structure and dynamics of parallel beta-sheets, hydrophobic core, and loops in Alzheimer's A beta fibrils. Biophys J. 2007;92(9):3032–9.

Cabaleiro-Lago C, et al. Inhibition of amyloid beta protein fibrillation by polymeric nanoparticles. J Am Chem Soc. 2008;130(46):15437–43.

Cabaleiro-Lago C, et al. Dual effect of amino modified polystyrene nanoparticles on amyloid beta protein fibrillation. ACS Chem Neurosci. 2010;1(4):279–87.

Cairo CW, et al. Affinity-based inhibition of beta-amyloid toxicity. Biochemistry. 2002;41 (27):8620–9.

Chang WE, et al. Molecular dynamics simulations of anti-aggregation effect of ibuprofen. Biophys J. 2010;98(11):2662–70.

Chiti F, Dobson CM. Protein misfolding, functional amyloid, and human disease. Annu Rev Biochem. 2006;75:333–66.

Cohen FE, Kelly JW. Therapeutic approaches to protein-misfolding diseases. Nature. 2003;426 (6968):905–9.

Cohen SIA, et al. Proliferation of amyloid-beta 42 aggregates occurs through a secondary nucleation mechanism. Proc Natl Acad Sci U S A. 2013;110(24):9758–63.

Cook NP, et al. Sensing amyloid-beta aggregation using luminescent dipyridophenazine ruthenium(II) complexes. J Am Chem Soc. 2011;133(29):11121–3.
Cook NP, et al. Unraveling the photoluminescence response of light-switching ruthenium (II) complexes bound to amyloid-beta. J Am Chem Soc. 2013;135(29):10810–6.
DeToma AS, et al. Misfolded proteins in Alzheimer's disease and type II diabetes. Chem Soc Rev. 2012;41(2):608–21.
Dobson CM. Protein folding and misfolding. Nature. 2003;426(6968):884–90.
Duan Y, et al. A point-charge force field for molecular mechanics simulations of proteins based on condensed-phase quantum mechanical calculations. J Comput Chem. 2003;24(16):1999–2012.
Dugan LL, et al. Carboxyfullerenes as neuroprotective agents. Proc Natl Acad Sci U S A. 1997;94 (17):9434–9.
Dugan LL, et al. Fullerene-based antioxidants and neurodegenerative disorders. Parkinsonism Relat Disord. 2001;7(3):243–6.
Ehrnhoefer DE, et al. EGCG redirects amyloidogenic polypeptides into unstructured, off-pathway oligomers. Nat Struct Mol Biol. 2008;15(6):558–66.
Eisenberg D, Jucker M. The amyloid state of proteins in human diseases. Cell. 2012;148 (6):1188–203.
Erkkila KE, Odom DT, Barton JK. Recognition and reaction of metallointercalators with DNA. Chem Rev. 1999;99(9):2777–95.
Esler WP, et al. Point substitution in the central hydrophobic cluster of a human beta-amyloid congener disrupts peptide folding and abolishes plaque competence. Biochemistry. 1996;35 (44):13914–21.
Fan HM, et al. Destabilization of Alzheimer's A beta 42 protofibrils with a novel drug candidate wgx-50 by molecular dynamics simulations. J Phys Chem B. 2015;119(34):11196–202.
Furumoto S, et al. Recent advances in the development of amyloid Imaging agents. Curr Top Med Chem. 2007;7(18):1773–89.
Gu RX, et al. Possible drug candidates for Alzheimer's disease deduced from studying their binding interactions with alpha7 nicotinic acetylcholine receptor. Med Chem. 2009;5 (3):250–62.
Guo J-P, Yu S, McGeer PL. Simple in vitro assays to identify amyloid-beta aggregation blockers for Alzheimer's disease therapy. J Alzheimers Dis. 2010;19(4):1359–70.
Haass C, Selkoe DJ. Soluble protein oligomers in neurodegeneration: lessons from the Alzheimer's amyloid beta-peptide. Nat Rev Mol Cell Biol. 2007;8(2):101–12.
Hamaguchi T, Ono K, Yamada M. Anti-amyloidogenic therapies: strategies for prevention and treatment of Alzheimer's disease. Cell Mol Life Sci. 2006;63(13):1538–52.
Hilbich C, et al. Substitutions of hydrophobic amino acids reduce the amyloidogenicity of Alzheimer's disease beta A4 peptides. J Mol Biol. 1992;228(2):460–73.
Hirohata M, et al. Non-steroidal anti-inflammatory drugs have anti-amyloidogenic effects for Alzheimer's beta-amyloid fibrils in vitro. Neuropharmacology. 2005;49(7):1088–99.
Ho L, et al. Heterogeneity in red wine polyphenolic contents differentially influences Alzheimer's disease-type neuropathology and cognitive deterioration. J Alzheimers Dis. 2009;16(1):59–72.
Jain D, Saha A, Marti AA. Non-covalent ruthenium polypyridyl complexes-carbon nanotubes composites: an alternative for functional dissolution of carbon nanotubes in solution. Chem Commun. 2011;47(8):2246–8.
Jarrett JT, Lansbury Jr PT. Seeding "one-dimensional crystallization" of amyloid: a pathogenic mechanism in Alzheimer's disease and scrapie? Cell. 1993;73(6):1055–8.
Jimenez-Hernandez ME, et al. A ruthenium probe for cell viability measurement using flow cytometry, confocal microscopy and time-resolved luminescence. Photochem Photobiol. 2000;72(1):28–34.
Jorgensen WL, et al. Comparison of simple potential functions for simulating liquid water. J Chem Phys. 1983;79(2):926–35.
Kim JE, Lee M. Fullerene inhibits beta-amyloid peptide aggregation. Biochem Biophys Res Commun. 2003;303(2):576–9.

Klunk WE, et al. Uncharged thioflavin-T derivatives bind to amyloid-beta protein with high affinity and readily enter the brain. Life Sci. 2001;69(13):1471–84.

Kollman PA, et al. Calculating structures and free energies of complex molecules: Combining molecular mechanics and continuum models. Acc Chem Res. 2000;33(12):889–97.

Lemkul JA, Bevan DR. Destabilizing Alzheimer's Abeta(42) protofibrils with morin: mechanistic insights from molecular dynamics simulations. Biochemistry. 2010;49(18):3935–46.

Linse S, et al. Nucleation of protein fibrillation by nanoparticles. Proc Natl Acad Sci U S A. 2007;104(21):8691–6.

Liu FF, et al. Molecular insight into conformational transition of amyloid beta-peptide 42 inhibited by (−)-epigallocatechin-3-gallate probed by molecular simulations. J Phys Chem B. 2011;115 (41):11879–87.

Lowe TL, et al. Structure-function relationships for inhibitors of beta-amyloid toxicity containing the recognition sequence KLVFF. Biochemistry. 2001;40(26):7882–9.

Luhrs T, et al. 3D structure of Alzheimer's amyloid-beta(1-42) fibrils. Proc Natl Acad Sci U S A. 2005;102(48):17342–7.

Masman MF, et al. In silico study of full-length amyloid beta 1-42 tri- and penta-oligomers in solution. J Phys Chem B. 2009;113(34):11710–9.

Mathis CA, et al. Synthesis and evaluation of C-11-labeled 6-substituted 2-arylbenzothiazoles as amyloid imaging agents. J Med Chem. 2003;46(13):2740–54.

McKee AC, et al. Ibuprofen reduces A beta, hyperphosphorylated tau and memory deficits in Alzheimer mice. Brain Res. 2008;1207:225–36.

Miller Y, Ma B, Nussinov R. Zinc ions promote Alzheimer A beta aggregation via population shift of polymorphic states. Proc Natl Acad Sci U S A. 2010;107(21):9490–5.

Monsonego A, et al. Increased T cell reactivity to amyloid beta protein in older humans and patients with Alzheimer disease. J Clin Investig. 2003;112(3):415–22.

Mucke L, et al. High-level neuronal expression of A beta(1-42) in wild-type human amyloid protein precursor transgenic mice: Synaptotoxicity without plaque formation. J Neurosci. 2000;20(11):4050–8.

Ono K, et al. Potent anti-amyloidogenic and fibril-destabilizing effects of polyphenols in vitro: implications for the prevention and therapeutics of Alzheimer's disease. J Neurochem. 2003;87 (1):172–81.

Podolski IY, et al. Effects of hydrated forms of C-60 fullerene on amyloid beta-peptide fibrillization in vitro and performance of the cognitive task. J Nanosci Nanotechnol. 2007;7 (4–5):1479–85.

Puckett CA, Barton JK. Mechanism of Cellular Uptake of a Ruthenium Polypyridyl Complex. Biochemistry. 2008;47(45):11711–6.

Raji CA, et al. Characterizing regional correlation, laterality and symmetry of amyloid deposition in mild cognitive impairment and Alzheimer's disease with Pittsburgh Compound B. J Neurosci Methods. 2008;172(2):277–82.

Raman EP, Takeda T, Klimov DK. Molecular dynamics simulations of Ibuprofen binding to Abeta peptides. Biophys J. 2009;97(7):2070–9.

Reinke AA, Gestwicki JE. Insight into amyloid structure using chemical probes. Chem Biol Drug Des. 2011;77(6):399–411.

Riviere C, et al. The polyphenol piceid destabilizes preformed amyloid fibrils and oligomers in vitro: hypothesis on possible molecular mechanisms. Neurochem Res. 2009;34(6):1120–8.

Selkoe DJ. Translating cell biology into therapeutic advances in Alzheimer's disease. Nature. 1999;399(6738):A23–31.

Soto C, et al. beta-sheet breaker peptides inhibit fibrillogenesis in a rat brain model of amyloidosis: Implications for Alzheimer's therapy. Nat Med. 1998;4(7):822–6.

Spoel D, et al. GROMACS: fast, flexible, and free. J Comput Chem. 2005;26(16):1701–18.

Sun Y, Xi W, Wei G. Atomic-level study of the effects of O-4 molecules on the structural properties of protofibrillar A beta trimer: beta-sheet stabilization, salt bridge protection, and binding mechanism. J Phys Chem B. 2015;119(7):2786–94.

Takeda T, et al. Nonsteroidal anti-inflammatory drug naproxen destabilizes Abeta amyloid fibrils: a molecular dynamics investigation. J Phys Chem B. 2010;114(46):15394–402.
Tang M, et al. A novel drug candidate for Alzheimer's disease treatment: gx-50 derived from Zanthoxylum bungeanum. J Alzheimers Dis. 2013;34(1):203–13.
Temussi PA, Masino L, Pastore A. From Alzheimer to Huntington: why is a structural understanding so difficult? EMBO J. 2003;22(3):355–61.
Vlad SC, et al. Protective effects of NSAIDs on the development of Alzheimer disease. Neurology. 2008;70(19):1672–7.
Wang J, et al. Grape-derived polyphenolics prevent Abeta oligomerization and attenuate cognitive deterioration in a mouse model of Alzheimer's disease. J Neurosci. 2008;28(25):6388–92.
Wu C, et al. The binding of thioflavin T and its neutral analog BTA-1 to protofibrils of the Alzheimer's disease A beta(16-22) peptide probed by molecular dynamics simulations. J Mol Biol. 2008;384(3):718–29.
Wu C, Bowers MT, Shea J-E. On the origin of the stronger binding of PIB over thioflavin T to protofibrils of the Alzheimer amyloid-beta peptide: a molecular dynamics study. Biophys J. 2011;100(5):1316–24.
Wu C, Scott J, Shea J-E. Binding of congo red to amyloid protofibrils of the Alzheimer A beta (9-40) peptide probed by molecular dynamics simulations. Biophys J. 2012;103(3):550–7.
Xia W. Amyloid inhibitors and Alzheimer's disease. Curr Opin Investig Drugs (London, England: 2000). 2003;4(1):55–9.
Zhang GB, et al. Multiple-peptide conjugates for binding beta-amyloid plaques of Alzheimer's disease. Bioconjug Chem. 2003;14(1):86–92.
Zhou X, et al. Interactions of a water-soluble fullerene derivative with amyloid-beta protofibrils: dynamics, binding mechanism, and the resulting SaltBridge disruption. J Phys Chem B. 2014;118(24):6733–41.

Chapter 7
Homology Modelling, Structure-Based Pharmacophore Modelling, High-Throughput Virtual Screening and Docking Studies of L-Type Calcium Channel for Cadmium Toxicity

Madhu Sudhana Saddala and A. Usha Rani

Abstract Cadmium (Cd) is a heavy metal present in air, water, soils and sediments. It is well known that long-term exposure to Cd causes various toxic effects in various organ systems such as cardiovascular, kidneys, liver, brain, lung, bones, immune/haemopoietic, endocrine and reproductive systems. Cd influx mediates voltage-gated L-type calcium channels (LCC) in excitable cells including mammalian neurons and also Cd uptake in non-excitable tissues. Therefore, LCC has been recognized as an attractive metal toxicity target. We construct a homology model of LCC in addition to the generated pharmacophore models then used to retrieve 50,500 molecules from Zinc database. There are 18 best reliable molecules mapped with core pharmacophore model of LCC. These hits were retrieved and further evaluated by molecular dynamics (MD) simulation, molecular docking and protein–ligand interactions, and binding affinity predictions as well as in silico ADMET properties were tested. Our work results focus on homology modelling, structure-based pharmacophore mapping, molecular docking, MD simulation, protein–ligand interactions and binding affinity predictions which were used in virtual screening strategy to spot new hits for blockade of LCC. Finally, the outcome results, priming the five best lead compounds, were expected to be the potential lead scaffolds for developing novel and potent blockers of LCC against metal toxicity.

Keywords Cadmium • LCC • Docking • Pharmacophore • MOE • ADMET

M.S. Saddala (✉) • A. Usha Rani
Division of Bioinformatics, DBT- Bioinformatics Centre, Department of Zoology, Sri Venkateswara University, Tirupati 517502, Andhra Pradesh, India
e-mail: madhubioinformatics@gmail.com

© Shanghai Jiao Tong University Press, Shanghai and Springer Science+Business Media Dordrecht 2017

D.-Q. Wei et al. (eds.), *Translational Bioinformatics and Its Application*, Translational Medicine Research, DOI 10.1007/978-94-024-1045-7_7

7.1 Introduction

Cadmium (Cd) is a heavy metal present in air, water, soils and sediments (Kocak and Akcil 2006). Cd is widely used in pigments, plastic stabilizers, electroplating, alloys, nickel–Cd batteries and welding industry and is also present in tobacco (Pappas et al. 2006). It is well known that long-term exposure to Cd causes various toxic effects in various organ systems such as cardiovascular, kidneys, liver, brain, lung, bones, immune/haemopoietic, endocrine and reproductive systems (Satarug et al. 2010). Several efforts are being made to find a substance which can significantly decrease the magnitude of metal toxicity when present in the biological system during heavy metal intoxication. Membrane damage caused by the reactive oxygen species (H_2O_2 and OH^- ions) generated from the exposure of living tissues to heavy metals may allow the entry of excess calcium into the cells with a subsequent biochemical cellular degradation and necrosis. Calcium channel blockers act on ion-conducting cell membrane channels. The 1,4-dihydropyridine moiety is commonly useful as calcium channel blockers and is used most frequently as drugs such as nifedipine, diltiazem, nicardipine and amlodipine (AD), which have been found as potent cardiovascular agents for the treatment of hypertension. Hence, this class of agents may be included in the search for protectors with a more favourable therapeutic index. Therefore, the present study is an attempt to find out the detoxifying action of calcium channel blockers against cadmium-induced toxicity in albino rats through computational tools.

In recent years, high-throughput virtual screening has been emerging as a complementary to high-throughput screening in an attempt to discover novel potential lead compounds in the process of drug discovery (Lyne 2002). Thus, to identify new and potent compounds that block the L-type calcium channel (LCC) model like AD, structure-based pharmacophore modelling and virtual screening may be considered as an effective approach. This study describes the structure-based pharmacophore modelling to identify the pharmacophoric features required for simultaneous inhibition of LCC for Cd toxicity by virtual screening: molecular docking, protein–ligand interaction fingerprints (PLIFs), binding energy calculations and binding affinity predictions.

7.2 Material and Methods

7.2.1 Homology Modelling

For unknown protein structures such as membrane proteins, homology modelling was introduced to construct the three-dimensional structure of a known atomic resolution model of the protein (target) and related homologous protein (template).

```
KcsA
M1              AGAATVLLVIVLLAGSYLA 47
CAC1C_HUMAN
IS5             IALLVLFVIIYAIIGLELF 290
IIS5            LLLLFLFIIIFSLLGMQLF 673
IIIS5           VIVTTLLQFMFACIGVQLF 1071
IVS5            ALLIVMLFFIYAVIGMQVF 1430

KcsA
P               ITYPRALWWSVETATTVGYGD 80
CAC1C_HUMAN
IP              DNFAFAMLTVFQCITMEGWTD 367
IIP             DNFPQSLLTVFQILTGEDWNS 710
IIIP            DNVLAAMMALFTVSTFEGWPE 1138
IVP             QTFPQAVLLLFRCATGEAWQE 1468

KcsA
M2              WGRCVAVVVMVAGITSFGLVTAALAT 112
CAC1C_HUMAN
IS6             WPWIYFVTLIIIGSFFVLNLVLGVLS 405
IIS6            LVCIYFIILFICGNYILLNVFLAIAV 753
IIIS6           VEISIFFIIYIIIIAFFMMNIFVGFV 1185
IVS6            FAVFYFISFYMLCAFLIINLFVAVIM 1524
```

Fig. 7.1 Pairwise alignment of CAC1C_HUMAN and KcsA sequences. The conserved key residues used to align the sequences are shown in *red boxes*. Residues reported to affect DHP antagonist binding and underscored and highlighted in *bold*

7.2.1.1 Construction of the Human LCC Model

The structural model of the human LCC was built using the recently reported 3.20 Å crystal structure of KcsA (Shaldam et al. 2014) (PDB entry code 1BL8) as a structural template. The sequence of the human LCC pore region α1c subunit (Cav1.2, CAC1C_HUMAN) was retrieved from the Swiss-Prot database (Shaldam et al. 2014) and aligned as described in the Results and Discussion section (Fig. 7.1). The construction of the transmembrane region of the model was achieved by the employment of the modeller 9.13.

The protocol used to develop the LCC model is divided into three phases: sequence alignment, model construction and model evaluation.

7.2.2 *Sequence Alignment*

The model was constructed using amino acid sequence of voltage-dependent LCC subunit alpha-1C (CAC1C HUMAN Q13936) obtained from UniProtKB/Swiss-Prot sequence database (Reyes et al. 1990; http://www.uniprot.org/unipro/Q13936). Coordinates of potassium channel KcsA atoms in their closed conformation were downloaded from the RCSB Protein Data Bank (PDB ID: 1BL8). Amino acid sequences of S5, S6 and P-loops in between for the four repeats (I–IV) (271–405, 654–753, 1052–1185 and 1411–1524, respectively) were used for sequence alignment with the amino acid sequence of KcsA as proposed by Zhorov et al. (2001) (Fig. 7.1). In order to favour valid superimposition of the residues, the sequence of each repeat was organized as S5, S6 and P-loop, allowing for a more flexible inspection of the results and easier corrections. The amino acid sequence of repeats I and III has a long extracellular loop which would decrease the quality of the generated model, so amino acid sequences were excluded from repeats I and III, respectively.

7.2.3 Construction of the LCC Model

The modelling procedure consisted of two steps: model construction from the template and refinement of loops. The above described sequence alignment file was used as input for the MODELLER 9.13 program (Sali et al. 1995) with the high-resolution NMR structure of potassium channel KcsA available in the RSCB Protein Data Bank (PDB ID: 1BL8) as a template for the 3D structure. Molecular modelling studies were performed using the MODELLER 9.13 running on Intel Core 2 Duo CPU personal computers. The model sequence, template structure and sequence alignment were used as input files to build the model. Loops can be defined automatically from the model to a template sequence alignment. The MODELER Loop Refinement-DOPE-Loop method (Shen and Sali 2006; Shaldam et al. 2014) was used for initial refinement of the loop conformation after model generation. The model side-chain conformation was optimized based on systematic searching of side-chain conformation and CHARMm energy minimization using the ChiRotor algorithm (Spassov et al. 2007; Shaldam et al. 2014). Five models were obtained from the first step of molecular modelling. These models were subjected to a comparison based on the best scores to reveal the differences among them. The model with the lowest energy and the lowest restraint violation was selected for the second step. Secondly, the loops between helices were subjected to refinement while keeping the start and end residues constrained. This procedure is based on the idea that transmembrane helices are much less flexible than loops, thus permitting to produce a sounder core alignment if the integrity of the helices is conserved. The more unpredictable loops can bear the more important differences. A CHARMm-based protocol (Spassov et al. 2008; Shaldam et al. 2014) that optimizes the conformation of a contiguous segment (i.e. a loop) of a protein structure was used for loop refinement. It is based on systematic conformational sampling of the loop backbone and CHARMm energy minimization. This approach can be used to refine a loop structure from a homology model as well as to optimize a segment of the protein experimental structure where the structure is poorly defined. The homology modelling (HM) phase was followed by the model evaluation phase. The stereochemical quality and structural integrity of the model were tested by RAMPAGE, ERRAT, MolProbity, ProSA and Verify3D software and target–template superimposition by PyMol (Eswar et al. 2008) (Fig. 7.2).

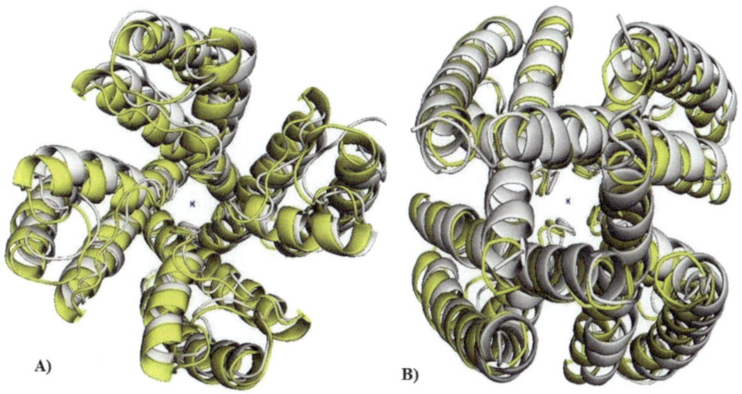

Fig. 7.2 Superimposition of the LCC model (*white*) and KcsA (*yellow*) (PDB: 3BPM). (**a**) Open conformation, (**b**) closed conformation

7.2.4 Active Site Identification

The active site of LCC model was identified using a CASTp server (Computer Atlas of Surface Topology of Proteins) (Dundas et al. 2006). A new program, CASTp, for automatically locating and measuring protein pockets and cavities, is based on precise computational geometry methods, including alpha shape and discrete flow theory. CASTp identification, measurements of surface accessible pockets as well as interior inaccessible cavities by locating, delineating and measuring concave surface regions on three-dimensional structure of proteins. The measurement includes the area and volume of pocket or void by solvent-accessible surface model (Richards' surface) and by molecular surface model (Connolly's surface), calculated analytically. It can also be used to study surface features and functional regions of proteins.

7.2.5 Generation of Structure-Based Pharmacophore Model

In the present study, the LCC modelled receptor complex with a channel blocker AD was used as starting structure for the generation of structure-based pharmacophore models (Abdul et al. 2012). LIGANDSCOUT (LS) is a tool that allows the automatic construction and visualization of 3D pharmacophore for structural data of macromolecule/ligand complexes. For the LS algorithm, chemical features include hydrogen bond donors and acceptors as directed vectors, and positive and negative ionizable regions as well as lipophilic areas are represented by spheres. Moreover, to increase the selectivity, the LS model includes spatial information regarding areas inaccessible to any potential ligand, thus reflecting possible steric restrictions. In particular, for excluded volume spheres placed in positions that are sterically forbidden, LS may also be used to construct pharmacophore of varying degrees of sophistication, suitable for export to different programs. In the present study, Molecular Operating Environment (MOE, version 2008, Chemical Computing Group Inc.)-compatible 3D pharmacophore model was first developed by LS using default parameters, and then, it was exported and converted into a MOE, pharmacophore query for virtual screening (http://www.chemcomp.com). Prior to the screening, it was necessary to make a number of adjustments, because feature interpretation differs slightly between the two programs. Those aromatic rings that LS classified simply as hydrophobic groups were classified as either aromatic or hydrophobic in MOE, using the PPCH_All scheme (which incorporates directionality of hydrogen bond donors and acceptors and orientation of aromatic rings). As in LS pharmacophore, the aromatic ring is not directly classified as such (because of the lack of detection of π–π stacking or cation–π interactions) but, rather as a set of hydrophobic atoms, can be interpreted in MOE in a manner that is useful in the prediction of right compounds in virtual screening.

7.2.6 Pharmacophore-Based Virtual Screening

The Zinc database (http://zinc.docking.org/), which allows the user to download compounds, structures from a variety of vendors as SDF files based on the structure-based amlodipine (AD) compound (Query), was used in this preliminary screen. Using MOE, the database was washed, and the 3D structure of each compound was built using the MMFF94x force field. Then for each compound, the low-energy conformers were generated using Conformation Import methodology implemented in MOE software. After assessing the pharmacophore query, virtual screening was carried out using the software MOE against the Zinc database. Because some changes may occur when the pharmacophore is exported from LS to MOE environment, therefore, the pharmacophore queries were validated before using it for virtual screening. To reduce the data of identified hits, they were docked into the recently identified binding pocket of LCC model, and the PLIFs were developed using MOE. Binding energies and binding affinities were calculated using LIGX (Chemical Computing Group, Montreal, Quebec, Canada) implemented in MOE to prioritize the final hits.

7.2.7 Molecular Docking

Docking is a computational method which predicts the preferred orientation of one molecule to a second when bound to each other to form a stable complex. Docking has been widely used to suggest the binding modes of protein inhibitors. Most docking algorithms are able to generate a large number of possible structures; thus, they also require a means to score each structure to identify those that of greatest interest. Docking was performed using AutoDock in PyRx Virtual Screening tool (Wolf 2009; Trott and Olson 2010).

Pharmacophoric hit compounds were docked into the active site of the refined LCC model. Lamarckian genetic algorithm was used as the number of individual population (150), max number of energy evaluation (25000000), max number of generation (27000) (Laskowski et al. 1993), gene mutation rate (0.02), crossover rate (0.8), Cauchy beta (1.0) and GA window size (10.0). The grid was set whole protein due to the multi-binding pocket at $X=3.42$, $Y=-56.23$, $Z=98.32$ and dimension (Å) at $X=89.92$, $Y=98.56$, $Z=98.32$ and exhaustiveness 8. The pose for a given ligands identified on the basis of highest binding energy. Only ligand flexibility was taken into account and the proteins were considered to be rigid bodies. The resulting complexes were clustered according to their root mean square deviation (RMSD) values and binding energies, which were calculated using the AutoDock scoring function. The PyMol molecular viewer (http://www.pymol.org/) was employed to analyse the docked structures.

7.2.8 Analysis of Drug Likeness

MolSoft Drug Likeness explorer (http://www.molsoft.com/mprop/) was used to analyse the drug likeness as per "Lipinski rule of 5" (Lipinski et al. 1997). According to "Lipinski rule of 5", a compound is more likely to be membrane permeable and easily absorbed by the body if its molecular weight is less than 500, its lipophilicity expressed as a quantity known as log P is less than 5, the number of groups in the molecule that can donate hydrogen atoms to hydrogen bonds is less than 5 and the number of groups that can accept hydrogen atoms to form hydrogen bonds is less than 10 (Leeson 2012).

7.2.9 ADMET Properties

The in silico pharmacokinetic properties and ADMET (absorption, distribution, metabolism, elimination and toxicity) analysis were predicted using OSIRIS property explorer (http://www.organic-chemistry.org/prog/peo/; Access date: September 23, 2014) which uses Chou and Jurs algorithm, based on computed atom contributions.

7.3 Results and Discussion

7.3.1 *Sequence Alignment*

Besides the choice of the reference, the accuracy of the alignment is the most crucial step in assuring the quality of the homology modelling. An accurate sequence alignment between the model and the template proteins is essential to achieve high-quality models. Voltage-gated LCC are members of a gene superfamily of transmembrane ion channel proteins that includes voltage-gated K^+ and Na^+ channels. LCC share structural similarities with K^+ and Na^+ channels in that they possess a pore-forming $\alpha 1$ subunit in four repeats of a domain with six transmembrane-spanning segments that include the voltage-sensing S4 segment and the pore-forming (P) region. As no atomic resolution images of calcium channel structures exist, much has been learnt about their structure since the recent determination of crystal structures of a number of potassium channels (Jiang et al. 2003; Long et al. 2005; Shaldam et al. 2014). The $\alpha 1$ subunit contains four repeated domains (I–IV), each of which includes six transmembrane segments (S1–S6) and a membrane-associated loop (the "P-loop") between segments S5 and S6. The four repeated domains are also remarkably similar to those known to form the voltage-gated potassium channels. However, potassium $\alpha 1$ subunit is homotetramer and calcium channel is heterotetramer. Potassium channel KcsA (PDB code 1BL8) has been selected to be the template. Amino acid sequences of S5, S6 and P-loops in between the four repeats (I–IV) (271–405, 654–753, 1052–1185 and 1411–1524, respectively) of voltage-dependent LCC subunit alpha-1C (CAC1C HUMAN Q13936) were used for sequence alignment with the amino acid sequence of KcsA as proposed by Zhorov et al. (2001), where S6 segments of LCC are aligned with M2 segments of KcsA in a manner similar to the alignment of the Na^+ channel with KcsA described by Lipkind and Fozzard (2000) and S5s were aligned with the M1 segment of KcsA as proposed by Huber et al. (2000) and the P-loops were aligned using MULTALIN server (Corpet 1988; Shaldam et al. 2014) (Fig. 7.1). Proteins that fold into similar structures can have large differences in the size and shape of residues at equivalent positions. These changes are tolerated not only because of replacements or movements in nearby side chains, as explored by Ponder and Richards, but also as a result of shifts in the backbone (Bowie et al. 1991; Shaldam et al. 2014). For a more flexible inspection, the sequence of each repeat was organized as S5, S6 and P-loop, allowing easier corrections. The amino acid sequence of repeats I and III has long extracellular loop which would reduce the quality of the generated model, so amino acid sequences were excluded from repeats I and III, respectively. Since the template is 88 residues shorter than the target protein, gaps were inserted to achieve best sequence similarity and identity without affecting sequence alignment proposed by Zhorov et al. (2001). The greatest attention was thus paid to the careful construction of transmembrane helices S5 and S6 and P-loop as well.

7.3.2 Construction of the LCC Model

Although the two proteins have low sequence identity of 9.5% and sequence similarity of 29.2%, the MODELLER program was applied to generate satisfactory models. As an integral process of model building, initial refinement of the loop conformation after model generation was automatically performed by MODELER Loop Refinement-DOPE-Loop method during the process. The model achieved from the alignments by Zhorov et al. (2001) was subjected to extensive loop optimization. This procedure is based on the idea that transmembrane helices are much less flexible than loops, thus permitting to produce a sounder core alignment if the integrity of the helices is conserved. On the contrary, the more volatile loops can bear the more important difference between the coordinates of the reference and the model. When a homology model is created, there are parts of the model sequence which are not aligned to any template structures. For these sections, no homology restraints (such as Cα–Cα distance restraints) can be applied. These parts of the structure generally have greater errors compared to the regions which are modelled based on a template structure. In attempts to reduce these errors, a CHARMm-based protocol that optimizes the conformation of a contiguous segment (i.e. a loop) of a protein structure called loop refinement was applied (Spassov et al. 2008; Shaldam et al. 2014). This is based on systematic conformational sampling of the loop backbone and CHARMm energy minimization. The algorithm goes through three stages: construction and optimization of loop backbone, construction of loop side chain and optimization of loop followed by reranking of the conformations. The model was then checked after a thorough energy minimization designed to reduce the steric clashes of the side chains without modifying the backbone of the protein to solve these contacts. To avoid modification of the backbone of the protein, the optimization of the geometry of side chain was performed with constraining the backbone. After the optimization, models were checked to assess the quality of their structure.

7.3.3 Model Evaluation

To assess stereochemical quality and structural integrity of the model, RAMPAGE (Lovell et al. 2003) (Fig. 7.3), ERRAT (Colovos and Yeates 1993; Shaldam et al. 2014), ProSA (Sippl 1993; Wiederstein and Sippl 2007) and Verify3D (Luthy et al. 1992; Shaldam et al. 2014) software were used. For comparison, these methods were also used to evaluate the template structure, and then each repeat was examined separately by means of ProSA. RAMPAGE is an offshoot of RAPPER which generates a Ramachandran plot using data derived by Lovell et al. (2003). It is recommended that it be used for this purpose in preference to PROCHECK, which is based on much older data. The Ramachandran diagram plots phi versus psi dihedral angels for each residue in the protein. The diagram is divided into

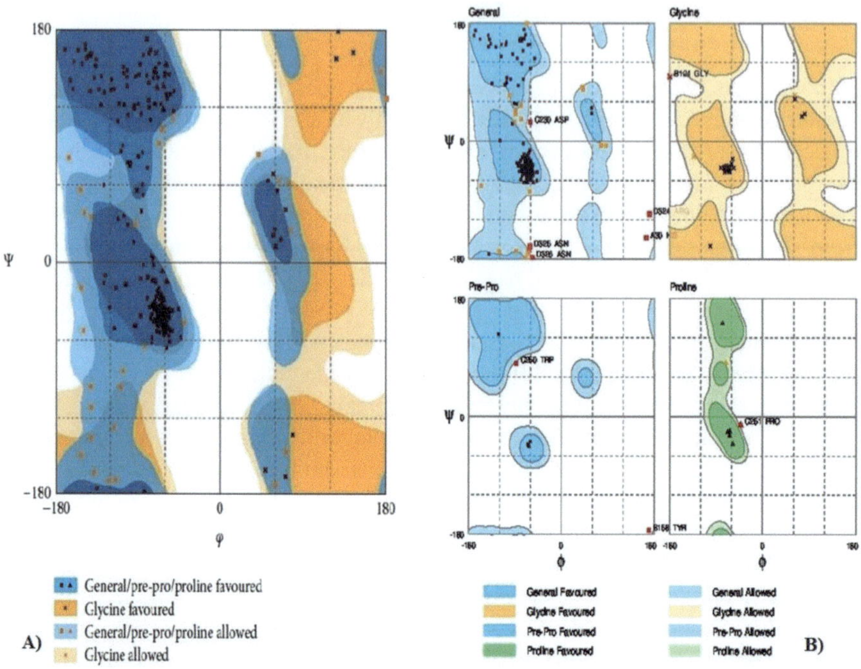

Fig. 7.3 Ramachandran plot. (**a**) The plot of LCC model shows that 92.3% of residues were found in the favoured, 7.7% in the allowed and none in the outlier regions. (**b**) The plot shows general, glycine, pre-proline and proline for LCC model

favoured, allowed and disallowed regions, whose contouring is based on density-dependent smoothing for 81,234 non-glycine, non-proline residues with $B < 30$ from 500 high-resolution protein structures. Regions are also defined for glycine, proline and pre-proline as shown in Fig. 7.3.

ERRAT is a protein structure verification algorithm, that is, especially well-suited for differentiating between correctly and incorrectly determined regions of protein structures based on characteristic atomic interactions (Colovos and Yeates 1993; Shaldam et al. 2014). Different types of atoms (C, N and O) are distributed nonrandomly with respect to each other in proteins because of energetic and geometric effects. Errors in model building lead to more randomized distributions of the different atom types, which can be distinguished from correct distributions by statistical methods. The program works by analysing the statistics of nonbonded interactions between different atom types. A single output plot is produced that gives the value of the error function versus position of a nine-residue sliding window. In comparison with statistics from highly refined structures, the error values have been calibrated to give confidence limits. The program provides an “overall quality factor” value which is defined as the percentage of the protein for which the calculated error value falls below the 95% statistical rejection limit. The ERRAT overall quality factor of the model is given in Table 7.1. This is not

Table 7.1 Assessment scores for the LCC receptor model

S. no.	Item	Model	Comment
1	ProSA	−3.89	ProSA Z-score as average of the four repeats
2	ERRAT	79.72	ERRAT overall quality factor
3	Verify3D	67.58% (W)	Percentage of residues with Verify3D average score > 0.2; verify3D overall assessment of the structure (P = pass, W= warning or F = fail) shown in parentheses

surprising since the model has longer loops than template. This method provides a useful tool for model building, structure verification and making decisions about reliability. A more reliable discrimination of incorrect regions would likely be obtained by combining the present analysis with others (Fig. 7.5).

ProSA and Verify3D are two methods that are sensitive in distinguishing between overall correct fold and those with an incorrect fold (Bhattacharya et al. 2008; Shaldam et al. 2014). ProSA (Protein Structure Analysis) program is a diagnostic tool that is based on the statistical analysis of all available protein structures (Wiederstein and Sippl 2007; Shaldam et al. 2014). It is a tool widely used to check 3D models of protein structures for potential errors. Its range of application includes error recognition in experimentally determined structures (Teilum et al. 2005; Llorca et al. 2006; Shaldam et al. 2014), theoretical models (Petrey and Honig 2005; Ginalski 2006; Shaldam et al. 2014) and protein engineering (Beissenhirtz et al. 2006; Mansfeld et al. 2006; Shaldam et al. 2014). The energy of the structure is evaluated using a distance-based pair potential and a potential that captures the solvent exposure of protein residues. From these energies, two characteristics are derived and displayed: Z-score and a plot of residue energies. The Z-score indicates overall model quality and measures the deviation of the total energy of the structure with respect to an energy distribution derived from random conformations. Z-scores outside a range characteristic of native proteins indicate erroneous structures. The overall quality score calculated by ProSA for a specific structure is displayed in a plot that shows the scores computed from all experimentally determined protein chains currently available in the Protein Data Bank (PDB). Structures which contain errors are likely to have Z-score outside the range of values characteristic of native proteins. Table 7.1 lists the Z-score calculated by ProSA (as average of the four repeats Z-score) for the model and compared against the template. The Z-scores for the model and template are much closer to the middle region of scores observed for experimentally determined protein structures in the PDB including the template structure. The energy plot shows the local model quality by plotting energies as a function of amino acid sequence. In general, positive values correspond to problematic or erroneous parts of the model (Fig. 7.4).

Verify3D analyses the compatibility of an atomic model (3D) with its own amino acid sequence (1D) and hence tests the accuracy of the model (Fig. 7.5). Each residue is assigned a structural class based on its location and environment. The environments are described by the area of the residue buried in the protein and inaccessible to solvent, the fraction of side chain area that is covered by polar atoms

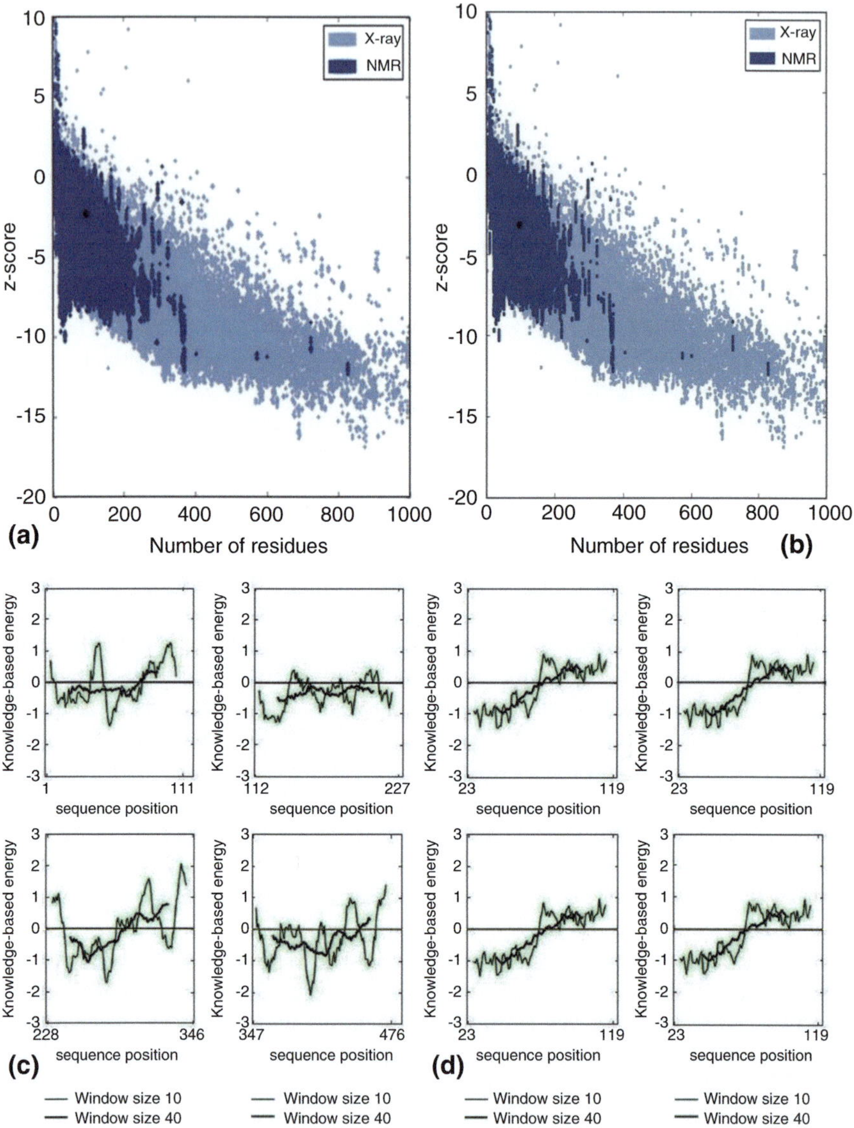

Fig 7.4 ProSA plot. Each repeat was examined separately. (**a**) ProSA Z-scores for LCC model; (**b**) ProSA Z-scores for template (KcsA) and *blue* and *sky blue dots* are Z-scores of PDB structures determined by X-ray crystallography and NMR, respectively; (**c**) ProSA energy profiles for LCC model (four repeats); (**d**) ProSA energy profiles for template (four repeats). Negative scores indicate a high-quality model

(O and N) and the local secondary structure. Based on these parameters, each residue position is categorized into an environmental class. In this manner, a 3D protein structure is converted into a 1D string, like a sequence, which represents the

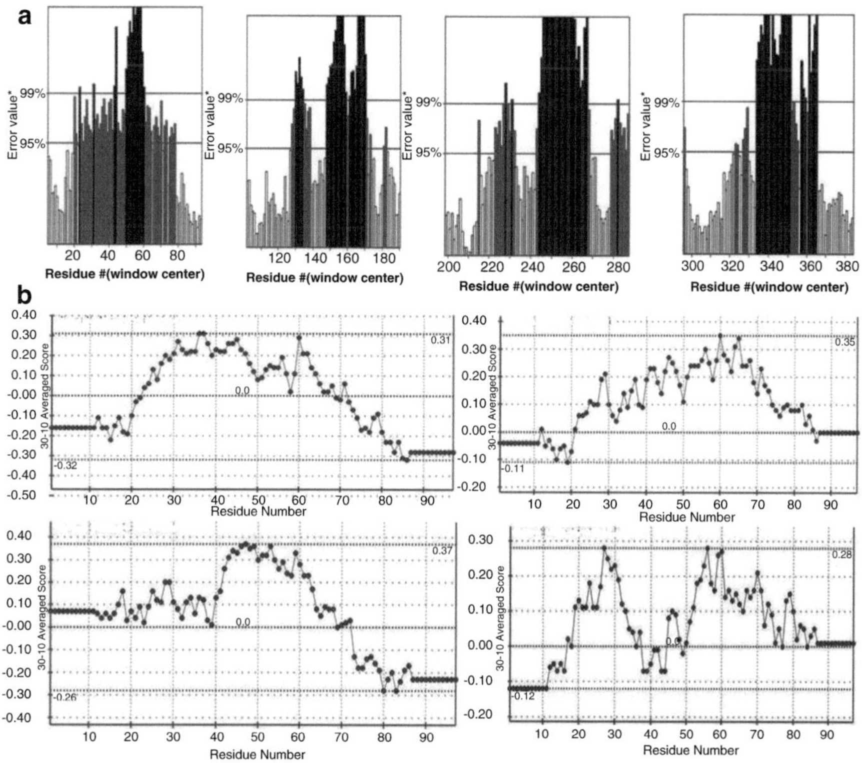

Fig. 7.5 (**a**) ERRAT score of the LCC model (four repeats). (**b**) Verify3Dscore profile calculated for LCC model. Scores over 0.2 indicate a high-quality model

environmental class of each residue in the folded protein structure. A collection of good structures is used as a reference to obtain a score for each of the 20 amino acids in this structural class. The scores of a sliding 21-residue window are added and plotted for individual residues. This method evaluates the fitness of a protein sequence in its current 3D environment. It can be applied to assess the quality of a theoretical model or to examine the characteristics of an experimental structure (Luthy et al. 1992; Shaldam et al. 2014). Table 7.1 shows the percentage of residues that had an average score > 0.2 and the Verify3D assessment of the structure (pass, warning or fail) for the model and template. Figure 7.5 shows the Verify3D profile for the model structure. Residues with a score over 0.2 should be considered reliable and the sequences exhibiting lower scores are those of extracellular loops.

Taken together, all of the above data indicate that the quality of the model is similar to that of the template. The model can be used for further computer-aided drug design (CADD) and it can be used in understanding how DHP work at the molecular level.

7.3.4 Generation of Structure-Based Pharmacophore Model

As shown in Fig. 7.6, the pharmacophore model automatically generated by the LS program includes four features: two hydrogen bond donors (HBD) (green colour) and three hydrophobic groups (yellow colour). Besides, the program automatically generated several excluded volumes (grey colour) in the model. The two HBD feature points are the amino group hydrogen atoms of the ligand towards the SER-78 and ILE-51, respectively. The three hydrophobic groups are located on the benzene group, chlorine atom on benzene and carboxy ethyl group of the ligand. The developed pharmacophore model was exported into MOE. Prior to screening, it was necessary to make a number of adjustments, as feature interpretation varies slightly between the two programs. As in LS pharmacophore, the aromatic ring of the compound in the complex was not classified as aromatic or hydrophobic features; thus, these were interpreted in MOE, using the PPCH_All scheme. Two modifications were made on this model to obtain appropriate model for virtual screening. The first modification is about the chlorobenzyl ring. It is clear that it is an aromatic group, but the LS could not interpret this ring as an aromatic group automatically. In MOE, additional features were developed using the MOE pharmacophore query editor. First, an aromatic feature was developed on the chlorobenzyl ring, and a hydrophobic feature was developed on the carboxy ethyl group of the ligand. This modified pharmacophore model was then validated by screening the test database. In the test database, we kept the compound (i.e. AD) present in complex structure. First, the AD was extracted, and then, hydrogen atoms were added and energy minimized by using MOE. The minimized structure of AD was added to the test database. After screening, the test compound was correctly mapped by the modified pharmacophore model as shown in Fig. 7.6. The result verified the validity of our modified pharmacophore model that can be used for the screening of large databases.

7.3.5 Pharmacophore-Based Virtual Screening

The modified validated pharmacophore model was then used as in silico filter to screen the Zinc database (http://zinc.docking.org/) of commercially available compounds. The Zinc database compounds in SDF format were loaded into MOE environment where the 3D structure of each compound was modelled using MMFF94x force field. The Conformation Import methodology was applied to generate low-energy conformations for each compound. All these compounds and their respective conformations were saved in MOE database. The conformers of each compound were then filtered by the pharmacophore model. To be considered as hit, the compound has to fit all the features of the pharmacophore. From the pharmacophore-based virtual screening, 18 hits (Fig. 7.7) were identified that mapped on the developed pharmacophore model (i.e. having the specified

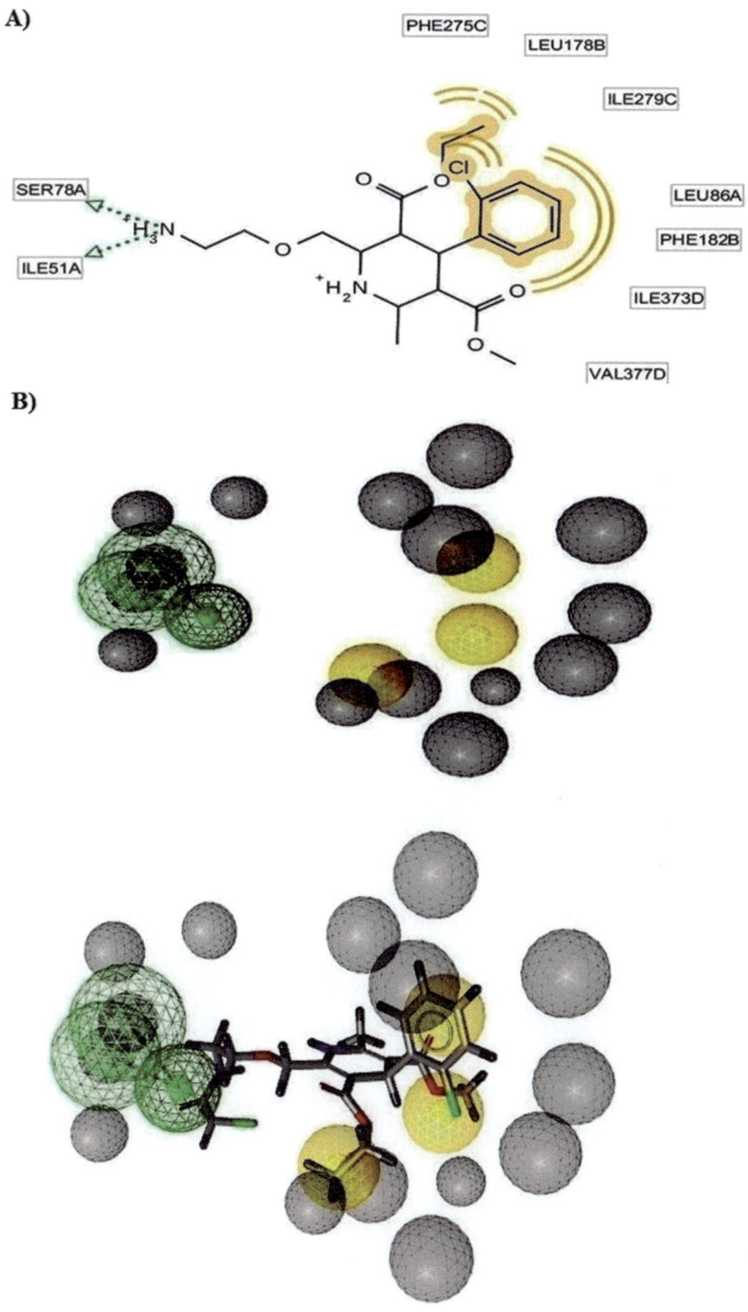

Fig. 7.6 (**a**) Two-dimensional pharmacophore model generated by LIGANDSCOUT from the complex structure of LCC and AD. The *dotted arrows* indicated the hydrogen bond donor (HBD) features. (**b**) The *yellow sphere* represented the HBD; the *yellow sphere* represented the hydrophobic feature in the ligand, whereas the *grey* colour spheres represented the excluded volumes

Fig. 7.7 The pharmacophore screened 18 hits from Zinc database

Table 7.2 AD and AD analogue compounds along with their respective interaction energies and their surrounding residues

S. no.	Compound	BE (kcal/mol)	Surrounding residues
1	Zinc59347487	−8.4	ILE-360 (**IP**), SER-393 (**IS6**), ASN-398 (**IS6**), ASN-740 (**IIS6**), ASN1517 (IVS6)'
2	Zinc20267861	−8.1	LEU-704 (**IIP**), ASN-740 (**IIS6**), ASN-1517 (**IVS6**)
3	Zinc59486248	−7.9	THR-361 (**IP**), MET-362 (**IP)**
4	Zinc59494792	−7.6	THR-361 (**IP**), ASN-740 (**IIS6**)
5	Zinc67664832	−7.2	THR-361 (**IP**), ILE-360 (**IP**), SER-393 (**IS6**), ASN-1178 (**IIIS6**), ASN-740 (**IIS6**)
6	Zinc19796039 (AD) (Query)	−5.4	ILE-360 (**IP**), SER-393 (**IS6**), ASN-1178 (**IIIS6**), ASN-740 (**IIS6**)

requirements). These initially identified hits were selected for further evaluation using docking studies.

7.3.6 Molecular Docking

In order to shed light on the molecular basis of the interactions between LCC and its ligands, docking simulations were undertaken on pharmacophoric hits of DHPs (dihydropyridines) on LCC model. Such calculations were conducted employing the automated docking program AutoDock which has proven to be really effective in reproducing the experimentally found posing of ligands into their binding site. As shown in Table 7.2, the predicted free energy of binding top five compounds. Docking of hits into active site of LCC model gave comparable binding solutions with the dihydropyridine ring fitting in the cleft formed by IIIS6, IIIS5 and IVS6 segments. The Zinc59347487 compounds bound -8.4 binding energy with ILE-360 (IP), SER-393 (IS6), ASN-398 (IS6), ASN-740 (IIS6) and ASN-1517 (IVS6) active site residues, respectively. The Zinc20267861 compounds bound -8.1 binding energy with LEU-704 (IIP), ASN-740 (IIS6) and ASN-1517 (IVS6) active site residues, respectively. The Zinc59486248 compounds bound -7.9 binding energy with THR-361 (IP) and MET-362 (IP) active site residues, respectively. The Zinc59494792 compounds bound -7.6 binding energy with THR-361 (IP) and ASN-740 (IIS6) active site residues, respectively. The Zinc67664832 compounds bound -7.2 binding energy with THR-361 (IP), ILE-360 (IP), SER-393 (IS6), ASN-1178 (IIIS6) and ASN-740 (IIS6) active site residues, respectively. The Zinc19796039 (AD) compounds bound -5.4 binding energy with ILE-360 (IP), SER-393 (IS6), ASN-1178 (IIIS6) and ASN-740 (IIS6) active site residues, respectively. The LCC model and best five screened compound interaction residues are shown in Table 7.2 and the graphical representation also shown in Fig. 7.8. The docking results showed that five compounds have best binding energies than AD compound (Table 7.2).

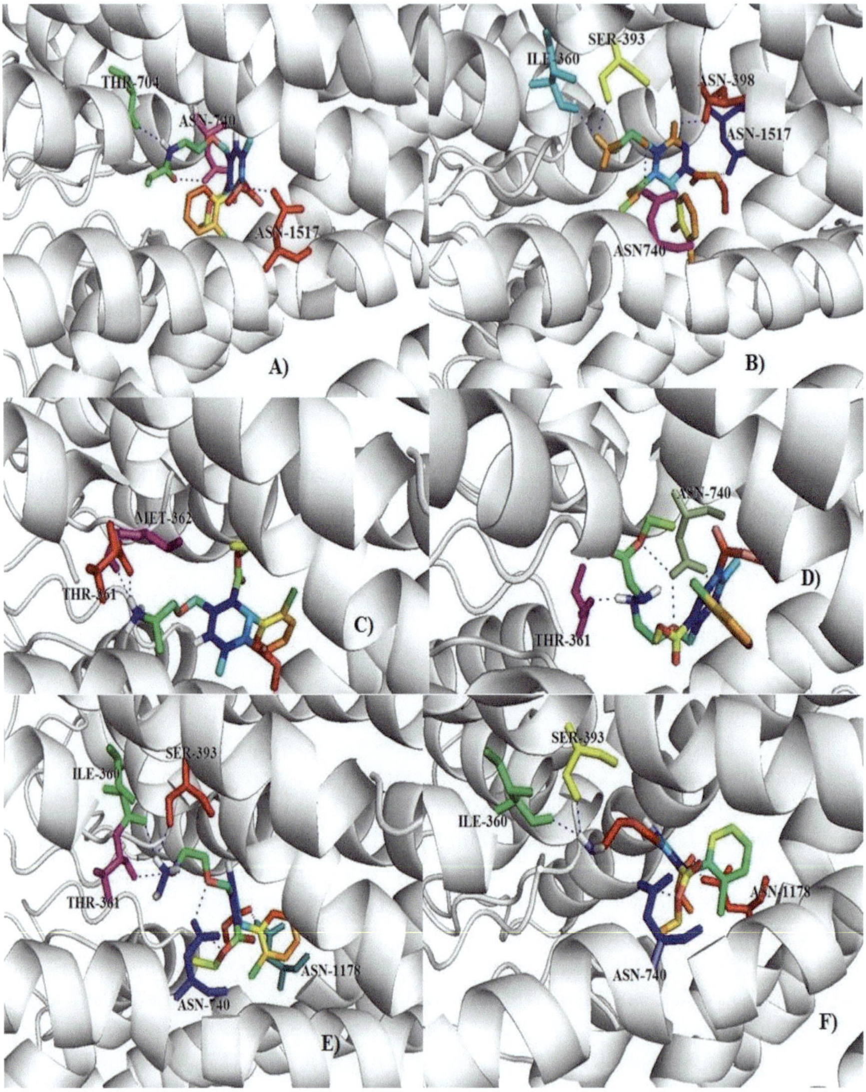

Fig. 7.8 Docked structures of Zinc20267861 (**a**), Zinc59347487 (**b**), Zinc59486248 (**c**), Zinc59494792 (**d**), Zinc67664832 (**e**) and AD (**f**) in model of LCC. DHPs are displayed as rainbow sticks, and key binding site residues are shown in *green*, *yellow*, *red*, *pink* and *blue*. Hydrogen bonds as represented with *dashed blue lines*

7.3.7 Analysis of Drug Likeness

All the compounds were tested for Lipinski "rule of 5", i.e. "drug-like" molecules have log P $\leq$5, molecular weight $\leq$500, number of hydrogen bond acceptors $\leq$10

Table 7.3 Molecular properties of compounds satisfying the Lipinski "rule of 5" by

Zinc ID	log S (moles/L)	Lipinski "rule of 5" Molecular weight ≤500	log P ≤5	HB acceptors ≤10	HB donors ≤5
Zinc59347487	−4.06	429.10	3.14	5	4
Zinc20267861	−4.18	484.12	3.03	6	2
Zinc59486248	−4.90	423.17	3.61	5	4
Zinc59494792	−4.30	499.19	2.43	5	3
Zinc67664832	−4.24	437.16	2.21	9	3
Amlodipine (Query)	−4.19	409.15	3.46	7	4

and number of hydrogen bond donors ≤5 (Table 7.3). Lipinski rule of 5 is a rule of thumb for evaluating the drug likeness or determining whether a chemical compound with a certain pharmacological or biological activity has properties that would make it a likely orally active drug in humans. Results showed that five compounds, i.e. (Zinc59347487), (Zinc20267861), (Zinc59486248), (Zinc59494792) and (Zinc67664832), satisfied the Lipinski "rule of 5". Their respective drug likeness properties are shown in Table 7.3.

7.3.8 *ADME Predicting Activity*

Although Lipinski "rule of 5" describes the molecular properties important for a drug's pharmacokinetics in the human body, including its ADME, it does not predict if a compound is pharmacologically active. Therefore, pharmacokinetic properties and toxicities were predicted for all the four compounds using OSIRIS property explorer. Results of pharmacokinetic properties and toxicity analysis are shown in Table 7.4. Solubility and partition coefficient were calculated for pharmacokinetic property, whereas mutagenicity, tumorigenicity, irritation effect and risk of reproductive effect were predicted for toxicity study. To determine the hydrophilicity, log P value was predicted. It is suggestive that a high log P value is associated with poor absorption or permeation and it must be less than 5 (Vyas et al. 2013). Results showed that all the five compounds confirmed to this limit, and among the five compounds, Zinc67664832 has a better cLog P value than others (Table 7.3). In general, a poor solubility is associated with bad absorption, and the aqueous solubility (log S) of a compound significantly affects its absorption and distribution characteristics. Results showed that Zinc67664832 has a better log S value than others (Table 7.3). In order to consider the compound overall potential as a drug candidate, drug score is calculated which combines drug likeness, cLog P, TPSA, molecular weight and toxicity risk parameters as shown in Table 7.4. Drug score showed that the compounds, Zinc59347487 and Zinc20267861, have higher scores of 0.56 and 0.47 compared to the others.

Table 7.4 In silico ADMET prediction by OSIRIS property explorer

Properties	Zinc59347487	Zinc20267861	Zinc59486248	Zinc59494792	Zinc67664832	Amlodipine
Mutagenic	–	+	–	–	–	–
Tumorigenic	–	+	–	–	–	–
Irritant	–	–	–	–	–	–
Reproductive effective	–	+	–	–	–	–
cLog P	1.18	1.71	2.43	0.98	1.99	2.07
Solubility	−3.08	−3.10	−3.68	−3.53	−3.18	−3.30
MW	429.0	485.0	422.0	495.0	438.0	408.0
TPSA	99.88	102.9	99.88	135.20	123.0	99.88
Drug likeness	−7.9	−5.54	−4.60	−2.14	−4.94	−6.2
Drug score	0.56	0.47	0.37	0.38	0.38	0.39

7.4 Conclusions

The point of present study was to produce a pharmacophore model to recognize vitally assorted lead hits. The recognized hits may be utilized for creating novel and strong inhibitors for VP-3. A structure-based pharmacophore was created situated in light of the complex structure of VP-3 and leupeptin. The created pharmacophore model was utilized for the screening of PubChem database. The recognized hits were further assessed by docking, MD simulation and binding energy forecast. Subsequently, five lead hits were accounted for that satisfied all the criteria for the outline of compounds that may go about as great leads for advancement of novel, intense and structurally diverse compounds for VP-3 inhibition. From the binding mode, anticipated by docking, it was observed that there are some particular groups that mimic the binding method of leupeptin and fit well to active site area of VP-3. The five leads likewise demonstrated the best binding energies among screened compounds. The MD simulations for the VP-3 five lead docking complexes were performed to comprehend conformational dependability, structural flexibility and molecular dynamics of the interaction in physiological environmental condition. RMSD investigation demonstrated that the molecular system was exceptionally steady in all trajectories. Therefore, five leads are proposed as the best potential inhibitor to begin with investigation acceptance towards outlining against VP-3 inhibitors.

Acknowledgements Author, Madhu Sudhana Saddala, is especially grateful to University Grants Commission, New Delhi, for their financial assistance with the award of BSR Meritorious Fellowship. This work was carried out in DBT – Bioinformatics Infrastructure Facility (BIF), Department of Zoology, Sri Venkateswara University, Tirupati (BT/BI/25/001/2006).

References

Abdul W, Abid Ali S, Sattar R, Arif Lodhi M, Ul-Haq Z. A novel pharmacophore model to identify leads for simultaneous inhibition of anti-coagulation and anti-inflammatory activities of snake venom phospholipase A_2. Chem Biol Drug Des. 2012;79:431–41.

Beissenhirtz MK, Scheller FW, Viezzoli MS, Lisdat F. Engineered superoxide dismutase monomers for superoxide biosensor applications. Anal Chem. 2006;78(3):928–35.

Bhattacharya A, Wunderlich Z, Monleon D, Tejero R, Montelione GT. Assessing model accuracy using the homology modeling automatically (HOMA) software. Proteins. 2008;70(1):105–18.

Bowie JU, Lüthy R, Eisenberg D. A method to identify protein sequences that fold into a known three-dimensional structure. Science. 1991;253(5016):164–70.

Colovos C, Yeates TO. Verification of protein structures: patterns of nonbonded atomic interactions. Protein Sci. 1993;2(9):1511–9.

Corpet F. Multiple sequence alignment with hierarchical clustering. Nucleic Acids Res. 1988;16 (22):10881–90.

Dundas J, Ouyang Z, Tseng J, Binkowski A, Turpaz Y, Liang J. CASTp: computed atlas of surface topography of proteins with structural and topographical mapping of functionally annotated residues. Nucleic Acids Res. 2006;34:W116–8.

Eswar N, Eramian D, Webb B, Shen MY, Sali A. Protein structure modeling with modeller. Methods Mol Biol. 2008;426:145–59.

Ginalski K. Comparative modeling for protein structure prediction. Curr Opin Struct Biol. 2006;16 (2):172–7.

Huber I, Wappl E, Herzog A, et al. Conserved Ca2+- antagonist-binding properties and putative folding structure of a recombinant high-affinity dihydropyridine-binding domain. Biochem J. 2000;347(3):829–36.

Jiang Y, Lee A, Chen J, et al. X-ray structure of a voltage-dependent K+ channel. Nature. 2003;423(6935):33–41.

Kocak M, Akcil E. The effects of chronic cadmium toxicity on the hemostatic system. Pathophysiol Haemost Thromb. 2006;35:411–6.

Laskowski RA, MacArthur MW, Moss DS, Thornton JM. PROCHECK: a program to check the stereo chemical quality of protein structures. J Appl Crystallogr. 1993;26:283–91.

Leeson P. Drug discovery: chemical beauty contest. Nature. 2012;481(15):455–6.

Lipinski CA, Lombardo F, Dominy BW, Feeny PJ. Experimental and computational approaches to estimate solubility and permeability in drug discovery and development settings. Adv Drug Deliv Rev. 1997;23:3–25.

Lipkind GM, Fozzard HA. KcsA crystal structure as framework for a molecular model of the Na+ channel pore. Biochemistry. 2000;39(28):8161–70.

Llorca O, Betti M, Gonźalez JM, Valencia A, Marquez AJ, Valpuesta JM. The three-dimensional structure of an eukaryotic glutamine synthetase: functional implications of its oligomeric structure. J Struct Biol. 2006;156(3):469–79.

Long SB, Campbell EB, MacKinnon R. Crystal structure of a mammalian voltage-dependent Shaker family K^+ channel. Science. 2005;309(5736):897–903.

Lovell SC, Davis IW, Arendall III WB, et al. Structure validation by Cα geometry: φ, and Cβ deviation. Proteins. 2003;50(3):437–50.

Luthy R, Bowie JU, Eisenberg D. Assessment of protein models with three-dimensional profiles. Nature. 1992;356(6364):83–5.

Lyne PD. Structure-based virtual screening: an overview. Drug Discov Today. 2002;7 (20):1047–55.

Mansfeld J, Gebauer S, Dathe K, Ulbrich-Hofmann R. Secretory phospholipase A2 from Arabidopsis thaliana: insights into the three-dimensional structure and the amino acids involved in catalysis. Biochemistry. 2006;45(18):5687–94.

Pappas RS, Polzin GM, Zhang L, Watson CH, Paschal DC, Ashley DL. Cadmium, lead, and thallium in mainstream tobacco smoke particulate. Food Chem Toxicol. 2006;44:714–23.

Perez-Reyes E, Wei X, Castellano A, Birnbaumer L. Molecular diversity of L-type calcium channels. "Evidence for alternative splicing of the transcripts of three non-allelic genes". J Biol Chem. 1990;265(33):20430–6.

Petrey D, Honig B. Protein structure prediction: inroads to biology. Mol Cell. 2005;20(6):811–9.

Sali A, Potterton L, Yuan F, Van Vlijmen H, Karplus M. Evaluation of comparative protein modeling by MODELLER. Proteins. 1995;23(3):318–26.

Satarug S, Scott Garrett H, Sens MA, Donald Sens A. Cadmium, environmental exposure, and health outcomes. Environ Health Perspect. 2010;118:182–90.

Shaldam MA, Elhamamsy MH, Esmat EA, El-Moselhy TF. 1,4-dihydropyridine calcium channel blockers: homology modeling of the receptor and assessment of structure activity relationship. ISRN Med Chem. 2014;2014(203518):1–14.

Shen M-Y, Sali A. Statistical potential for assessment and prediction of protein structures. Protein Sci. 2006;15(11):2507–24.

Sippl MJ. Recognition of errors in three-dimensional structures of proteins. Proteins. 1993;17 (4):355–62.

Spassov VZ, Yan L, Flook PK. The dominant role of side-chain backbone interactions in structural realization of amino acid code. ChiRotor: a side-chain prediction algorithm based on side-chain backbone interactions. Protein Sci. 2007;16(3):494–506.

Spassov VZ, Flook PK, Yan L. LOOPER: a molecular mechanics-based algorithm for protein loop prediction. Protein Eng Des Sel. 2008;21(2):91–100.

Teilum K, Hoch JC, Goffin V, Kinet S, Martial JA, Kragelund BB. Solution structure of human prolactin. J Mol Biol. 2005;351(4):810–23.

Trott O, Olson AJ. AutoDockVina: improving the speed and accuracy of docking with a new scoring function, efficient optimization, and multithreading. J Comput Chem. 2010;31:455–61.

Vyas VK, Ghate M, Goel A. Pharmacophore modeling virtual screening docking and *in silico* ADMET analysis of protein kinase B (PKB beta) inhibitors. J Mol Graph Model. 2013;42 (13):17–25.

Wiederstein M, Sippl MJ. ProSA-web: interactive web service for the recognition of errors in three-dimensional structures of proteins. Nucleic Acids Res. 2007;35:W407–10.

Wolf LK. PyRx. Chem Eng News. 2009;87:31.

Zhorov BS, Folkman EV, Ananthanarayanan VS. Homology model of dihydropyridine receptor: implications for L-type Ca^{2+} channel modulation by agonists and antagonists. Arch Biochem Biophys. 2001;393(1):22–41.

Chapter 8
Natural Compounds Are Smart Players in Context to Anticancer Potential of Receptor Tyrosine Kinases: An In Silico and In Vitro Advancement

Pushpendra Singh, Shashank Kumar, and Felix Bast

Abstract Cancer is the ruling cause of mortality worldwide. Chemotherapeutic toxicity and drug resistance have provided impulsion for the formulation of new anticancer agents. Receptor tyrosine kinases (RTKs) are the most activated cell surface receptors for copious polypeptide growth factors, cytokines, and hormones that play a considerable role in cancer initiation, promotion, and progression. Natural products are a prime source of new anticancer drugs and their leads. The objective of computer-aided drug design (CADD) is to enhance the set of compounds with prudent active drug-like properties and eliminate inactive, toxic, poor absorption, distribution, metabolism, and excretion toxicity (ADME/T) compounds. In the present chapter, in silico advancement of anticancer natural compounds and molecular mechanisms of action of flavonoids, viz., genistein, myricetin, quercetin, luteolin, morin, kaempferol, catechin, and epigallocatechin gallate (EGCG), on RTK and PI3K signaling pathway attributing to their potential anticancer activity have been discussed.

Keywords Receptor tyrosine kinases • Cancer • Natural compounds • Computer-aided drug design

P. Singh (✉) • F. Bast
Centre for Biosciences, School of Basic and Applied Sciences, Central University of Punjab, Bathinda 151001, Punjab, India
e-mail: pushsingh02@rediffmail.com

S. Kumar
Center for Biochemistry and Microbial Sciences, School of Basic and Applied Sciences, Central University of Punjab, Bathinda 151001, Punjab, India

© Shanghai Jiao Tong University Press, Shanghai and Springer Science+Business Media Dordrecht 2017

D.-Q. Wei et al. (eds.), *Translational Bioinformatics and Its Application*, Translational Medicine Research, DOI 10.1007/978-94-024-1045-7_8

8.1 Natural Products: Promising Resource for Cancer Drug Discovery

Cancer is the ruling cause of mortality worldwide, especially breast and prostate cancer. They are regarded as the most frequent cancer in women and men, respectively, second to skin cancer (Jakowlew 2006). The toxicity associated with drug resistance and poor prognosis in the current chemotherapeutics has provided the much-needed impulsion for the formulation of new anticancer agents (Biswas et al. 2006; Martin et al. 2011). Natural products are a prime source of new anticancer drugs and their leads. Anticancer drug development of natural origin including plants (vincristine, vinblastine, etoposide, and paclitaxel), marine organisms (cytarabine and aplidine), and microorganisms (dactinomycin and doxorubicin) added a new concept for drug discovery. Furthermore, various compounds recognized from fruits and vegetables have been used as anticancer therapy. Moreover, curcumin, resveratrol, genistein, and diallyl sulfide had a most promising anticancer activity in the different model and entered into the clinical trial. Traditional medicines owe their capability to exhibit various biological activities including anticancer potential. Furthermore, synthetic analogs of natural compounds with improved potency and safety might be prepared, thus portraying them as the beacon for cancer drug discovery. In fact, natural products are an inspiration for the majority of US Food and Drug Administration (FDA)-approved drugs. Another remarkable character is that natural products can also be prepared by synthesis and have played a mid-role in the drug development by providing challenging synthetic targets. Plants have large reservoir of potent, novel, and highly varied structures that are dubious to be synthesized in laboratories. Over the past many years, plants have been known to be a cornucopia of biologically active compounds including cocaine, digitalis, quinine, and muscarine (Kumar et al. 2013). Many of these active compounds are useful drugs such as anticancer agent paclitaxel (Taxol) from the yew tree and antimalarial agent artemisinin from *Artemisia annua*. Flavonoids comprise a significant group of polyphenolic compounds, which are primarily benzo-α-pyrone (phenyl chromone) derivatives, structurally diverse low molecular mass molecules (Kumar and Pandey 2013). Bioactive flavonoids have been found to be indispensable for the growth and development of plants, additionally providing the physical environment that proves to be essential for plant survival under stress circumstances. Among the various natural products, flavonoids have attracted more attention owing to their remarkable spectrum of pharmacological activities such as antioxidant, antiangiogenic, anti-inflammatory, and anticancer activity (Mishra et al. 2013; Kumar et al. 2014).

Lim et al. (2008) reported seven *Aspidosperma* indole alkaloids (jerantinine A to G) that were extracted from the *Tabernaemontana corymbosa* leaf (Lim et al. 2008). Jerantinine A has been reported for its potent cytotoxic activity against vincristine-resistant nasopharyngeal carcinoma cells and has the capability to inhibit cell cycle at G2M stage and polymerization of tubulin (Raja 2015; Raja et al. 2014). Furthermore, jerantinine B and E are also reported for their potent anticancer activity with a variety of mechanisms including disruption of microtubule organization and induction of apoptosis in different human cancer cell lines (Frei et al. 2013, Qazzaz et al. 2016). Jerantinine B, δ-tocotrienol, and combined low-dose treatments induced a dose-dependent growth inhibition against U87MG and HT-29 cells indubitably disrupted the microtubule networks (Abubakar et al. 2016).

Astragalus membranaceus is an adaptogenic herb that belongs to Leguminosae family originating in Northern China and has been used to treat a range of disorders including chronic illnesses, metabolic disorders, compromised immunity, inflammation, and cancer. Furthermore, treatment with *A. membranaceus* supplemented injection with current chemotherapy was found to hamper the tumor growth, decrease the unavoidable side effect of chemotherapy, restore the impaired T cell functions, and improve the drug sensitivity of tumor cells (Cho and Chen 2009; Zou and Liu 2003). Moreover, *A. membranaceus* injection might efficiently encourage the immune response of tumor-bearing host that led to improve the anti-metastasis activity of dendritic cells in vivo (Dong and Dong 2005). Further, it was also reported that *A. membranaceus*-based medicine might augment the usefulness of platinum-based chemotherapy for advanced non-small cell lung cancer (McCulloch et al. 2006). A polysaccharide isolated from the radix of *A. membranaceus* was reported to increase tumor sensitivity and reduce chemotherapeutic toxicity. Further, it was reported that treatment of *A*. polysaccharide integrated with vinorelbine and cisplatin had appreciably enhanced QOL in patients with advanced NSCLC compared with vinorelbine and cisplatin alone (Guo et al. 2012). Cho and Leung (2007a, b) reported that administration of *A. membranaceus* root fraction in tumor-bearing mice and cyclophosphamide-treated mice (in vivo) could reestablish the depressed immune functions. Thus, *A. membranaceus* could reveal immunomodulating and immune-restorative effects, both in vitro and in vivo (Cho and Leung 2007a, b, c). Furthermore, it was reported that the root of *A. membranaceus* was proficient to induce monocytic differentiation of both human and murine cells. Moreover, in vivo administration of *A. membranaceus* fraction could reestablish the depressed mitogenic response in tumor-bearing mice. Moreover, roots of this plant have polysaccharides (UV-absorbing compounds) which may have potential in protecting against solar-induced skin damage (Curnow and Owen 2016).

Identification and development of anticancer agents from the natural product by computer-aided high-throughput virtual screening (HTVS) and extra precision (XP) molecular docking has been well documented. In silico screening approach is the primary technique for identification of natural products as inhibitors of target protein and predicting their interactions. Examples of natural compounds that have been reported as anticancer properties identified by using HTVS and XP molecular docking are represented in Table 8.1.

Table 8.1 Anticancer natural compounds identified by HTVS and XP

Agents	Targets	References
Wortmannin	Wild-type and mutant PIK3CA	Kuete et al. (2015) and Dan et al. (2010)
Noscapine derivatives	Microtubule	Santoshi et al. (2014) and Naik et al. (2012)
Hinokiflavone	MMP-9	Kalva et al. (2014)
Combretastatin	Microtubule-destabilizing	Do et al. (2014) and Abolhasani et al. (2015)
Xanthone derivatives	DNA topoisomerase IIα	Alam and Khan (2014) and Verbanac et al. (2012)
Camptothecin	RAD9	(Prasad et al. (2013) and Yamazaki et al. (2004)
Linarin	CDK4	Meshram et al. (2012)
Violacein	Estrogen receptor	Meshram et al. (2012)
Hydroxycinnamic acid	MMP-2 and MMP-9	Wang et al. (2012a)
S-Adenosylmethionine	S-Adenosylmethionine	Taylor et al. (2009)
De novo drug design	c-Met tyrosine kinase	Chen (2008)
Sulfobenzoic acid	Transformylase	Xu et al. (2004)
Genistein	Acetylcholinesterase	Fang et al. (2014)
Quercetin	Inducible nitric oxide synthase	Singh and Konwar (2012)
Rutin and myricetin	α-Glucosidase	Hee and JuSung (2014)

8.2 RTK Signaling Inhibitors as Promising Anticancer Agent

RTKs play a prominent role in blood cancer and solid tumors as they are the most activated cell surface receptors for numerous growth factors, cytokines, and hormones (Robinson et al. 2000). They have been shown not just to be the key regulators of normal cellular processes but additionally to play a critical role in the growth and development of various cancers (Singh and Bast 2014a). There are two types of RTK family, first containing the transmembrane domain and second not possessing transmembrane domains (Hubbard and Till 2000). Overexpression of RTKs has been reported in numerous cancers, including non-small cell lung cancer and breast and prostate cancers. RTKs comprise of many protein molecules including EGFR, insulin receptor (IR) and insulin-like growth factor 1 (IGF1R), and vascular endothelial growth factor receptors (VEGFR). They have covalently bound heterotetramer protein consisting of two extracellular α-subunits and two transmembrane β-subunits.

Ligand-receptor interactions induce conformational changes that led to activate autophosphorylation of a cascade of tyrosine residues, ultimately resulting in activation of the PI3K pathway and rat sarcoma (RAS) pathway that is participated

Table 8.2 Examples of anticancer natural compounds as inhibitors of RTK signaling proteins

Natural compounds	Targets	References
Cyclopentyl-pyrimidine	IGF1R	Aware et al. (2015)
Oxindole-based inhibitors	FGFR1	
Platycodin D	VEGFR2	Luan et al. (2014)
ZINC natural database	VEGFR2	Li et al. (2014)
Genistein	EGFR	Yuan et al. (2008)
Quercetin	PDK1, PI3K, and mTOR	Singh and Bast (2014b)
Quercetin	EGFR and mutated EGFR	Singh and Bast (2014a)
Curcuminoid analogs	HER2	Yim-Im et al. (2014)
Alkaloids/flavonoids	PI3K	Jackson and Setzer (2013)
Curcumin derivatives	STAT3	Kumar and Bora (2012)
Natural compounds	STAT3	Liu et al. (2014)
Morin, myricetin, and EGCG	STAT3, IR, EGFR, and AR/ER	Singh and Bast (2014a, b and 2015a, b, c)

in cellular growth and metabolism. PI3K, Akt, PDK1, and mTOR are activated by a number of cellular processes including expression of oncogenes and inactivation of tumor suppressor genes, tyrosine kinase receptors, and G-protein coupled receptors (Frasca et al. 2008). Numerous anticancer natural compounds that have been reported as inhibitors of RTK signaling proteins demonstrated by using CADD are given in Table 8.2.

Flavonoids exhibit anticancer activity by synchronizing the expression of EGFR, VEGF, and matrix metalloproteinases (MMPs) in addition to inhibiting NF-κβ and PI3K signaling pathways (Gu et al. 2013). VEGFR activation led to angiogenesis which is closely linked to the development of cancer including prostate, breast, lung, and hepatocellular carcinoma (Chu et al. 2013; Folkman 2002; Hicklin and Ellis 2005; Huang et al. 2011; Tanno et al. 2004). Encouraging strategy for combating cancer by inhibiting abnormal angiogenesis and employing monoclonal antibodies, ribozymes, and TRK inhibitors are currently in clinical trials (Arora and Scholar 2005; Ferrara et al. 2004; Saini and Hurwitz 2008). Oral tyrosine kinase inhibitors, namely, sorafenib, sunitinib, and pazopanib, have been endorsed by the USFDA for the treatment of diverse cancer (Wang et al. 2012b). Various in vitro, in vivo, and preclinical findings convincingly proclaim the use of dietary products in the prevention and treatment of cancer also (Amin et al. 2009).

Cancer is an extremely heterogeneous malignancy, with its signal pathways evince a complex array of cross signaling pathways. Appropriately, when blocking the key signal transduction pathways, the single-targeted drugs can also activate the other pathways that led to increasing cell proliferation. Therefore, multitargeted drugs have the better option for future drug discovery. RTK inhibitors have played an increasingly significant role in the treatment of various cancers. Recently, published phase clinical III trials have exposed potential efficacies of these drugs. Multitargeted TKIs have been regarded as promising agents for various cancers due to their potential antitumor mechanisms. Single-targeted drugs have poor efficiencies for most cancer patients, while they may be highly productive in certain cancer patients. Thus, it is imperative to identify populations that are suitable for TKIs (Zhou 2012). Moreover, synergistic action by multi-targeting compounds produces a new strategy for discovering anticancer drugs for cancer drug resistance (Zhang et al. 2014a). In this context, obstruction of many essential kinases at the level of receptors or downstream serine/threonine kinases may assist to optimize the most anticancer therapeutic sake.

8.3 Multidrug Resistance Development in Cancer

Different critical factors are responsible for the development of cancer multidrug resistance such as (1) mutations in target proteins, (2) augmented action of drug efflux pumps (ATP-binding cassette superfamily), (3) decreased drug influx, and (4) distorted expression of apoptosis and (5) anti-apoptotic proteins (Costantino and Barlocco 2013). ABC transport molecules are expressed on the membranes of cellular vesicles and affect the biochemical and biophysical properties, i.e., ADME/T of chemotherapeutics. Mechanism such as insensitivity to drug-induced apoptosis and induction of drug detoxification perhaps play a vital role in earning of anticancer drug resistance. Overactivity of ABC transporters in cancer cells modulates anticancer drug resistance. In this context, an ongoing effort to succeed therapies could either block or inactivate these transporters. This may lead to increase the anticancer drug concentration within the cells. Bioactive flavonoids have been found to be indispensable for the growth and development of plants, additionally providing the natural environment that proves to be essential for plant survival under stress condition. Among the various natural products, flavonoids have attracted more attention owing to their remarkable spectrum of pharmacological activities such as antioxidant, antiangiogenic, anti-inflammatory, and anticancer activity.

8.4 CADD

It is broadly accepted that drug discovery and development are risky, costly, and time- and resource-consuming processes. A variety of cancer drugs are small compounds designed to bind and modulate the biological action of the receptors. Molecular docking inheres in three key consecutive goals: pose prediction, virtual screening, and binding affinity evaluation. There is an ever improved endeavor to apply the computational method for drug design, development, and optimization in the field of chemical and biological sciences. In the modern arena, computer-aided or in silico molecular drug development is being utilized to accelerate and facilitate hit identification and optimization of the absorption, distribution, metabolism, excretion, and toxicity profile. Recently, researchers are dynamically involved in the development of more sophisticated computational tools that will ameliorate potency and efficiency of the drug development process, decrease the use of animals, and increase accuracy of pose predictability. The rapid expansion of CADD by the advancement of computational software (AutoDock, DOCK, GOLD, and Maestro), identification of molecular targets, and an expanded database of the publicly accessible target crystal structure of the protein provided the preeminent environment for drug discovery. CADD is being exploited to identify hits, pick leads, and optimize leads, i.e., transform biologically active compounds into good drugs by enhancing their physicochemical, pharmaceutical, and ADME/T properties. HTVS is used to discover novel agents from different chemical scaffolds by searching commercial, public, and private databases. It is deliberated to reduce the size of chemical space and thereby allow cornerstone on more promising candidates for lead discovery and optimization. The aim of CADD is to enhance the set of compounds with drug-like properties and eliminate compounds with inactive, toxic, poor ADME/T. In other words, in silico modeling is used noticeably to minimize time and resource necessities of chemical synthesis and biological in vitro and in vivo testing (Guedes et al. 2014; Kapetanovic 2008).

Natural products are a significant source of bioactive compounds for drug breakthrough. However, their utilization in drug discovery has somehow diminished because of barriers to the screening of natural products against anticancer targets. In another study, Kapetanovic (2008) reported that the estimated time and cost of new drug bringing to market differ, by 7–12 years and $ 1.2 billion. Also, 5 out of 40,000 compounds experimentally validated in animals reach primary human testing. Furthermore, only one of five compounds achieves approval for clinical studies (expected). Taking in these barriers for drug development, here we discuss the strategies for in silico screening of natural compounds that strap up the current technology that may help to abridge these barriers (Harvey et al. 2015). Commonly used computational approaches for screening of natural products against anticancer targets include (1) target identification and validation (reverse docking, protein structure prediction, target druggability, probe design, and chemical sensing) and (2) lead discovery and optimization (molecular docking, de novo design, designs virtual library based on pharmacophore, quantitative structure-activity relationship models, and sequence-based method for phosphorylation site prediction). Approaches used for target identification, validation, lead discovery, and optimization are depicted in Fig. 8.1.

1. *Target identification and validation*

- Reverse docking

Due to increased number of well-known protein structures (NMR and 3D crystallographic), a new molecular docking method called reverse docking comes in a picture, in which docking is carried out by probing a protein database instead of a compound database. Reverse docking is proving to be an influential tool for identification and validation of small molecules into a set of target proteins, in addition to the lead discovery and optimization stages of the drug development cycle (Chen and Ung 2001), for example, Indock, a reverse docking platform to study drug toxicity, and TarFisDock, used to identify drug targets (Li et al. 2006).

Approches used for Target identification, validation, lead discovery and optimization

Target identification and validation
a) Reverse docking
b) Protein structure prediction
c) Target druggability
d) Probe Design and Chemical Sensing

Lead discovery and optimization
a) Molecular docking
c) De novo Design
d) Designs virtual library based on Pharmacophore
e) Quantitative structure-activity relationship models (QSAR models)
f) Sequence-based method for phosphorylation site prediction

↓

***In silico* prediction of ADME\T and pharmacokinetics**

↓

Preclinical tests

Fig. 8.1 Outline of CADD-based drug designing and development

- Protein structure prediction

Over the past genomic decade, the whole genome sequencing projects have produced a huge quantity of protein sequence data, which led to fill the gap between protein sequence and structure. Furthermore, in vitro experimental determination of a protein structure and function is rigorous, time-consuming, and expensive. Therefore, the use of computational tools for conveying structure to a protein represents the most proficient option for experimental methods (Neerincx and Leunissen 2005). To overcome this problem, a plethora of computerized methods are accessible (online servers and software) to predict protein primary, secondary, and tertiary structure from the amino acid sequence (Fischer 2006; Pavlopoulou and Michalopoulos 2011). A variety of protein databases provided information regarding amino acid sequences derived from nucleotide databases such as GenPept, RefSeq, the protein information resource, and the UniProt knowledgebase (Bairoch et al. 2005; Pruitt et al. 2007; Wu et al. 2002).

- Druggability

Druggability is the property of target molecules (proteins and nucleic acids) that elicits a positive clinical response when bind with a compound. It is known that best drug target should have the following properties: approving capability for high-throughput screening and capability to change a disease physiology and differential expression of target molecules (Bakheet and Doig 2009). Due to the lack of knowledge about the molecular mechanism of disease and target identification, experimentally an assessment of proteins for their druggability is a discouraging job and makes the convoluted situation. In this context, with the aid of progress information such as protein-protein interaction and metabolic and gene regulatory networks, computational models can predict drug targets with high sensitivity and in lesser time (Costa et al. 2010; Kandoi et al. 2015).

- Chemical probes

Chemical probes (fluorescence resonance energy transfer-based probes and MRI probes) are crucial tools for evaluation of biochemical processes and detection of hazardous compounds in cells. Therefore, the development of chemical probes provided a lot of information regarding appreciation of disease marker. Recently, fluorescent-based probes have the best consideration because they are easy and more sensitive to predict protein targets (Jun et al. 2011; Kikuchi 2010).

2. *Lead discovery and optimization*

- Molecular docking

The discovery of potent drug targets has regular increases in the last few decades due to the expansion of genomic and proteomics techniques. Experimental and computational tools are dynamically applied to lead identification and optimization. The lead molecules are capable of modulating the biological function of the target proteins. Various molecular docking techniques such as HTVS, XP, and induced molecular docking technology prompt identification of drug-like leads.

- Designs virtual library based on pharmacophore

Pharmacophore models are a geometrical description of the chemical functionalities and can be generated using two different approaches depending on the input data employed for model construction (Güner and Bowen 2014). (1) Structure-based modeling and the interaction pattern of a molecule and its targets are extracted from experimentally determined ligand-protein interactions (Kaserer et al. 2015). (2) In the case of ligand-based modeling, 3D structures of two or more known compounds are aligned, and pharmacophore character is shared among these training set molecules.

- De novo design

Biochemical and organic model builder is used to develop molecules by adding layers of substituents to a core molecule that has been positioned in a binding site.

- Quantitative structure-activity relationship (QSAR)

The aims of quantitative structure-activity relationship (QSAR) analysis are (1) to predict biological activity (biological/toxicological) and physicochemical properties of compounds (2) and to rationalize the mechanisms of action within a series of chemicals employing the interdisciplinary information of chemistry, mathematics, and biology. Numerous studies have attempted to correlate mathematically the property of molecules using different computationally derived quantitative parameters termed as descriptors. There are two types of QSAR used in drug discovery: (1) 2D-QSAR and (2) 3D-QSAR (Divakar and Hariharan 2015).

- Sequence-based method for phosphorylation site prediction

Kinase-mediated phosphorylation is one of the imperative posttranslational modifications. Cell signaling defects linked with protein phosphorylation are associated with cancer initiation and progression. Therefore, identification of protein phosphorylation sites is essential for studying disease finding. However, experimental recognition of phosphorylation sites is costly and labor intensive. Computational methods are helpful tools for phosphorylation site identification (sequence-based method for serine, threonine, and tyrosine phosphorylation site) and afford information regarding cell signaling (He et al. 2012; Trost and Kusalik 2011).

- ADME/T modeling for drug design

In recent decades, in silico absorption, distribution, metabolism, excretion (ADME), and toxicity modeling are used as a tool for computer-aided drug design in pharmaceutical research. Recently, various ADME/T-related prediction models have been reported by many software and online predictors. Due to easy compound screening, low-cost nature of these models permits more rationalized drug identification and their structural optimization in addition to the parallel investigation of bioavailability and activity. However, the modern in silico approaches still need additional progress (Wang et al. 2015).

8.5 Flavonoids: Molecular Mode of Action

8.5.1 Genistein

Genistein is a phytoestrogen soy product belonging to the class of isoflavones predominantly found as glycosylated form in plants. Its effect has been reported for copious biological processes such as growth and development that were found to result in metabolic alterations at the cellular level. Furthermore, it was established that ingestion of dietary genistein also led to changes in metabolic hormones including insulin, leptin, thyroid, adrenocorticotropic, cortisol, and corticosterone (Takeda et al. 1997; Zhou et al. 2014). Experimental evidences collected over the past few decades have upheld the information that inhibition of cancer cell growth by genistein is conciliated via the inflection of RTK signaling pathways that result in control of cell cycle and apoptosis (Chen et al. 2013; Chung et al. 1997; Niu et al. 1999; Shim et al. 2010; Siddiqui et al. 2004; Walker et al. 2000). Consequently, it has been observed that antiproliferative activity of daidzein and genistein may be linked with the oncogene products such as estrogen receptor α and TK c-erbB-2 expression in breast cancer cells. As evident in several reports,

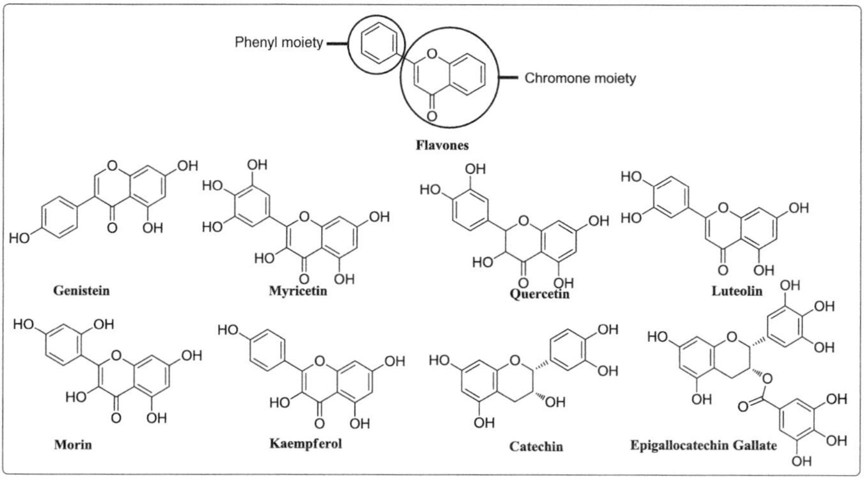

Fig. 8.2 Molecular structure of flavonoids

genistein impedes the activation of PI3K and MAPK signaling molecules which are known to perpetuate a homeostatic equilibrium between cell growth and apoptosis in nasopharyngeal carcinoma cells (Ding et al. 2003; Kerns and Di 2010; Mechoulam and Pierce 2005). It is reported that genistein effectively inhibits the activity of downstream targets such as Src, Akt, and glycogen synthase kinase-3β (Müller et al. 2001). Molecular structures of active multitargeted RTK signaling inhibitor flavonoid are represented in Fig. 8.2.

8.5.2 *Quercetin*

Quercetin belongs to flavonoids that play an important role in cancer prevention. Quercetin inhibited EGFR cell signaling with an activation of a forkhead family of transcription factor FOXO1 activation. Many reports confirm that quercetin is an efficient anticancer agent that induces apoptosis and decreases cell proliferation in oral cancer cells overexpressing EGFR (Huang et al. 2013). Furthermore, it was also observed that quercetin may induce apoptosis and reduce cell proliferation in HeLa cells via the AMPK-induced HSP70 and downregulation of EGFR (Jung et al. 2010). In vitro and in vivo experiments revealed that quercetin could inhibit the proliferation and induce apoptosis in cancer cells (Huang et al. 2013; Lyne 2002). Moreover, quercetin is observed to arrest the cell cycle at the G0/G1 phase and induce apoptosis in PC3 cells via intrinsic apoptotic stimuli that led to DNA damage (Liu et al. 2012a, b).

Genistein, quercetin, luteolin, morin, and kaempferol anticancer properties in different cancer cells are represented in Table 8.3.

Table 8.3 Genistein, quercetin, luteolin, morin, and kaempferol anticancer properties in different cancer cells

Agents	Tumors	References
Genistein	Breast cancer	Parra et al. (2016)
	Breast cancer	Fang et al. (2016)
	Colorectal cancer	Qin et al. (2015)
	Pancreatic cancer	Suzuki et al. (2014)
	Breast cancer	Xie et al. (2014)
Quercetin	Liver cancer cells	Wu et al. (2014)
	Liver cancer cells	Olayinka et al. (2014)
	Anticancer	Pandey et al. (2015)
	Anticancer	Brito et al. (2015)
	Colon cancer	Refolo et al. (2015)
Morin	Colon cancer	Hyun et al. (2015)
	Leukemia cell	Karimi et al. (2013)
	Anticancer	Neves and Kwok (2015)
	Leukemia cell	Park et al. (2014a)
	Osteoblast and breast tumor	Naso et al. (2013)
Luteolin	Head and neck cancer	Majumdar et al. (2014)
	Anticancer	Lin et al. (2008)
	Lung cancer cells	Ma et al. (2015)
	Gastric cancer	Lu et al. (2015)
	Anticancer	Sak (2014)
Kaempferol	Cervical cancer	Tu et al. (2016a, b)
	Esophageal cancer	Tu et al. (2016a, b)
	Anticancer	Kadioglu et al. (2015)
	Lung cancer	Park et al. (2014b)
	Anticancer	Batra and Sharma (2013)

8.5.3 Morin

Morin is a flavone that exhibits antiproliferative, antioxidant, and anti-inflammatory activity by modulating NF-κβ signaling pathway. It has been observed to inhibit the NF-κβ-dependent gene expression activated by tumor necrosis factor (TNF) and the p65 subunit of NF-κβ. It enhances apoptosis and reduced invasion via downregulation of MMP2 and MMP9. These effects were correlated with enhancement of apoptosis induced by TNF and chemotherapeutic agents (Wang et al. 2009). Furthermore, in vitro and in vivo findings indicate that morin possesses anti-inflammatory, anti-angiogenesis, and antiproliferative activity by supporting suppression of diethylnitrosamine-induced hepatocellular carcinoma cells via downregulation of MMP2 and MMP9 (Masuda et al. 2003). It was also observed that morin induces apoptosis in HL-60 and hepatocellular cells by activation of the cysteine-aspartic acid protease-3 (caspase-3) (Kuo et al. 2007; Luo et al. 2001; Sivaramakrishnan and Devaraj 2010).

It has been suggested that morin prevents acute liver damage by inhibiting the production of the pro-inflammatory cytokine (Park et al. 2010).

8.5.4 *Luteolin*

Luteolin is one of the most widespread naturally occurring flavonoids present in edible plants. It has been reported that luteolin suppresses VEGF-induced phosphorylation of VEGF2R in prostate cancer cells (Liu et al. 2012a, b). Luteolin exhibited cyclin-dependent kinase cell cycle arrest in breast cancer cells. Further, it has been reported that luteolin down-regulate the EGFR mRNA expression followed by the inhibiting MAPK activation (Morales and Haza 2012). Luteolin exhibited cyclin-dependent kinase cell cycle arrest in breast cancer cells. Further, it has been reported that luteolin down-regulate the EGFR mRNA expression followed by the inhibition of MAPK activation (Azevedo et al. 2015; Kim et al. 2012; Lin et al. 2008; Lopez-Lazaro 2009; Maggioni et al. 2014; Phillips et al. 2011; Sak 2014; F. Sun et al. 2012; Xu et al. 2013; Zhang et al. 2010, 2014a, b, 2015). Luteolin could sensitize cancer cells to inhibit cell proliferation and induce apoptosis and cell cycle arrest through suppressing cell survival pathways such as PI3K and MAPK in colon cancer cell, epithelioid cancer, pancreatic cancer cells, hepatoma cells, breast cancer cells, lung cancer xenograft models, and gastric carcinoma xenografts in nude mice (Azevedo et al. 2015). Luteolin also inhibited hypoxia-induced cell proliferation, motility, and adhesion via inhibiting the expression of integrin β1 and focal adhesion kinase (Wang et al. 2014).

8.5.5 *Kaempferol*

Kaempferol is a yellow crystalline solid, slightly water-soluble natural polyphenol belonging to the group of antioxidant flavonoids. Moreover, numerous studies showed that consumption of kaempferol containing foods led to reduced risk of cancer and cardiovascular diseases. Furthermore, numerous preclinical studies have demonstrated that kaempferol and some glycosides of kaempferol have extensive pharmacological activities including antioxidant, anti-inflammatory, anticancer, neuroprotective, antidiabetic, analgesic, and antiallergic (Prasad et al. 2013). Moreover, the cytotoxicity and resistance of anticancer drugs that conjugate with glutathione may be influenced by long-term intake of kaempferol (Sivashanmugam et al. 2013). It is reported that oxidative stress induced by kaempferol in chronic myelogenous leukemia cells (K562) and promyelocytic leukemia cells (U937, K562, and U937) affects the inactivation of PI3K signaling pathways which may lead to cell death. Kaempferol has been reported to induce apoptosis via endoplasmic reticulum stress and mitochondria-dependent pathway in osteosarcoma U2OS cells. In vivo efficacy of kaempferol was assessed in BALB/nu/nu mice inoculated with U2OS cells and indicated inhibition of tumor growth by reducing cell proliferation and inducing apoptosis. Furthermore, it inhibits the growth of human osteosarcoma cells both in vivo and in vitro (Meshram et al. 2012; Prasad et al. 2013). Kaempferol and quercetin were known to induce apoptosis and cell cycle arrest in various oral cancer cell lines including SCC1483, SCC-25, and SCC-QLL1 via caspase-3-dependent activity. Kaempferol induces apoptosis through oxidative stress and induces G2/M cell cycle arrest in glioblastoma, HeLa, and leukemic cell lines (Taylor et al. 2009; Xu et al. 2004). Kaempferol is known to reduce cell proliferation through downregulation of oncoprotein c-Myc and promoting apoptosis and cell cycle arrest in ovarian cancer cells (Luo et al. 2010).

8.5.6 Green Tea

Green tea is a wonderful beverage with potential health benefits made from the leaves of *Camellia sinensis* via minimum oxidation processing with abundant polyphenols, including epigallocatechin gallate (EGCG), epicatechin gallate (ECG), epigallate catechin (EGC), and epicatechin (EC) targeting multiple signaling pathways that lead to anti-oxidative and anticarcinogenic potential (Kadioglu et al. 2014; Khan et al. 2006). Multitargeted anticancer activities of EGCG have been demonstrated by using in silico, in vitro, and in vivo study represented in Table 8.4.

EGCG affect numerous molecular targets involved in cancer cell proliferation and survival; however, polyphenolic catechins such as EGCG exhibit poor oral bioavailability. The consumption of green tea has been recommended for chemoprotective activity. Previous studies have established that active anticancer constituent in green tea is EGCG with the anticancer activity highlighted in various in vitro and in vivo studies. The anticancer activity of EGCG may be accredited to the combinatory effects on multiple targets that are determinant for cell proliferation and apoptosis (Alam and Khan 2014; Dennler et al. 2002; Kalva et al. 2014; Kuete et al. 2015; Santoshi et al. 2014; Xu et al. 2004). A number of reports confirm that green tea reduces cell proliferation and sensitizes the cell to apoptosis, ultimately leading to cancer cell growth inhibition in diverse cancer cells including colorectal and hepatocellular carcinoma, SW480 colon cancer, SV40 virally transformed WI38 human fibroblasts (WI38VA), prostate cancer, and Ishikawa cells (Li et al. 2014; Liu et al. 2014; Mayer and Gustafson 2004, 2008; Sun et al. 2014; Yim-Im et al. 2014). Interestingly, green tea showed anticancer property in colorectal and hepatocellular carcinoma cells via regulating the activation of tyrosine kinase EGFR (Khan et al. 2006; Shimizu et al. 2011; Shimizu et al. 2008). Furthermore, it was reported that EGCG has potential inhibitory effects on tumor angiogenesis, induced by IGF1 in non-small cell lung cancer cells via downregulation of HIF-1α and VEGF expression. EGCG inhibiting tumor invasion and angiogenesis underlines the role of green tea as a cancer chemopreventive agent (Jackson and Setzer 2013). EGCG treatment was found to inhibit UV-induced

Table 8.4 EGCG anticancer properties in different cancer cells

Agent	Tumors and tumor cells	References
(−)-EGCG	Nasopharyngeal carcinoma cells	Fang et al. (2015)
(−)-EGCG	Lung cancer	Zhang et al. (2015)
(−)-EGCG	Squamous cell carcinoma	Irimie et al. (2015)
(−)-EGCG	HL-60 promyelocytic leukemia cells	Saiko et al. (2015)
(−)-EGCG	Head and neck tumor	Masuda et al. (2003)
(−)-EGCG	Mouse embryonic fibroblast cells	Yagiz et al. (2007)
(−)-EGCG	Human osteogenic sarcoma (HOS) cells	Ji et al. (2006)
(−)-EGCG	Laryngeal squamous carcinoma cells	X. Wang et al. (2009)
(−)-EGCG	Nasopharyngeal carcinoma cells	Luo et al. (2001)
(−)-EGCG	Renal cell carcinoma	Gu et al. (2009)
(−)-EGCG	Hypopharyngeal carcinoma cells	Park et al. (2010)
(−)-EGCG	Pancreatic cancer cells	Z. Wang et al. (2008)

epidermal lipid peroxidation, decrease antioxidant enzyme glutathione peroxidase activity, and increase catalase activity against exposures to UV light in the human skin (Kumar and Bora 2012). Catechin induces apoptosis and cell cycle arrest and reduces cell proliferation via deviated antioxidant parameters including superoxide dismutase, catalase, and lipid peroxidation in HepG2 cells (Amaral et al. 2014).

8.5.7 *Myricetin*

Myricetin is a naturally occurring phenolic flavonol found in red wine, fruits, vegetables, and herbs possessing antioxidant and anti-inflammatory activity. In vitro investigations demonstrate that in high concentrations, it can improve lipoproteins such as low-density lipoprotein cholesterol. Myricetin was observed to ameliorate inoperative insulin signaling via β-endorphin signaling in the skeletal muscles of fructose-fed rats. It enhances the secretion β-endorphin, followed by peripheral μ-opioid receptor activation which leads to amelioration of impaired insulin receptor signaling (Lee et al. 2004; Yamaguchi et al. 1995). JAK1/STAT3 pathway activated by cytokine and growth factor including insulin, IGF1, and EGF has been recommended to play a significant role in cell proliferation, differentiation, and cell migration (Simon et al. 1998; Vela et al. 2015). It was reported that myricetin directly binds to JAK1/STAT3 molecules to inhibit cell transformation in EGF-activated mouse $JB6P^{+}$ cells (Kumamoto et al. 2009). The multitargeted anticancer activity of myricetin has been demonstrated by using in silico, in vitro, and in vivo study represented in Table 8.5.

Table 8.5 Myricetin anticancer potential in different cancer cells

Agent	Tumors and tumor cells	References
Myricetin	Esophageal carcinoma	Wang et al. (2014)
Myricetin	breast cancer	Zhang et al. (2014a, b)
Myricetin	Lung cancer	Zhang et al. (2014a, b)
Myricetin	Colon cancer cells	Kim et al. (2014)
Myricetin	Squamous cell carcinoma	Maggioni et al. (2014)
Myricetin	Prostate cancer cells	Xu et al. (2013)
Myricetin	Pancreatic cancer cells	Phillips et al. (2011)
Myricetin	Hepatocellular carcinoma	Zhang et al. (2010)
Myricetin	Gastric and ovarian cancers	Sak (2014)
Myricetin	Leukemia	Morales and Haza (2012)
Myricetin	Bladder cancer	Sun et al. (2012)
Myricetin	Glioblastoma cells	Siegelin et al. (2009)

8.6 Conclusions and Outlook

Insulin, IGF, EGF, and VEGF growth factors are crucial to the growth and regulation of cancer cells. Receptors for these growth factors are a striking target to combat cancer. Moreover, binding of ligands to receptor molecules induces conformational changes and activates autophosphorylation of a cascade of tyrosine residues of small protein molecules such as STAT3. PI3K pathway is one of the most habitually activated signal transduction pathways distant from RAS that plays a significant role in cellular growth and metabolism. PI3K, Akt, PDK1, and mTOR are activated by a number of biological processes including expression of oncogenes and inactivation of tumor suppressor genes.

A vast scientific data has accumulated, elucidating the molecular mechanisms of cancer development and the action of anticancer agents in cancer prevention. These research findings have provided the basis for the identification of molecular mechanisms for cancer prevention and treatment. Notably, these discoveries have identified key molecular targets for screening and testing novel natural anticancer drugs that have fewer adverse side effects. However, despite increasing advances in drug discovery and preclinical testing, anticancer drug development remains a laborious, time-consuming process with limited success. This suggests a critical need to differentiate at an earlier stage of development between promising candidates and those less likely to be effective. Even though progress has been made in identifying important molecular targets and potential nontoxic anticancer agents, transitioning preclinical results into the clinic has been extremely challenging. Unfortunately very few compounds have shown real promise in clinical trials. The combination of two or more compounds that target multiple pathways simultaneously is a strategy that is rapidly gaining widespread acceptance. Researchers suggested that combinations of drugs that could block heterogeneous cancers by inhibiting multiple signaling pathways have also been revealed beneficial in clinical trials.

Furthermore, combinations of agents with natural compounds will probably require a lower dose of each compound which led to less toxicity and fewer side effects. In silico screening uses molecular docking programs that target molecules into the active site and then ranks molecules by their aptitude to interact with the target protein. Such computer-identified drug targets can be validated in vitro and in vivo using cell-based biochemical assays and animal studies as well.

Acknowledgments We would like to thank Vice Chancellor, Central University of Punjab, Bathinda, Punjab, (India) for supporting this study with infrastructural requirements. We also thank Professor P. Ramarao (Dean, Academic Affairs), Central University of Punjab, Bathinda, Punjab, India, for his suggestions during the course that tremendously helped to improve this article. This study was also supported by Senior Research Fellowship Grant-in-Aid from Indian Council of Medical Research (ICMR) awarded to PS.

Conflict of Interest The authors declare that there is no conflict of interests regarding the publication of this article.

References

Abolhasani H, Zarghi A, Hamzeh-Mivehroud M, Alizadeh AA, Mojarrad JS, Dastmalchi S. *In-silico* investigation of tubulin binding modes of a series of novel antiproliferative spiroisoxazoline compounds using docking studies. Iranian J Pharm Res: IJPR. 2015;14 (1):141–8.

Abubakar IB, Lim K-H, Kam T-S, Loh H-S. Synergistic cytotoxic effects of combined δ-tocotrienol and jerantinine B on human brain and colon cancers. J Ethnopharmacol. 2016;184:107–18.

Alam S, Khan F. QSAR and docking studies on xanthone derivatives for anticancer activity targeting DNA topoisomerase IIα. Drug Des Dev Ther. 2014;8:183–95.

Amin AR, Kucuk O, Khuri FR, Shin DM. Perspectives for cancer prevention with natural compounds. J Clin Oncol. 2009;27(16):2712–25.

Arora A, Scholar EM. Role of tyrosine kinase inhibitors in cancer therapy. J Pharmacol Exp Ther. 2005;315(3):971–9.

Aware V, Gaikwad N, Chavan S, Manohar S, Bose J, Khanna S, Chandrika BR, Dixit N, Singh KS, Damre A. Cyclopentyl-pyrimidine based analogues as novel and potent IGF-1R inhibitor. Eur J Med Chem. 2015;92:246–56.

Azevedo C, Correia-Branco A, Araújo JR, Guimarães JT, Keating E, Martel F. The chemopreventive effect of the dietary compound kaempferol on the mcf-7 human breast cancer cell line is dependent on inhibition of glucose cellular uptake. Nutr Cancer. 2015;67(3):504–13.

Bairoch AM, Apweiler R, Wu CH, Barker WC, Boeckmann B, Ferro Rojas S, Gasteiger E, Huang H, Lopez R, Magrane M. The universal protein resource (UniProt). Nucleic Acids Res. 2005;33 (Database issue):D154–9.

Bakheet TM, Doig AJ. Properties and identification of human protein drug targets. Bioinformatics. 2009;25(4):451–7.

Batra P, Sharma AK. Anti-cancer potential of flavonoids: recent trends and future perspectives. Biotech. 2013;3(6):439–59.

Biswas S, Criswell TL, Wang SE, Arteaga CL. Inhibition of transforming growth factor-β signaling in human cancer: targeting a tumor suppressor network as a therapeutic strategy. Clin Cancer Res. 2006;12(14):4142–6.

Chen CY-C. Discovery of novel inhibitors for c-Met by virtual screening and pharmacophore analysis. J Chin Inst Chem Eng. 2008;39(6):617–24.

Chen Y, Ung C. Prediction of potential toxicity and side effect protein targets of a small molecule by a ligand–protein inverse docking approach. J Mol Graph Model. 2001;20(3):199–218.

Chen Y-J, Cheng Y-J, Hung AC, Wu Y-C, Hou M-F, Tyan Y-C, Yuan S-SF. The synthetic flavonoid WYC02-9 inhibits cervical cancer cell migration/invasion and angiogenesis via MAPK14 signaling. Gynecol Oncol. 2013;131(3):734–43.

Cho WC, Leung KN. In vitro and in vivo anti-tumor effects of Astragalus membranaceus. Cancer Lett. 2007a;252(1):43–54.

Cho WC, Leung KN. In vitro and in vivo immunomodulating and immunorestorative effects of Astragalus membranaceus. J Ethnopharmacol. 2007b;113(1):132–41.

Cho WC, Leung KN. *In vitro* and *in vivo* immunomodulating and immunorestorative effects of Astragalus membranaceus in murine models and in tumor-bearing mice. Cancer Res. 2007c;67 (9 Suppl):227–227.

Cho WC, Chen HY. Clinical efficacy of traditional Chinese medicine as a concomitant therapy for nasopharyngeal carcinoma: a systematic review and meta-analysis. Cancer Investig. 2009;27 (3):334–344.

Chu JS, Ge FJ, Zhang B, Wang Y, Silvestris N, Liu LJ, Zhao CH, Lin L, Brunetti AE, Fu YL. Expression and prognostic value of VEGFR-2, PDGFR-β, and c-Met in advanced hepatocellular carcinoma. J Exp Clin Cancer Res. 2013;32(1):16.

Chung CD, Liao J, Liu B, Rao X, Jay P, Berta P, Shuai K. Specific inhibition of Stat3 signal transduction by PIAS3. Science. 1997;278(5344):1803–5.

Costa PR, Acencio ML, Lemke N. A machine learning approach for genome-wide prediction of morbid and druggable human genes based on systems-level data. BMC Genomics. 2010;11 (5):1.

Costantino L, Barlocco D. Challenges in the design of multitarget drugs against multifactorial pathologies: a new life for medicinal chemistry? Future Med. Chem. 2013;5(1):5–7.

Curnow A, Owen S. An evaluation of root phytochemicals derived from Althea officinalis (Marshmallow) and Astragalus membranaceus as potential natural components of UV-protecting dermatological formulations. Oxid Med Cell Longev. 2016; Article ID 7053897.

Dan S, Okamura M, Seki M, Yamazaki K, Sugita H, Okui M, Mukai Y, Nishimura H, Asaka R, Nomura K. Correlating phosphatidylinositol 3-kinase inhibitor efficacy with signaling pathway status: *in silico* and biological evaluations. Cancer Res. 2010;70(12):4982–94.

de la Parra C, Castillo-Pichardo L, Cruz-Collazo A, Cubano L, Redis R, Calin GA, Dharmawardhane S. Soy isoflavone genistein-mediated downregulation of miR-155 contributes to the anticancer effects of genistein. Nutr Cancer. 2016;68(1):154–64.

Dennler S, Goumans M-J, Ten Dijke P. Transforming growth factor β signal transduction. J Leukoc Biol. 2002;71(5):731–40.

Ding Y, Shimada Y, Maeda M, Kawabe A, Kaganoi J, Komoto I, Hashimoto Y, Miyake M, Hashida H, Imamura M. Association of CC chemokine receptor 7 with lymph node metastasis of esophageal squamous cell carcinoma. Clin Cancer Res. 2003;9(9):3406–12.

Divakar S, Hariharan S. 3D-QSAR studies on plasmodium falciparam proteins: a mini-review. Comb Chem High Throughput Screen. 2015;18(2):188–98.

Do Amaral DN, Cavalcanti BC, Bezerra DP, Ferreira PMP, de Paula Castro R, Sabino JR, Machado CML, Chammas R, Pessoa C, Sant'Anna CM. Docking, synthesis and antiproliferative activity of n-acylhydrazone derivatives designed as combretastatin a4 analogues. PLoS One. 2014;9(3):e85380.

Dong J, Dong X. Comparative study on effect of astragulus injection and interleukin-2 in enhancing anti-tumor metastasis action of dendrite cells. Chinese J Integ Tradit West Med. 2005;25(3):236–9.

Fang J, Wu P, Yang R, Gao L, Li C, Wang D, Wu S, Liu A-L, Du G-H. Inhibition of acetylcholinesterase by two genistein derivatives: kinetic analysis, molecular docking and molecular dynamics simulation. Acta Pharm Sin B. 2014;4(6):430–7.

Fang C-Y, Wu C-C, Hsu H-Y, Chuang H-Y, Huang S-Y, Tsai C-H, Chang Y, Tsao GS-W, Chen C-L, Chen J-Y. EGCG inhibits proliferation, invasiveness and tumor growth by up-regulation of adhesion molecules, suppression of gelatinases activity, and induction of apoptosis in nasopharyngeal carcinoma cells. Int J Mol Sci. 2015;16(2):2530–58.

Fang Y, Zhang Q, Wang X, Yang X, Wang X, Huang Z, Jiao Y, Wang J. Quantitative phosphoproteomics reveals genistein as a modulator of cell cycle and DNA damage response pathways in triple-negative breast cancer cells. Int J Oncol. 2016;48(3):1016–28.

Ferrara N, Hillan KJ, Gerber H-P, Novotny W. Discovery and development of bevacizumab, an anti-VEGF antibody for treating cancer. Nat Rev Drug Discov. 2004;3(5):391–400.

Filipa Brito A, Ribeiro M, Margarida Abrantes A, Salome Pires A, Jorge Teixo R, Guilherme Tralhao J, Filomena Botelho M. Quercetin in cancer treatment, alone or in combination with conventional therapeutics? Curr Med Chem. 2015;22(26):3025–39.

Fischer D. Servers for protein structure prediction. Curr Opin Struct Biol. 2006;16(2):178–82.

Folkman J. Role of angiogenesis in tumor growth and metastasis. Semin Oncol. 2002;29(6 Suppl 16):15–8.

Frasca F, Pandini G, Sciacca L, Pezzino V, Squatrito S, Belfiore A, Vigneri R. The role of insulin receptors and IGF-I receptors in cancer and other diseases. Arch Physiol Biochem. 2008;114 (1):23–37.

Frei R, Staedler D, Raja A, Franke R, Sasse F, Gerber-Lemaire S, Waser J. Total synthesis and biological evaluation of jerantinine E. Angew Chem. 2013;125(50):13615–8.

Gu B, Ding Q, Xia G, Fang Z. EGCG inhibits growth and induces apoptosis in renal cell carcinoma through TFPI-2 overexpression. Oncol Rep. 2009;21(3):635–40.

Gu J-W, Makey KL, Tucker KB, Chinchar E, Mao X, Pei I, Thomas E, Miele L. EGCG, a major green tea catechin suppresses breast tumor angiogenesis and growth via inhibiting the activation of HIF-1α and NFκB, and VEGF expression. Vasc Cell. 2013;5(1):9.

Guedes IA, de Magalhães CS, Dardenne LE. Receptor–ligand molecular docking. Biophys Rev. 2014;6(1):75–87.

Güner OF, Bowen JP. Setting the record straight: the origin of the pharmacophore concept. J Chem Inf Model. 2014;54(5):1269–83.

Guo L, Bai S-P, Zhao L, Wang X-H. Astragalus polysaccharide injection integrated with vinorelbine and cisplatin for patients with advanced non-small cell lung cancer: effects on quality of life and survival. Med Oncol. 2012;29(3):1656–62.

Harvey AL, Edrada-Ebel R, Quinn RJ. The re-emergence of natural products for drug discovery in the genomics era. Nat Rev Drug Discov. 2015;14(2):111–29.

He Z-S, Shi X-H, Kong X-Y, Zhu Y-B, Chou K-C. A novel sequence-based method for phosphorylation site prediction with feature selection and analysis. Protein Pept Lett. 2012;19 (1):70–8.

Hicklin DJ, Ellis LM. Role of the vascular endothelial growth factor pathway in tumor growth and angiogenesis. J Clin Oncol. 2005;23(5):1011–27.

Huang J, Zhang X, Tang Q, Zhang F, Li Y, Feng Z, Zhu J. Prognostic significance and potential therapeutic target of VEGFR2 in hepatocellular carcinoma. J Clin Pathol. 2011;64(4):343–8.

Huang C-Y, Chan C-Y, Chou I-T, Lien C-H, Hung H-C, Lee M-F. Quercetin induces growth arrest through activation of FOXO1 transcription factor in EGFR-overexpressing oral cancer cells. J Nutr Biochem. 2013;24(9):1596–603.

Hubbard SR, Till JH. Protein tyrosine kinase structure and function. Annu Rev Biochem. 2000;69 (1):373–98.

Hyun H-B, Lee WS, Go S-I, Nagappan A, Park C, Han MH, Hong SH, Kim G, Kim GY, Cheong J. The flavonoid morin from Moraceae induces apoptosis by modulation of Bcl-2 family members and Fas receptor in HCT 116 cells. Int J Oncol. 2015;46(6):2670–8.

Irimie AI, Braicu C, Zanoaga O, Pileczki V, Gherman C, Berindan-Neagoe I, Campian RS. Epigallocatechin-3-gallate suppresses cell proliferation and promotes apoptosis and autophagy in oral cancer SSC-4 cells. Onco Targets Ther. 2015;8:461.

Jackson DA, Setzer WN. Selective phosphoinositide 3-kinase inhibition by natural products: a molecular docking study. Der Pharma Chemica. 2013;5(6):303–11.

Jakowlew SB. Transforming growth factor-β in cancer and metastasis. Cancer Metastasis Rev. 2006;25(3):435–57.

Ji SJ, Han DH, Kim JH. Inhibition of proliferation and induction of apoptosis by EGCG in human osteogenic sarcoma (HOS) cells. Arch Pharm Res. 2006;29(5):363–8.

Jorgensen WL. The many roles of computation in drug discovery. Science. 2004;303(5665): 1813–8.

Jun ME, Roy B. Ahn KH: “Turn-on” fluorescent sensing with “reactive” probes. Chem Commun. 2011;47(27):7583–601.

Jung JH, Lee JO, Kim JH, Lee SK, You GY, Park SH, Park JM, Kim EK, Suh PG, An JK. Quercetin suppresses HeLa cell viability via AMPK-induced HSP70 and EGFR down-regulation. J Cell Physiol. 2010;223(2):408–14.

Kadioglu O, Cao J, Saeed ME, Greten HJ, Efferth T. Targeting epidermal growth factor receptors and downstream signaling pathways in cancer by phytochemicals. Target Oncol. 2014;10 (3):337–53.

Kadioglu O, Nass J, Saeed ME, Schuler B, Efferth T. Kaempferol is an anti-inflammatory compound with activity towards NF-κB pathway proteins. Anticancer Res. 2015;35 (5):2645–50.

Kalva S, Singam EA, Rajapandian V, Saleena LM, Subramanian V. Discovery of potent inhibitor for matrix metalloproteinase-9 by pharmacophore based modeling and dynamics simulation studies. J Mol Graph Model. 2014;49:25–37.

Kandoi G, Acencio ML, Lemke N. Prediction of druggable proteins using machine learning and systems biology: a mini-review. Front Physiol. 2015;6.

Kapetanovic I. Computer-aided drug discovery and development (CADDD): *in silico*-chemico-biological approach. Chem Biol Interact. 2008;171(2):165–76.

Karimi R, Parivar K, Roudbari NH, Sadeghi SV, Hashemi M, Hayat P. Anti-proliferative and apoptotic effects of morin in human Leukemia cell lines (HUT-78). Int J Cell Mol Biotechnol. 2013; Article ID ijcmb-00001, 1–13.

Kaserer T, Beck KR, Akram M, Odermatt A, Schuster D. Pharmacophore models and pharmacophore-based virtual screening: concepts and applications exemplified on hydroxysteroid dehydrogenases. Molecules. 2015;20(12):22799–832.

Kerns E, Di L. Drug-like properties: concepts, structure design and methods: from ADME to toxicity optimization. Amsterdam: Academic Press; 2010.

Khan N, Afaq F, Saleem M, Ahmad N, Mukhtar H. Targeting multiple signaling pathways by green tea polyphenol (−)-epigallocatechin-3-gallate. Cancer Res. 2006;66(5):2500–5.

Kikuchi K. Design, synthesis and biological application of chemical probes for bio-imaging. Chem Soc Rev. 2010;39(6):2048–53.

Kim HY, Jung SK, Byun S, Son JE, Oh MH, Lee J, Kang MJ, Heo YS, Lee KW, Lee HJ. Raf and PI3K are the molecular targets for the anti-metastatic effect of luteolin. Phytother Res. 2012;27 (10):1481–8.

Kim ME, Ha TK, Yoon JH, Lee JS. Myricetin induces cell death of human colon cancer cells via BAX/BCL2-dependent pathway. Anticancer Res. 2014;34(2):701–6.

Kuete V, Saeed ME, Kadioglu O, Börtzler J, Khalid H, Greten HJ, Efferth T. Pharmacogenomic and molecular docking studies on the cytotoxicity of the natural steroid wortmannin against multidrug-resistant tumor cells. Phytomedicine. 2015;22(1):120–7.

Kumamoto T, Fujii M, Hou D-X. Myricetin directly targets JAK1 to inhibit cell transformation. Cancer Lett. 2009;275(1):17–26.

Kumar A, Bora U. Molecular docking studies on inhibition of Stat3 dimerization by curcumin natural derivatives and its conjugates with amino acids. Bioinformation. 2012;8(20):988.

Kumar S, Pandey AK. Chemistry and biological activities of flavonoids: an overview. Sci World J. 2013, Article ID162750; (2013).

Kumar S, Chashoo G, Saxena AK Pandey AK. *Parthenium hysterophorus*: a probable source of anticancer, antioxidant and anti-HIV agent. BioMed Res Int 2013:Article ID810734; (2013).

Kumar S, Pandey S, Pandey A K. *In vitro* antibacterial, antioxidant, and cytotoxic activities of *Parthenium hysterophorus* and characterization of extracts by LC-MS analysis. BioMed Res Int. 2014:Article ID 495154; (2014).

Kuo H-M, Chang L-S, Lin Y-L, Lu H-F, Yang J-S, Lee J-H, Chung J-G. Morin inhibits the growth of human leukemia HL-60 cells via cell cycle arrest and induction of apoptosis through mitochondria dependent pathway. Anticancer Res. 2007;27(1A):395–405.

Lee B-C, Lee T-H, Avraham S, Avraham HK. Involvement of the chemokine receptor CXCR4 and its ligand stromal cell-derived factor 1α in breast cancer cell migration. Mol Cancer Res. 2004;2(6):327–38.

Li H, Gao Z, Kang L, Zhang H, Yang K, Yu K, Luo X, Zhu W, Chen K, Shen J. TarFisDock: a web server for identifying drug targets with docking approach. Nucleic Acids Res. 2006;34 suppl 2: W219–24.

Li J, Zhou N, Luo K, Zhang W, Li X, Wu C, Bao J. *In silico* discovery of potential VEGFR-2 inhibitors from natural derivatives for anti-angiogenesis therapy. Int J Mol Sci. 2014;15 (9):15994–6011.

Lim K-H, Hiraku O, Komiyama K, Kam T-S. Jerantinines A– G, cytotoxic Aspidosperma alkaloids from Tabernaemontana corymbosa. J Nat Prod. 2008;71(9):1591–4.

Lin Y, Shi R, Wang X, Shen H-M. Luteolin, a flavonoid with potentials for cancer prevention and therapy. Curr Cancer Drug Targets. 2008;8(7):634.

Liu KC, Yen CY, Wu RSC, Yang JS, Lu HF, Lu KW, Lo C, Chen HY, Tang NY, Wu CC. The roles of endoplasmic reticulum stress and mitochondrial apoptotic signaling pathway in quercetin-mediated cell death of human prostate cancer PC-3 cells. Environ Toxicol. 2012a;27(4):428–39.

Liu L-C, Tsao TC-Y, Hsu S-R, Wang H-C, Tsai T-C, Kao J-Y, Way T-D. EGCG inhibits transforming growth factor-β-mediated epithelial-to-mesenchymal transition via the inhibition of Smad2 and Erk1/2 signaling pathways in nonsmall cell lung cancer cells. J Agric Food Chem. 2012b;60(39):9863–73.

Liu L, Leung K, Chan DS, Wang Y, Ma D, Leung C. Identification of a natural product-like STAT3 dimerization inhibitor by structure-based virtual screening. Cell Death Dis. 2014;5(6): e1293.

Lopez-Lazaro M. Distribution and biological activities of the flavonoid luteolin. Mini-Rev Med Chem. 2009;9(1):31–59.

Lu J, Li G, He K, Jiang W, Xu C, Li Z, Wang H, Wang W, Wang H, Teng X. Luteolin exerts a marked antitumor effect in cMet-overexpressing patient-derived tumor xenograft models of gastric cancer. J Transl Med. 2015;13(1):42.

Luan X, Gao Y-G, Guan Y-Y, Xu J-R, Lu Q, Zhao M, Liu Y-R, Liu H-J, Fang C, Chen H-Z. Platycodin D inhibits tumor growth by antiangiogenic activity via blocking VEGFR2-mediated signaling pathway. Toxicol Appl Pharmacol. 2014;281(1):118–24.

Luo F-J, Hu Z, Deng X-Y, Zhao Y, Zeng L, Dong Z-G, Yi W, Cao Y. Effect of tea polyphenols and EGCG on nasopharyngeal carcinoma cell proliferation and the mechanisms involved. Chin J Cancer Res. 2001;13(4):235–42.

Luo H, Daddysman MK, Rankin GO, Jiang B-H, Chen YC. Kaempferol enhances cisplatin's effect on ovarian cancer cells through promoting apoptosis caused by down regulation of cMyc. Cancer Cell Int. 2010;10:16.

Lyne PD. Structure-based virtual screening: an overview. Drug Discov Today. 2002;7 (20):1047–55.

Ma L, Peng H, Li K, Zhao R, Li L, Yu Y, Wang X, Han Z. Luteolin exerts an anticancer effect on NCI-H460 human non-small cell lung cancer cells through the induction of Sirt1-mediated apoptosis. Mol Med Rep. 2015;12(3):4196–202.

Maggioni D, Nicolini G, Rigolio R, Biffi L, Pignataro L, Gaini R, Garavello W. Myricetin and naringenin inhibit human squamous cell carcinoma proliferation and migration *in vitro*. Nutr Cancer. 2014;66(7):1257–67.

Majumdar D, Jung K-H, Zhang H, Nannapaneni S, Wang X, Amin AR, Chen Z, Shin DM. Luteolin nanoparticle in chemoprevention: *in vitro* and in vivo anticancer activity. Cancer Prev Res. 2014;7(1):65–73.

Martin S, Wu Z-H, Gehlhaus K, Jones T, Zhang Y-W, Guha R, Miyamoto S, Pommier Y, Caplen N. RNAi screening identifies TAK1 as a potential target for the enhanced efficacy of topoisomerase inhibitors. Curr Cancer Drug Targets. 2011;11(8):976.

Masuda M, Suzui M, Lim JT, Weinstein IB. Epigallocatechin-3-gallate inhibits activation of HER-2/neu and downstream signaling pathways in human head and neck and breast carcinoma cells. Clin Cancer Res. 2003;9(9):3486–91.

Mayer AM, Gustafson KR. Marine pharmacology in 2001–2: antitumour and cytotoxic compounds. Eur J Cancer. 2004;40(18):2676–704.

Mayer AM, Gustafson KR. Marine pharmacology in 2005–2006: antitumour and cytotoxic compounds. Eur J Cancer. 2008;44(16):2357–87.

McCulloch M, See C, Shu X-J, Broffman M, Kramer A, Fan W-Y, Gao J, Lieb W, Shieh K, Colford JM. Astragalus-based Chinese herbs and platinum-based chemotherapy for advanced non–small-cell lung cancer: meta-analysis of randomized trials. J Clin Oncol. 2006;24 (3):419–30.

Mechoulam H, Pierce EA. Expression and activation of STAT3 in ischemia-induced retinopathy. Investig Ophthalmol Vis Sci. 2005;46(12):4409–16.

Meshram RJ, Bhiogade NH, Gacche RN, Jangle SN. Virtual screening and docking exploration on estrogen receptor: an *in silico* approach to decipher novel anticancer agents. Indian J Biotechnol. 2012;11(4):389–95.

Mishra A, Sharma AK, Kumar S, Saxena AK, Pandey AK. *Bauhinia variegata* leaf extracts exhibit considerable antibacterial, antioxidant and anticancer activities. BioMed Res Int. 2013;2013:915436.

Morales P, Haza AI. Selective apoptotic effects of piceatannol and myricetin in human cancer cells. J Appl Toxicol. 2012;32(12):986–93.

Müller A, Homey B, Soto H, Ge N, Catron D, Buchanan ME, McClanahan T, Murphy E, Yuan W, Wagner SN. Involvement of chemokine receptors in breast cancer metastasis. Nature. 2001;410(6824):50–6.

Naik PK, Lopus M, Aneja R, Vangapandu SN, Joshi HC. *In silico* inspired design and synthesis of a novel tubulin-binding anti-cancer drug: folate conjugated noscapine (Targetin). J Comput Aided Mol Des. 2012;26(2):233–47.

Naso LG, Lezama L, Rojo T, Etcheverry SB, Valcarcel M, Roura M, Salado C, Ferrer EG, Williams PA. Biological evaluation of morin and its new oxovanadium (IV) complex as antioxidant and specific anti-cancer agents. Chem Biol Interact. 2013;206(2):289–301.

Neerincx PB, Leunissen JA. Evolution of web services in bioinformatics. Brief Bioinform. 2005;6 (2):178–88.

Neves H, Kwok HF. Recent advances in the field of anti-cancer immunotherapy. BBA Clin. 2015;3:280–8.

Niu G, Heller R, Catlett-Falcone R, Coppola D, Jaroszeski M, Dalton W, Jove R, Yu H. Gene therapy with dominant-negative Stat3 suppresses growth of the murine melanoma B16 tumor *in vivo*. Cancer Res. 1999;59(20):5059–63.

Olayinka ET, Ore A, Ola OS, Adeyemo OA. Protective effect of quercetin on melphalan-induced oxidative stress and impaired renal and hepatic functions in rat. Chemotherap Res Pract. (2014); Article ID 936526.

Pandey SK, Patel DK, Thakur R, Mishra DP, Maiti P, Haldar C. Anti-cancer evaluation of quercetin embedded PLA nanoparticles synthesized by emulsified nanoprecipitation. Int J Biol Macromol. 2015;75:521–9.

Park J-H, Yoon J-H, Kim S-A, Ahn S-G, Yoon J-H. (−)-Epigallocatechin-3-gallate inhibits invasion and migration of salivary gland adenocarcinoma cells. Oncol Rep. 2010;23 (2):585–90.

Park J-H, Yoon J-H, Kim S-A, Ahn S-G, Yoon J-H. Morin, a flavonoid from Moraceae, induces apoptosis by induction of BAD protein in human leukemic cells. Int J Mol Sci. 2014a;16 (1):645–59.

Park C, Lee WS, Go SI, Nagappan A, Han MH, Hong SH, Kim GS, Kim GY, Kwon TK, Ryu CH, et al. Morin, a flavonoid from Moraceae, induces apoptosis by induction of BAD protein in human leukemic cells. Int J Mol Sci. 2014b;16(1):645–59.

Pavlopoulou A, Michalopoulos I. State-of-the-art bioinformatics protein structure prediction tools (Review). Int J Mol Med. 2011;28(3):295–310.

Phillips P, Sangwan V, Borja-Cacho D, Dudeja V, Vickers S, Saluja A. Myricetin induces pancreatic cancer cell death via the induction of apoptosis and inhibition of the phosphatidylinositol 3-kinase (PI3K) signaling pathway. Cancer Lett. 2011;308(2):181–8.

Prasad NK, Kanakaveti V, Eadlapalli S, Vadde R, Meetei A P, Vindal V. Ligand-based pharmacophore modeling and virtual screening of RAD9 inhibitors. J Chem. (2013); Article ID 679459.

Pruitt KD, Tatusova T, Maglott DR. NCBI reference sequences (RefSeq): a curated non-redundant sequence database of genomes, transcripts and proteins. Nucleic Acids Res. 2007;35 suppl 1: D61–5.

Qazzaz ME, Raja VJ, Lim K-H, Kam T-S, Lee JB, Gershkovich P, Bradshaw TD. *In vitro* anticancer properties and biological evaluation of novel natural alkaloid jerantinine B. Cancer Lett. 2016;370(2):185–97.

Qin J, Chen JX, Zhu Z, Teng JA. Genistein inhibits human colorectal cancer growth and suppresses MiR-95, Akt and SGK1. Cell Physiol Biochem. 2015;35(5):2069–77.

Raja VJ (2015) Biological characterisation of a novel and naturally isolated indole alkaloid. University of Nottingham, PhD thesis.

Raja VJ, Lim K-H, Leong C-O, Kam T-S, Bradshaw TD. Novel antitumour indole alkaloid, Jerantinine A, evokes potent G2/M cell cycle arrest targeting microtubules. Investig New Drugs. 2014;32(5):838–50.

Refolo MG, D'Alessandro R, Malerba N, Laezza C, Bifulco M, Messa C, Caruso MG, Notarnicola M, Tutino V. Anti proliferative and pro apoptotic effects of flavonoid quercetin are mediated by CB1 receptor in human colon cancer cell lines. J Cell Physiol. 2015;230 (12):2973–80.

Robinson DR, Wu Y-M, Lin S-F. The protein tyrosine kinase family of the human genome. Oncogene. 2000;19(49):5548–57.

Saiko P, Steinmann M-T, Schuster H, Graser G, Bressler S, Giessrigl B, Lackner A, Grusch M, Krupitza G, Bago-Horvath Z. Epigallocatechin gallate, ellagic acid, and rosmarinic acid perturb dNTP pools and inhibit de novo DNA synthesis and proliferation of human HL-60 promyelocytic leukemia cells: synergism with arabinofuranosylcytosine. Phytomedicine. 2015;22(1):213–22.

Saini S, Hurwitz, H. VEGF inhibition for cancer therapy. In: Molecular targeting in oncology. Springer; (2008). pp. 573–584.

Sak K. Cytotoxicity of dietary flavonoids on different human cancer types. Pharmacogn Rev. 2014;8(16):122–46.

Santoshi S, Manchukonda NK, Suri C, Sharma M, Sridhar B, Joseph S, Lopus M, Kantevari S, Baitharu I, Naik PK. Rational design of biaryl pharmacophore inserted noscapine derivatives as potent tubulin binding anticancer agents. J Comput Aided Mol Des. 2014;29(3):249–70.

SeungHee E, JuSung K. Molecular docking studies for discovery of plant-derived α-glucosidase inhibitors. Plant Omics. 2014;7(3):166–70.

Shim J-H, Su Z-Y, Chae J-I, Kim DJ, Zhu F, Ma W-Y, Bode AM, Yang CS, Dong Z. Epigallocatechin gallate suppresses lung cancer cell growth through Ras–GTPase-activating protein SH3 domain-binding protein 1. Cancer Prev Res. 2010;3(5):670–9.

Shimizu M, Shirakami Y, Moriwaki H. Targeting receptor tyrosine kinases for chemoprevention by green tea catechin, EGCG. Int J Mol Sci. 2008;9(6):1034–49.

Shimizu M, Adachi S, Masuda M, Kozawa O, Moriwaki H. Cancer chemoprevention with green tea catechins by targeting receptor tyrosine kinases. Mol Nutr Food Res. 2011;55(6):832–43.

Siddiqui IA, Adhami VM, Afaq F, Ahmad N, Mukhtar H. Modulation of phosphatidylinositol-3-kinase/protein kinase B-and mitogen-activated protein kinase-pathways by tea polyphenols in human prostate cancer cells. J Cell Biochem. 2004;91(2):232–42.

Siegelin M, Gaiser T, Habel A, Siegelin Y. Myricetin sensitizes malignant glioma cells to TRAIL-mediated apoptosis by down-regulation of the short isoform of FLIP and bcl-2. Cancer Lett. 2009;283(2):230–8.

Simon AR, Rai U, Fanburg BL, Cochran BH. Activation of the JAK-STAT pathway by reactive oxygen species. Am J Phys Cell Phys. 1998;275(6):C1640–52.

Singh P, Bast F. *In silico* molecular docking study of natural compounds on wild and mutated epidermal growth factor receptor. Med Chem Res. 2014a;23(9):5074–85.

Singh P, Bast F. Multitargeted molecular docking study of plant-derived natural products on phosphoinositide-3 kinase pathway components. Med Chem Res. 2014b;23(4):1690–700.

Singh P, Bast F. High-throughput virtual screening, identification and *in vitro* biological evaluation of novel inhibitors of signal transducer and activator of transcription 3. Med Chem Res. 2015a;24(6):2694–708.

Singh P, Bast F. Screening and biological evaluation of myricetin as a multiple target inhibitor insulin, epidermal growth factor, and androgen receptor; *in silico* and *in vitro*. Investig New Drugs. 2015b;33(3):575–93.

Singh P, Bast F. Screening of multitargeted natural compounds for receptor tyrosine kinases inhibitors and biological evaluation on cancer cell lines; *in silico* and *in vitro*. Med Oncol. 2015c;32(9):233.

Singh SP, Konwar BK. Molecular docking studies of quercetin and its analogues against human inducible nitric oxide synthase. SpringerPlus. 2012;1(1):1–10.

Sivaramakrishnan V, Devaraj SN. Morin fosters apoptosis in experimental hepatocellular carcinogenesis model. Chem Biol Interact. 2010;183(2):284–92.

Sivashanmugam M, Raghunath C, Vetrivel U. Virtual screening studies reveal linarin as a potential natural inhibitor targeting CDK4 in retinoblastoma. J Pharmacol Pharmacother. 2013;4(4):256.

Sun F, Zheng XY, Ye J, Wu TT, Wang J l, Chen W. Potential anticancer activity of myricetin in human T24 bladder cancer cells both *in vitro* and *in vivo*. Nutr Cancer. 2012;64(4):599–606.

Sun X-Q, Chen L, Li Y-Z, Li W-H, Liu G-X, Tu Y-Q, Tang Y. Structure-based ensemble-QSAR model: a novel approach to the study of the EGFR tyrosine kinase and its inhibitors. Acta Pharmacol Sin. 2014;35(2):301–10.

Suzuki R, Kang Y a, Li X, Roife D, Zhang R, Fleming JB. Genistein potentiates the antitumor effect of 5-fluorouracil by inducing apoptosis and autophagy in human pancreatic cancer cells. Anticancer Res. 2014;34(9):4685–92.

Takeda K, Noguchi K, Shi W, Tanaka T, Matsumoto M, Yoshida N, Kishimoto T, Akira S. Targeted disruption of the mouse Stat3 gene leads to early embryonic lethality. Proc Natl Acad Sci. 1997;94(8):3801–4.

Tanno S, Ohsaki Y, Nakanishi K, Toyoshima E, Kikuchi K. Human small cell lung cancer cells express functional VEGF receptors, VEGFR-2 and VEGFR-3. Lung Cancer. 2004;46(1):11–9.

Taylor JC, Bock CW, Takusagawa F, Markham GD. Discovery of novel types of inhibitors of S-adenosylmethionine synthesis by virtual screening. J Med Chem. 2009;52(19):5967–73.

Trost B, Kusalik A. Computational prediction of eukaryotic phosphorylation sites. Bioinformatics. 2011;27(21):2927–35.

Tu LY, Bai HH, Cai JY, Deng SP. The mechanism of kaempferol induced apoptosis and inhibited proliferation in human cervical cancer SiHa cell: From macro to nano. Scanning. 2016a; doi:10.1002/sca.21312.

Tu L-Y, Pi J, Jin H, Cai J-Y, Deng S-P Synthesis, characterization and anticancer activity of kaempferol-zinc (II) complex. Bioorg Med Chem Lett. 2016b; S0960-894X (16):30323–30327.

Vela M, Aris M, Llorente M, Garcia-Sanz J, Kremer L. Chemokine receptor specific antibodies in cancer immunotherapy: achievements and challenges. Name: Front Immunol. 2015;6:12.

Verbanac D, Jain SC, Jain N, Chand M, Paljetak HČ, Matijašić M, Perić M, Stepanić V, Saso L. An efficient and convenient microwave-assisted chemical synthesis of (thio) xanthones with additional *in vitro* and *in silico* characterization. Bioorg Med Chem. 2012;20(10):3180–85.

Walker EH, Pacold ME, Perisic O, Stephens L, Hawkins PT, Wymann MP, Williams RL. Structural determinants of phosphoinositide 3-kinase inhibition by wortmannin, LY294002, quercetin, myricetin, and staurosporine. Mol Cell. 2000;6(4):909–19.

Wang Z, Desmoulin S, Banerjee S, Kong D, Li Y, Deraniyagala RL, Abbruzzese J, Sarkar FH. Synergistic effects of multiple natural products in pancreatic cancer cells. Life Sci. 2008;83(7):293–300.

Wang X, Hao M-W, Dong K, Lin F, Ren J-H, Zhang H-Z. Apoptosis induction effects of EGCG in laryngeal squamous cell carcinoma cells through telomerase repression. Arch Pharm Res. 2009;32(9):1263–9.

Wang L, Li X, Zhang S, Lu W, Liao S, Liu X, Shan L, Shen X, Jiang H, Zhang W. Natural products as a gold mine for selective matrix metalloproteinases inhibitors. Bioorg Med Chem. 2012a;20 (13):4164–71.

Wang P, Heber D, Henning SM. Quercetin increased the antiproliferative activity of green tea polyphenol (−)-epigallocatechin gallate in prostate cancer cells. Nutr Cancer. 2012b;64(4):580–7.

Wang L, Feng J, Chen X, Guo W, Du Y, Wang Y, Zang W, Zhang S, Zhao G. Myricetin enhance chemosensitivity of 5-fluorouracil on esophageal carcinoma *in vitro* and *in vivo*. Cancer Cell Int. 2014;14(1):71.

Wang Y, Xing J, Xu Y, Zhou N, Peng J, Xiong Z, Liu X, Luo X, Luo C, Chen K. In silico ADME/T modelling for rational drug design. Q Rev Biophys. 2015;48(04):488–515.

Wu CH, Huang H, Arminski L, Castro-Alvear J, Chen Y, Hu Z-Z, Ledley RS, Lewis KC, Mewes H-W, Orcutt BC. The protein information resource: an integrated public resource of functional annotation of proteins. Nucleic Acids Res. 2002;30(1):35–7.

Wu C, Shi L, Wu C, Guo D, Selke M, Wang X. Enhanced *in vitro* anticancer activity of quercetin mediated by functionalized CdTe QDs. Science China Chem. 2014;57(11):1579–88.

Xie Q, Bai Q, Zou LY, Zhang QY, Zhou Y, Chang H, Yi L, Zhu JD, Mi MT. Genistein inhibits DNA methylation and increases expression of tumor suppressor genes in human breast cancer cells. Genes Chromosom Cancer. 2014;53(5):422–31.

Xu L, Li C, Olson AJ, Wilson IA. Crystal structure of avian aminoimidazole-4-carboxamide ribonucleotide transformylase in complex with a novel non-folate inhibitor identified by virtual ligand screening. J Biol Chem. 2004;279(48):50555–65.

Xu R, Zhang Y, Ye X, Xue S, Shi J, Pan J, Chen Q. Inhibition effects and induction of apoptosis of flavonoids on the prostate cancer cell line PC-3 *in vitro*. Food Chem. 2013;138(1):48–53.

Yagiz K, Wu LY, Kuntz CP, James Morre D, Morré DM. Mouse embryonic fibroblast cells from transgenic mice overexpressing tNOX exhibit an altered growth and drug response phenotype. J Cell Biochem. 2007;101(2):295–306.

Yamaguchi K, Shirakabe K, Shibuya H, Irie K, Oishi I, Ueno N, Taniguchi T, Nishida E, Matsumoto K. Identification of a member of the MAPKKK family as a potential mediator of TGF-β signal transduction. Science. 1995;270(5244):2008–11.

Yamazaki Y, Kitajima M, Arita M, Takayama H, Sudo H, Yamazaki M, Aimi N, Saito K. Biosynthesis of camptothecin. *In silico* and in vivo tracer study from [1-13C] glucose. Plant Physiol. 2004;134(1):161–70.

Yim-Im W, Sawatdichaikul O, Semsri S, Horata N, Mokmak W, Tongsima S, Suksamrarn A. Computational analyses of curcuminoid analogs against kinase domain of HER2. BMC Bioinform. 2014;15(1):1–13.

Yuan J, Liu H, Kang X, Zou G. Molecular docking of epidermal growth factor receptor tyramine kinase domain and its inhibitor genistein. Chin J Biotechnol. 2008;24(10):1813–7.

Zhang X, Ling Y, Yu H, Ji Y. Studies on mechanism of myricetin-induced apoptosis in human hepatocellular carcinoma HepG-2 cells. China J Chin Materia Medica. 2010;35(8):1046–50.

Zhang C, Zhai S, Li X, Zhang Q, Wu L, Liu Y, Jiang C, Zhou H, Li F, Zhang S. Synergistic action by multi-targeting compounds produces a potent compound combination for human NSCLC both *in vitro* and *in vivo*. Cell Death Dis. 2014a;5(3):e1138.

Zhang S, Wang L, Liu H, Zhao G, Ming L. Enhancement of recombinant myricetin on the radiosensitivity of lung cancer A549 and H1299 cells. Diagn Pathol. 2014b;9:68.

Zhang Y, Wang X, Han L, Zhou Y, Sun S. Green tea polyphenol EGCG reverse cisplatin resistance of A549/DDP cell line through candidate genes demethylation. Biomed Pharmacother. 2015;69:285–90.

Zhou C. Multi-targeted tyrosine kinase inhibitors for the treatment of non-small cell lung cancer: an era of individualized therapy. Transl Lung Cancer Res. 2012;1(1):72–7.

Zhou J, Farah BL, Sinha RA, Wu Y, Singh BK, Bay B-H, Yang CS, Yen PM. Epigallocatechin-3-Gallate (EGCG), a green tea polyphenol, stimulates hepatic autophagy and lipid clearance. PLoS One. 2014;9(1):e87161.

Zou Y, Liu X. Effect of astragalus injection combined with chemotherapy on quality of life in patients with advanced non-small cell lung cancer. Chinese J Integr Tradit Western Med. 2003;23(10):733–5.

Part III
Omics for Precision Medicine

Chapter 9
Genome-Wide Association Studies: A Comprehensive Tool to Explore Comparative Genomic Variations and Interactions

Aruni Wilson

Abstract In the recent past, significant advances in sequencing technologies have led to genome-wide association studies (GWASs) that had revealed substantial insight into the genetic architecture of human phenotypes. The technique involves rapid scanning of markers across the whole genome of many people in order to find genetic variants that can be attributed to a particular disease condition. The enormous contributions of genetic information to common disease conditions and newer algorithms have set a stage for rapid and efficient screening of the data. Such information in the future will enable a tailor-made disease prevention program through selection of treatment options. An archive of data from genome-wide association studies on a variety of diseases and conditions already can be accessed through an NCBI Web site, called the Database of Genotype and Phenotype (dbGaP) located at http://www.ncbi.nlm.nih.gov/entrez/query.fcgi?db=gap.

Keywords GWAS • PheWAS • Genome bioinformatics • Microbiome variations

9.1 Introduction

Rapid development of sequencing technology and computational methods has now paved the way for GWAS that has now become a powerful tool to detect natural variations underlying complex traits. The GWAS methodology became well established in human genetics during a decade. This method of study follows an approach that involves rapidly scanning markers across the entire set of DNA in comparison with many subjects to find genetic variations associated with a particular

A. Wilson (✉)
Division of Microbiology and Molecular Genetics, School of Medicine, Loma Linda University, 92350 Loma Linda, CA, USA

Stem Cell Center, University of California, Riverside, CA, USA

School of Biosciences and Technology, VIT University, Vellore, India
e-mail: aruniwilson@llu.edu

© Shanghai Jiao Tong University Press, Shanghai and Springer Science+Business Media Dordrecht 2017
D.-Q. Wei et al. (eds.), *Translational Bioinformatics and Its Application*, Translational Medicine Research, DOI 10.1007/978-94-024-1045-7_9

disease or trait or to study variations in the environmental influence. Currently, over 1000 GWASs have been published linking nearly 4000 statistically significant loci to over 500 human traits and diseases (Hindorff et al. 2013). GWAS has identified thousands of statistically significant genetic variants that are associated with a number of human conditions; however, the main drawback is that most GWASs do not identify clinically significant associations. Furthermore, identification of biologically significant variants by GWAS also presents a significant challenge. While GWAS is a phenotype–genotype approach, an alternative is to study the genotype–phenotype approach that is currently called the "phenome-wide association studies (PheWASs) (Hebbring et al. 2013). Millions of SNPs were identified in human populations by which a high-density haplotype map of human genome could be constructed (Abecasis et al. 2012). Furthermore, many commercial arrays designed for large-scale genotyping has led to the use of this technique in identifying genes involved in human disease (2118). The HapMap Project showed that common but minute variations in human DNA occur about once in very 1000 base pairs of DNA across the human genome, which contains about three billion base pairs. These variations, called single nucleotide polymorphisms (SNPs), can be used to identify genetic contributions to common diseases. Recent GWASs have identified about 500,000 of these SNPs in each individual. The advent of genome-wide association (GWA) technology has transformed the landscape of human genetic research. It has enabled those in the field to move beyond the limitations of small-scale candidate gene studies, and well over 200 loci influencing a wide range of complex phenotypes have now been identified. The National Center for Biotechnology Information (NCBI), a part of NIH's National Library of Medicine, is developing databases for use by the research community. An archive of data from genome-wide association studies on a variety of diseases and conditions already can be accessed through an NCBI Web site, called the Database of Genotype and Phenotype (dbGaP), located at http://www.ncbi.nlm.nih.gov/entrez/query.fcgi?db=gap.

Three of the major diseases in which there has been the greatest yield of novel complex trait-susceptibility genes from GWA studies include metabolic/cardiovascular, autoimmune, and cancer.

Over 50 novel loci now known to modify individual risk of type 2 diabetes and cardiovascular disease have been described to influence circulating levels of lipids or to alter energy balance and thereby body mass index and potential for obesity (Mohlke et al. 2008). There has been a similar explosion in respect of autoimmune diseases, and Chen and Shapiro (2015) summarize how the total numbers of loci implicated in predisposition to celiac disease, inflammatory bowel disease, multiple sclerosis, rheumatoid arthritis, lupus, or type 1 diabetes have more than quadrupled in the past 3 years, with several of those loci predisposing to more than one autoimmune disease. GWASs into cancer predisposition tell the same story: Easton (Abecasis et al. 2012) documents over 20 new loci for breast cancer, prostate cancer, colorectal cancer, or melanoma discovered in the last 2 years. The findings in cancer also provide an early glimpse of the possible complexity of susceptibility loci, with multiple alleles at the same 8q24 locus influencing the risk of different combinations of cancers.

9.2 Background

The primary purpose of GWAS is to identify single nucleotide polymorphisms (SNPs) that are associated with phenotypic traits, typically those associated with a particular disease. Nearly half of the disease-associated SNPs from published GWAS are not located in or near genes (Welter et al. 2014). Therefore, despite the fact that significant associations are often found between complex traits and SNPs in gene deserts (i.e., genomic regions of $>$ 500 kb that lack annotated genes or protein-coding sequences, the possibility of the SNPs within such genomic regions regulating the unlinked genes lies within the twists and turns that form when 3 m of human DNA (chromosomes) is packaged within a roughly spherical nucleus that is only approximately 10 μm in diameter (Schierding et al. 2014). Within the hierarchy of folding necessary to package the genome within the eukaryotic nucleus, regions of each chromosome contact other chromosomes to form an intricate three-dimensional DNA network. Therefore, while two regions of DNA (loci) may be distant on a linear scale, DNA folding provides a mechanism for these two loci to become spatially close together. Implicit in this concept is the idea that all genetic functions (regulation, reading, repair, and replication) are influenced by this three-dimensional architecture, generating the cell's morphology and function (Misteli 2001). Intracellular DNA structure cannot be divorced from its functions. Therefore, it is possible that intergenic SNPs associated with diseases are indeed involved in the regulation of genes and pathways through spatial associations with different genes.

9.3 GWAS in Human Disease Study

GWAS approach is through studying the phenotype to genotype variations (Fig. 9.1a). While there is every effort to catalog human variation, the most recent versions of dbSNP and the human gene mutation database contain 38,072,522 validated variants (Sherry et al. 2001) and ~100,000 mutations in nuclear genes (Stenson et al. 2014) that are associated with complex human traits, respectively. However, the associations between common variants (SNPs) and phenotypic traits or diseases held in these databases, and others like them, only describe a small fraction of the overall heritability of complex disease traits (Frazer et al. 2009). Thus, our ability to elucidate functional pathways related to these SNPs has been limited. Also, it is important to determine if SNPs located outside of genes contribute to disease phenotypes through alterations to spatial regulatory interactions.

One caveat to the study of SNPs within non-genic regions is that while it is known that common SNPs explain a substantial portion of heritability, not all SNPs contribute equally to the heritability of a trait. Despite this, it remains possible that SNPs located outside of coding regions represent a new class of regulatory SNPs that make an important contribution toward explaining heritability.

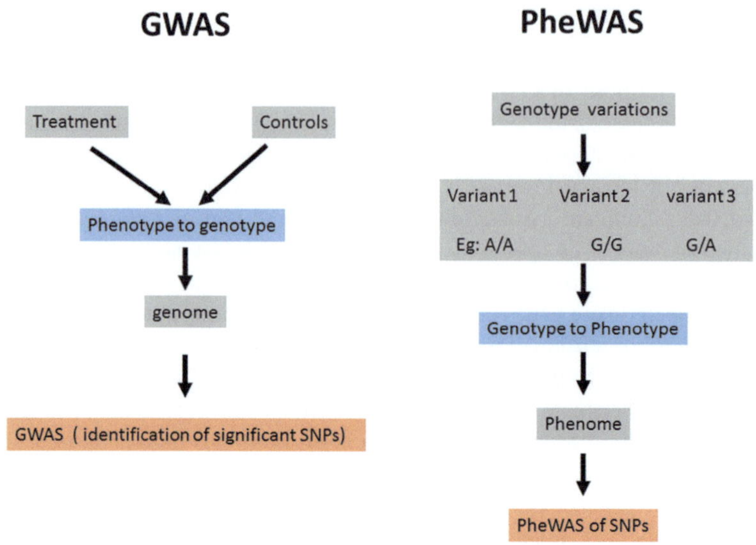

Fig. 9.1 Illustrative representation of GWAS (**a**) and PheWAS (**b**)

Recently, many methods that try to explain the roles of these SNPs in the context of 3D structure have recently been developed. A recently developed database provides functional annotations of SNPs using actual long-range interaction data sets (Wang et al. 2012). By going beyond conservation information and incorporating information from multiple different sources (e.g., HapMap, ENCODE), the GWAS 3D database has branded itself as an efficient solution to interpret the regulatory role of genetic variation in the noncoding regions, associating SNPs with 3D structure changes.

9.4 Planning and Executing GWAS

GWASs are fast becoming the default study design to study and discover new genetic variants that may influence an expressed trait or a phenotype. The major planning and execution of GWAS rely upon a proper ethical consideration, a well-balanced study design, selection of a phenotype or phenotypes, power analysis of the sample size, sample tracking and storage of big data, and genotyping product selection. Furthermore, more importantly, due consideration is given to DNA quantity and preparation, genotyping methods, quality control and checking the genotype data and above all analysis part through in silico genotyping (imputing), study of the test of association, and replication of association signals.

The major ethical considerations and consent documents and sharing of GWAS data under biobank policies are available at http://grants.nih.gov/grants/gwas/, and the deposition of data could be retrieved from the NIH database of genotype and phenotype (dbGaP) (Tryka et al. 2014). However, it is always best to consult local experts in ethics and genetics on their local review board.

Although GWAS uses a family design, the resulting data can be analyzed under the test of association of familial traits (Martin et al. 2000; Chen and Abecasis 2007). However, many of the earlier GWASs have originated from case–control study designs. This is advantageous in independent sampling units and ease of analysis. Also, such approach could contribute to power of tests of common genetic variants that contribute only moderate to low relative risk groups. Nevertheless, matching of samples is more important in this approach to avoid biases.

It has been earlier shown that many studies have demonstrated the utility of genotyping a common set of population controls for analyses of multiple traits. The Wellcome Trust Case Control Consortium (WTCCC) (Wellcome Trust Case Control Consortium 2007) genotyped 500,000 SNPs in a common set of 3000 controls drawn from the 1958 British birth cohort and the UK Blood Services collection and has used these for a number of disease-based GWAS. Tests of association using a Cochran–Armitage additive trend statistic showed a high degree of concordance of the separately ascertained, but ethnically matched, UK controls. The use of previously genotyped population controls ascertained from an independent study can result in a considerable cost saving or increase in power (Zondervan and Cardon 2007), but the population must be substantially free of the disease or phenotype under investigation and must be ethnically well matched. Many of the related issues have been reviewed previously (Zondervan and Cardon 2007; McCarthy et al. 2008; Cooper et al. 2008; Amos 2007).

One of the most neglected aspects of GWAS is the choice of phenotype definition and method of measurement of primary phenotypes and their potential cofounders. Hence, a large-scale study such as a well-designed prospective cohort is often advantageous. Given the need for a large-scale study, often international and collaborative efforts to establish replication or identify alleles could help in developing standardized phenotypic protocols. Such studies will facilitate comparable cross analyses and analysis of metadata.

The power to detect association is a function of the effect sample size (number of cases and controls or families) and the tested association disease model. These factors are influenced by the prevalence of the disease, disease allele frequency, and the genotypic relative risk (GRR). For a typical study design that plans to genotype 1000 cases and 1000 controls for 300,000 markers, a disease-predisposing variant with GRR $= 1.415$ under a multiplicative model, with prevalence 0.1 and risk allele frequency 0.5, can be detected with 80% power (Klein 2007; Mukherjee et al. 2011). To reduce the high cost of genotyping, a two-stage design has been proposed where a proportion of samples are genotyped on every marker in stage 1, and a proportion of these markers are later followed up by genotyping them on the remaining samples in stage 2 (Nguyen et al. 2009). For the above example, nearly the same power (77%) can be achieved with only 34% as many genotypes by using 30% of samples in stage 1 and 5% markers in stage 2. Software packages such as CaTS and other statistical packages (Table 9.1) can be used to plan the sample size and power for their studies (Skol et al. 2006).

Table 9.1 Bioinformatics tools in GWAS

Function	Tool	PubMed reference/Tool url
Quality control	Quality control of data sets is the important prerequisite for GWAS analysis and analysis of metadata	
	GTOOL	http://www.well.ox.ac.uk/~cfreeman/software/gwas/gtool.html
	GWAtool box	http://www.ncbi.nlm.nih.gov/pubmed/22155946
		http://www.eurac.edu/en/research/health/biomed/services/Pages/GWAtoolbox.aspx
	PhenoMan	http://www.ncbi.nlm.nih.gov/pubmed/24336645
		https://code.google.com/p/phenoman/
	QCGWAS	http://www.ncbi.nlm.nih.gov/pubmed/24395754
		http://cran.r-project.org/web/packages/QCGWAS/
	QCTOOL	http://www.well.ox.ac.uk/~gav/qctool/#overview
Association mapping	The analysis relies upon detection of linkage disequilibrium (LD) between genetic markers and genes controlling the phenotype of interest through exploring recombination events that have accumulated over generations	
	PLINK	http://www.ncbi.nlm.nih.gov/pubmed/17701901
		http://pngu.mgh.harvard.edu/~purcell/plink/
	EMMA (efficient mixed-model association)	http://www.ncbi.nlm.nih.gov/pubmed/18385116
		http://mouse.cs.ucla.edu/emma/
	GAPIT (genome association and prediction integrated tool)	http://www.maizegenetics.net/#!gapit/cmkv
	GenABEL	http://www.ncbi.nlm.nih.gov/pubmed/17384015
		http://www.genabel.org/packages/GenABEL
	GLOGC (genome-wide LOGistic mixed model/score) set	http://www.ncbi.nlm.nih.gov/pubmed/22522135
		http://www.bioinformatics.org/~stanhope/GLOGS/
	GWAPP	http://www.ncbi.nlm.nih.gov/pubmed/23277364
		http://gwapp.gmi.oeaw.ac.at/index.html#!homePage
	GWASpi	http://www.ncbi.nlm.nih.gov/pubmed/21586520
		http://www.gwaspi.org/

(continued)

Table 9.1 (continued)

Function	Tool	PubMed reference/Tool url
	Matapax	http://www.ncbi.nlm.nih.gov/pubmed/22353578
		http://matapax.mpimp-golm.mpg.de/
	Software engineering the mixed model for genome-wide association studies on large samples	http://www.ncbi.nlm.nih.gov/pubmed/19933212
		http://www.ncbi.nlm.nih.gov/pubmed/19933212
	TASSEL	http://www.ncbi.nlm.nih.gov/pubmed/17586829
		http://www.maizegenetics.net/#!tassel/c17q9
Complex trait prediction	The success of genome-wide association studies (GWASs) has led to increasing interest in making predictions of complex trait phenotypes, including disease, from genotype data. Rigorous assessment of the value of predictors is crucial before implementation	
	ATHENA (analysis tool for heritable and environmental network association)	http://www.ncbi.nlm.nih.gov/pubmed/24149050
		http://ritchielab.psu.edu/software/athena-downloads
	GCTA (genome-wide complex trait analysis)	http://www.ncbi.nlm.nih.gov/pubmed/21167468
		http://www.complextraitgenomics.com/software/gcta/
	GVCBLUP	http://www.ncbi.nlm.nih.gov/pubmed/25107495
		http://animalgene.umn.edu/gvcblup_win/index.html
	MultiBLUP	http://www.ncbi.nlm.nih.gov/pubmed/24963154
		http://dougspeed.com/multiblup/
Regulatory SNP prediction	Genome-wide association studies revealed that most disease-associated single nucleotide polymorphisms (SNPs) are located in regulatory regions within introns or in regions between genes. Regulatory SNPs (rSNPs) are such SNPs that affect gene regulation by changing transcription factor (TF) binding affinities to genomic sequences. Identifying potential rSNPs is crucial for understanding disease mechanisms	
	Is-Rsnp (In silico regulatory SNP detection)	http://www.ncbi.nlm.nih.gov/pubmed/20823317
		http://bioinformatics.research.nicta.com.au/software/is-rsnp/
	AtSNP (affinity testing for regulatory SNPs)	http://www.ncbi.nlm.nih.gov/pubmed/26092860
		https://github.com/chandlerzuo/atsnp
	Bayes PI-BAR (Bayesian method for protein–DNA interaction with binding affinity ranking)	http://doi.org/10.1093/nar/gkv733
		http://folk.uio.no/junbaiw/BayesPI-BAR/

(continued)

Table 9.1 (continued)

Function	Tool	PubMed reference/Tool url
SNP trait association database	Genome-wide association studies (GWASs) have identified numerous single nucleotide polymorphisms (SNPs) that are associated with development of multifactorial diseases, such as coronary artery disease, rheumatoid arthritis, type 2 diabetes mellitus, and cancers	
	DbGaP (database of genotypes and phenotypes)	http://www.ncbi.nlm.nih.gov/pubmed/24297256
		http://www.ncbi.nlm.nih.gov/gap
	GRASP (Genome-Wide Repository of Associations between SNPs and Phenotypes)	http://www.ncbi.nlm.nih.gov/pubmed/25428361
		http://apps.nhlbi.nih.gov/Grasp/Overview.aspx
	GWAS central	http://www.ncbi.nlm.nih.gov/pubmed/24301061
		http://www.gwascentral.org/
	HuGE Navigator	http://www.ncbi.nlm.nih.gov/pubmed/18227866
		http://www.hugenavigator.net/HuGENavigator/home.do
	LincSNP	http://www.ncbi.nlm.nih.gov/pubmed/24885522
		http://210.46.85.180:8080/LincSNP/
	NHGRI GWAS catalog	http://www.ncbi.nlm.nih.gov/pubmed/24316577
		http://www.genome.gov/gwastudies/
Rare variant association analysis	Although genome-wide association studies have been successful in detecting associations with common variants, there is currently an increasing interest in identifying low frequency and rare variants associated with complex traits	
	EPACTS (Efficient and Parallelizable Association Container Toolbox)	http://genome.sph.umich.edu/wiki/EPACTS
	FamFLM	http://www.ncbi.nlm.nih.gov/pubmed/26111046
		http://mga.bionet.nsc.ru/soft/famFLM/
	FamLBL (family-triad-based logistic Bayesian Lasso)	http://www.ncbi.nlm.nih.gov/pubmed/24849576
		http://www.stat.osu.edu/~statgen/SOFTWARE/LBL/
	FARVAT (family-based rare variant association test)	http://www.ncbi.nlm.nih.gov/pubmed/25075118
		http://healthstat.snu.ac.kr/software/farvat/
	SCORE-Seq	http://www.ncbi.nlm.nih.gov/pubmed/21885029
		http://dlin.web.unc.edu/software/SCORE-Seq/

(continued)

Table 9.1 (continued)

Function	Tool	PubMed reference/Tool url
	VAT (variant association tools)	http://www.ncbi.nlm.nih.gov/pubmed/24791902
		http://varianttools.sourceforge.net/Association/HomePage
Linkage disequilibrium software tools	Assessing linkage disequilibrium (LD) across ancestral populations is a powerful approach for investigating population specific genetic structure as well as functionally mapping regions of disease susceptibility	
	Haploview	http://www.ncbi.nlm.nih.gov/pubmed/15297300
		http://www.broadinstitute.org/scientific-community/science/programs/medical-and-population-genetics/haploview/haploview
	SNAP (SNP annotation and proxy search)	http://www.broadinstitute.org/mpg/snap/
Pathway analysis software tools	Genome-wide association (GWA) studies have typically focused on the analysis of single markers, which often lacks the power to uncover the relatively small effect sizes conferred by most genetic variants. Recently, pathway-based approaches have been developed, which use prior biological knowledge on gene function to facilitate more powerful analysis of GWA study data sets. These approaches typically examine whether a group of related genes in the same functional pathway are jointly associated with a trait of interest	
	PLINK	http://www.ncbi.nlm.nih.gov/pubmed/17701901
		http://pngu.mgh.harvard.edu/~purcell/plink/
	ALIGATOR	http://www.ncbi.nlm.nih.gov/pubmed/19539887
		http://x004.psycm.uwcm.ac.uk/~peter/
	EW_dmGWAS	http://www.ncbi.nlm.nih.gov/pubmed/25805723
		http://bioinfo.mc.vanderbilt.edu/dmGWAS/
	GenGen	http://www.ncbi.nlm.nih.gov/pubmed/17966091
		http://www.openbioinformatics.org/gengen/
	GESBAP	http://www.ncbi.nlm.nih.gov/pubmed/19502494
Unclassified GWAS tools		
	Birdsuite	http://www.ncbi.nlm.nih.gov/pubmed/18776909
		http://www.broadinstitute.org/scientific-community/science/programs/

(continued)

Table 9.1 (continued)

Function	Tool	PubMed reference/Tool url
		medical-and-population-genetics/birdsuite/birdsuite
	Genome track analyzer	http://www.ncbi.nlm.nih.gov/pubmed/25627242
		http://ancorr.eimb.ru/
	PrediXcan	http://dx.doi.org/10.1101/020164
		https://github.com/hakyimlab/PrediXcan

A known potential source of error for any GWAS is sample handling within the laboratory. A number of sample tracking and evaluation steps can be put in place to reduce the potential for sample mishandling or mislabeling. The National Cancer Institute's Office of Biorepositories and Biospecimen Research (http://biospecimens.cancer.gov/) is a useful resource for best practices and policies for biospecimen storage and tracking and has also developed a suite of informatics tools available through the cancer Biomedical Informatics Grid (caBIG). A most effective way of sample tracking and monitoring could be through the generation of a "mini-fingerprint" of highly polymorphic genetic variants (SNPs or microsatellites) on all incoming samples which can serve as a more specific sample reference.

The Illumina Infinium assay (Illumina Inc., San Diego, CA) (Steemers et al. 2006) and other genotyping platforms are available for GWAS for genotyping. The most commonly used alternative is the Affymetrix platform (Affymetrix Inc., Santa Clara, CA) and Illumina BeadChip.

The main factors for genotyping success are DNA quality and quantity. The 260/280 nm ratio, although a good measure of nucleic acid contamination of protein, is a poor measure of DNA contamination by protein. DNA of 400 ng for "Duo" products that process two samples per BeadChip, or 200 ng for "Quads" (four DNAs per BeadChip), has been shown by earlier workers to be efficient (Sale et al. 2009).

Illumina's Infinium assay (Steemers et al. 2006) is capable of multiplexing approximately 6000 to 1 million SNPs/CNVs, either using fixed content products for GWAS or customizable focused-content products (termed iSelect). At present, fixed content products for GWAS in humans range from approximately 370,000 to over 1 million markers per sample. In brief, Illumina's Infinium assay (Steemers et al. 2006) consists of four modular components: (a) a single-tube whole-genome amplification step, (b) an array-based hybridization capture step, (c) an "on array" enzymatic single-base extension (SBE) step, and (d) an amplified-signal detection step. SBE uses a single 50 bp probe designed to hybridize adjacent to the SNP query site. After hybridization of target DNA to the BeadChip (a microelectromechanical systems (MEMS)-patterned substrate on silica slides), the SNP locus-specific primers, attached to 3-micron silica beads, are extended in the presence of hapten-labeled dideoxynucleotides. Biotin-labeled ddCTP, 2,4-dinitrophenol (DNP)-labeled ddATP and ddUTP are efficiently incorporated planning and executing a genome-wide association study 409 by polymerases and allow detection with a dual-color, orthogonal, multilayer immunohistochemical sandwich assay. Biotin and DNP are simultaneously detected by staining with a combination of Alexa555-labeled streptavidin (SA) and Alexa647-labeled rabbit primary antibody against DNP, counterstaining with biotinylated anti-SA and DNP-labeled goat anti-rabbit secondary antibody (Sale et al. 2009).

Rigorous quality control is a crucial component of any GWAS since subtle biases in raw data can lead to hundreds or thousands of false positive results, confounding efforts to validate lead SNPs at the replication stage (Sale et al. 2009). Quality control steps to reject SNPs or samples are necessarily a trade-off between stringency to prevent type 1 error against loss of data, reducing power. The thresholds used in the individual steps reflect common values that are currently in use but can be modified to be more or less tolerant of type 1 error. This decision will depend on study design, availability, and size of replication study samples and willingness to include downstream manual steps to review cluster patterns of many SNP loci that appear to show significant association (Skol et al. 2006).

Although a variety of approaches can be used to analyze a GWAS, some widely used applications, as well as a method uniquely capable of handling multivariate data, are available as open source and also as standalone software (Table 9.1). For example, PLINK, a GWAS analysis software (Skol et al. 2006), has been developed specifically for the analysis of GWAS data for single SNP analyses in case–control data sets. Also, a multivariate trait GWA algorithm has been implemented the software package Ghost *(*people.virginia.edu/wc9c/ghost/*)*. This implementation can help systematically identify genetic variants that are responsible for multiple traits. More elaborate reviews on GWAS planning and execution could be seen (Zondervan and Cardon 2007; Sale et al. 2009; Distefano and Taverna 2011).

9.5 GWAS for Bacteria

In the recent past, the application of GWAS approach explored the horizon to study the bacterial host preference, antibiotic resistance, and virulence (Chen and Shapiro 2015). The study also threw light on the bacterial and human genome dynamics. GWASs in bacteria had boomed in the recent years and provide a genetic basis for bacterial phenotypes. Hence, any measurable bacterial phenotype can be dissected with this approach. An extension of such conventional GWASs has now shown to explore the environmentally and industrially relevant phenotypes. The two primary requirements for GWAS are genotypic and phenotypic measurements from a sample of organisms. Such phenotypes should be measurable that can be related to genotypic measurement through high-throughput screening. At the genotypic level, the bacterial genomes can be broken to core and accessory genome that are composed of elements that are more commonly present in the strains and the genes individually involved in a given trait such as environmental adaptation (Lapierre and Gogarten 2009; Vernikos et al. 2015). Hence, this may be a genetic variant of the core by either presence or absence of a polymorphism such as SNP or a variation on a large piece of genome including gene clusters or operons. Since many GWASs to date have either used SNPs or the presence of gene or its absence as the basic criterion, recent methods using DNA–"mer" counts as the basic unit of association to study the core and flexible genome are emerging (Sheppard et al. 2013). Since bacterial genome diversity is influenced by population stratification which is nothing but the close relationship of related subgroups than the wider population (Balding 2006), studies to explore the impact of clonal frames and population stratification using GWAS on mycobacterium tuberculosis genomes compared to phylogenetic convergence studies have shown *M. tuberculosis* possess an extensive linage disequilibrium and strong population structure making them challenging subjects for traditional GWAS (Farhat et al. 2013). Bacteria such as *Streptococcus pneumoniae* that have an extensive recombining efficiency have less long-range linkage disequilibrium (LD) and are more localized blocks that facilitate GWAS (Chewapreecha et al. 2014). Thus, the important first step before performing a bacterial GWAS is to characterize the linkage disequilibrium. Hence the major key obstacles of bacterial genome GWAS being the long-range LD within the clonal frame and extensive bacterial population stratification which reduces the ability to zero in on the causal mutations with confidence. However, the relative strength of positive selection provides an opportunity for increased resolution in bacterial GWAS hits. Hence, the success of bacterial genome GWASs focuses on performing the genome-wide selection scanning of specific genomes that are putatively under positive selection and, also, performing a targeted association study on these positive selection genomic regions.

To date, there are no genome-wide studies that attempt to characterize specific genes and pathways in the human genome that shape the composition of the microbiome. Human Microbiome Studies in relation to human host have also shown a clear evidence for the influence of environmental factors that support host genetic components in structuring of these human microbial communities (Spor et al. 2011). SNPs in the *MEFV* genes are associated with changes in gut bacterial communities (Khachatryan et al. 2008). Furthermore, irritable bowel syndrome (IBS) risk loci are associated with changes in the gut microbiome composition (Li et al. 2012). Many new findings such as the loss of function polymorphism and its relation to Crohn's disease through gene *FUT2* and NOD2 risk allele count correlation with increase in the relative abundance of *Enterobacteriaceae* (Knights et al. 2014) were some of the findings through extensive GWASs.

In addition to candidate gene approaches, researchers have also used host genome-wide genetic variation to find important interactions with the human microbiome. In a study using 416 twin pairs to assess the heritability of the microbiome, microbial taxa for which relative abundance is more similar in monozygotic compared to dizygotic twins are identified (Goodrich et al. 2014). In the laboratory mouse, quantitative trait locus (QTL) mapping approaches have found multiple loci associated with gut microbial community composition, some of which overlap genes involved in immune response (Benson et al. 2010). Furthermore, it is shown that host mitochondrial DNA haplogroups are correlated with the structure of microbiome communities (Ma et al. 2014). Genome-wide analysis study to identify human genes and correlate pathways with microbiome composition was carried out using data generated by the Human Microbiome Project (HMP) as HMP has sampled and cataloged the microbial diversity across multiple body sites in a few hundred individuals (Human Microbiome Project Consortium 2012). A more elaborate review can be found by Chen et al. (Chen and Shapiro 2015).

9.6 Phenome-Wide Association Studies (PheWASs)

Over the past decade, GWASs have been used to identify thousands of statistically significant variants that are associated with many human conditions including the more complex immunological etiologies such as rheumatoid arthritis, multiple sclerosis, Alzheimer's disease, etc. However, unfortunately most of the GWASs fail to identify the clinical significance of association. Hence, identifying such biologically significant variants always poses a challenge through GWAS. As a complementary alternative approach to GWAS, many studies have begun to exploit the genotype-to-phenotype approach through the phenome-wide association studies (PheWASs) (Fig. 9.1b). With its fast improvements of this technique, it has already demonstrated its capacity to rediscover many human diseases and their relations. Furthermore, PHeWAS has been shown to be capable of exploring the genetic variants with pleotropic properties. With its first publication (Denny et al. 2010), this study associated only five genetic targets to curate the phenome that was later refined with clinical expertise. The international classification of disease version 9 (ICD9) spectrum was used to define the phenome in PheWAS (Denny et al. 2010).

In another study, curated phenome using the Electronic Medical Records and Genomics (eMERGE) network (McCarty et al. 2011). In this study, GWAS was used to inform PheWAS within the same cohort. One of the significant findings was the common SNP near FOXE1 (rs965513) associated with risk of hypothyroidism. ICD9 codes that define hypothyroidism were significantly associated with this rs965513 genotype by PheWAS (Denny et al. 2011). Recently, this concept of GWAS-informed PheWAS approach has also been applied to the study of platelet phenotypes. Using a similar eMERGE population as described above, this study identified 81 GWAS-significant SNPs including 56 SNPs associated with platelet count, 29 SNPs associated with platelet volume, and four SNPs associated with both. Many of these SNPs validate previously published GWAS results. Each of the 81 SNPs was then individually associated with the phenome. Similar to GWAS PheWAS is a hypothesis-generating approach that is challenged by multiple comparison testing (Hebbring 2014). While there are inherent limitation to PheWAS, like GWAS, differences across populations may affect the ability to validate findings. At the SNP level, it is more likely to see very different GWAS results.

For example, from a population with European ancestry compared with a population with African ancestry could result due to significant differences in the linkage disequilibrium structure and allele frequencies between the two populations. In a genetically driven PheWAS, there is often one SNP associated across the phenome. If the SNP genotyped is not the functional variant, and/or observed in multiple populations, replicating PheWAS results could be difficult.

On the contrary the advantages of PheWASs are more, and the selection of a phenotype for GWAS is important for the success of any GWAS. So far, the PheWASs focusing on genetic targets have concentrated on SNPs that were already identified by GWAS (Denny et al. 2010; Shameer et al. 2014). Even with the complex challenges described above, PheWAS has demonstrated its capacity to identify expected associations when going in the opposite direction compared with GWAS. More exemplified advantages and limitations of PheWAS can be found by Hebbings et al. (Hebbring 2014).

9.7 Bioinformatics Tools in GWAS Analysis

The genetic complexity of studying the genetic variants through GWAS underpinning the correlation for most of the common diseases remains largely unexplained. Traditional GWAS focuses on one single nucleotide polymorphism at a time and has failed to account for the complexity of many genotype–phenotype relationships that are very heterogeneous due to gene–gene and gene–environment interactions. Bioinformatics tools are necessary to uncover nonlinear genetic predictors of these common diseases. These data mining and machine learning methods increase the power of discovering genetic predictors of common diseases. Furthermore, filter and wrapper algorithms are necessary to limit the number of examined attributes making the analysis more powerful, and hence computations can become practical. However, prior biological knowledge can improve the analysis and interpretation of GWAS data. Many powerful and intuitive software packages exist (Table 9.1 provides some of the commonly used software) for effective interrogation and analysis of the GWAS data through crunching the complex data sets to more useful information.

With more variable approaches to methodology, PheWAS is always limited by how well the phenome can be defined. Efforts to reliably define phenotypes using electronic medical record (EMR) data have been limited to specific phenotypes (McCarty et al. 2011). Although these disease-specific methods can discriminate between cases and controls, they often do not necessarily provide a high-throughput mechanism to define the thousands of phenotypes within a phenome such as the genome. Automated medical informatics tools capable of reliably defining the phenotypes within a phenome will be always required to refine the PheWAS potential.

9.8 Future Directions

A catalog of published GWAS can be seen at the National Human Genome Research Institute Web site (*http://www.genome.gov/gwastudies/*).

As the sequencing technology becomes cheaper, there is an explosion of targeted gene sequencing studies looking for rarer risk. Recent advances such as the second generation of GWAS performed using new chips targeting variants throughout the genome at ever-lower frequencies are becoming popular. This could lead to a complete blend of whole-genome sequencing of hundreds to thousands of disease patients and controls.

GWAS and its recent variants along with PheWAS will have a harvest of rare disease-associated variants with much stronger effects on risk than the common variants. This will lead to new insights into disease pathways and more importantly predicts individual risk. Hence, the emergence of these variants will make personal genomics vastly more useful for health predictions.

The future of the study lies on the fact that this will be an invaluable tool to study the role on the effect of one gene over other modifier genes in a combined genetic background that arise due to interactions between or within them. Hence, understanding the roles of "epistasis" (gene–gene interactions) involves studying both "functional epistasis," which will address the molecular interactions that proteins (and other genetic elements) have with one another and will convey whether they operate within the same pathway or consist of proteins who directly complex with one another, and "compositional epistasis" intended to study the blocking of one allelic effect by an allele at another locus (Boone et al. 2007). Furthermore, GWAS and their variants will be used to study the gene–environment interactions, copy number variants, and epigenetic phenomena which are anticipated to provide additional insights into our understanding of complex human disorders through genome-wide studies taking into consideration other variables such as environment and epigenetic modifications.

References

Abecasis GR, Auton A, Brooks LD, DePristo MA, Durbin RM, Handsaker RE, Kang HM, Marth GT, McVean GA. An integrated map of genetic variation from 1,092 human genomes. Nature. 2012;491:56–65.

Amos CI. Successful design and conduct of genome-wide association studies. Hum Mol Genet. 2007;16 (2):R220–5. Epub;%2007 Jun 27.: R220–5.

Balding DJ. A tutorial on statistical methods for population association studies. Nat. Rev. Genet. 2006;7:781–91.

Benson AK, Kelly SA, Legge R, Ma F, Low SJ, Kim J, Zhang M, Oh PL, Nehrenberg D, Hua K, Kachman SD, Moriyama EN, Walter J, Peterson DA, Pomp D. Individuality in gut microbiota composition is a complex polygenic trait shaped by multiple environmental and host genetic factors. Proc Natl Acad Sci U S A. 2010;107:18933–8.

Boone C, Bussey H, Andrews BJ. Exploring genetic interactions and networks with yeast. Nat. Rev. Genet. 2007;8:437–49.

Chen WM, Abecasis GR. Family-based association tests for genome wide association scans. Am J Hum Genet. 2007;81:913–26.

Chen PE, Shapiro BJ. The advent of genome-wide association studies for bacteria. Curr Opin Microbiol. 2015;25:17–24. doi:10.1016/j.mib.2015.03.002. . Epub;%2015 Mar 31.

Chewapreecha C, Marttinen P, Croucher NJ, Salter SJ, Harris SR, Mather AE, Hanage WP, Goldblatt D, Nosten FH, Turner C, Turner P, Bentley SD, Parkhill J. Comprehensive identification of single nucleotide polymorphisms associated with beta-lactam resistance within pneumococcal mosaic genes. PLoS Genet. 2014;10:e1004547.

Cooper JD, Smyth DJ, Smiles AM, Plagnol V, Walker NM, Allen JE, Downes K, Barrett JC, Healy BC, Mychaleckyj JC, Warram JH, Todd JA. Meta-analysis of genome-wide association study data identifies additional type 1 diabetes risk loci. Nat Genet. 2008;40:1399–401.

Denny JC, Ritchie MD, Basford MA, Pulley JM, Bastarache L, Brown-Gentry K, Wang D, Masys DR, Roden DM, Crawford DC. PheWAS: demonstrating the feasibility of a phenome-wide scan to discover gene-disease associations. Bioinformatics. 2010;26:1205–10.

Denny JC, Crawford DC, Ritchie MD, Bielinski SJ, Basford MA, Bradford Y, Chai HS, Bastarache L, Zuvich R, Peissig P, Carrell D, Ramirez AH, Pathak J, Wilke RA, Rasmussen L, Wang X, Pacheco JA, Kho AN, Hayes MG, Weston N, Matsumoto M, Kopp PA, Newton KM, Jarvik GP, Li R, Manolio TA, Kullo IJ, Chute CG, Chisholm RL, Larson EB, McCarty CA, Masys DR, Roden DM, de AM. Variants near FOXE1 are associated with hypothyroidism and other thyroid conditions: using electronic medical records for genome- and phenome-wide studies. Am J Hum Genet. 2011;89:529–42.

Distefano JK, Taverna DM. Technological issues and experimental design of gene association studies. Methods Mol Biol. 2011;700:3–16. doi:10.1007/978-1-61737-954-3_1.

Farhat MR, Shapiro BJ, Kieser KJ, Sultana R, Jacobson KR, Victor TC, Warren RM, Streicher EM, Calver A, Sloutsky A, Kaur D, Posey JE, Plikaytis B, Oggioni MR, Gardy JL, Johnston JC, Rodrigues M, Tang PK, Kato-Maeda M, Borowsky ML, Muddukrishna B, Kreiswirth BN, Kurepina N, Galagan J, Gagneux S, Birren B, Rubin EJ, Lander ES, Sabeti PC, Murray M. Genomic analysis identifies targets of convergent positive selection in drug-resistant Mycobacterium tuberculosis. Nat Genet. 2013;45:1183–9.

Frazer KA, Murray SS, Schork NJ, Topol EJ. Human genetic variation and its contribution to complex traits. Nat. Rev. Genet. 2009;10:241–51.

Goodrich JK, Waters JL, Poole AC, Sutter JL, Koren O, Blekhman R, Beaumont M, Van TW, Knight R, Bell JT, Spector TD, Clark AG, Ley RE. Human genetics shape the gut microbiome. Cell. 2014;159:789–99.

Hebbring SJ. The challenges, advantages and future of phenome-wide association studies. Immunology. 2014;141:157–65.

Hebbring SJ, Schrodi SJ, Ye Z, Zhou Z, Page D, Brilliant MH. A PheWAS approach in studying HLA-DRB1*1501. Genes Immun. 2013;14:187–91.

Hindorff LA, MacArthur J, Morales J, Junkins HA, Hall PN, Klemm AK, Manolio TA. A catalog of published genome-wide association studies. 2013.

Human Microbiome Project Consortium. A framework for human microbiome research. Nature. 2012;486:215–21.

Khachatryan ZA, Ktsoyan ZA, Manukyan GP, Kelly D, Ghazaryan KA, Aminov RI. Predominant role of host genetics in controlling the composition of gut microbiota. PLoS One. 2008;3: e3064.

Klein RJ. Power analysis for genome-wide association studies. BMC Genet. 2007;8:58.

Knights D, Silverberg MS, Weersma RK, Gevers D, Dijkstra G, Huang H, Tyler AD, van SS, Imhann F, Stempak JM, Huang H, Vangay P, Al-Ghalith GA, Russell C, Sauk J, Knight J, Daly MJ, Huttenhower C, Xavier RJ. Complex host genetics influence the microbiome in inflammatory bowel disease. Genome Med. 2014;6:107–0107.

Lapierre P, Gogarten JP. Estimating the size of the bacterial pan-genome. Trends Genet. 2009;25:107–10.

Li E, Hamm CM, Gulati AS, Sartor RB, Chen H, Wu X, Zhang T, Rohlf FJ, Zhu W, Gu C, Robertson CE, Pace NR, Boedeker EC, Harpaz N, Yuan J, Weinstock GM, Sodergren E, Frank DN. Inflammatory bowel diseases phenotype, *C. difficile* and NOD2 genotype are associated with shifts in human ileum associated microbial composition. PLoS One. 2012;7:e26284.

Ma J, Coarfa C, Qin X, Bonnen PE, Milosavljevic A, Versalovic J, Aagaard K. mtDNA haplogroup and single nucleotide polymorphisms structure human microbiome communities. BMC Genomics. 2014;15:257–15. doi:10.1186/1471-2164-15-257.

Martin ER, Monks SA, Warren LL, Kaplan NL. A test for linkage and association in general pedigrees: the pedigree disequilibrium test. Am J Hum Genet. 2000;67:146–54.

McCarthy MI, Abecasis GR, Cardon LR, Goldstein DB, Little J, Ioannidis JP, Hirschhorn JN. - Genome-wide association studies for complex traits: consensus, uncertainty and challenges. Nat Rev Genet. 2008;9:356–69.

McCarty CA, Chisholm RL, Chute CG, Kullo IJ, Jarvik GP, Larson EB, Li R, Masys DR, Ritchie MD, Roden DM, Struewing JP, Wolf WA. The eMERGE Network: a consortium of biorepositories linked to electronic medical records data for conducting genomic studies. BMC Med Genet. 2011;4:13. doi:10.1186/1755-8794-4-13.

Misteli T. The concept of self-organization in cellular architecture. J Cell Biol. 2001;155:181–5.

Mohlke KL, Boehnke M, Abecasis GR. Metabolic and cardiovascular traits: an abundance of recently identified common genetic variants. Hum Mol Genet. 2008;17:R102–8.

Mukherjee S, Simon J, Bayuga S, Ludwig E, Yoo S, Orlow I, Viale A, Offit K, Kurtz RC, Olson SH, Klein RJ. Including additional controls from public databases improves the power of a genome-wide association study. Hum Hered. 2011;72:21–34.

Nguyen TT, Pahl R, Schafer H. Optimal robust two-stage designs for genome-wide association studies. Ann Hum Genet. 2009;73:638–51.

Sale MM, Mychaleckyj JC, Chen WM. Planning and executing a genome wide association study (GWAS). Methods Mol Biol. 2009;590:403–18. doi:10.1007/978-1-60327-378-7_25.

Schierding W, Cutfield WS, O'Sullivan JM. The missing story behind Genome Wide Association Studies: single nucleotide polymorphisms in gene deserts have a story to tell. Front Genet. 2014;5:39. doi:10.3389/fgene.2014.00039. eCollection;%2014.

Shameer K, Denny JC, Ding K, Jouni H, Crosslin DR, de AM, Chute CG, Peissig P, Pacheco JA, Li R, Bastarache L, Kho AN, Ritchie MD, Masys DR, Chisholm RL, Larson EB, McCarty CA, Roden DM, Jarvik GP, Kullo IJ. A genome- and phenome-wide association study to identify genetic variants influencing platelet count and volume and their pleiotropic effects. Hum Genet. 2014;133:95–109.

Sheppard SK, Didelot X, Meric G, Torralbo A, Jolley KA, Kelly DJ, Bentley SD, Maiden MC, Parkhill J, Falush D. Genome-wide association study identifies vitamin B5 biosynthesis as a host specificity factor in Campylobacter. Proc Natl Acad Sci U S A. 2013;110:11923–7.

Sherry ST, Ward MH, Kholodov M, Baker J, Phan L, Smigielski EM, Sirotkin K. dbSNP: the NCBI database of genetic variation. Nucleic Acids Res. 2001;29:308–11.

Skol AD, Scott LJ, Abecasis GR, Boehnke M. Joint analysis is more efficient than replication-based analysis for two-stage genome-wide association studies. Nat Genet. 2006;38:209–13.

Spor A, Koren O, Ley R. Unravelling the effects of the environment and host genotype on the gut microbiome. Nat Rev Microbiol. 2011;9:279–90.

Steemers FJ, Chang W, Lee G, Barker DL, Shen R, Gunderson KL. Whole-genome genotyping with the single-base extension assay. Nat Methods. 2006;3:31–3.

Stenson PD, Mort M, Ball EV, Shaw K, Phillips A, Cooper DN. The Human Gene Mutation Database: building a comprehensive mutation repository for clinical and molecular genetics, diagnostic testing and personalized genomic medicine. Hum Genet. 2014;133:1–9.

Tryka KA, Hao L, Sturcke A, Jin Y, Wang ZY, Ziyabari L, Lee M, Popova N, Sharopova N, Kimura M, Feolo M. NCBI's database of genotypes and phenotypes: dbGaP. Nucleic Acids Res. 2014;42:D975–9.

Vernikos G, Medini D, Riley DR, Tettelin H. Ten years of pan-genome analyses. Curr Opin Microbiol. 2015;23:148–54. doi: 10.1016/j.mib.2014.11.016. Epub;%2014 Dec 5.

Wang MC, Chen FC, Chen YZ, Huang YT, Chuang TJ. LDGIdb: a database of gene interactions inferred from long-range strong linkage disequilibrium between pairs of SNPs. BMC Res Notes. 2012;5:212–5. doi:10.1186/1756-0500-5-212.

Wellcome Trust Case Control Consortium. Genome-wide association study of 14,000 cases of seven common diseases and 3,000 shared controls. Nature. 2007;447:661–8.

Welter D, MacArthur J, Morales J, Burdett T, Hall P, Junkins H, Klemm A, Flicek P, Manolio T, Hindorff L, Parkinson H. The NHGRI GWAS Catalog, a curated resource of SNP-trait associations. Nucleic Acids Res. 2014;42:D1001–6.

Zondervan KT, Cardon LR. Designing candidate gene and genome-wide case-control association studies. Nat Protoc. 2007;2:2492–501.

Chapter 10
A Survey of Bioinformatics-Based Tools in RNA-Sequencing (RNA-Seq) Data Analysis

Pallavi Gaur and Anoop Chaturvedi

Abstract The capability of next-generation sequencing can be understood by one of its techniques like RNA sequencing (RNA-Seq) that deals with the transcriptome complexity in a powerful and cost-effective way. This technique has emerged as a revolutionary tool with high sensitivity and accuracy over old techniques. Additionally, this technique also gives unprecedented ability to detect novel mRNA transcripts as well as ncRNAs and analyze alternative splicing. Being a high-throughput sequencing technique, it poses a great demand for bioinformatics-based analysis of the generated data. Here, we explain how RNA-Seq data can be analyzed, discuss its challenges, and provide an overview of the data analysis methods/tools. We discuss strategies for quality check, mapping, and differential expression in transcriptomic data along with discussions on lately developed strategies for alternative splicing and isoform quantification. We also mention some useful R/Bioconductor packages for aforementioned tasks.

Keywords RNA-Seq • Mapping • Differential expression • Bioconductor • Galaxy

10.1 Introduction

RNA-Seq is one of the most advanced techniques which use the platform of high-throughput sequencing (HTS) also called the next-generation sequencing (NGS) technologies to decipher the transcriptome. Transcriptome comprises the complete set of transcripts in a tissue, organism, or a specific cell for a given physiological

P. Gaur (✉)
Center of Bioinformatics, Institute of Inter Disciplinary Studies, Nehru Science Center, University of Allahabad, Allahabad 211002, Uttar Pradesh, India
e-mail: palbioinfor@gmail.com

A. Chaturvedi
Department of Statistics, Nehru Science Center, University of Allahabad, Allahabad 211002, Uttar Pradesh, India
e-mail: anoopchaturv@gmail.com

© Shanghai Jiao Tong University Press, Shanghai and Springer Science+Business Media Dordrecht 2017

D.-Q. Wei et al. (eds.), *Translational Bioinformatics and Its Application*, Translational Medicine Research, DOI 10.1007/978-94-024-1045-7_10

condition. Transcripts include protein-coding messenger RNA (mRNA) and non-coding RNA like ribosomal RNA (rRNA), transfer RNA (tRNA), and other ncRNAs (Lindberg and Lundeberg 2010; Okazaki et al. 2002). RNA-Seq basically helps us in looking at the regions of genome being transcribed in a sample and quantifying the expression of such transcripts. Transcriptome has the tendency to vary with different physiological conditions that make transcriptomics a significant field of study, thus turning RNA-Seq a powerful tool for dissecting and understanding many biological phenomena like underlying mechanism and pathways controlling disease initiation, development, and progression.

Over the years, several technologies have come to the existence to study transcriptome, but lately developed RNA-Seq has the ability to characterize the transcriptome in a more global and relatively better way than microarrays and other traditional strategies. RNA-Seq uses cDNA sequencing, from RNA sample of interest (Wilhelm et al. 2008). Basically, RNA-Seq starts by library construction, followed by sequencing on a specific NGS platform and subsequent bioinformatic analysis. In a nutshell, library construction requires isolation of RNA which is randomly fragmented into smaller pieces, followed by reverse transcription. Reverse transcription converts RNA fragments into cDNA with ligation of adapter sequences to either one or both ends for amplification. Fragmentation of RNA can be done prior to reverse transcription, or reverse transcription can be done first followed by cDNA fragmentation (Roberts et al. 2011; Wang et al. 2009). This choice plays an important role because it mostly causes a bias in final results. Especially, cDNA fragmentation generates an under-representation of the 5′ of the transcripts, while RNA fragmentation allows a better representation of the transcript body although somehow may end up in delivering depleted transcript end (Mortazavi et al. 2008). Basic steps and strategy executed by RNA-sequencing experiment are almost the same for every platform which is shown in Fig. 10.1.

Fragment size selection and priming the sequence reaction along with the above steps can vary with the implementation of the protocol and introduce some technical biases in the resulting data. The final sequencing step relies on the NGS platform like 454 pyrosequencing system (a subsidiary of Roche), the AB SOLiD system (Life Technologies), and the Illumina Genome Analyzer (Illumina) (Liu et al. 2012; Marguerat and Bahler 2010; Ansorge 2009), each having its own library construction method. Both the 454 and the SOLiD systems employ an innovative emulsion polymerase chain reaction (emulsion PCR) method for clonal amplification. In emulsion PCR, the cDNA fragments from a library are attached to beads followed by compartmentalization in the aqueous droplets called water-in-oil emulsion. This way, each droplet contains a single DNA molecule as well as the segregated template fragments. These fragments are then amplified in very small emulsified aqueous droplets (Dressman et al. 2003).

The Illumina Genome Analyzer (GA) utilizes the strategy of "bridge PCR" amplification where the adapter-linked single-stranded fragments of cDNA are immobilized on a glass slide by oligonucleotide hybridization in a bridging way, followed by clonal PCR amplification (Fedurco et al. 2006). A population of identical templates is resulted from clonal amplification, but it may introduce a

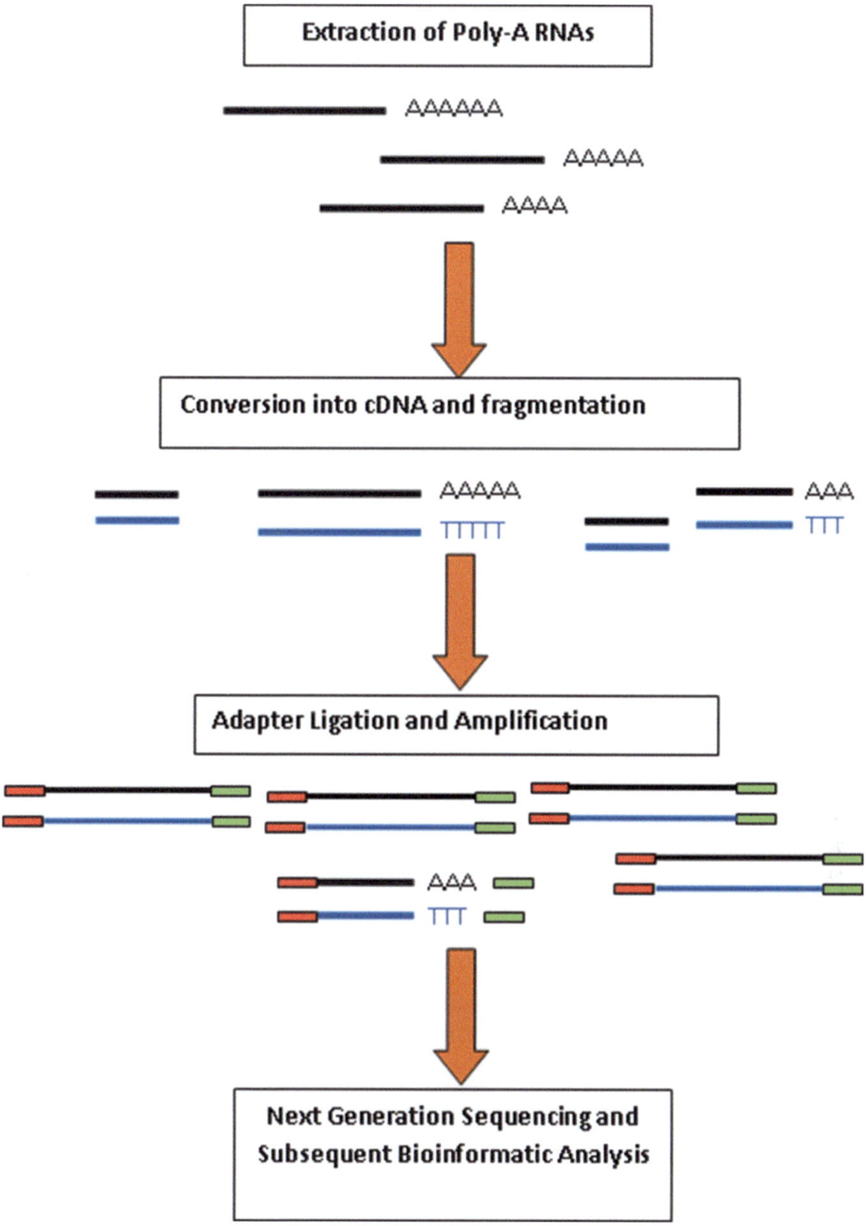

Fig. 10.1 A basic layout of RNA-sequencing experiment

bias in the RNA-Seq result due to PCR artifacts. That is why performances on different biological replicates are needed to determine whether the same short reads are present in different replicates (Wang et al. 2009). Different NGS platforms use different sequencing strategies (Metzker 2009), and several reviews can be found describing details including mechanisms and comparisons of these NGS technologies (Liu et al. 2012a; Metzker 2009; Shendure and Ji 2008; Ansorge 2009). Sequencing can produce single-end or paired-end reads. In paired-end sequencing, a fragment is sequenced from both ends, while in single-end sequencing, only one end is used. Having the advantage of sequencing from both ends, paired-end sequencing generates data of comparatively high quality.

Since the advent of RNA-Seq in 2008, it has emerged as a superior technique to study transcriptome over traditional methods which were either hybridization (microarray) or sequence based (SAGE, CAGE). Being superior in resolution at the single-base level, this technique can effectively measure the expression level of thousands of genes simultaneously in addition to information on alternative splicing, unannotated exons, allele-specific expression (Heap et al. 2010), microRNAs, variants like SNPs (Quinn et al. 2013), and novel transcripts (gene or noncoding RNAs). Additionally, many significant phenomena such as detection of differential alternative splicing and isoform abundance can be studied in detail with RNA-Seq technique (Park et al. 2013).

Although RNA-Seq is clearly more informative and advantageous, the data produced by this technique are still complex and huge. NGS platforms generate high-throughput data in the form of millions of short sequences termed as "reads." These reads are associated with their base-call quality scores that indicate the reliability of each base call. The length of these short reads depends on the type of NGS platform used for sequencing, but generally they fall within a length of 25–450 bp. The resulting reads are categorized into three types: exonic reads, exon–intron junction reads, and poly(A) reads (Wang et al. 2009). The analysis of this kind of data is not a straightforward task and is usually a bottleneck to deal with. Fortunately, continuous progress in the area of bioinformatics has eased the way to deal with RNA-Seq data. There are now various bioinformatic tools/software, web servers, as well as whole pipelines to tackle and analyze RNA-Seq data. Also, various strategies applicable to RNA-Seq data analysis can be implemented in Bioconductor (Huber et al. 2015; Gentleman et al. 2004) through statistical language "R" (https://www.r-project.org). Bioconductor is free, is open-source, and can deal with analysis of not only RNA-Seq data but other high-throughput genomic data as well. Bioconductor basically works on the basis of different "packages" dedicated to different types of tasks. There are many Bioconductor packages dedicated to the whole RNA-Seq data analysis executable with even a little proficiency in R. Many tools can be combined for analysis of RNA-Seq data, and researchers may form their own custom data analysis pipelines according to their objectives.

Bioinformatic analysis of RNA-Seq data can be divided into several stages. The very first step is experiment/technology dependent, and choice of the methods for downstream analysis is made on the basis of the type of experiment. During sequencing only, the first step of bioinformatic analysis gets started with the transformation of fluorescent measurements into associated nucleotide bases with their quality scores. Base quality score is usually a value representing the confidence of the called bases. The final output of this base-calling step is the short reads (raw data) in FASTQ (FAST-All with quality score) format. The next task is to map

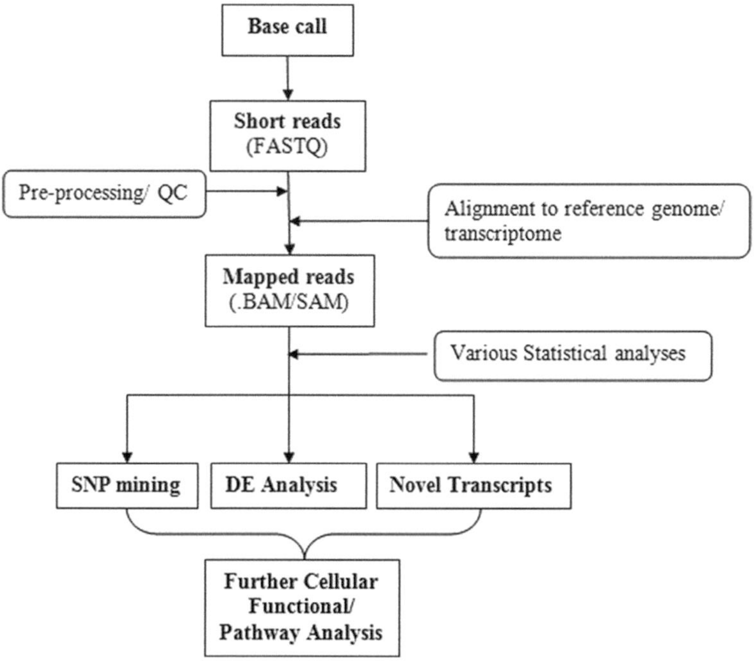

Fig. 10.2 A usual flow chart of bioinformatics-based analysis of RNA-Seq data

these short reads to reference genome (or transcriptome in case of transcriptomic data) in case it's already available or otherwise firstly assemble them de novo. After mapping, further downstream analysis may proceed according to research goals, though a usual work flow of bioinformatics-based analysis associated with RNA-Seq data is shown in the flowchart (Fig. 10.2). During the analysis, different tools/software or strategies may be applied at different steps.

It would not be inappropriate to say that RNA sequencing has a variety of different applications and data analysis strategies depending on the organism under study and research objectives. RNA-Seq has the power of identifying transcripts and quantifying gene expression which is the key to decipher more knowledge on the relationship between genome and proteome. Elucidating RNA isoform expression, alternative splicing, and ncRNA levels are other applications of RNA-Seq having great importance in molecular biology.

10.2 Data Format, Quality Check, and Preprocessing

Raw reads (FASTQ format) obtained after the base-calling step contain nucleotides associated with quality scores. Although different NGS platforms have their own methods of base calling (base-calling software) to evaluate base quality, various third party groups have also put efforts in developing base-calling methods. The most profitable and notable example is the enhanced ABI base caller, Phred, which played an important role in the Human Genome Project (Ewing and Green 1998; Ewing et al. 1998). Nowadays, most NGS platforms provide the user with a Phred-like score value (Ewing et al. 1998) for base quality evaluation which is based on a logarithmic scale encoding the probability of error in the corresponding base call. This base-calling step is particularly important because its accuracy affects the downstream analysis. The resulting format of base-calling algorithm, i.e., FASTQ, is a FASTA (FAST-All) standard format of biological sequences like format but comes with associated quality score for each nucleotide, usually Phred score.

Reads may be represented in other formats like FASTA and standard flowgram format (SFF) that may be converted to one another, but generally FASTQ format is the most frequent one that can be used as input in many applications. FASTQ files may be so huge in size and also consist of contaminations that need to be eliminated before downstream analysis because contaminated input directly affects the outcome. Preprocessing of data is thus a very important and necessary step before jumping onto the downstream analysis. Preprocessing includes steps like checking the Phred scores, length of reads per base, and read quality and trimming the reads to remove adapters, low-quality sequences, duplicate sequences, and Ns (means no base assigned during the base call). Various available preprocessing tools may be in the form of stand-alone software or accessed with different whole data analysis pipelines, web servers like Galaxy (https://galaxyproject.org/), language platforms like R/Bioconductor, or simply based on command lines.

Some popular tools for quality check and preprocessing of RNA-Seq data are FastQC (Andrews 2010) (http://www.bioinformatics.babraham.ac.uk/projects/fastqc), FASTX-Toolkit (http://hannonlab.cshl.edu/fastx_toolkit), Cutadapt (https://cutadapt.readthedocs.org/en/stable) (Martin 2011), and Trimmomatic (http://www.usadellab.org/cms/index.php?page=trimmomatic) (Bolger et al. 2014). These tasks are also achievable through some R/Bioconductor packages like "ShortRead" (Morgan et al. 2009). We present a list of some recently developed tools for data quality check and preprocessing (Table 10.1).

10.3 Mapping

Mapping is the most important step in way of analyzing any NGS data. "Mapping" makes each read correspond to a particular position in genome/transcriptome. Since RNA-Seq data may produce reads either from single exon without accessing the exon-exon boundary (unspliced) or from a pair of exon where a read would span the intronic region (spliced), the mapping strategy demands a deeper lookout. If we empirically align the RNA-Seq reads using methods like Burrows–Wheeler transform, we have to consider both the aligned and unaligned reads. Fully aligned reads may be unspliced, but the reads which fail to align may be truly spliced reads spanning an intron. Today, we have many aligners for NGS data using different

Table 10.1 List of recently developed tools/software for data QC and preprocessing

Tool/package	A brief introduction	Input	References
Category: data QC			
AuPairWise	Implemented in R scripts. Measures RNA-Seq replicability by modeling the effects of noise	Expression data	Ballouz and Gillis (2016)
ClinQC	Analysis pipeline. Analyzes both; Sanger and NGS data	Raw reads in any native file format of their sequencing platforms	Pandey et al. (2016)
SinQC	Software tool. Detects technical artifacts in single-cell RNA-seq. Python and R based. R package – ROCR	Gene expression patterns	Jiang et al. Jiang et al. (2016)
TIN (transcript integrity number)	Based on python. Measures RNA degradation	RNA-Seq datasets	Wang et al. (2016)
dupRadar	R package for plotting and analyzing duplication rates dependent on expression levels	BAM file with mapped and duplicate marked reads and a gene model in GTF format	Sayols and Klein (2015)
HTSeq	Python script-based tool	FASTQ, BAM	Anders et al. (2015)
mRIN	Perl- and R-based package. Assess mRNA integrity directly from RNA-Seq data	Coverage profile	Feng et al. (2015)
NOISeq	Bioconductor package. Includes modeling noise distribution of count	Raw and mapped data	Tarazona et al. (2015)
Qualimap 2	Java- and R-based GUI as well as command line interface. Supports multi-sample QC	BAM/SAM, GTF/GFF/ BED and read counts table	Okonechnikov et al. (2015)
Rcorrector	Corrects error for Illumina RNA-Seq reads (k-mer-based method). Written in C, C++, and Perl	k-mers based on input reads and counts	Song and Florea (2015)
deepTools	Galaxy-based server	BAM, SAM	Ramirez et al. (2014)
FIXSEQ	R based. Corrects over-dispersed read-count distribution	Read counts	Hashimoto et al. (2014)
QuaCRS	An integrated quality control pipeline for RNA-Seq data. Command line interface	FASTQ, BAM, additional metadata	Kroll et al. (2014)
BlackOPs	Blacklist mismapping in RNA-Seq. Written in Perl	Aligned data	Cabanski et al. (2013)

(continued)

Table 10.1 (continued)

Tool/package	A brief introduction	Input	References
GeneScissors	Detects and corrects spurious transcriptome features leading misalignment. Written in C++, Python, and BamTools	Can be added to any standard pipeline before mapping	Zhang et al. (2013)
HTQC	Toolkit implemented in C++. For graphics – Perl is used	FASTQ	Yang et al. (2013)
IDCheck	RNA-Seq sample identity check	BAM	Huang et al. (2013)
Kraken	Tool package. Pipeline written in Perl and R	FASTQ	Davis et al. (2013)
SEECER	Command line interface. Uses HMMs. Applicable to de novo RNA-Seq	Raw reads	Le et al. (2013)
BM-Map	Software package. Allocates multireads in RNA-Seq data. C++ based	SAM	Yuan et al. (2012)
RSeQC	Python-script-based package. Visualization facilitated through genome browsers like UCSC, IGB, IGV and also using R scripts	SAM, BAM, FASTA, BED or chromosome size file	Wang et al. (2012)
RNA-SeQC	Java based (no installation required). Also integrated in "GenePattern" web interface	One/more BAM	Deluca et al. (2012)
ArrayExpressHTS/AEHTS	R/Bioconductor-based pipeline	Raw reads	Goncalves et al. (2011)
BIGpre	Stand-alone/integrated in Galaxy	FASTQ	Zhang et al. (2011)
NGSQC	Cross platform QC analysis pipeline	FASTQ (IlluQC) or FASTA (454QC)	Dai et al. (2010)
SAMStat	C language-based tool package	SAM, BAM, FASTA, FASTQ	Lassmann et al. (2010)
Category: Trimmers and adapter removers			
ADEPT	Written in Perl5.8. Command line based	One or more FASTQ files	Feng et al. (2016)
Cookiecutter	k-mer-based algorithm. Command line based. Implemented in C++	One or more FASTQ files and a list of k-mers (user provided or cookiecutter generated from FASTA)	Starostina et al. (2015)
NxTrim	For Illumina Nextera Mate Pair (NMP) reads, Command line interface	Raw reads	O'Connell et al. (2015)

(continued)

Table 10.1 (continued)

Tool/package	A brief introduction	Input	References
PEAT	Specifically for paired-end sequencing. Command line interface	FASTQ, no adapter sequence required	Li et al. (2015b)
leeHom	Based on Bayesian maximum a posteriori probability approach. Command-line-based package	One or more FASTQ files, unaligned BAM, adapter sequence	Renaud et al. (2014)
ngsShoRT	Software package written in Perl	FASTQ or Illumina's native QSEQ format	Chen et al. (2014)
QTrim	Stand-alone command line based (python version) as well as a web interface	FASTQ or a FASTA file with its associated quality file (.qual)	Shrestha et al. (2014)
Skewer	"Bit-masked k-difference matching algorithm" based	FASTQ	Jiang et al. (2014)
AlienTrimmer	Command line based	One or more FASTQ files	Criscuolo and Brisse (2013)
NGS QC Toolkit	Implemented in Perl. Command line based, web based	FASTQ, FASTA	Patel and Jain (2012)

Most of the tools shown in table are attributed to RNA-Seq data, but some lately developed tools for NGS data QC and preprocessing are also included in the table. Many data QC tools given in the table are not only limited to raw data QC but to advance stages also like mapping. A brief about basic property of each tool is also included in the table

approaches like seed based (e.g., SHRiMP2;David et al. 2011) , BFAST (Homer et al. 2009), SeqMap (Jiang and Wong 2008), CUSHAW3 (Liu et al. 2014), SOAP (Li et al. 2008a), MAQ (Li et al. 2008b), STAMPY (Lunter and Goodson 2011) or hash based (e.g., MOSAIK; Lee et al. 2014), and HIVE hexagon (Santana et al. 2014). Additionally, a popularly used algorithm in data compression technique, the Burrows–Wheeler transform (BWT), also contributes in providing some excellent mapping tools like BWA (Li and Durbin 2009d), SOAP2 (Li et al. 2009a), and Bowtie (Langmead 2010). Several tools such as TopHat (Trapnell et al. 2009), STAR (Dobin et al. 2013), SpliceMap (Au et al. 2010), and MapSplice (Wang et al. 2010) are available today that perform mapping while considering both the exonic and splicing events.

Mapping refers to locating the short reads onto reference genome/transcriptome which is comparatively feasible with the availability of a reference genome/ transcriptome; otherwise a de novo assembly is required to proceed further. Without a reference genome or transcriptome, mapping is not feasible as in such case a de novo assembly of RNA-Seq reads would be required to generate full transcript sequences (Robertson et al. 2010). De novo assembly is usually complex in nature that involves construction of de Bruijn graphs using k-mers. There are many tools for de novo assembly for RNA-Seq data like Trinity (Haas et al. 2013), Velvet (Zerbino and Birney 2008), Bridger (Chang et al. 2015), SOAPdenovo (Li et al.

2010), and Trans-ABySS (Simpson et al. 2009). Here we discuss some useful assemblers for de novo assembly and mappers that are very efficient in RNA-Seq reads mapping.

10.3.1 Trinity

Trinity (Haas et al. 2013) is the first method designed specifically for transcriptome assembly and works on the basis of de Bruijn graphs. It comprises three independent software modules, Inchworm, Chrysalis, and Butterfly, which are used sequentially to produce transcripts. Inchworm assembles the RNA-Seq data into transcript sequences, Chrysalis clusters the Inchworm contigs and constructs complete de Bruijn graphs for each cluster, and then Butterfly processes the individual graphs in parallel to trace the paths of reads within the graph, ultimately reporting full-length transcripts.

10.3.2 Bridger

Bridger is a newer framework for de novo transcript assembly (Chang et al. 2015). It is so named as if to build a bridge between the basic keys of two popular assemblers: Cufflinks (the reference-based assembler (Trapnell et al. 2012)) and Trinity (the de novo assembler (Haas et al. 2013)). It has some advantages over other de novo aligners like it allows the use of different k-mer lengths for different data, while trinity has a fixed k-mer length of 25. It also has a lower false-positive rate and uses less memory and run time compared with Trinity.

On the other hand, the presence of reference genome/transcriptome makes mapping process relatively faster and easier to implement with some web-based/command-line-based tools. In mapping, the problem of multimapping is also usually seen and needs to be taken care of. Generally, mapping utilizes a heuristic first step to find likely candidates followed by local alignment, but alignment is not sufficient for mapping moderate- to large-sized genomes. Thus, the strategy used by most of the aligners/mappers is to somehow enable a fast heuristic method so that the smaller number of local alignments has to be performed. As aforementioned, RNA-Seq mappers should be able to consider the spliced alignment problem, i.e., they should be able to place spliced read across introns and correctly determine exon–intron boundaries. In the present scenario of RNA-Seq research, many aligners work well in this kind of mapping, among which Bowtie2 (Langmead et al. 2009) is a popular one. We discuss a few other tools that have proven their worth.

10.3.3 TopHat

TopHat is a program that aligns RNA-Seq reads to a genome/transcriptome while considering splice junction mapping (Trapnell et al. 2009). It uses the ultrahigh-throughput short read aligner Bowtie and then analyzes the mapping results to identify splice junctions between exons. Using this initial mapping information from Bowtie, TopHat builds a database of possible splice junctions and then again maps the reads against these junctions to confirm them. It runs on Linux and MacOS X and was originally designed to work with reads produced by the Illumina Genome Analyzer, although it is successfully applied with reads from other technologies as well. It also can be implemented in R using some Bioconductor packages as well as on Galaxy server. Moreover, mapping can be visualized through Integrated Genome Viewer (https://www.broadinstitute.org/igv/) (Robinson et al. 2011).

Before performing further downstream analysis, it is also recommended to check the quality of mapping as it greatly influences the downstream analysis. A list of data QC and preprocessing tools capable of checking and processing the data at many stages (including mapping) of data analysis is provided in Table 10.1. Tools like SAMStat (Lassmann et al. 2010) and dupRadar (Sayols and Klein 2015) (R package for QC) are easily accessible and very useful in checking and dealing with mapping quality issues.

10.3.4 STAR

STAR (Spliced Transcripts Alignment to Reference) (Dobin et al. 2013) is one of the important alignment tools that are capable of identifying the alternative splicing junctions in RNA-Seq reads. It is a free, open-source software (under GPLv3 license) that can be downloaded from http://code.google.com/p/rna-star/. It works by indexing the reference genome first, followed by producing a suffix array index to accelerate the alignment step in further processing. STAR has high accuracy like TopHat with comparatively less time consumption. While it can fairly handle single- or paired-end reads, it also increases its accuracy if provided with an annotation (.gtf) file. Advantageously, STAR was not developed as an extension of a short read mapper but a stand-alone C++ code. Being capable of running parallel threads on multi-core systems, STAR is faster in comparison with other tools.

Visualization of mapped reads in a graphical or preferably and advantageously in interactive mode is necessary to closely look at the mapped regions and other factors. There are various tools/software packages such as “SAMtools tview” (Li et al. 2009b), “MapView” (Bao et al. 2009), “Tablet” (Milne et al. 2013), “IGV” (Thorvaldsdóttir et al. 2013), and “Bambino” (Edmonson et al. 2011) that enable the visualization of mapped reads.

In NGS data analysis, the factor of quality control is significant at every single step. Since mapping is the basis for further analysis of data, it is mandatory to check the quality of mapped files to assure the error-free results. Among already available NGS data manipulators like Picard (http://picard.sourceforge.net.) and SAMtools (Li et al. 2009b), some lately developed powerful tools like RseQC and QoRTs assist in quality control, data processing, and management to an excellent level. These tools are included in a package of various utilities that handle the data at different levels.

QoRTs (Hartley and Mullikin 2015) is a fast and portable multifunction toolkit that easily handles cross-comparison of replicates (biological/experimental) and detection of errors, artifacts, and biases. Additionally it can produce count data that can be used in Bioconductor package such as DESeq, DESeq2, and edgeR.

On the other hand, RSeQC (Wang et al. 2012), a comprehensive package of python programs, provides a number of modules to evaluate RNA-Seq data from different aspects. Quality check of raw reads for properties like sequence quality, PCR bias, nucleotide composition bias, and GC bias can be checked with its "basic modules," while "RNA-Seq specific modules" evaluate the quality/status of sequencing saturation of both splice junction detection and expression estimation. RSeQC is written in Python and C and is freely available at http://code.google.com/p/rseqc/.

Mapping is also fundamental in many versatile applications of RNA-Seq like transcript identification and characterization, gene expression quantification, detection of alternatively spliced isoforms, detection of allele-specific expression (ASE), and differential gene expression. Programs like HTSeq-count (Anders et al. 2015) and featureCounts (Liao et al. 2014) use the raw counts of mapped reads for gene quantification. Gene quantification also utilizes a gene transfer format (GTF) file containing the genome coordinates of exons and genes. The number of reads mapped to transcript reference is also the most important information in estimating gene and transcript expression. For expression analysis, only read counts are not sufficient because of other factors like sequence biases, number of reads, and transcript length. These factors are handled by various normalization methods like RPKM (reads per kilobase per million mapped reads) (Mortazavi et al. 2008), FPKM (fragments per kilobase of transcript per million mapped reads) (Trapnell et al. 2010), and TPM (transcripts per million) which are elaborated later in other sections. "Cufflinks" (Trapnell et al. 2012) is a widely used program for estimating transcript level expression from mapping using an EM (expectation–maximization) approach while taking into account biases like nonuniform distribution of reads along the gene length.

The power of identification and quantification of an overall expression of RNAs in a sample is facilitated by RNA-Seq by enabling the genome-wide studies of alternative pre-mRNA splicing which is an important factor to understand the differential expression. Since alternative splicing produces multiple isoforms by skipping or differential joining of exons or introns within a pre-mRNA transcript during transcription (Fig. 10.3), it delivers functional diversity of a gene during posttranscriptional processing and affects gene regulation.

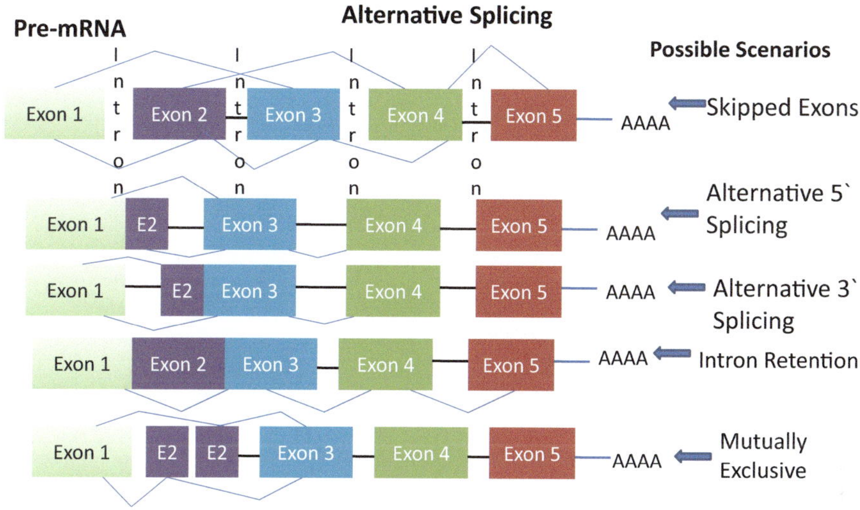

Fig. 10.3 A graphical illustration of alternative splicing event that eventually results in isoforms

Analyzing expression of transcripts at the isoform level is very important in order to understand differential expression. Since many genes may have multiple isoforms, deciphering isoform-specific expression is definitely not straightforward because it is not simple to assign some reads to a particular isoform. The basic approach for dealing with this difficult task was to quantify the transcript isoforms using only those sequences which were unique to particular isoforms (Filichkin et al. 2010). This approach worked on the basis of some already known or predicted transcript isoforms for a given gene that were used to form a set of sequences which in turn could differentiate one isoform from others. Then the mapping of reads to such a set of sequences elaborated the corresponding isoform expression precisely.

Similarly ALEXA-seq (Griffith et al. 2010) method used only those reads that mapped uniquely to one isoform to estimate isoform-specific expression, but these kinds of approaches usually are limited. This is because many isoforms are mostly nonunique or may have minor sequence differences, and also these approaches demand a prior knowledge of precise annotation of splice variants.

The tools related to isoform identification, quantification, abundance estimation, pre-mRNA alternative splicing discovery, and mapping/alignment are already widespread, and the development of new methods is progressing at a very accelerating speed. We present a list (Table 10.2) consisting some recently developed methods/tools dedicated to these tasks along with a brief description of each tool.

Lately, some algorithms like Sailfish, Kallisto, and Salmon have come into existence that use an alignment-free approach to deal with gene/isoform quantification task. These algorithms are considered to be lightweight algorithms that are faster than traditional mapping steps. A succinct overview of all three algorithms is briefed below.

Table 10.2 Recently developed methods/tools for isoform discovery, quantification, abundance estimation, alternative splicing discovery, assembling transcriptome, and alignment of RNA-Seq reads

Tool	A brief description of utility	URL	References
CIDANE	Transcript reconstruction, isoform discovery, and abundance estimation	http://ccb.jhu.edu/software/cidane/.	Canzar et al. (2016)
CLASS CLASS2	Transcriptome assembly. Alternative splicing discovery	http://sourceforge.net/projects/Splicebox	Song and Florea (2013), Song et al. (2016)
Rail-RNA	A cloud-enabled spliced aligner. Analyzes many samples at once. For many samples, Rail-RNA is more accurate than annotation-assisted aligners	http://rail.bio	Nellore et al. (2015)
Rockhopper 2	De novo assembly of bacterial transcriptomes	http://cs.wellesley.edu/~btjaden/Rockhopper	McClure et al. (2013), Tjaden (2015)
JAGuaR	An alignment protocol for RNA-Seq reads. Does not detect novel junctions	http://www.bcgsc.ca/platform/bioinfo/software/jaguar	Butterfield et al. (2014)
MaLTA	Simultaneous transcriptome assembly and quantification from Ion Torrent RNA-Seq data	http://alan.cs.gsu.edu/NGS/?q=malta	Mangul et al. (2014)
HSA	An effective spliced aligner of RNA-Seq reads. Better call rate and efficiency but little less accurate at some attributes	https://github.com/vlcc/HSA	Bu et al. (2013)

10.3.5 Sailfish

Sailfish (Patro et al. 2014) is a free and open-source software, available at http://www.cs.cmu.edu/~ckingsf/software/sailfish. It is a much faster in silico method facilitating the quantification of RNA-isoform abundance by totally avoiding the time-consuming mapping step. Instead of mapping, it inspects k-mers in reads to observe transcript coverage that results in a fast processing of reads. It also maintains the accuracy up to the mark by incorporating an EM procedure that brings a statistical coupling between k-mers. It discards k-mers that overlap inaccurate bases to handle sequencing errors. Overall, it has only a single explicit parameter the k-mer length to rely on. Longer k-mers tend to resolve their origin easier than short k-mers but may be more affected by errors for which Sailfish has implemented an error handling EM procedure. Process wise, Sailfish first builds an index from a FASTA reference transcript file and a chosen k-mer length. Data structures like minimal perfect hash function 9 in the index file play an important role in mapping each k-mer in reference transcript to an identifier in such a way that no two k-mers share an identifier. There is no need to change or rebuild the index unless the reference or the choice of k changes. Next to building index files, the step of quantification is proceeded that takes index and a set of RNA-Seq reads as input to estimate the isoform abundance, measured in RPKM, KPKM (k-mers per kilobase per million mapped k-mers), and TPM. Sailfish can also be used for non-model organisms in de novo mode. Since Sailfish has an overall parameter of the k-mer counts, it is also computationally efficient that can effectively exploit many CPU cores.

10.3.6 Kallisto

Kallisto (Bray et al. 2016) was developed by Pachter lab with the same lightweight algorithm approach as Sailfish to quantify transcript abundance but improves it with a "pseudoalignment" process. It is a fast software program written mainly in C++. It is considered to be near optimal in speed along with accuracy and tested successfully by its developers in analyzing 30 million unaligned paired-end RNA-Seq reads in less than 5 min on a standard desktop. This software is widely popular because of its accuracy as compared to those of the already existing tools. It does not work on the basis of position in transcript where a read aligns but the compatibility of read with a particular transcript that takes a lot less time than the traditional alignment process.

10.3.7 Salmon

Salmon (Patro et al. 2015) is an open-source software under the GPL v3 license and available at http://combine-lab.github.io/salmon/. Its developers call it a wicked-fast transcript quantification software that requires a set of target transcripts for quantification task and may be run in two modes: the quasi-mapping-based mode and the alignment-based mode. The quasi-mapping-based mode like Sailfish incorporates two phases, indexing and quantification, while the alignment-based mode uses the alignment file (SAM/BAM) provided by the user along with reference transcript FASTA file and does not require indexing.

10.4 Differential Expression

An important application of RNA-Seq technique is to identify genes that change in abundance between conditions, i.e., they differ in counts in different conditions. Differential expression (DE) is simply to compare expression levels of genes between two conditions, e.g., stimulated versus unstimulated or wild type versus mutant or normal versus treated. If there is a statistically significant difference or change in read counts between two conditions, a gene can be affirmed as a differentially expressed gene. The aforementioned steps of data preprocessing and mapping are mandatory for analysis of differential expression. Also, for differential expression, it is necessary to analyze read-count distributions, typically represented as a matrix N of n × m where N_{ij} is the number of reads assigned to gene in sequencing experiment/condition j. Bioconductor has many packages to support DE analysis of RNA-Seq data. Many packages like DESeq2 (Love et al. 2014), edgeR (Robinson et al. 2010), limma (Ritchie et al. 2015), and baySeq (Hardcastle 2012) have whole RNA-Seq data analysis pipelines which can be of great use. Most of the packages for DE analysis expect input data in the form of matrix of integer values. To prepare the count matrix, SAM/BAM alignment file along with a file specifying the genomic features, e.g., a GFF3 or GTF, can be used. For this, we may use other packages of Bioconductor like Rsubread (Liao et al. 2013) and GenomicAlignments (Lawrence et al. 2013).

Two most popular packages for DE analysis are DESeq2 (Love et al. 2014) and edgeR (Robinson et al. 2010). They are modular in nature that means there are many entry points in the package from where the package can be used. They often give freedom to use an alternative aligner or a different strategy or tool to obtain read counts and then use the package for rest of the analysis. Since there is not any universal standard for DE analysis, it may somewhat be objective oriented and heavily dependent on external data like reference assemblies and annotation. Thus, we can't expect that two different analysis strategies of the same data will end up with the same results, similarity is still expected though.

It is also worth mentioning about the importance of normalization which is a very significant step in the analysis of DE. Normalization is necessary to correct for biases which can arise from technical biases like between-sample differences that denote library size and within-sample gene-specific effects that may be related to gene length and GC-content (Oshlack and Wakefi 2009). There are various normalization methods for DE analysis including Total Count (TC), Upper Quartile (UQ), Median (Med), the DESeq normalization implemented in the DESeq Bioconductor package, Trimmed Mean of M values (TMM) implemented in the edgeR Bioconductor package, Quantile (Q), and RPKM normalization. FPKM normalization is also a popular method and is used by tools like cufflinks (Trapnell et al. 2010). FPKM is analogous to RPKM but does not use read counts.

This overview of DE analysis is superficial and descriptive of basics only used in DE analysis. There are actually a huge number of parameters in each step that can change results. Every step including preprocessing and mapping affects the analysis of subsequent steps. Like other tasks of RNA-Seq data analysis where newer algorithms and tools are making a mark, task of differential expression has also opened up the way for the development of newer and different algorithms/tools. BitSeq (Hensman et al. 2015; Glaus et al. 2012), deGPS (Chu et al. 2015), NOISeq (Tarazona et al. 2015), and XBSeq (Chen et al. 2015) are some of the recently developed tools which are really different in their algorithm and performance and give a broader spectrum to differential expression analysis in RNA-Seq data.

Although in this chapter we elaborate on different approaches and tools for analysis of RNA-Seq data, continuous research in this field has provided us some great whole analysis pipelines to also deal with RNA-Seq data. Since RNA-Seq technique has unprecedented ability to study transcriptome to a much greater extent than previous technologies, the analyses of ncRNAs have also become more accessible and feasible. Here we succinctly present a list of recently developed pipelines dedicated to RNA-Seq data and also some tools/pipelines dedicated to analyses of ncRNAs obtained through RNA-Seq technique (Table 10.3).

Table 10.3 List of some popular and recently developed pipelines dedicated to whole RNA-Seq and ncRNA data (obtained through RNA-Seq technique) analysis

Tool	A brief introduction of utility	URL	References
Category: RNA-Seq data analysis pipelines			
CANEapp	GUI and an automated server-side analysis pipeline for RNA-Seq	http://psychiatry.med.miami.edu/research/laboratory-of-translational-rna-genomics/CANE-app	Velmeshev et al. (2016)
QuickRNASeq	A pipeline for large-scale RNA-Seq data analyses and visualization	http://sourceforge.net/projects/quickrnaseq/	Shanrong Zhao et al. (2016)
TRAPLINE	Pipeline for RNA sequencing data analysis, evaluation, and annotation	https://usegalaxy.org/u/mwolfien/w/rnaseq-wolfien-pipeline	Wolfien et al. (2016)
BioWardrobe	Integrated pipeline. Analyzes epigenomics and transcriptomic data	https://biowardrobe.com/	Kartashov and Barski (2015)
QuickNGS	Pipeline that analyzes data from multiple NGS projects at a time. Parallel computing resources	http://bifacility.uni-koeln.de/quickngs/web/	Wagle et al. (2015)
RAP	A cloud-computing web application for RNA-Seq analysis	https://bioinformatics.cineca.it/rap/	D'Antonio et al. (2015)
RNAMiner	A multilevel bioinformatics protocol and pipeline for RNA-Seq	http://calla.rnet.missouri.edu/rnaminer/index.html	Li et al. (2015a)
Category: ncRNA analysis tools/pipelines			
isomiR-SEA	Details miRNAs, isomiRs, and conserved miRNA: mRNA interaction. Specialized alignment algorithm	http://eda.polito.it/isomir-sea/	Urgese et al. (2016)
Chimira	An online tool (pipeline) for analyzing large amounts of small RNA-Seq data	http://wwwdev.ebi.ac.uk/enright-dev/chimira/	Vitsios and Enright (2015)
iSRAP	A one-touch integrated small RNA analysis pipeline	http://israp.sourceforge.net/	Quek et al. (2015)
miRA	ncRNA identification tool. Identifies miRNA precursors in plants	https://github.com/mhuttner/miRA	Evers et al. (2015)
mirPRo	A stand-alone pipeline that quantifies known miRNAs and predicts novel miRNAs	http://sourceforge.net/projects/mirpro/	Shi et al. (2015)

(continued)

Table 10.3 (continued)

Tool	A brief introduction of utility	URL	References
miRge	Ultrafast, small RNA-Seq solution pipeline. Decreases computational requirements	http://atlas.pathology.jhu.edu/baras/miRge.html	Baras et al. (2015)
Oasis	Fast and flexible web application. Facilitates online analysis of small-RNA-Seq (smRNA-Seq) data	https://oasis.dzne.de/	Capece et al. (2015)
segmentSeq	Bioconductor package. Identifies robust sets of siRNA precursors	http://www.bioconductor.org/packages/release/bioc/html/segmentSeq.html	Hardcastle (2015), Hardcastle et al. (2012)
sRNAtoolbox	smRNA analysis pipeline. Collection of small RNA research tools	http://bioinfo5.ugr.es/srnatoolbox	Rueda et al. (2015)
SMiRK	Automated pipeline for miRNA analysis	https://github.com/smirkpipeline/SMiRK	Milholland et al. 2015
Tailor	Read aligner for small silencing RNAs. Also captures the tailing events directly from the alignments without extensive post-processing	https://github.com/jhhung/Tailor	Chou et al. (2015)
tDRmapper	t-RNA derived RNA annotation tool. Maps and quantifies tRNA-derived RNAs (tDRs). Includes graphical visualization that facilitates the discovery of novel tRNA.	https://github.com/sararselitsky/tDRmapper	Selitsky and Sethupathy (2015)
YM500v2	A small RNA sequencing (smRNA-Seq) database for human cancer miRNome research	http://ngs.ym.edu.tw/ym500v2/index.php	Cheng et al. (2015), Cheng et al. (2013)
BioVLAB-MMIA-NGS	A whole software pipeline for microRNA-mRNA integrated analysis using high-throughput sequencing data	http://epigenomics.snu.ac.kr/biovlab_mmia_ngs/	Chae et al. (2014)
CAP-miRSeq	Whole pipeline for microRNA sequencing data	http://bioinformaticstools.mayo.edu/research/cap-mirseq/	Sun et al. (2014)
MAGI	Fast microRNA-Seq data analysis in a GPU infrastructure	http://elgar.ucsd.edu/software/magi/	Kim et al. (2014)
mrSNP	Predicts the impact of a SNP in a 3UTR on miRNA binding	http://mrsnp.osu.edu/	Deveci et al. (2014)
piClust	Finds piRNA clusters and transcripts from small RNA-Seq data	http://epigenomics.snu.ac.kr/piclustweb/	Jung et al. (2014)

(continued)

Table 10.3 (continued)

Tool	A brief introduction of utility	URL	References
CoRAL	ncRNA identification tool. Predicts the precursor class of small RNAs present in RNA-sequencing dataset	http://wanglab.pcbi.upenn.edu/coral/	Leung et al. (2013)
ISRNA	Software pipeline designed for storage, visualization, and analysis of small RNA sequencing data	http://omicslab.genetics.ac.cn/resources.php	Luo et al. (2014)
iMir	A modular pipeline for comprehensive analysis of small RNA-Seq data	http://www.labmedmolge.unisa.it/inglese/research/imir	Giurato et al. (2013)
miReader	Detects mature miRNAs directly from next-generation sequencing read data, without any need of reference/genomic sequences	http://scbb.ihbt.res.in/2810-12/miReader.php	Jha and Shankar (2013)
miRDeep	An integrated application tool for miRNA identification from RNA sequencing data	http://sourceforge.net/projects/mirdeepstar/	An et al. (2013)
ShortStack	Processes and analyzes small RNA-Seq data with respect to a reference genome and outputs a comprehensive and informative annotation of all discovered small RNA genes	http://axtell-lab-psu.weebly.com/shortstack.html	Axtell (2013)
SHRiMP2	Software package for aligning genomic reads against a target genome. Works great with small RNA mapping	http://compbio.cs.toronto.edu/shrimp/	David et al. (2011)

10.5 Summary

Today, RNA-Seq is the mainstream tool for analysis of transcriptomes that are so rich in information and progressing day by day. This technique has its wide applications in various areas like clinical diagnostics, pharmacogenomics, and drug development. It can find novel transcripts and identify drug-related genes and microRNAs. Although RNA-Seq technology is still in progressive and developmental stage, yet it has made substantial contributions to our understanding of many transcriptomes from those of simple unicellular organisms to complex mammalian cells, as well as in tissues in normal and disease states. Still, the data from RNA-Seq is complex to analyze and very sensitive to technical biases. This chapter focused mainly with some tools/software for RNA-Seq data analysis and some interesting platforms like R/Bioconductor and Galaxy web server where many of these tools can be accessed and data can be analyzed. It is worth noting that many tools mentioned in this chapter are not restricted only to RNA-Seq data and may be used for other kinds of NGS data as well. Also, there are several other tools, software, whole analysis pipeline, and statistical strategies for analyzing RNA-Seq data, but they are not discussed here. Still, bioinformatics-based tools are progressing rapidly, and there is a wide opportunity of building new tools and strategies for analyzing RNA-Seq data as well as data derived from other NGS technologies. As NGS technologies are continuously evolving, we can hope for RNA-Seq having more technical and analytical developments with lower cost in the near future.

Acknowledgments The authors would like to thank University Grant Commission, India for the support. The authors express their gratitude to Nimisha Chaturvedi, Dr. Raghvendra Singh, and Swadha Singh for giving valuable suggestions regarding the improvement of this chapter.

References

An J, Lai J, Lehman ML, Nelson CC. miRDeep*: an integrated application tool for miRNA identification from RNA sequencing data. Nucleic Acids Res. 2013. PMID: 23221645.

Anders S, Pyl PT, Huber W. HTSeq-a Python framework to work with high-throughput sequencing data. Bioinformatics. 2015;31:166–9.

Andrews S. Fast QC: a quality control tool for high throughput sequence data. 2010. Available online at: http://www.bioinformatics.babraham.ac.uk/projects/fastqc

Ansorge WJ. Next-generation DNA sequencing techniques. N Biotechnol. 2009;25:195–203. Bioinformatics 25:1754–60.

Au KF, Jiang H, Lin L, Xing Y, Wong WH. Detection of splice junctions from paired-end RNA-seq data by Splice Map. Nucleic Acids Res. 2010;38:4570–8.

Axtell MJ. ShortStack: comprehensive annotation and quantification of small RNA genes. RNA. 2013. PMID: 23610128.

Ballouz S, Gillis J. AuPairWise: a method to estimate RNA-seq replicability through co-expression. bioRxiv. 2016; doi:10.1101/044669.

Bao H, Guo H, Wang J, Zhou R, Lu X, Shi S. MapView: visualization of short reads alignment on a desktop computer. Bioinformatics. 2009. PMID: 19369497.

Baras AS, Mitchell CJ, Myers JR, Gupta S, Weng LC, Ashton JM et al. miRge – a multiplexed method of processing small RNA-Seq data to determine microRNA entropy. PloS one. 2015. PMID: 26571139.

Bolger AM, Lohse M, Usadel B. Trimmomatic: a flexible trimmer for Illumina sequence data. Bioinformatics. 2014;30:2114–20. doi:10.1093/bioinformatics/btu170.

Bray NL, Pimentel H, Melsted P, Pachter L. Near-optimal probabilistic RNA-Seq quantification. Nat Biotechnol. 2016; doi:10.1038/nbt.3519.

Bu J, Chi X, Jin Z. HSA: a heuristic splice alignment tool. BMC Systems Biol. 2013. PMID: 24564867.

Butterfield YS, Kreitzman M, Thiessen N, Corbett RD, Li Y, Pang J et al. JAGuaR: junction alignments to genome for RNA-seq reads. PloS one. 2014. PMID: 25062255.

Cabanski CR, Wilkerson MD, Soloway M, Parker JS, Liu J, Prins JF, et al. BlackOPs: increasing confidence in variant detection through mappability filtering. Nucleic Acids Res. 2013. PMID: 23935067.

Canzar S, Andreotti S, Weese D, Reinert K, Klau GW. CIDANE: comprehensive isoform discovery and abundance estimation. Genome Biol. 2016. PMID: 26831908.

Capece V, Garcia Vizcaino JC, Vidal R, Rahman RU, Pena Centeno T, Shomroni O et al. Oasis: online analysis of small RNA deep sequencing data. Bioinformatics. 2015. PMID: 25701573.

Chae H, Rhee S, Nephew KP, Kim S. BioVLAB-MMIA-NGS: microRNA-mRNA integrated analysis using high throughput sequencing data. Bioinformatics. 2014. PMID: 25270639.

Chang Z, Li G, Liu J, Zhang Y, Ashby C, Liu D, et al. Bridger: a new framework for de novo transcriptome assembly using RNA-seq data. Genome Biol. 2015.

Chen C, Khaleel SS, Huang H, Wu CH. Software for pre-processing Illumina next-generation sequencing short read sequences. Source code for biology and medicine. 2014. PMID: 24955109.

Chen HH, Liu Y, Zou Y, Lai Z, Sarkar D, Huang Y, et al. Differential expression analysis of RNA sequencing data by incorporating non-exonic mapped reads. BMC Genomics. 2015; doi:10.1186/1471-2164-16-S7-S14.

Cheng WC, Chung IF, Huang TS, Chang ST, Sun HJ, Tsai CF, et al. YM500: a small RNA sequencing (smRNA-seq) database for microRNA research. Nucleic Acids Res. 2013. PMID: 23203880.

Cheng WC, Chung IF, Tsai CF, Huang TS, Chen CY, Wang SC, et al. YM500v2: a small RNA sequencing (smRNA-seq) database for human cancer miRNome research. Nucleic Acids Res. 2015. PMID: 25398902.

Chou MT, Han BW, Hsiao CP, Zamore PD, Weng Z, Hung JH. Tailor: a computational framework for detecting non-templated tailing of small silencing RNAs. Nucleic Acids Res. 2015. PMID: 26007652.

Chu C, Fang Z, Hua X, Yang Y, Chen E, CowleyJr AW, et al. deGPS is a powerful tool for detecting differential expression in RNA-sequencing studies. BMC Genomics. 2015. doi: 10.1186/s12864-015-1676-0.

Criscuolo A, Brisse S. AlienTrimmer: a tool to quickly and accurately trim off multiple short contaminant sequences from high-throughput sequencing reads. Genomics. 2013;102:500–6.

Dai M, Thompson RC, Maher C, Contreras-Galindo R, Kaplan MH, Markovitz DM, et al. NGSQC: cross-platform quality analysis pipeline for deep sequencing data. BMC Genomics. 2010. PMID: 21143816.

D'Antonio M, D'Onorio De Meo P, Pallocca M, Picardi E, D'Erchia AM, Calogero RA, et al. RAP: RNA-Seq Analysis Pipeline, a new cloud-based NGS web application. BMC Genomics. 2015. PMID: 26046471.

David M, Dzamba M, Lister D, Ilie L, Brudno M. SHRiMP2: sensitive yet practical SHort Read Mapping. Bioinformatics. 2011. PMID: 21278192.

Davis MPA, Dongen SV, Goodger CA, Bartonicek N, Enright AJ. Kraken: A set of tools for quality control and analysis of high-throughput sequence data. Methods. 2013;63(1): 41–9. doi:10.1016/j.ymeth.2013.06.027. PMID 23816787.

Deluca DS, Levin JZ, Sivachenko A, Fennell T, Nazaire MD, Williams C, et al. RNA-SeQC: RNA-seq metrics for quality control and process optimization. Bioinformatics. 2012;28(11):1530–2. doi:10.1093/bioinformatics/bts196.

Deveci M, Catalyürek UV, Toland AE. mrSNP: software to detect SNP effects on microRNA binding. BMC Bioinf. 2014. PMID: 24629096.

Dobin A, Davis CA, Schlesinger F, Drenkow J, Zaleski C, Jha S, et al. STAR: ultrafast universal RNA-seq aligner. Bioinformatics. 2013. PMID: 23104886.

Dressman D, Yan H, Traverso G, Kinzler KW, Vogelstein B. Transforming single DNA molecules into fluorescent magnetic particles for detection and enumeration of genetic variations. Proc Natl Acad Sci U S A. 2003;100:8817–22.

Edmonson MN, Zhang J, Yan C, Finney RP, Meerzaman DM, Buetow KH. Bambino: a variant detector and alignment viewer for next-generation sequencing data in the SAM/BAM format. Bioinformatics. 2011. PMID: 21278191.

Evers M, Huttner M, Dueck A, Meister G, Engelmann JC. miRA: adaptable novel miRNA identification in plants using small RNA sequencing data. BMC Bioinf. 2015. PMID: 26542525.

Ewing B, Green P. Base-calling of automated sequencer traces using Phred II error probabilities. Genome Res. 1998;8(3):186–94.

Ewing B, Hillier L, Wendl MC, Green P. Base-calling of automated sequencer traces using phred. I Accuracy assessment. Genome Res. 1998;8(3):175–85.

Fedurco M, Romieu A, Williams S, Lawrence I, Turcatti G. BTA, a novel reagent for DNA attachment on glass and efficient generation of solid-phase amplified DNA colonies. Nucleic Acids Res. 2006;34:e22.

Feng H, Zhang X, Zhang C. mRIN for direct assessment of genome-wide and gene-specific mRNA integrity from large-scale RNA sequencing data. Nat Commun. 2015;6(7816) doi:10.1038/ncomms8816.

Feng S, Lo CC, Li PE, Chain PS. ADEPT, a dynamic next generation sequencing data error-detection program with trimming. BMC Bioinf. 2016; doi:10.1186/s12859-016-0967-z.

Filichkin SA, Priest HD, Givan SA, Shen R, Bryant DW, Fox SE, et al. Genome-wide mapping of alternative splicing in *Arabidopsis thaliana*. Genome Res. 2010;20:45–58.

Gentleman RC, Carey VJ, Bates DM, Bolstad B, Dettling M, Dudoit S, et al. Bioconductor: open software development for computational biology and bioinformatics. Genome Biol. 2004;5: R80.

Giurato G, De Filippo MR, Rinaldi A, Hashim A, Nassa G, Ravo M, et al. iMir: An Integrated pipeline for high-throughput analysis of small non-coding RNA data obtained by smallRNA-Seq. BMC Bioinf. 2013. PMID: 24330401.

Glaus P, Honkela A, Rattray M. Identifying differentially expressed transcripts from RNA-seq data with biological variation. Bioinformatics. 2012. PMID: 22563066.

Goncalves A, Tikhonov A, Brazma A, Kapushesky M. A pipeline for RNA-seq data processing and quality assessment. Bioinformatics. 2011. PMID: 21233166.

Griffith M, Griffith OL, Mwenifumbo J, Goya R, Morrissy AS, Morin RD, et al. Alternative expression analysis by RNA sequencing. Nat Methods. 2010;7:843–7.

Haas BJ, Papanicolaou A, Yassour M, Grabherr M, Blood PD, Bowden J, et al. De novo transcript sequence reconstruction from RNA-seq using the Trinity platform for reference generation and analysis. Nat Protoc. 2013;8:1494–512.

Hardcastle TJ. Discovery of methylation loci and analyses of differential methylation from replicated high-throughput sequencing data. bioRxiv. 2015; doi:10.1101/021436.

Hardcastle TJ. baySeq: eEmpirical Bayesian analysis of patterns of differential expression in count data. R package version 2.8.0. 2012.

Hardcastle TJ, Kelly KA and Baulcombe DC. Identifying small interfering RNA loci from high-throughput sequencing data. Bioinformatics. 2012. PMID: 22171331.

Hartley SW, Mullikin JC. QoRTs: a comprehensive toolset for quality control and data processing of RNA-Seq experiments. BMC Bioinf. 2015; doi:10.1186/s12859-015-0670-5.

Hashimoto TB, Edwards MD, Gifford DK. Universal count correction for high-throughput sequencing. PLoS Comput Biol. 2014. PMID: 24603409.

Heap GA, Yang JHM, Downes K, Healy BC, Hunt KA, et al. Genome-wide analysis of allelic expression imbalance in human primary cells by high-throughput transcriptome resequencing. Hum Mol Genet. 2010;19:122–34.

Hensman J, Papastamoulis P, Glaus P, Honkela A, Rattray M. Fast and accurate approximate inference of transcript expression from RNA-seq data. Bioinformatics. 2015. PMID: 26315907.

Homer N, Merriman B, Nelson SF. BFAST: an alignment tool for large scale genome resequencing. PLoS One. 2009;4:e7767.

Huang J, Chen J, Lathrop M, Liang L. A tool for RNA sequencing sample identity check. Bioinformatics. 2013. PMID: 23559639.

Huber W, Carey VJ, Gentleman R, Anders S, Carlson M, Carvalho BS, et al. Orchestrating high-throughput genomic analysis with Bioconductor. Nat Methods. 2015;12(2):115–21.

Jha A, Shankar R. miReader: discovering novel miRNAs in species without sequenced genome. PloS one. 2013. PMID: 23805282.

Jiang H, Wong WH. SeqMap: mapping massive amount of oligonucleotides to the genome. Bioinformatics. 2008;24:2395–6.

Jiang H, Lei R, Ding SW, Zhu S. Skewer: a fast and accurate adapter trimmer for next-generation sequencing paired-end reads. BMC Bioinf. 2014. PMID: 24925680.

Jiang P, Thomson JA, Stewart R. Quality Control of Single-cell RNA-seq by SinQC. Bioinformatics. 2016; doi:10.1093/bioinformatics/btw176.

Jung I, Park JC, Kim S. piClust: a density based piRNA clustering algorithm. Comput Biol Chem. 2014. PMID: 24656595.

Kartashov AV, Barski A. BioWardrobe: an integrated platform for analysis of epigenomics and transcriptomics data. Genome Biol. 2015. PMID: 26248465.

Kim J, Levy E, Ferbrache A, Stepanowsky P, Farcas C, Wang S, et al. MAGI: a Node.js web service for fast MicroRNA-Seq analysis in a GPU infrastructure. Bioinformatics. 2014. PMID: 24907367.

Kroll KW, Mokaram NE, Pelletier AR, Frankhouser DE, Westphal MS, Stump PA, et al. Quality Control for RNA-Seq (QuaCRS): an integrated quality control pipeline. Cancer Inf. 2014. PMID: 25368506.

Langmead B. Aligning short sequencing reads with Bowtie. Curr Protoc Bioinf Chapter 11, Unit 11.7. 2010.

Langmead B, Trapnell C, Pop M, Salzberg SL. Ultrafast and memory-efficient alignment of short DNA sequences to the human genome. Genome Biol. 2009;10:R25.

Lassmann T, Hayashizaki Y, Daub CO. SAMStat: monitoring biases in next generation sequencing data. Bioinformatics. 2010;27(1):130–1. doi:10.1093/bioinformatics/btq614. PMID 21088025.

Lawrence M, Huber W, Pagès H, Aboyoun P, Carlson M, Gentleman R, Morgan M, Carey V. Software for computing and annotating genomic RANGES. PLoS Comput Biol 2013;9.

Le HS, Schulz MH, McCauley BM, Hinman VF, Bar-Joseph Z. Probabilistic error correction for RNA sequencing. Nucleic Acids Res. 2013. PMID: 23558750.

Lee WP, Stromberg MP, Ward A, Stewart C, Garrison EP, Marth GT. MOSAIK: a hash-based algorithm for accurate next-generation sequencing short-read mapping. PloS one. 2014. PMID: 24599324.

Leung YY, Ryvkin P, Ungar LH, Gregory BD, Wang LS. CoRAL: predicting non-coding RNAs from small RNA-sequencing data. Nucleic Acids Res. 2013. PMID: 23700308

Li H, Durbin R. Fast and accurate short read alignment with Burrows-Wheeler transform. Bioinformatics. 2009d;25(14):1754–60.

Li R, Li Y, Kristiansen K, Wang J. SOAP: short oligonucleotide alignment program. Bioinformatics. 2008a;24:713–4.

Li H, Ruan J, Durbin R. Mapping short DNA sequencing reads and calling variants using mapping quality scores. Genome Res. 2008b. PMID: 18714091.

Li R, Yu C, Li Y, Lam T-W, Yiu S-M, Kristiansen K, et al. SOAP2: an improved ultra-fast tool for short read alignment. Bioinformatics. 2009a;25:1966–7.

Li H, Handsaker B, Wysoker A, Fennell T, Ruan J, Homer N, et al. The sequence alignment/map format and SAMtools. Bioinformatics. 2009b. PMID: 19505943.

Li H, Handsaker B, Wysoker A, Fennell T, Ruan J, Homer N, et al. 1000 genome project data processing subgroup. 2009c.

Li R, Fan W, Tian G, Zhu H, He L, Cai J, et al. The sequence and de novo assembly of the giant panda genome. Nature. 2010;463(7279):311–7.

Li J, Hou J, Sun L, Wilkins JM, Lu Y, Niederhuth CE, et al.. From gigabyte to kilobyte: A bioinformatics protocol for mining large RNA-Seq transcriptomics data. PloS one. 2015a. PMID: 25902288.

Li YL, Weng JC, Hsiao CC, Chou MT, Tseng CW, Hung JH. PEAT: an intelligent and efficient paired-end sequencing adapter trimming algorithm. BMC Bioinf. 2015b. PMID: 25707528

Liao Y, Smyth GK, Shi W. The Subread aligner: fast, accurate and scalable read mapping by seed-and-vote. Nucleic Acids Res. 2013;41:e108.

Liao Y, Smyth GK, Shi W. Feature counts: an efficient general-purpose read summarization program. Bioinformatics. 2014;30:923–30.

Lindberg J, Lundeberg J. The plasticity of the mammalian transcriptome. Genomics. 2010;95:1–6.

Liu L, Li Y, Li S, Hu N, He Y, Pong R, et al. Comparison of next-generation sequencing systems. J Biomed Biotechnol. 2012;2012:251364.

Liu Y, Popp B, Schmidt B. CUSHAW3: sensitive and accurate base-space and color-space short-read alignment with hybrid seeding. PloS one. 2014. PMID: 24466273.

Love MI, Huber W, Anders S. Moderated estimation of fold change and dispersionfor RNA-seq data with DESeq2. Genome Biol. 2014;15:550.

Lunter G, Goodson M. Stampy: a statistical algorithm for sensitive and fast mapping of Illumina sequence reads. Genome Res. 2011. PMID: 20980556.

Luo GZ, Yang W, Ma YK, Wang XJ. ISRNA: an integrative online toolkit for short reads from high-throughput sequencing data. Bioinformatics. 2014. PMID: 24300438.

Mangul S, Caciula A, Al Seesi S, Brinza D, Măndoiu I, Zelikovsky A. Transcriptome assembly and quantification from Ion Torrent RNA-Seq data. BMC Genomics. 2014. PMID: 25082147.

Marguerat S, Bähler J. RNA-seq: from technology to biology. Cell Mol Life Sci. 2010;67:569–79.

Martin M. Cutadapt removes adapter sequences from high-throughput sequencing reads. EMBnet J. 2011;17(1):10–2.

McClure R, Balasubramanian D, Sun Y, Bobrovskyy M, Sumby P, Genco CA, et al. Computational analysis of bacterial RNA-Seq data. Nucleic Acids Res. 2013. PMID: 23716638.

Metzker ML. Sequencing technologies – the next generation. Nat Rev Genet. 2009;11:31–46.

Milholland B, Gombar S, Suh Y. SMiRK: an automated pipeline for miRNA analysis. J Genomics. 2015. PMID: 26613105.

Milne I, Stephen G, Bayer M, Cock PJ, Pritchard L, Cardle L, et al. Using Tablet for visual exploration of second-generation sequencing data. Brief Bioinf. 2013. PMID: 22445902.

Morgan M, Anders S, Lawrence M, Aboyoun P, Pagès H, Gentleman R. Short read: a Bioconductor package for input, quality assessment and exploration of high-throughput sequence data. Bioinformatics. 2009;25:2607–8.

Mortazavi A, Williams BA, McCue K, Schaeffer L, Wold B. Mapping and quantifying mammalian transcriptomes by RNA-Seq. Nature Methods. 2008. PMID: 18516045.

Nellore A, Collado-Torres L, Jaffe AE, Morton J, Pritt J, Alquicira-Hernández J, et al. Rail-RNA: Scalable analysis of RNA-seq splicing and coverage. bioRxiv. 2015. doi:10.1101/019067.

O'Connell J, Schulz-Trieglaff O, Carlson E, Hims MM, Gormley NA, Cox AJ. NxTrim: optimized trimming of Illumina mate pair reads. Bioinformatics. 2015. PMID: 25661542.

Okazaki Y, Furuno M, Kasukawa T, et al. Analysis of the mouse transcriptome based on functional annotation of 60,770 full-length cDNAs. Nature. 2002;420:563–73.

Okonechnikov K, et al. Qualimap 2: advanced multi-sample quality control for high-throughput sequencing data. Bioinformatics. 2015. PMID: 26428292.

Oshlack A, Wakefield MJ. Transcript length bias in RNA-seq data confounds systems biology. Biol Direct. 2009;4:14.

Pandey RV, Pabinger S, Kriegner A, Weinhäusel A. ClinQC: a tool for quality control and cleaning of Sanger and NGS data in clinical research. BMC Bioinf. 2016; doi:10.1186/s12859-016-0915.

Park JW, Tokheim C, Shen S, Xing Y. Identifying differential alternative splicing events from RNA sequencing data using RNASeq-MATS. Methods Mol Biol. 2013. PMID: 23872975.

Patel RK, Jain M. NGS QC Toolkit: a toolkit for quality control of next generation sequencing data. PloS one. 2012. PMID: 22312429.

Patro R, Mount SM, Kingsford C. Sailfish enables alignment-free isoform quantification from RNA-seq reads using lightweight algorithms. Nat Biotechnol. 2014;32:462–4. PMID: 23912058

Patro R, Duggal G, Kingsford C. Salmon: accurate, versatile and ultrafast quantification from RNA-seq data using lightweight-alignment. bioRxiv. 2015. http://dx.doi.org/10.1101/021592

Quek C, Jung CH, Bellingham SA, Lonie A, Hill AF. iSRAP – a one-touch research tool for rapid profiling of small RNA-seq data. J Extracell Vesicles. 2015. PMID: 26561006.

Quinn EM, Cormican P, Kenny EM, et al. Development of strategies for SNP detection in RNA-seq data: application to lymphoblastoid cell lines and evaluation using 1000 Genomes data. PLoS One. 2013;8(3):e58815.

Ramirez F, Dündar F, Diehl S, Grüning BA, Manke T. deepTools: a flexible platform for exploring deep-sequencing data. Nucleic Acids Res. 2014. PMID: 24799436.

Renaud G, Stenzel U, Kelso J. leeHom: adaptor trimming and merging for Illumina sequencing reads. Nucleic Acids Res. 2014. PMID: 25100869.

Ritchie ME, Phipson B, Wu D, Hu Y, Law CW, Shi W, Smyth GK. Limma powers differential expression analyses for RNA-sequencing and microarray studies. Nucleic Acids Res. 2015;43 (7):e47.

Roberts A, Trapnell C, Donaghey J, Rinn JL, Pachter L. Improving RNA-Seq expression estimates by correcting for fragment bias. Genome Biol. 2011;12:R22.

Robertson G, Schein J, Chiu R, et al. De novo assembly and analysis of RNA-seq data. Nat Methods. 2010;7:909–12.

Robinson MD, McCarthy DJ, Smyth GK. edgeR: a Bioconductor package for differential expression analysis of digital gene expression data. Bioinformatics. 2010;26:139–40.

Robinson JT, Thorvaldsdóttir H, Winckler W, Guttman M, Lander ES, et al. Integrative Genomics Viewer. Nat Biotechnol. 2011;29:24–6.

Rueda A, Barturen G, Lebrón R, Gómez-Martín C, Alganza Á, Oliver JL, et al. sRNAtoolbox: an integrated collection of small RNA research tools. Nucleic Acids Res. 2015. PMID: 26019179.

Santana-Quintero L, Dingerdissen H, Thierry-Mieg J, Mazumder R, Simonyan V. HIVE-hexagon: high-performance, parallelized sequence alignment for next-generation sequencing data analysis. PLoS ONE. 2014;9(6):e99033. doi:10.1371/journal.pone.0099033.

Sayols S, Klein H. dupRadar: assessment of duplication rates in RNA-Seq datasets. R package version 1.1.0. 2015.

Selitsky SR, Sethupathy P. tDRmapper: challenges and solutions to mapping, naming, and quantifying tRNA-derived RNAs from human small RNA-sequencing data. BMC Bioinf. 2015; doi:10.1186/s12859-015-0800-0.

Shendure J, Ji H. Next-generation DNA sequencing. Nat Biotechnol. 2008;26:1135–45.

Shi J, Dong M, Li L, Liu L, Luz-Madrigal A, Tsonis PA et al. mirPRo-a novel standalone program for differential expression and variation analysis of miRNAs. Scientific Rep. 2015. PMID: 26434581.

Shrestha RK, Lubinsky B, Bansode VB, Moinz MB, McCormack GP and Travers SA. QTrim: a novel tool for the quality trimming of sequence reads generated using the Roche/454 sequencing platform. BMC Bioinf. 2014. PMID: 24479419.

Simpson JT, Wong K, Jackman SD, et al. ABySS: a parallel assembler for short read sequence data. Genome Res. 2009;19:1117–23.

Song L, Florea L. CLASS: constrained transcript assembly of RNA-seq reads. BMC Bioinf. 2013. PMID: 23734605.

Song L, Florea L. Rcorrector: efficient and accurate error correction for Illumina RNA-seq reads. GigaScience. 2015; doi:10.1186/s13742-015-0089-y.

Song L, Sabunciyan S, Florea L. CLASS2: accurate and efficient splice variant annotation from RNA-seq reads. Nucleic Acids Res. 2016. PMID: 26975657.

Starostina E, Tamazian G, Dobrynin P, O'Brien S, Komissarov A. Cookiecutter: a tool for kmer-based read filtering and extraction. bioRxiv. 2015. doi:10.1101/024679.

Sun Z, Evans J, Bhagwate A, Middha S, Bockol M, Yan H, et al. CAP-miRSeq: a comprehensive analysis pipeline for microRNA sequencing data. BMC Genomics. 2014. PMID: 24894665.

Tarazona S, Furió-Taríl P, Turrà D, Pietro AD, José Nueda M, Ferrer A, et al. Data quality aware analysis of differential expression in RNA-seq with NOISeq R/Bioc package. Nucleic Acids Res. 2015; doi:10.1093/nar/gkv711.

Thorvaldsdóttir H, Robinson JT, Mesirov JP. Integrative Genomics Viewer (IGV): high-performance genomics data visualization and exploration. Brief Bioinform. 2013;14:178–92.

Tjaden B. De novo assembly of bacterial transcriptomes from RNA-seq data. Genome Biol. 2015. PMID: 25583448.

Trapnell C, Pachter L, Salzberg SL. TopHat: discovering splice junctions with RNA-Seq. Bioinformatics. 2009;25:1105–11.

Trapnell C, Williams BA, Pertea G, Mortazavi A, Kwan G, van Baren MJ, et al.. Transcript assembly and quantification by RNA-Seq reveals unannotated transcripts and isoform switching during cell differentiation. Nat Biotechnol. 2010;28:511–5.

Trapnell C, Roberts A, Goff L, Pertea G, Kim D, Kelley DR, et al. Differential gene and transcript expression analysis of RNA-seq experiments with TopHat and Cufflinks. Nat Protoc. 2012;7:562–78.

Urgese G, Paciello G, Acquaviva A, Ficarra E. isomiR-SEA: an RNA-Seq analysis tool for miRNAs/isomiRs expression level profiling and miRNA-mRNA interaction sites evaluation. BMC Bioinf. 2016. PMID: 27036505.

Velmeshev D, Lally P, Magistri M, Faghihi MA. CANEapp: a user-friendly application for automated next generation transcriptomic data analysis. BMC Genomics. 2016. PMID: 26758513.

Vitsios DM, Enright AJ. Chimira: analysis of small RNA sequencing data and microRNA modifications. Bioinformatics. 2015. PMID: 26093149.

Wagle P, Nikolić M, Frommolt P. QuickNGS elevates next-generation sequencing data analysis to a new level of automation. BMC Genomics. 2015. PMID: 26126663.

Wang Z, Gerstein M, Snyder M. RNA-Seq: A revolutionary tool for transcriptomics. Nat Rev Genet. 2009;10:57–63.

Wang K, Singh D, Zeng Z, Coleman SJ, Huang Y, Savich GL, et al.. MapSplice: accurate mapping of RNA-seq reads for splice junction discovery. Nucleic Acids Res. 2010. PMID: 20802226.

Wang, L, Wang, S, Li W. RSeQC: quality control of RNA-seq experiments. Bioinformatics. 2012;28(16): 2184–2185. http://doi.org/10.1093/bioinformatics/bts356

Wang L, Nie J, Sicotte H, Li Y, Eckel-Passow JE, Dasari S, et al. Measure transcript integrity using RNA-seq data. BMC Bioinf. 2016;17(1):1–16. http://doi.org/10.1186/s12859-016-0922-z Rseqc

Wilhelm BT, Marguerat S, Watt S, et al. Dynamic repertoire of a eukaryotic transcriptome surveyed at single-nucleotide resolution. Nature. 2008;453:1239–43.

Wolfien M, Rimmbach C, Schmitz U, Jung JJ, Krebs S, Steinhoff G, et al.. TRAPLINE: a standardized and automated pipeline for RNA sequencing data analysis, evaluation and annotation. BMC Bioinf. 2016. PMID: 26738481

Yang X, Liu D, Liu F, Wu J, Zou J, Xiao X, et al.. HTQC: a fast quality control toolkit for Illumina sequencing data. BMC Bioinf. 2013. PMID: 23363224.

Yuan Y, Norris C, Xu Y, Tsui KW, Ji Y and Liang H. BM-Map: an efficient software package for accurately allocating multireads of RNA-sequencing data. BMC Genomics. 2012. PMID: 23281802.

Zerbino DR, Birney E. Velvet: algorithms for de novo short read assembly using de Bruijn graphs. Genome Res. 2008;18:821–9.

Zhang T, Luo Y, Liu K, Pan L, Zhang B, Yu J, et al. BIGpre: a quality assessment package for next-generation sequencing data. Genom Proteom Bioinform. 2011;9:238–44. PMID: 22289480.

Zhang Z, Huang S, Wang J, Zhang X, Pardo Manuel de Villena F, McMillan L, et al. GeneScissors: a comprehensive approach to detecting and correcting spurious transcriptome inference owing to RNA-seq reads misalignment. Bioinformatics. 2013;29:i291–9. . PMID: 23812996

Zhao S, Xi L, Quan J, Xi H, Zhang Y, Schack DV, et al. QuickRNASeq lifts large-scale RNA-seq data analyses to the next level of automation and interactive visualization. BMC Genomics. 2016; doi:10.1186/s12864-015-2356-9.

Chapter 11
Epigenetics and Its Role in Human Cancer

Utkarsh Raj and Pritish Kumar Varadwaj

Abstract Cancer is often associated with heritable epigenetic changes, which are characterized by the change in gene expression profile without changing the underlying DNA sequence. The most prominent epigenetic modification is methylation of DNA, which to a large extent is connected to modifications of histone proteins. Epigenetic modifications resulting in a normal gene are reversible, thus endow functional flexibility and diversity to the genome, and these modifications can be cured with selective epigenetic target inhibitors. The role of epigenetics in human cancer has been vastly studied and reported in recent decade with emerging evidences about the significance of epigenetic alterations to comprehend various cellular mechanisms. The cellular mechanisms which are crucial for controlling the growth and progression were seen to be impaired by epigenetic changes, which result into development of various human cancer diseases. Although several targets for cancer epigenetics have been identified and annotated in recent past, the development of novel anticancer treatments for these targets is still in nascent stage. By recognizing the spectrum of cancer epigenetics, an array of new drug discoveries has been possible these days. In this chapter, we presented an overview of such epigenetic modifications which occurs and resulted into human cancer and the relationship between those epigenetic enzyme classes and cancer types, with a note on preclinical utilizations of inhibitors for the treatment of such cancer types. This chapter focuses on the practical understanding of human cancer epigenetics and its perspective use for drug designing.

Keywords Cancer epigenetics • DNA methylation • Histone modifications • CpG methylation • miRNAs

U. Raj • P.K. Varadwaj (✉)
Department of Bioinformatics, Indian Institute of Information Technology, CC2-4203, IIIT-Allahabad, Allahabad 211012, India
e-mail: pritish@iiita.ac.in

© Shanghai Jiao Tong University Press, Shanghai and Springer Science+Business Media Dordrecht 2017

D.-Q. Wei et al. (eds.), *Translational Bioinformatics and Its Application*, Translational Medicine Research, DOI 10.1007/978-94-024-1045-7_11

11.1 Introduction

The essential role of epigenetics in various cellular processes of normal and cancerous cells has drawn considerable attentions in recent years. It has been reported that the selective expression of gene resulted due to epigenetic modifications was instrumental in deciding the fate of proteins involved in binding of chromatin and the related machinery of transcription. These findings had revealed critical infection-related epigenetic components and pathways which are crucial for discovery of novel therapeutics. A major fraction of such reported cases comprised of epigenetic misregulation related to human cancer. Cancer epigenetics is the investigation of epigenetic alterations to the genome of tumor cells that don't essentially include a change or variation in the nucleotide succession.

The earliest indications of an epigenetic connection to cancer were resultant of various studies on gene expression and methylation of DNA. The quantum of such studies was well discussed elsewhere in a survey article by Feinberg enumerating the historical backdrop of growth of epigenetics (Feinberg and Tycko 2004). The International Cancer Genome Consortium (ICGC) has significantly strengthened these early observations. The whole genome sequencing in an immeasurable cluster of cancers has given an index of recurrent somatic mutations in several epigenetic controllers (Forbes et al. 2011; Stratton et al. 2009). Epigenetic data is contained in the cell in various forms that incorporate methylation of DNA, modification of histones (methylation, phosphorylation, acetylation, and so forth), positioning of nucleosome, and microRNA expression; these data together constitute the epigenome (Campbell and Tummino 2014). All these modifications in the chromatin structure lead to the activation or silencing of the expression of genes (Herceg and Ushijima 2010; Baylin 2008; Jones and Baylin 2007; Kouzarides 2007a; Kelly et al. 2010). Although an exhaustive understanding of epigenomic dysregulation in specific type of cancer has not been clarified yet, there exists a comprehension of tumor-specific types of modification which occurs in human cancer (Baylin and Jones 2011a; Croce 2009). The remodeling of chromatin is carried out with the help of two important mechanisms: the cytosine residue methylation in DNA and an array of posttranslational modifications (PTMs) occurring at the N-terminal ends of histone proteins. These PTMs comprise of methylation, acetylation, ubiquitylation, phosphorylation, glycosylation, sumoylation, ADP-ribosylation, citrullination, biotinylation, and carbonylation (Sidoli et al. 2012; Gardner et al. 2011). Among all such PTMs, the lysine amino acid residues of histone tails are reported to be methylated, acetylated, or ubiquitylated; also the arginine amino acid residues are found to be methylated, whereas threonine and serine amino acid residues were seen to undergo phosphorylation (Cosgrove et al. 2004; Cruickshank et al. 2010; Imhof 2006; Weake and Workman 2008; De Koning et al. 2007). These covalent alterations have the propensity to bring cross talk, which is known as the histone code that can be positively or negatively associated with specific states of transcription or chromatin organization (Cruickshank et al. 2010; Sippl and Jung 2010; Chi et al. 2010). Human tumors are considered fundamentally to be a disease of

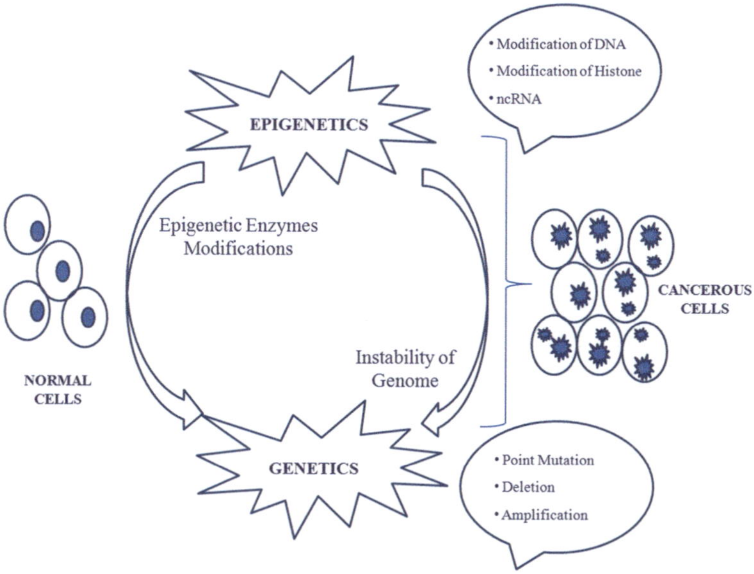

Fig. 11.1 Oncogenic mutation in normal cells due to epigenetic modifications

genetic level, where several genes get mutated or abnormally proliferated during the formation of cancer (Martin 2004; Vogelstein and Kinzler 2004). In the meantime, epigenetic modifications such as methylation of DNA, histone modifications, and microRNAs (miRNAs) lead to abnormal expression of genes (Chen et al. 2014), which induce instability of genome as explained in Fig. 11.1. Hence, an epigenetic can be safely characterized as a steadily inherited phenotype resulting out of progressions in a chromosome without apparent modifications in the DNA arrangement. In fact, all the various cellular pathways contributing to the neoplastic phenotype are affected by epigenetic genes in cancer (Jiang et al. 2015; Fornaro et al. 2016; Delpu et al. 2013). They are being investigated as biomarkers in clinical use for early detection of disease, tumor classification, and response to treatment with classical chemotherapy agents, target compounds, and epigenetic drugs (Mack et al. 2015; Andreol et al. 2013). These sorts of subtle adjustments are fundamental for ordinary cell physiology and function, aiding in the initiation or restraint of essential qualities in different phases of advancement. There are occurrences, however, in which the changes can be modified to actuate sporadic transcription of gene. In these cases, the results can incite different types of tumors in humans, with two key zones of modification, viz., methylation of DNA sequences and changes on the histones encompassing DNA. Since the disclosure of their association in the change of expression of the gene, modification of histones and methylation of DNA have been involved in sicknesses other than malignancy. One paramount part of epigenetic methylation is its reversibility; this key property has made a guaranteeing field of epigenetic treatment, which has prompted the improvement of a few FDA sanction drugs for treatment of tumors. It has likewise produced a few new and energizing thoughts for future ways of treatment.

11.2 Epigenetics and Cancer Types

In this segment, we portray the present understanding about different types of cancer with their associated epigenetic enzyme classes, taking into account that established cause-consequence might not so much specific that these receptor targets can be accepted for anticancer drug discovery. In Table 11.1, we enumerated the associations between the major types of cancer and diverse epigenetic targets classes, which can be so much informative to fetch relevant drug discovery information (Andreol et al. 2013).

11.2.1 Breast Cancer

Epigenetic modifications including methylation of DNA and remodeling of chromatin play an important role in the development of breast cancer. In the similar manner, altered expression of microRNAs has also been reported to control important genes in the breast cancer development and progression (Veeck and Esteller 2010). Besides, various synthetic drugs based on epigenetic therapy which can decrease hypermethylation of DNA and deacetylation of histones are currently in preclinical and clinical trials (Lustberg and Ramaswamy 2010).

Table 11.1 Cancer types with their associated epigenetic enzyme classes

Epigenetic enzymatic classes	Cancer type
Methyltransferases	Breast cancer, colorectal cancer, leukemia, ovarian cancer, liver cancer, prostate cancer
Deacetylases	Breast cancer
Deacetylases (classes I, II, and IV)	Colorectal cancer, leukemia, ovarian cancer, gastric cancer, prostate cancer, liver cancer
K and R methyltransferases	Breast cancer, leukemia, myeloma, ovarian cancer, prostate cancer
Acetyltransferases	Leukemia, prostate cancer
Demethylases	Kidney cancer
miRNAs regulating proteins	Breast cancer, colorectal cancer, leukemia, myeloma, lung cancer, ovarian cancer, liver cancer
Kinases/phosphatases	Liver cancer

Epigenetic enzyme classes shown as bold characters are validated targets for the associated cancer types, whereas epigenetic enzyme classes shown as normal characters are partially validated targets

11.2.2 Ovarian Cancer

Epigenetic alterations, viz., aberrant methylation of DNA and unregulated distinct microRNAs expression, have resulted in altered expression of gene favoring survival of cells (Asadollahi et al. 2010). With reference to other cancerous diseases, the therapeutic improvement went for turning around oncogenic chromatin abnormalities which have been principally examined with DNA methyltransferase and histone deacetylase inhibitors. Moreover, the examination of various epigenetic events in which there is posttranscriptional gene regulation by small noncoding microRNAs has also been done (Ahluwalia et al. 2001).

11.2.3 Colorectal Cancer

Modification or extensive loss of DNA methylation patterns at several steps involved in the progression of colorectal cancer contributes fundamentally to epigenetic dysregulation (Kim and Deng 2007). In addition to this, epigenetically miRNAs modification has also been established to perform an important part in colorectal cancer (Grady and Markowitz 2002). Since major pathways of colorectal carcinogenesis are closely linked with changes in epigenetics, emerging evidence demonstrates that the risk of colorectal cancer can be impacted by lifestyle and factors affecting environment (Nyström 2009).

11.2.4 Prostate Cancer

In this cancer, epigenetic modifications come into view earlier and more frequently than the mutations occurring at genetic level. The identification of the silencing of multiple genes due to epigenetic alterations has been done (Chin et al. 2011). Preclinical confirmation including the epigenome as a key go-between in this cancer type involved preliminary clinical tests with epigenetic drugs, viz., histone deacetylase inhibitors (Kim and Deng 2007).

11.2.5 Leukemia

DNA and histone posttranslational modifications have been exhibited to be connected with a few changes in epigenetic targets for distinctive hematologic malignancies (Bishton et al. 2007). Biological players that are being used for clinical applications comprise deacetylases (Altucci and Minucci 2009), histone and DNA methyltransferases (Rodríguez-Paredes and Esteller 2011), and miRNA

(Florean et al. 2011). In this type of cancer, the function of different epigenetic enzyme classes is being studied primarily for acute promyelocytic leukemia (Petrie et al. 2009) and acute myeloid leukemia (Florean et al. 2011).

11.2.6 Gastric Cancer

The abnormal changes that occurred due to acetylation of histones which is regulated by histone acetyltransferases and histone deacetylases have been associated with gastric cancer (W-jian et al. 2012). Despite the fact that different connections between gastrointestinal malignancy and histone acetyltransferases and histone deacetylases have been distinguished, contrasting with other cancers, fewer advances have been accounted for to treat gastrointestinal carcinogenesis with epigenetic drugs.

11.2.7 Myeloma and Lymphomas

The importance in the modulation of epigenetic enzymes has been significantly raised for the treatment of myelomas and lymphomas, mainly as combination therapies (Mahadevan and Fisher 2011). For example, histone deacetylase inhibitors and DNA methyltransferase inhibitors are being already investigated for the cure of non-Hodgkin's lymphomas (Cotto et al. 2010; Yoshimi and Kurokawa 2011).

11.2.8 Liver Cancer

Methylation of DNA and RNA interference, as well as several modifications in histones, has been established as epigenetic events which contribute to the progression of hepatocellular carcinoma (Herceg and Paliwal 2011; Tischoff 2008). At present only histone deacetylase inhibitors have been studied for the treatment of such type of carcinoma (Lachenmayer et al. 2010).

11.2.9 Lung Cancer

Epigenetic changes, viz., methylation of DNA and covalent modifications of histone and chromatin with the help of epigenetic enzymes and miRNAs, are involved in the silencing of tumor suppressor genes and in enhancing the oncogene expression level (Yang 2011; Heller et al. 2010; Herman 2004). The restoration in the expression level of silenced genes involved in epigenetics with novel targeted strategies and combined therapy with entinostat and azacitidine, as well as DNA methyltransferase inhibitors and histone deacetylase inhibitors, was examined in phase I/II clinical trials for the treatment of non-small-cell lung carcinoma (Heller et al. 2010).

11.2.10 Kidney Cancer

The modifications occurring due to methylation of DNA at an early stage of cancer may expose renal tissue to various changes taking place both at the epigenetic and genetic level, producing more cancerous growth (Dressler 2008). Currently, there are some clinical trials of phase I/II for testing inhibitors involved in deacetylation of histones which can lead to advanced renal cell carcinomas (Gan et al. 2009).

Most of the structural information about the enzyme classes involved in cancer epigenetics is well known and is used in the application of targetable molecules as mentioned in Table 11.2. First-generation epigenetic inhibitors such as histone deacetylase inhibitors and DNA methyltransferase inhibitors have as of now been affirmed for treatment of cancer. Extensive efforts have been made in current drug development that mainly focused to investigate more selective inhibitors which can be useful in multi-targeted approach therapy for the treatment of cancer.

11.2.11 Mechanisms of Epigenetic Regulation of Cancer

There are numerous chemical alterations that influence DNA, as well as RNA and proteins, and make diverse epigenetic layers. Out of these alterations, DNA methylation is the most well-studied epigenetic modification; in any case, it turns out to be progressively acknowledged that DNA methylation does not work alone yet rather is connected to different alterations, for example, histone modifications. As evident from Fig. 11.2, studies focusing on the methylation of the DNA cover almost half of the cancer epigenetic research (Razvi 2013). miRNA studies also become an integral part of the epigenetic research as nearly a quarter of the research community working on cancer epigenetics focuses on it.

11.2.11.1 DNA Methylation

Methylation of DNA is a prevalent alteration in bacteria, plants, and mammals. DNA methylation which occurs during the replication of DNA is a stable gene-silencing mechanism. It involves the addition of a –CH3 group to 5′ end of the CpG dinucleotide of the cytosine ring. Catalyzation of this reaction is being carried out by the DNA methyltransferase (DNMT) family, which comprises of DNMT1,

Table 11.2 Information about epigenetic enzyme classes and their connections with drug discovery for the cancer treatment

Epigenetic enzyme classes	Structural data	Clinical trials	Approved drugs	Known ligands
DNA methyltransferases	✓	✓	✓	✓
Histone acetylation				
Deacetylases class I, II, IV	✓	✓	✓	✓
Deacetylases class III	✓			✓
Acetyltransferases	✓	✓		✓
Histone ADP-ribosylation				
Mono-ADP-ribosyltransferases	✓			✓
Poly-ADP-ribosyltransferases	✓	✓		✓
Histone biotinylation				
Biotin ligase	✓			
Histone citrullination				
Deiminases	✓			✓
Histone glycosylation				
Glycosyltransferases/glycosidase	✓			✓
Histone methylation				
K and R methyltransferases	✓	✓		✓
Demethylases	✓	✓		✓
Histone phosphorylation				
Kinases/phosphatases	✓			
Histone ubiquitination and sumoylation				
E1, E2, and E3 enzymes	✓			✓
microRNA expression				
miRNA-regulating proteins	✓			✓

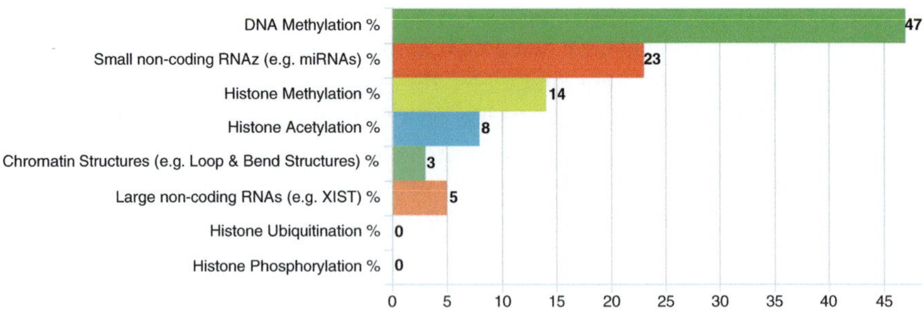

Fig. 11.2 Breakout of epigenetic research on the basis of epigenetic modifications

DNMT3A, and DNMT3B. Methylation of DNA is performed by DNMT3A and DNMT3B, which further results in the formation of 5-methylcytosine from cytosine residues of CpG dinucleotides during the formation of embryo, while DNMT1 is involved in maintaining the status of methylation during the process of embryo

Table 11.3 Aberrant DNA-methylated genes with associated cancer types

Gene	Cancer type
DNA methyltransferase	
DNMT 1	Ovarian and colorectal cancer
DNMT3b	Colon, colorectal, breast, ovarian, squamous cell carcinoma, esophageal cancers
Methyl-CpG-binding proteins	
MBD1	Prostate, lung, and colon cancer
MBD2	Prostate, lung, and colon cancer
MBD3	Colon and lung cancer
MBD4	Stomach, colon, endometrium cancers
MeCP2	Rett syndrome and prostate cancer
Kaiso	Lung, intestinal, and colon cancer

These genes can be overexpressed or silenced and cause cancer when their methylation activity is affected

formation. Conversion of 5-hydroxymethyl-2′-deoxycytidine from 5-methylcytosine is further carried out by the ten-eleven translocation (TET) family enzymes (Tahiliani et al. 2009). CpG islands are referred to the regions of DNA in the genome of human normally ranging from 0.5 to 5 kb in size and frequently occur in the promoter region of genes. Although the process of methylation of DNA in 5′ promoter region has been thoroughly investigated in various studies and has exhibited suppression of gene expression, the significance of 5-hydroxymethylation is still unclear and under investigation. Recent studies reported that the methylation of DNA occurs downstream in the promoter region (both intra- and intergenic) of genes as well as in regions with low CpG density neighboring CpG islands (Maunakea et al. 2010; Hansen et al. 2011). The following is the list of some DNA methylation genes that get altered in different human cancer types as mentioned in Table 11.3.

11.2.11.2 Histone Modification

The fundamental structure of nucleosome is comprised of histones, namely, H2A, H2B, H3, and H4, which together form the histone octamer around it (Luger et al. 1997). The N-terminals of histones protrude outward from the core of the nucleosome, whereas an array of covalent modifications, viz., methylation, acetylation, phosphorylation, sumoylation, ubiquitination, etc., occurs in the amino acid residues of this terminal. These covalent modifications can change the structure of chromatin from an open to a closed, condensed form and vice versa. The mono-, di-, or trimethylation of histones occurs at the ϵ-NH2 group of lysine amino acid residues, followed by mono- or dimethylation at arginine amino acid residues. In addition to other abovesaid covalent modifications, histone protein methylations are thought to represent an epigenetic code by the creation of binding interfaces for

Table 11.4 Aberrant histone-modified genes and their cancer-causing diseases (these histone-modifying genes can also be silenced or overexpressed by aberrant activity to cause cancer)

Gene	Cancer
SIRT1	Colon cancer
SIRT2	Glioma and gastric cancer
SIRT3	Breast cancer
SIRT4	Acute myeloid cancer
SIRT5	Breast cancer
SIRT6	Prostate, breast cancer
SIRT7	Thyroid carcinoma, breast cancer
HDAC1	Colorectal, cervical dysplasia, gastric, colon, stromal sarcomas, and prostate cancer
HDAC2	Colon and multiple gastric carcinomas
HDAC3	Prostate, colon cancer
HDAC4	Breast, prostate, and colon cancer
HDAC5	Acute myeloid cancer, colon cancer
HDAC6	Breast and acute myeloid cancer
HDAC7	Colon cancer
HDAC8	Colon cancer
HDAC9	Breast, lung cancer
HDAC10	Gastric cancer
P300	Ovarian, breast, oral, colorectal, hepatocellular cancers
CBP	Breast, colon, acute myeloid cancer, ovarian cancer
MOZ	Neurogenic progenitors, hematopoietic, leukemia cancer
PCAF	Colon cancer
MORF	Uterine, leiomyomata
Tip60	Prostate, colorectal cancer
DOT1L	Mixed lineage leukemia
MLL1	Cervical tumor
EHMT1, EHMT2	Esophageal squamous cell carcinoma

proteins involved in the regulation of chromatin (Sharma et al. 2010; Suzuki and Bird 2008).

The modifications of histones have a great impact on several biological processes, viz., transcriptional repression, activation of genes, and repair of DNA, with the exception of the packaging of chromatin. Based on the function, there are three classes of histone interacting proteins: (i) the writers which place the modification of histones, (ii) the erasers which remove these histone alterations, (iii) and the readers that recognize these alterations and may provide histone, nucleosome, or DNA-modifying enzymes (Kouzarides 2007b). The following is the list of a few histone-modified genes which get altered in different types of human cancers as mentioned in Table 11.4.

Some other posttranslational modifications are as follows:

Phosphorylation: Phosphorylation of histone plays an important role in DNA repair, gene silencing, cell cycle control, signal transduction pathway, cellular differentiation, and chromatin structure. It basically occurs on threonine, serine, and tyrosine amino acid residues.

ADP-ribosylation: ADP-ribosylation is the process in which one or more ADP-ribose molecules are added to a protein. It has been observed that histone protein is described to be mono-and poly-ADP ribosylated; thus they have a connection between codes. It is involved in different processes like cell signaling, gene regulation, and DNA repair; thus improper functioning causes disease like cancer.

Biotinylation: It is the process of attachment of biotin to a protein, nucleic acid, or other kind of molecule. This process is described in various histone variants and involved in various biological processes like cellular response to damage DNA, gene silencing, and cell proliferation.

Acetylation and Deacetylation: Acetylation and deacetylation of histone are the processes in which lysine residues of N-terminal tail of the nucleosome are acetylated or deacetylated, and it takes part in gene regulation. These processes are essential for gene regulation and catalyzed by enzymes like HDACs and HATs. Increased activity or overexpression of these enzymes can lead to formation of metastasis and tumor (cancer).

Citrullination/Deimination: This is the process of conversion of arginine residue in a protein into citrulline. In this process primary ketimine group (=NH) is replaced by a ketone group (=O) by the activity of enzymes like PADs (petideylarginine deiminases).

Carbonylation: It is the process in which RCS (reactive carbonyl species) covalently modifies cysteine residues in histone. RCS are produced with the help of enzymes like tyrosine kinases/phosphatases, transcription factors like p53, Nrf2, NFkB, and peroxiredoxins by redox signaling process. There is also a worse condition that some abnormal changes in redox signaling process can lead to the formation of malignant cells or cancerous cells.

Ubiquitination/Sumoylation: Ubiquitination and sumoylation are two very important PTM processes that play their roles in protein trafficking, cell survival, DNA damage response, signaling regulation, and cancer. Deregulation of these two processes causes abnormal activity of proteins; thus it contributes to disease like cancer.

Other PTMs: There are also kinds of posttranslational modifications such as histone proline isomerization and histone tail clipping. Histone tail clipping process removes N-terminal tail of histone molecule during transcription process and in other process which involve process like chromatin remodeling. Proline isomerization is a specific posttranslational modification that does not include covalent modifications, but it does include isomerization of proline residue.

11.2.11.3 microRNAs

microRNAs are short, endogenous, and noncoding RNAs normally 19–25 nucleotides in length and are conserved throughout evolution. These miRNAs mainly belong to the 3′ untranslated regions (3′ UTR) of target mRNA to control the expression of genes in two ways: (i) silencing of posttranscriptional process (ii) and target mRNA degradation (Rouhi et al. 2008). The relationship between several epigenetic mechanisms and these miRNAs is a quite convoluted and complex regulatory network (Iorio et al. 2010). miRNA expression is tissue specific and is controlled by various epigenetic changes, viz., methylation of DNA and alterations of histones (Friedman et al. 2009). miRNAs can also have an effect on epigenetic mechanisms which control the transcription of genes and the capability to target posttranscriptional silencing of genes (Kasinski and Slack 2011). Convincing proof now demonstrates that miRNAs are liable to both hypo- and hypermethylation in a tumor as well as tissue-specific manner (Wee et al. 2014). Recent studies also suggest that hypermethylation can mimic small chromosomal deletions or loss of heterozygosity with the help of long-range epigenetic silencing (Malkhosyan et al. 1996; Duval et al. 2001). The concept of long-range epigenetic silencing is no longer "one methylated CpG island – one silent gene" but rather involves large regions which may include several genes (Perucho 1996).

11.2.12 Identification Methods for Epigenetic Modifications Involved in Cancer

Earlier, the profiles associated with epigenetics were restricted to individual genes only, but these days, the researchers have adopted a whole genomic approach in order to find out an entire genomic profile for cancer cells versus normal cells.

Prominent methodologies for measuring CpG methylation in cells consist of:

- Bisulfite sequencing: This type of sequencing involves the use of bisulfite treatment of DNA to determine its methylation patterns. Treatment of DNA with bisulfite converts cytosine residues to uracil but leaves 5-methylcytosine residues unaffected. Subsequently, bisulfite treatment brings about specific changes in the sequence of DNA that rely on upon the methylation status of individual cytosine residues, yielding single-nucleotide determination data about the methylation status of a DNA segment.

  Different investigations can be performed on the altered sequence to retrieve this information. The target of this examination is consequently reduced to differentiating between single-nucleotide polymorphisms (cytosines and thymidine) coming about because of bisulfite conversion.
- MethyLight: It is a highly sensitive assay, equipped for recognizing methylated alleles within the sight of a 10,000-fold excess of unmethylated alleles. The test is likewise exceedingly quantitative and can very precisely decide the relative

prevalence of a particular pattern of methylation of the DNA. The most striking point of interest of MethyLight, when contrasted with existing procedures, is its capability to permit the fast screening of hundreds to thousands of samples.

- Pyrosequencing: It is a strategy for DNA sequencing (determining the order of nucleotides in DNA) based on the "sequencing by synthesis" principle. It contrasts from Sanger sequencing, in that it depends on the detection of pyrophosphate release on nucleotide incorporation, rather than chain termination with dideoxynucleotides.
- Arbitrary primed PCR: A deoxyribonucleic acid (DNA) fingerprinting technique in which one short arbitrary primer is used to amplify multiple DNA fragments of different length, which yield a fingerprint after separation in gel electrophoresis. It is also known as random amplification.
- Combined bisulfite restriction analysis (COBRA): A molecular biology technique that takes into account the sensitive quantification of levels of DNA methylation at a specific genomic locus on a DNA sequence in a small sample of genomic DNA. This method is a modification of bisulfite sequencing, which combines bisulfite conversion-based polymerase chain reaction with restriction digestion.
- Restriction landmark genomic scanning: It is a genome investigation technique that takes into consideration fast concurrent visualization of thousands of landmarks or restriction sites. By utilizing a combination of restriction enzymes, some of which are specific to modifications of DNA, the method can be used to visualize differences in the levels of methylation across the genome of a given organism.
- Chromatin immunoprecipitation (ChIP): An immunoprecipitation experimental technique used to investigate the interaction between proteins and DNA in the cell. It intends to figure out whether specific proteins are associated with specific genomic regions, such as transcription factors on promoters or other DNA-binding sites and potentially characterizing cistromes. It also aims to determine the specific location in the genome where various histone modifications are linked with, demonstrating the target of the histone modifiers.
- HELP assay (Hpall tiny fragment enrichment by ligation-mediated PCR): This is one of a few procedures utilized for figuring out if DNA has been methylated. The system can be adjusted to look at DNA methylation within and around individual genes, or it can be extended to inspect methylation in a whole genome.
- Methylated DNA immunoprecipitation: MeDIP or mDIP is a large-scale (chromosome- or genome-wide) purification technique which is used to enrich for methylated DNA sequences. It consists of isolating methylated DNA fragments via an antibody raised against 5-methylcytosine (5mC). In any case, comprehension of the methylome stays simple; its study is entangled by the way that, as other epigenetic properties, patterns vary from cell type to cell type.
- Profiling of the expression of genes using DNA microarray: Comparing mRNA levels from diseased cell lines prior and then afterward treatment with a demethylating agent.

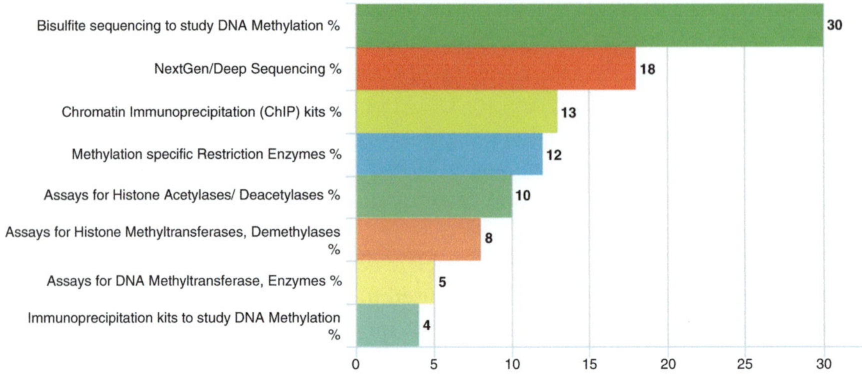

Fig. 11.3 Segmentation of the epigenetic research on the basis of deployment of assay classes

In view of the fact that bisulfite sequencing is an important marker strategy to measure CpG methylation, when one of alternate methods is utilized, results are normally affirmed utilizing this technique. The bisulfite conversion kits to study the methylation of DNA cover around 30% of the epigenetic research market as clear from the Fig. 11.3. The ChIP utilizing antibodies specific for methyl-CpG-binding domain proteins also occupies around 13% of the research based on the cancer epigenetics (Razvi 2013). The well-known methodologies for determining profiles of histone modification in healthy versus cancerous cells comprise of the two techniques, i.e., mass spectrometry and chromatin immunoprecipitation assay.

11.2.13 Clinical Use of Epigenetics

Currently, there are two noteworthy zones of interest for the clinical use of epigenetics, to be specific, biomarkers and therapeutics.

1. *Cancer Biomarkers:* The methylated genomic DNA has a wide range of properties, which makes it an alluring molecule for biomarker utility. Initially, it is steady in biofluids, for example, blood, saliva, and urine. Secondly, in most of the cases, methylation in CpG is obtained amid malignant transformation and hence specific to neoplasia. Lastly, the systems utilized for detection of methylated DNA are promptly manageable to automation.
2. *Cancer Therapeutics:* Both epigenetic proteins and protein markers are great focuses for the improvement of new anticancer medications. The verification of idea for epigenetic treatments is the FDA and EMEA approval of demethylating agents and histone acetylase (HDAC) inhibitors for the treatment of MDS, AML, and certain types of lymphomas, respectively. In any case, we ought not to overlook that these agents are nonselective without having their side effects clearly known.

11.2.14 Future Perspective for Cancer Epigenetics Therapy

Various reported studies on genome-wide mapping suggested the information about how normal genomes are developed significantly illuminating our perspective of epigenetic variations in cancer. From a period that started with recognizing cancer-specific abnormalities in the methylation of DNA, both gains and losses, we now comprehend that these must not just be connected to characterize key-related changes in chromatin but also viewed in the perspective that all genomic regions are not equal for susceptibility to these modifications (Baylin and Jones 2011b; Berman et al. 2012). An important example is the disclosure that both the gains and losses of methylation of DNA in cancer can be biased to different genomic regions associated with nuclear lamin, late-replicating DNA, i.e., enriched for low-transcription developmental genes with bivalent chromatin in the promoter region (Hegi et al. 2009). In both the embryonic and adult stem cells, such chromatin is necessary for maintenance of the state of stem cells and appears susceptible to evolve epigenetic variations during progression of tumor. This susceptibility may seriously include stresses, viz., increased ROS that intensely shifts a complex of proteins, comprising DNA methyltransferases and polycomb proteins into CpG islands. The confinement of such proteins may lead to aberrant methylation of DNA. Several stimulating examples of the effectiveness of such methodologies have emerged, and these will without a doubt increment significantly in the near future.

The use of epigenetic drugs with an intention to reestablish sensitivity to hormonal as well as cytotoxic drugs is a big challenge in cancer therapy. The restoration of the hormonal sensitivity in breast cancer is of highest medical significance and has come under serious consideration in various reported studies of the most recent decades (Baylin and Jones 2011b; Berman et al. 2012). Altogether 25% of breast cancers have the repressed estrogen receptor alpha (ERalpha) because of hypermethylation of the ER promoter and don't react to endocrine treatment. Recent reported studies established that decitabine and histone deacetylase inhibitors, viz., trichostatin A, entinostat, and scriptaid, can restore ER mRNA expression (Raha et al. 2011).

Striking advancement has been made in the last few years on the methylation of DNA and modifications of histones involved in the transcription of genes; however, the significance of these phenomena in epigenetic regulation of cancer has not been fully clarified. On the other hand, a lot of research advancement has been made in context to improve drugs associated with cancer epigenetics which can target chromatin and enzymes taking part in the modification of histones. Numerous epigenetic medications, comprising a histone deacetylase inhibitor and two DNA methyltransferase enzyme inhibitors, have been sanctioned by the FDA as viable medications for the treatment of cancer. In the meantime, different inhibitor drugs, for example, SAHA (Marks 2007), MS-275 (Bracker et al. 2009), and FK228 (Saijo et al. 2012), have as of now been the prime focus and are in step III clinical tests.

Therefore, more specific and potent inhibitors should be developed to diminish undesirable side effects. Studies on understanding of the impact of epigenetic changes occurring in cancer and tumor pathology are likely to improve the capability to detect and treat cancer (Mack et al. 2015; Wu et al. 2016).

11.3 Conclusion

The utmost challenge for researchers working on cancer therapy is to integrate the available data to understand the translational prospective of specific expression profile. However, various studies on epidemiology have acknowledged both the environmental and dietary factors as also related with cancer; animal models are capable to recognize the mechanisms as well as correlation between these environmental factors and carcinogenesis. In spite of the fact that the viability of epigenetic treatment for cancer therapy remains unproven till now, there is a strong urge to consider the use of epigenetic agents, perhaps informed by epigenetic profiling of individual patient, which may facilitate the therapeutic window for personalized medication. In addition, such studies will also facilitate the identification of specific subtypes of cancer which are more prone to chemotherapy (Lv et al. 2015; Cho 2011). This will also help in effective use of various epigenetic target inhibitors, comprising DNA-demethylating agents, histone deacetylase inhibitors, or several other promising therapies which are undergoing preclinical and clinical tests. The plethora of genetic modifications in epigenetic regulators offers numerous conceivable focuses for drug discovery and will probably draw in the consideration of the pharmaceutical business. Therefore, the characterization of the progression of tumor at the molecular level, involving both genetic and epigenetic profiles, is considered to be an important step in assessment of the progress of individualized treatment modalities as well as personalized therapies available for such cancers.

References

Ahluwalia A, Hurteau JA, Bigsby RM, Nephew KP. DNA methylation in ovarian cancer. II. Expression of DNA methyltransferases in ovarian cancer cell lines and normal ovarian epithelial cells. Gynecol Oncol. 2001;82(2):299–304.

Altucci L, Minucci S. Epigenetic therapies in haematological malignancies: searching for true targets. Eur J Cancer. 2009;45(7):1137–45.

Andreol F, Barbosa AJM, Daniele Parenti M, Rio AD. Modulation of epigenetic targets for anticancer therapy: clinicopathological relevance, structural data and drug discovery perspectives. Curr Pharm Des. 2013;19(4):578–613.

Asadollahi R, CAC H, Zhong XY. Epigenetics of ovarian cancer: from the lab to the clinic. Gynecol Oncol. 2010;118(1):81–7.

Baylin SB. Epigenetics and cancer. Mol Basis Cancer. 2008:57–65.

Baylin SB, Jones PA. A decade of exploring the cancer epigenome – biological and translational implications. Nat Rev Cancer. 2011a;11(10):726–34.

Baylin SB, Jones PA. A decade of exploring the cancer epigenome – biological and translational implications. Nat Rev Cancer. 2011b;11:726–34.

Berman BP, Weisenberger DJ, Aman JF, Hinoue T, Ramjan Z, Liu Y, Noushmehr H, Lange CPE, Dijk CMV, Tollenaar RAEM, Berg DVD, Laird PW. Regions of focal DNA hypermethylation and long-range hypomethylation in colorectal cancer coincide with nuclear lamina-associated domains. Nat Genet. 2012;44(1):40–6.

Bishton M, Kenealy M, Johnstone R, Rasheed W, Prince HM. Epigenetic targets in hematological malignancies: combination therapies with HDAC is and demethylating agents. Expert Rev Anticancer Ther. 2007;7(10):1439–49.

Bracker TU, Sommer A, Fichtner I, Faus H, Haendler B, Hess-Stumpp H. Efficacy of MS-275, a selective inhibitor of class I histone deacetylases, in human colon cancer models. Int J Oncol. 2009;35(4):909–20.

Campbell RM, Tummino PJ. Cancer epigenetics drug discovery and development: the challenge of hitting the mark. J Clin Invest. 2014;124(1):64–9.

Chen QW, Zhu XY, Li YY, Meng ZQ. Epigenetic regulation and cancer (Review). Oncol Rep. 2014;31(2):523–32.

Chi P, Allis CD, Wang GG. Covalent histone modifications–miswritten, misinterpreted and mis-erased in human cancers. Nat Rev Cancer. 2010;10(7):457–69.

Chin SP, Dickinson JL, Holloway AF. Epigenetic regulation of prostate cancer. Clin Epigenet. 2011;2(2):151–69.

Cho WC. Epigenetic alteration of microRNAs in feces of colorectal cancer and its clinical significance. Expert Rev Mol Diagn. 2011;11(7):691–4.

Cosgrove MS, Boeke JD, Wolberger C. Regulated nucleosome mobility and the histone code. Nat Struct Mol Biol. 2004;11(11):1037–43.

Cotto M, Cabanillas F, Tirado M, García MV, Pacheco E. Epigenetic therapy of lymphoma using histone deacetylase inhibitors. Clin Transl Oncol. 2010;12(6):401–9.

Croce CM. Causes and consequences of microRNA dysregulation in cancer. Nat Rev Genet. 2009;10(10):704–14.

Cruickshank MN, Besant P, Ulgiati D. The impact of histone posttranslational modifications on developmental gene regulation. Amino Acids. 2010;39(5):1087–105.

De Koning L, Corpet A, Haber JE, Almouzni G. Histone chaperones: an escort network regulating histone traffic. Nat Struct Mol Biol. 2007;14(11):997–1007.

Delpu Y, Cordelier P, Cho WC, Torrisani J. DNA methylation and cancer diagnosis. Int J Mol Sci. 2013;14(7):15029–58.

Dressler GR. Epigenetics, development, and the kidney. J Am Soc Nephrol JASN. 2008;19(11):2060–7.

Duval A, Rolland S, Compoint A, Tubacher E, Iacopetta B, Thomas G, Hamelin R. Evolution of instability at coding and non-coding repeat sequences in human MSI-H colorectal cancers. Hum Mol Genet. 2001;10:513–8.

Feinberg AP, Tycko B. The history of cancer epigenetics. Nat Rev Cancer. 2004;4:143–53.

Florean C, Schnekenburger M, Grandjenette C, Dicato M, Diederich M. Epigenomics of leukemia: from mechanisms to therapeutic applications. Epigenomics. 2011;3(5):581–609.

Forbes SA, Bindal N, Bamford S, Cole C, Kok CY, Beare D, Jia M, Shepherd R, Leung K, Menzies A, et al. COSMIC: mining complete cancer genomes in the catalogue of somatic mutations in cancer. Nucleic Acids Res. 2011;39(Database issue):D945–50.

Fornaro L, Vivaldi C, Caparello C, Musettini G, Baldini E, Masi G, Falcone A. Pharmacoepigenetics in gastrointestinal tumors: MGMT methylation and beyond. Front Biosci (Elite edition). 2016;8:170–80.

Friedman JM, Liang G, Liu CC, Wolff EM, Tsai YC, Ye W, Zhou X, Jones PA. The putative tumor suppressor microRNA-101 modulates the cancer epigenome by repressing the polycomb group protein EZH2. Cancer Res. 2009;69(6):2623–9.

Gan HK, Seruga B, Knox JJ. Targeted therapies for renal cell carcinoma – more gains from using them again. Curr Oncol. 2009;16(S1):45–51.

Gardner KE, Allis CD, Strahl BD. Operating on chromatin, a colorful language where context matters. J Mol Biol. 2011;409(1):36–46.
Grady WM, Markowitz SD. Genetic and epigenetic alterations in colon cancer. Annu Rev Genomics Hum Genet. 2002;3(37):101–28.
Hansen KD, Timp W, Bravo HC, Sabunciyan S, Langmead B, McDonald OG, Wen B, Wu H, Liu Y, Diep D, Briem E, Zhang K, Irizarry RA, Feinberg AP. Increased methylation variation in epigenetic domains across cancer types. Nat Genet. 2011;43(8):768–75.
Hegi ME, Sciuscio D, Murat A, Levivier M, Stupp R. Epigenetic deregulation of DNA repair and its potential for therapy. Clin Cancer Res. 2009;15(16):5026–31.
Heller G, Zielinski CC, Zöchbauer-Müller S. Lung cancer: from single-gene methylation to methylome profiling. Cancer Metastasis Rev. 2010;29(1):95–107.
Herceg Z, Paliwal A. Epigenetic mechanisms in hepatocellular carcinoma: how environmental factors influence the epigenome. Mutat Res/Rev Mutat Res. 2011;727(3):55–61.
Herceg Z, Ushijima T. Introduction: epigenetics and cancer. Adv Genet. 2010;70:1–23.
Herman JG. Epigenetics in lung cancer: focus on progression and early lesions. Chest. 2004;125(5 Suppl):119S–22S.
Imhof A. Epigenetic regulators and histone modification. Brief Funct Genom Proteom. 2006;5(3):222–7.
Iorio MV, Piovan C, Croce CM. Interplay between microRNAs and the epigenetic machinery: an intricate network. Biochimica et Biophysica Acta (BBA)-Gene Regul Mech. 2010;1799:694–701.
Jiang W, Liu N, Chen XZ, Sun Y, Li B, Ren XY, Qin WF, Jiang N, Xu YF, Li YQ, Ren J, Cho WC, Yun JP, Zeng J, Liu LZ, Li L, Guo Y, Mai HQ, Zeng MS, Kang TB, Jia WH, Shao JY, Ma J. Genome-wide identification of a methylation gene panel as a prognostic biomarker in nasopharyngeal carcinoma. Mol Cancer Ther. 2015;14(12):2864–73.
Jones PA, Baylin SB. The epigenomics cancer. Cell. 2007;128(4):683–92.
Kasinski AL, Slack FJ. Epigenetics and genetics. MicroRNAs en route to the clinic: progress in validating and targeting microRNAs for cancer therapy. Nat Rev Cancer. 2011;11(12):849–64.
Kelly TK, De Carvalho DD, Jones PA. Epigenetic modifications as therapeutic targets. Nat Biotechnol. 2010;28(10):1069–78.
Kim YS, Deng G. Epigenetic changes (aberrant DNA methylation) in colorectal neoplasia. Gut Liver. 2007;1(1):1–11.
Kouzarides T. Chromatin modifications and their function. Cell. 2007a;128(4):693–705.
Kouzarides T. Chromatin modifications and their function. Cell. 2007b;128:693–705.
Lachenmayer A, Alsinet C, Chang CY, Llovet JM. Molecular approaches to treatment of hepatocellular carcinoma. Dig Liver Dis. 2010;42(Suppl 3):S264–72.
Luger K, Mader AW, Richmond RK, Sargent DF, Richmond TJ. Crystal structure of the nucleosome core particle at 2.8 Å resolution. Nature. 1997;389:251–60.
Lustberg MB, Ramaswamy B. Epigenetic therapy in breast cancer. Curr Breast Cancer Rep. 2010;3(1):34–43.
Lv JF, Hu L, Zhuo W, Zhang CM, Zhou HH, Fan L. Epigenetic alternations and cancer chemotherapy response. Cancer Chemother Pharmacol. 2015;77(4):673–84.
Mack SC, Hubert CG, Miller TE, Taylor MD, Rich JN. An epigenetic gateway to brain tumor cell identity. Nat Neurosci. 2015;19(1):10–9
Mahadevan D, Fisher RI. Novel therapeutics for aggressive non- Hodgkin's lymphoma. J Clin Oncol. 2011;29(14):1876–84.
Malkhosyan S, Rampino N, Yamamoto H, Perucho M. Frameshift mutator mutations. Nature. 1996;382(6591):499–500.
Marks PA. Discovery and development of SAHA as an anticancer agent. Oncogene. 2007;26(9):1351–6.
Martin GS. The road to Src. Oncogene. 2004;23:7910–7.
Maunakea AK, Nagarajan RP, Bilenky M, Ballinger TJ, D'Souza C, Fouse SD, Johnson BE, Hong C, Nielsen C, Zhao Y, Turecki G, Delaney A, Varhol R, Thiessen N, Shchors K, Heine

VM, Rowitch DH, Xing X, Fiore C, Schillebeeckx M, Jones SJM, Haussler D, Marra MA, Wang T, Costello JF. Conserved role of intragenic DNA methylation in regulating alternative promoters. Nature. 2010;466(7303):253–7.

Nyström M. Diet and epigenetics in colon cancer. World J Gastroenterol. 2009;15(3):257.

Perucho M. Microsatellite instability: the mutator that mutates the other mutator. Nat Med. 1996;2:630–1.

Petrie K, Zelent A, Waxman S. Differentiation therapy of acute myeloid leukemia: past, present and future. Curr Opin Hematol. 2009;16(2):84–91.

Raha P, Thomas S, Munster PN. Epigenetic modulation: a novel therapeutic target for overcoming hormonal therapy resistance. Epigenomics. 2011;3(4):451–70.

Razvi E. Epigenetic research classes and assay trends. GEN Reports: Market & Tech Analysis. 2013; 1–10.

Rodríguez-Paredes M, Esteller M. Cancer epigenetics reaches mainstream oncology. Nat Med. 2011;17(3):330–9.

Rouhi A, Mager DL, Humphries RK, Kuchenbauer F. MiRNAs, epigenetics, and cancer. Mamm Genome. 2008;19:517–25.

Saijo K, Katoh T, Shimodaira H, Oda A, Takahashi O, Ishioka C. Romidepsin (FK228) and its analogs directly inhibit phosphatidylinositol 3-kinase activity and potently induce apoptosis as histone deacetylase/phosphatidylinositol 3-kinase dual inhibitors. Cancer Sci. 2012;103(11):1994–2001.

Sharma S, Kelly TK, Jones PA. Epigenetics in cancer. Carcinogenesis. 2010;31:27–36.

Sidoli S, Cheng L, Jensen ON. Proteomics in chromatin biology and epigenetics: elucidation of post-translational modifications of histone proteins by mass spectrometry. J Proteome. 2012;75(12):3419–33.

Sippl W, Jung M. Epigenetic targets in drug discovery. Chem List. 2010;104:131–2.

Stratton MR, Campbel PJ, Futreal PA. The cancer genome. Nature. 2009;458:719–24.

Suzuki MM, Bird A. DNA methylation landscapes: provocative insights from epigenomics. Nat Rev Genet. 2008;9:465–76.

Tahiliani M, Koh KP, Shen Y, Pastor WA, Bandukwala H, Brudno Y, Agarwal S, Iyer LM, Aravind L, Rao A. Conversion of 5-methylcytosine to 5-hydroxymethylcytosine in mammalian DNA by MLL partner TET1. Science. 2009;324(5929):930–5.

Tischoff I. DNA methylation in hepatocellular carcinoma. World J Gastroenterol. 2008;14(11):1741.

Veeck J, Esteller M. Breast cancer epigenetics: from DNA methylation to microRNAs. J Mammary Gland Biol Neoplasia. 2010;15(1):5–17.

Vogelstein B, Kinzler KW. Cancer genes and the pathways they control. Nat Med. 2004;10:789–99.

Weake VM, Workman JL. Histone ubiquitination: triggering gene activity. Mol Cell. 2008;29(6):653–63.

Wee S, Dhanak D, Li H, Armstrong SA, Copeland RA, Sims R, Baylin SB, Liu XS, Schweizer L. Targeting epigenetic regulators for cancer therapy. Ann N Y Acad Sci. 2014;1309(1):30–6.

W-jian S, Zhou X, Zheng J-hang LMD, Nie JY, Yang XJ, Zheng ZQ. Histone acetyltransferases and deacetylases: molecular and clinical implications to gastrointestinal carcinogenesis. Acta Biochim Biophys Sin. 2012;44(1):80–91.

Wu K, Sharma S, Venkat S, Liu K, Zhou X, Watabe K. Non-coding RNAs in cancer brain metastasis. Front Biosci (Scholar edition). 2016;8:187–202.

Yang IV, Schwartz DA. Epigenetic control of gene expression in the lung. Am J Respir Crit Care Med. 2011;183(10):1295–301.

Yoshimi A, Kurokawa M. Key roles of histone methyltransferase and demethylase in leukemogenesis. J Cell Biochem. 2011;112(2):415–24.

Chapter 12
Methods for Microbiome Analysis

Kalibulla Syed Ibrahim and Nachimuthu Senthil Kumar

Abstract Metagenomics is gaining importance as an invaluable tool as it attempts to determine directly the whole collection of genes and analyze from microbes in a particular environment where they interact with each other by exchanging nutrients, metabolites, and signaling molecules. The development of affordable next-generation sequencers has led to democratization of sequencing, but their ever-growing throughput is making data analysis increasingly complex. This has introduced a plethora of challenges with respect to design of experiments, bioinformatics, and downstream processing. This chapter aims to provide an overview of the currently available methodologies and tools for performing every individual step of a typical metagenomic data set analysis and expected to serve as a useful resource for microbial ecologists and bioinformaticians.

Keywords 16S • Analysis pipeline • Bioinformatics • Genome annotation • Human microbiome • Metagenomics • Metatranscriptomics • Next-generation sequencing

12.1 Introduction

Microorganisms make up only 1 to 2% of the mass of the body of a healthy human, but they are suggested to outnumber human cells by 10 to 1 and to outnumber human genes by 100 to 1. The majority of microbes were identified to inhabit the gut and have profound influence on human well-being (Bäckhed et al. 2005). It has been recognized that microbes play major roles in maintaining health and causing illness, but relatively little is known about the role that microbial communities play in human health and disease (Cho and Blaser 2012; Lampe 2008). The knowledge about the human microbiome that we currently possess is from culture-based approaches using the 16S rRNA technology. However, it has to be noted around 20–60% of the microbiome associated with human is uncultivable (Peterson et al.

K.S. Ibrahim • N.S. Kumar (✉)
Department of Biotechnology, Mizoram University, Aizawl, Mizoram 796 004, India
e-mail: syedibrahim.k@gmail.com; nskmzu@gmail.com

© Shanghai Jiao Tong University Press, Shanghai and Springer Science+Business Media Dordrecht 2017

D.-Q. Wei et al. (eds.), *Translational Bioinformatics and Its Application*, Translational Medicine Research, DOI 10.1007/978-94-024-1045-7_12

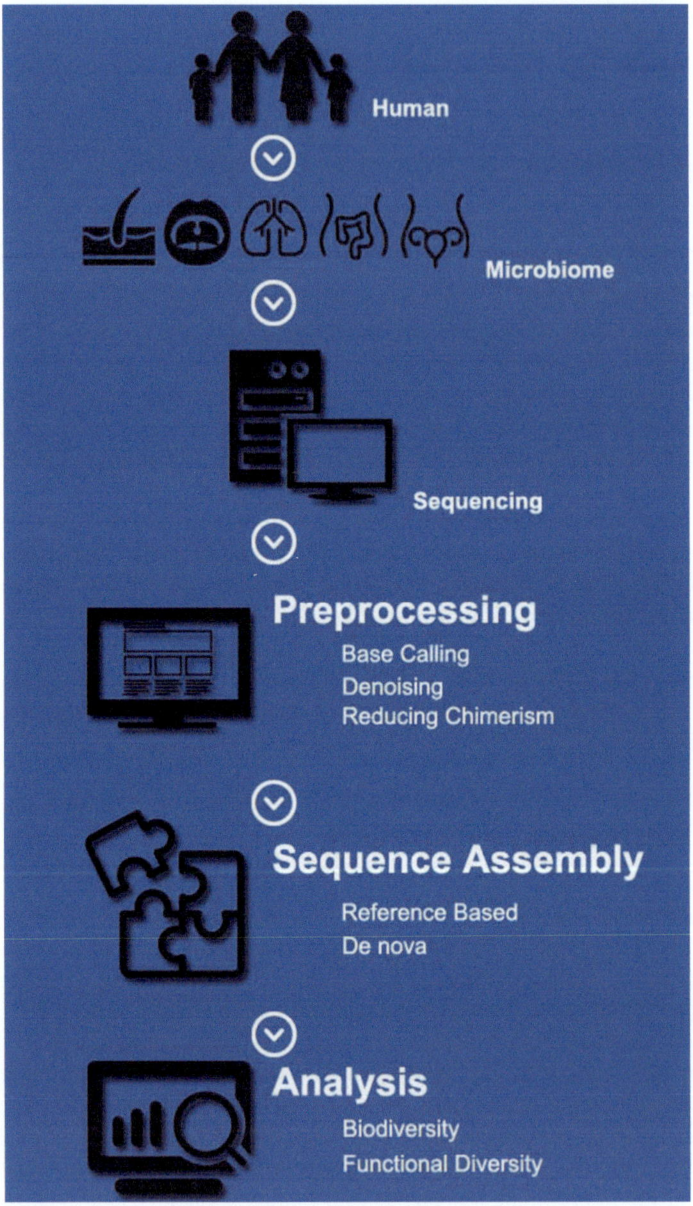

Fig. 12.1 Overall workflow of human microbiome analysis

2009). Projects such as Human Microbiome Project and MetaHIT (Qin et al. 2010) were launched with an intention to generate resources to enable a comprehensive characterization of the human microbiota and analysis of its role in human health and disease. Figure 12.1 provides an overview of the methods involved in human microbiome analysis.

Metagenomics, the term coined by Handelsman et al. (1998), made it possible for direct genetic analysis of species that are refractory to culturing methods. Using metagenomics, several types of ecosystems including extreme environments and low-diversity environments have been studied so far (Oulas et al. 2015). Decoding the metagenome and its comprehensive genetic information can also be used to understand the functional properties of the microbial community besides studying population ecology. This has provided an infinite capacity for bioprospecting that allowed the discovery of novel compounds of biotechnological commercialization (Segata et al. 2011). Initially metagenomics was used mainly to identify novel biomolecules from environmental microbial assemblages (Chistoserdova 2010). But the advent of next-generation sequencing techniques at affordable costs has allowed for more comprehensive examination of microbial communities such as comparative community metagenomics, metatranscriptomics, and metaproteomics (Simon and Daniel 2010).

In order to disentangle complex ecosystem functions of the microbial communities and fulfill the promise of metagenomics, the comprehensive data sets derived from the next-generation sequencing technologies require intensive analyses (Scholz et al. 2011). This demand has created the need for more powerful tools and software that have unprecedented potential to shed light on ecosystem functions of microbial communities and evolutionary processes.

12.2 Sequence Processing

Compared to conventional Sanger sequencing, several next-generation sequencing platforms provide huge data at much lower recurring cost. Though these technologies include a number of methods like template preparation, sequencing and imaging, and data analysis in common, it is the unique combination of specific protocols that distinguishes one technology from another. Besides that, it also determines the type of data produced from each platform, posing challenges when comparing platforms based on data quality and cost. As these new sequencing technologies produce hundreds of megabases of data at affordable costs, metagenomics is within the reach of many laboratories. The metagenomic analysis workflow begins with sampling and metadata collection and then proceeds with DNA extraction, library construction, sequencing, read preprocessing, and assembly. Either for reads, contigs, or both, binning is applied. Community composition analysis is made using databases. Some details of the workflow will be different in different sequencing facilities.

One has to take greater care when processing sequences of metagenomic data sets than when processing genomic data sets because in the later there is no fixed end point and lacks many of the quality assurance procedures (Kunin et al. 2008).

12.2.1 Preprocessing

Preprocessing of sequence reads is a critical and largely overlooked aspect of metagenomic analysis. Preprocessing comprises the base calling of raw data coming off the sequencing machines, vector screening to remove cloning vector sequence, quality trimming to remove low-quality bases (as determined by base calling), and contaminant screening to remove verifiable sequence contaminants. Errors in each of these steps can have greater downstream consequences in metagenomes.

12.2.2 Sources of Bias and Error in 16S rRNA Gene Sequencing and Reducing Sequencing Error Rates

Irrespective of the technologies used, the scientist needs to understand the quality of their data and how to reduce errors that affect downstream analyses. Two main categories of errors that are commonly observed with 16S sequencing are due to misrepresentation of the relative abundances of microbial populations in a sample (bias) and misrepresentation of an actual sequence itself due to PCR amplification and sequencing (error) (Schloss et al. 2011). Misrepresentation of the relative abundances might be due to DNA extraction method (Miller et al. 1999), PCR primer and cycling conditions, 16S rRNA gene copy number, and the actual community composition in the original sample (Hansen et al. 1998). On the other hand, error due to misrepresentation of an actual sequence is due to PCR polymerases that typically have error rates of one substitution per 105–106 bases (Cline et al. 1996), risk of chimera formation (Haas et al. 2011), and errors introduced by sequencers (Margulies et al. 2005). Because of their relative rates, sequencing errors and chimeras are of the most concern (Schloss et al. 2011).

Sequencing errors can be reduced by the following ways: removing sequence associated with low-quality scores, removing ambiguous base calls, removing mismatches to the PCR primer, or removing sequences that were shorter or longer than expected. Besides these, using denoising and removing sequences that cannot be taxonomically classified are also followed. But the later generally reduce the number of spurious OTUs and phylotypes and do not minimize the actual error rate. Laehnemann et al. (2015) has reported an extensive survey of the errors that are generated during sequencing by the commonly used high-throughput sequencing platforms.

12.2.3 Base Calling and Quality Trimming

Base calling involves identifying DNA bases from the readout of a sequencing machine. Popular base caller widely used is Phred (Ewing et al. 1998). The quality score, q, assigned to a base is related to the estimated probability, p, of erroneously calling the base by the following formula: $q = -10 \times \log^{10}(p)$. Thus, a Phred quality score of 20 corresponds to an error probability of 1%. Paracel's TraceTuner (www.paracel.com) and ABI's KB (www.appliedbiosystems.com) are the other two frequently used base callers, which behave very similar to Phred by converting raw data into accuracy probability base calls. Since metagenomic assemblies have lower coverage than genomes, errors are more likely to propagate to the consensus. Some post-processing pipelines ignore base quality scores associated with reads and contigs, and few take positional sequence depth into account as a weighting factor for consensus reliability. Because of this, for an average user, low-quality data will be indistinguishable from the rest of the data set. When poor-quality read that inadvertently passed through to gene prediction it may pass into public repositories. Hence, quality trimming is highly recommended.

12.2.4 Denoising

Denoising is a computationally intensive process that removes problematic reads and increases the accuracy of the taxonomic analysis. This is critically important for 16S metagenomic data analysis as it may give rise to erroneous OTUs, and it is sequencing platform-specific too. Illumina require less denoising than others. Though generally a considerable number of sequences is lost, it usually results in high-quality sequences (Gaspar and Thomas 2013) at certain level of stringency (Bakker et al. 2012). Notable software packages that are commonly used to correct amplicon pyrosequencing errors include Denoiser (Reeder and Knight 2010), AmpliconNoise (Quince et al. 2011), Acacia (Bragg et al. 2012), DRISEE (duplicate read inferred sequencing error estimation) (Keegan et al. 2012), JATAC (Balzer et al. 2013), and CorQ (Iyer et al. 2013). Denoiser uses frequency-based heuristics rather than statistical modeling to cluster reads and makes more accurate assessments of alpha diversity when combined with chimera-checking methods. AmpliconNoise is highly effective but is computationally intensive and applies an approximate likelihood using empirically derived error distributions to remove pyrosequencing noise from reads. These two tools do not modify individual reads; rather they both select an "error-free" read to represent reads in a given cluster. Acacia, on the other hand, is an error-correction tool, reduces the number and complexity of alignments, and uses a quicker but less sensitive statistical approach to distinguish between error and genuine sequence differences. DRISEE assess sequencing quality and provides positional error estimates that can be used to inform read trimming within a sample. JATAC algorithm identifies duplicate reads based on the flowgram that has been shown to be superior for noise removal in metagenomics amplicon data and also allows for a more effective removal of artificial duplicates. CorQ corrects homopolymer and non-homopolymer insertion and deletion (indel) errors by utilizing inherent base quality in a sequence-specific context.

12.2.5 Reducing Chimerism

Chimeras are fusion products that are formed between multiple parent sequences. These are falsely interpreted as novel organisms. These are not sequencing errors as they are not derived from a single reference sequence to which it can be mapped. Few commonly used programs for combating chimerism are Bellerephon, Pintail (Ashelford et al. 2005), ChimeraSlayer (Haas et al. 2011), Perseus (Quince et al. 2011), and Uchime (Edgar et al. 2011). The two algorithms most widely used for 16S chimera detection are Pintail and Bellerophon. The former is used by the databases like the RDP (Cole et al. 2009) and SILVA (Pruesse et al. 2007) and the latter is used by the GreenGenes 16S rRNA sequence collection (DeSantis et al. 2006). Pintail is generally visualized as 16S anomaly detection tool rather than a chimera detection tool. But interestingly most anomalies detected by Pintail were chimeras (Ashelford et al. 2005). Perseus, unlike Pintail and Bellerophon, does not use a reference database, but does require a training set of sequences similar to the sequences for characterization. Uchime outperformed ChimeraSlayer, especially in cases where the chimera has more than two parents and its performance was comparable to that of Perseus.

12.3 Sequence Assembly

The shotgun sequencing generates sequences for multiple small fragments separately which are then combined into a reconstruction of the original genome using computer programs called genome assemblers. These programs assemble shorter reads first into contigs, and these are then oriented into scaffolds that provide a more compact and concise view of the sequenced community. New challenges for the assembly process are posed by recent advances in genome sequencing technologies in terms of volume of data generated, length of the fragments, and new types of sequencing errors especially in metagenomics (Pop 2009). Earlier metagenomic data assemblies used tools that were originally designed for conventional whole-genome shotgun sequence (WGS) projects with minor parameter modifications (Wooley and Ye 2009). But recent ones have evolved as more robust specifically in handling samples containing multiple genomes. The assembly process can be approached either as reference-based assembly or as de novo assembly.

12.3.1 Reference-Based Assembly

In reference-based assembly, contigs are created by mapping on one or more reference genomes that belong to a particular species or genus, or sequences from closely related organism would have already been deposited in online data repositories and databases. Reference-based assembly tools are not computationally intensive and can perform well when metagenomic samples are derived from the areas that are extensively studied. Tools like GS Reference Mapper (Roche), MIRA 4 (Chevreux et al. 2004) or AMOS, and MetaAMOS (Treangen et al. 2013) are commonly used in metagenomics applications. The assemblies can be visualized using tools such as Tablet (Milne et al. 2009), EagleView (Huang and Marth 2008), and MapView (Bao et al. 2009). Gaps in the query genome(s) of the resulting assembly indicate that the assembly is incomplete or that the reference genomes used are too distantly related to the community under investigation.

12.3.2 De Novo Assembly

On the other hand, de novo assembly is a computationally expensive process requiring hundreds of gigabytes of memory and has long execution times, which assembles the contigs based on the de Bruijn graphs without any reference genome (Miller et al. 2010). Though tools such as EULER (Pevzner et al. 2001), FragmentGluer (Pevzner et al. 2004), Velvet (Zerbino and Birney 2008), SOAP (Li et al. 2008), ABySS (Simpson et al. 2009), and ALLPATHS (Maccallum et al. 2009) were built for assembling a single genome, even today they are used for metagenomics applications. EULER and ALLPATHS attempt to correct errors in reads prior to assembly, while Velvet and FragmentGluer deal with errors by editing the graphs. These often underperform when used for metagenome assemblies due to problems coming from variation between similar subspecies and genomic sequence similarity between different species. Besides that, difference in abundance for species in a sample was also affected by different sequencing depths for individual species. Tools like Genova (Laserson et al. 2011), MAP (Lai et al. 2012), MetaVelvet (Namiki et al. 2012), MetaVelvet-SL (Afiahayati and Sakakibara 2014), and Meta-IDBA (Peng et al. 2011) managed to create more accurate assemblies especially from data sets containing a mixture of multiple genomes by making use of k-mer frequencies to detect kinks in the de Bruijn graph. Using k-mer thresholds, they decompose the graph into subgraphs and further assemble contigs and scaffolds based on the decomposed subgraphs. The IDBA-UD algorithm (Peng et al. 2012) additionally address the issue of metagenomic sequencing technologies with uneven sequencing depths by making use of multiple depth-relative k-mer thresholds in order to remove erroneous k-mers in both low-depth and high-depth regions.

12.4 Analyzing Community Biodiversity

12.4.1 The Marker Gene

Microbial community fundamentally is a collection of individual cells, with distinct genomic DNA. In order to describe the community, it is impractical to fully sequence every genome in every cell. Hence, microbial ecology has defined a number of unique tags to distinct genomes called molecular markers. A marker is a small segment of DNA sequence that identifies the genome that contains it, eliminating the need to sequence the entire genome. Despite its numerous varieties, there are some which are desirable for properties for a good marker like it should be present in every member of a population and discriminate individuals with distinct genomes and, ideally, should differ proportionally to the evolutionary distance between distinct genomes.

By far the most ubiquitous and significant (Lane et al. 1985) is the small or 16S ribosomal RNA subunit gene (Tringe and Hugenholtz 2008) as the preferred target marker gene for bacteria and archaea. But in case of fungi and eukaryotes, the preferred marker genes are the internal transcribed spacer (ITS) and 18S rRNA gene, respectively (Oulas et al. 2015). The gold standard (Nilakanta et al. 2014) for the 16S data analysis is QIIME (Caporaso et al. 2010). Yet another popular tool is Mothur (Schloss et al. 2009) which provides the user with a variety of choices by incorporating software such as DOTUR (Schloss and Handelsman 2005), SONS (Schloss and Handelsman 2006a), Treeclimber (Schloss and Handelsman 2006b), and many more algorithms. Other tools include SILVAngs (Quast et al. 2012) and MEGAN (Huson et al. 2007). These marker gene analyses generally involve searching a reference database to find the closest match to an OTU from which a taxonomic lineage is inferred. Some widely utilized databases for 16S rRNA gene analysis include GreenGenes (DeSantis et al. 2006) and Ribosomal Database Project (Cole et al. 2007; Cole et al. 2009). Besides 16S, SILVA (Pruesse et al. 2007) also supports analysis of 18S in case of fungi and eukaryotes. Unite (Koljalg et al. 2013) can be used for analyzing ITS.

Unfortunately, not much databases are available for analyzing extremely diverse protists and viruses for which considerably less sequence information is available compared to bacteria. Humans are not only reported to carry viral particles consisting mainly of bacteriophages (Haynes and Rohwer 2011) but also a substantial number of eukaryotic viruses (Virgin et al. 2009). Like bacterial microbiota, viromes show similar patterns in different stages of human (Caporaso et al. 2011; Koenig et al. 2010), but the effects of these patterns in the human virome are mostly not understood, although certain bacteriophages in other animals are beneficial to the host (Oliver et al. 2009). The lack of a universal gene that is present in all virus makes amplicon-based studies difficult for characterizing the virome in its totality.

12.5 Analyzing Functional Diversity

This generally involves identifying protein coding sequences from the metagenomic reads and comparing the coding sequence to a database (for which some functional information is identified) to infer the function based on its similarity to sequences in the database. Besides picturing the functional composition of the community (Looft et al. 2012) or functions that associate with specific environmental or host-physiological variables (Morgan et al. 2012), they may also reveal the presence of novel genes (Nacke et al. 2011) or provide insight into the ecological conditions associated with those genes for which the function is currently unknown (Buttigieg et al. 2013). Functional annotation of metagenome involves two non-mutually exclusive steps: gene prediction and gene annotation.

12.5.1 Gene Prediction

This can be done on assembled or unassembled metagenomic sequences. Metagenomic reads/contigs are scanned for identifying protein coding genes (CDSs), as well as CRISPR repeats, noncoding RNAs, and tRNA. Predicting CDSs from metagenomic reads is a fundamental step for annotation. Gene prediction for metagenomic sequences can be performed in three ways: first, by mapping the metagenomic reads or contigs to a database of gene sequences; second, based on protein family classification; and, third, by de novo gene prediction.

Mapping the metagenomic reads or contigs to a database of gene sequences is a straightforward method of identifying coding sequences in a metagenome. This method of gene prediction can simultaneously provide functional annotation, if functional annotation of the gene is available. It comes under high-throughput gene prediction procedure as the mapping algorithms assess rapidly whether a genomic fragment is nearly identical to a database sequence or not. This method is generally useful for cataloging the specific genes present in the metagenome but not appropriate from predicting novel or highly divergent genes due to underrepresentation of genomes in sequence databases.

The second method is the most frequently used gene prediction procedure where each metagenomic read is translated into all six possible protein coding frames and each of the resulting peptides is compared to a database of protein sequences. Tools like transeq (Rice et al. 2000), USEARCH (Edgar 2010), RAPsearch (Zhao et al. 2011), and lastp (Kielbasa et al. 2011) translate reads prior to conducting protein sequence alignment. On the other hand, algorithms like blastx (Altschul et al. 1997), USEARCH with the ublast option, or lastx (Kielbasa et al. 2011) translate nucleic acid sequences on the fly. As this also relies on database, it can reveal only diverged homologues of known proteins and not useful for identifying novel types of proteins. Common functional databases includes SMART (Schultz et al. 1998), SEED (Overbeek et al. 2005), NCBI nr (Pruitt et al. 2011), the KEGG Orthology

(Kanehisa and Goto 2000), COGs (Tatusov et al. 1997), MetaCyc (Caspi et al. 2012), eggNOGs (Powell et al. 2011), and PFAM (Punta et al. 2011). Integrated pipelines with integrated functional annotation like MG-RAST (Meyer et al. 2008), MEtaGenome ANalyzer (MEGAN) (Huson et al. 2007), and HUMAnN (Abubucker et al. 2012) are also available to automate these tasks.

Contrary to the above two methods, de novo gene prediction does not rely on a reference database for identifying sequence similarity. Rather, gene prediction systems are trained by evaluating various properties of microbial genes like length of the gene, codon usage, GC bias, etc. Hence this method can potentially identify novel genes, but it is difficult to determine if the predicted gene is real or spurious. Tools like MetaGene (Noguchi et al. 2006), MetaGeneAnnotator (Meyer et al. 2008), Glimmer-MG (Kelley et al. 2011), MetaGeneMark (Zhu et al. 2010), FragGeneScan (Rho et al. 2010), Orphelia (Hoff et al. 2009), and MetaGun (Liu et al. 2013) can be used for de novo gene prediction. Yok and Rosen (2011) recommended that gene prediction in metagenomes can be improved when multiple methods are applied to the same data like following a consensus approach. Though time-consuming, this method tends to be more discriminating than 6-frame translation while annotating (Trimble et al. 2012).

RNA genes (tRNA and rRNA) can be predicted using tools like tRNAscan (Lowe and Eddy 1997). Predictions of tRNA predictions are quite reliable, but not the rRNA genes. Other types of noncoding RNA (ncRNA) genes can be detected by comparison to covariance models (Griffiths-Jones et al. 2005) and sequence-structure motifs (Macke et al. 2001). These methods are computationally intensive and take long time for metagenomic data sets. Predicting ncRNAs are usually excluded from downstream analyses because of the complexity due to lack conservation and reliable "ab initio" methods even for isolated genomes.

Errors in gene prediction mainly occur due to chimeric assemblies or frameshifts (Mavromatis et al. 2007). Hence, the quality of the gene prediction normally relies on the quality of read preprocessing and assembly. Though gene prediction can be performed with both assembled reads (contigs) and unassembled reads, it is advised to perform gene calling on both reads and contigs. It was observed that gene prediction methods used on accurately assembled sequences predicted more than 90% when compared to predictions made on unassembled reads which exhibited lower accuracy (~70%) (Mavromatis et al. 2007).

12.5.2 Functional Annotation

Functional annotation of metagenomic data sets are made by comparing predicted genes to existing, previously annotated sequences or by context annotation. Metagenomic data will have complications when predicted proteins are short and lack homologues. Databases that are used for comparing protein sequences include alignment of profiles from the protein families in TIGRFAMs (Selengut et al. 2007), PFAM (Finn et al. 2008), COGs (Tatusov et al. 1997), and RPS-BLAST

(Markowitz et al. 2006). PFAMs allow the identification and annotation of protein domains. TIGRFAM database include models for both domain and full-length proteins. Though COGs also allow the annotation of the full-length proteins, it is not frequently updated like PFAMs and TIGRFAMs. It is also recommended not to assign protein function solely based on BLAST results as there is a potential for error propagation through databases (Kyrpides and Ouzounis 1999). Context-based annotation methods include genomic neighborhood (Overbeek et al. 1999), gene fusion (Marcotte et al. 1999b), phylogenetic profiles (Pellegrini et al. 1999), and coexpression (Marcotte et al. 1999a). It was observed that neighborhood analysis was performed on metagenomic data, which, combined with homology searches, inferred specific functions for 76% of the metagenomic data sets (83% when nonspecific functions are considered) (Harrington et al. 2007) and is expected to be used in predicting protein function in metagenomic data in the future.

12.6 Metatranscriptomic Analysis

Metatranscriptome sequencing has been recently employed to identify RNA-based regulation and expression in human microbiome (Markowitz et al. 2008). Accessing metatranscriptome of the microbiome through metatranscriptomic shotgun sequencing (RNAseq) has led to the discovery and characterization of new genes from uncultivated microorganisms under different conditions. Few investigations (Bikel et al. 2015; Franzosa et al. 2014; Gosalbes et al. 2011; Jorth et al. 2014; Knudsen et al. 2016) have been performed on metatranscriptomics combined with metagenomics. Several technical issues affecting large-scale application of metatranscriptomics are discussed by Bikel et al. (2015). Though metagenomic and metatranscriptomic data provide extensive information about microbiota diversity, gene content, and their potential functions, it is very difficult to say whether DNA comes from viable cells or whether the predicted genes are expressed at all and, if so, under what conditions and to what extent (Gosalbes et al. 2011).

The bioinformatics pipeline for analyzing the data obtained from a metatranscriptomic experiment is similar to the one used in metagenomics. Basically this is also divided in two strategies: mapping sequence reads to reference genomes or pathways to identify the taxonomical classification of active microorganism and the functionality of their expressed genes and de novo assembly of new transcriptomes. For de novo assembly, there are several programs like SOAPdenovo (Li et al. 2009), ABySS (Birol et al. 2009), and Velvet-Oases (Schulz et al. 2012) that have been reported to be successfully applied to the metatranscriptome assembly (Ghaffari et al. 2014; Ness et al. 2011; Schulz et al. 2012; Shi et al. 2011). A program specially developed for de novo transcriptome assembly from short-read RNAseq data, Trinity (Haas et al. 2013), is one of the most used bioinformatics tools to assemble de novo transcriptomes of different species. It is a very efficient and sensitive in recovering full-length transcripts and isoforms (Ghaffari et al. 2014; Luria et al. 2014).

Metatranscriptome analyses involves stepwise approach for detecting the different RNA types, such as rRNAs, mRNAs, and other noncoding RNAs, facilitating the researchers to study them individually. The reads can be then compared against the small subunit rRNA reference database (SSUrdb), and later, the remaining unassigned reads can be analyzed with the large subunit rRNA reference database (LSUrdb)—the databases compiled from SILVA (Pruesse et al. 2007) or RDP II (Cole et al. 2009). The non-rRNA representation can be then identified from subtracting the LSU rRNA and SSU rRNA reads from the total reads obtained. The non-rRNAs are finally carried forward for functional analyses.

The functional diversity of the microbiome can be predicted by annotating metatranscriptomic sequences with known functions. cDNA sequences with no significant homology with any of the rRNA databases can be searched against the NCBI nr protein database using BLASTX (Altschul et al. 1997). The sequence reads that contain protein coding genes are identified, and their sequences are compared to the coding sequences of protein databases like the Kyoto Encyclopedia of Genes and Genomes (KEGG), protein family annotations (Pfam), gene ontologies (GO), and clusters of orthologous groups (COG). Thus, the function of the query sequence is assigned based on its homology to sequences functionally annotated in all the above mentioned databases.

Pipelines for combined metatranscriptomics with metagenomics include INFERNAL, a powerful tool for predicting small RNA in the metagenomic data (Nawrocki and Eddy 2013). HUMAnN is another automated pipeline, an offline platform, to determine the presence/absence and abundance of microbial pathways and gene families in a community directly from metagenomic sequence. This is done by converting sequence reads into coverage and abundance and finally summarizes the gene families and pathways in a microbial community (Abubucker et al. 2012). Other offline platforms used to analyze metagenomic data include MEGAN (Huson et al. 2007), IMG/M server (Markowitz et al. 2008), RAST (MG-RAST) (Meyer et al. 2008), and JCVI Metagenomics Reports (METAREP) (Goll et al. 2010).

12.7 Statistical Analysis in Metagenomics

Statistical analysis plays critical role in analyzing and interpreting metagenomic data. Even simple metagenomic analysis like estimate of species diversity seems not so straightforward and obviously needs statistical attention due to the artifacts created during the sequencing (discussed earlier).

Often critical statistical analysis precedes with normalization (i.e., normalization to a reference sample), a step that reduces the systematic variance and improves the overall performance for downstream statistical analysis. These include methods like centering, autoscaling, pareto scaling, range scaling, vast scaling, log transformation, and power transformation. Appropriate selection of data pretreatment methods and its significance have been by van den Berg et al. (2006).

Robust data processing algorithms for wide range of analysis are mostly created using repositories available from the open-source R-project (http://www.R-project.org) and the R-based bioconductor project (https://www.bioconductor.org/). These are widely considered to be the most complete collection of up-to-date statistical and machine learning algorithms (Xia et al. 2009). Common statistical analysis includes missing value estimation, diversity analysis, and univariate and multivariate analysis like directions of variance, cluster analysis, etc.

Missing value exclusion, missing value replacement, and missing value imputation can be identified by probabilistic PCA (PPCA), Bayesian PCA (BPCA), and singular value decomposition imputation (SVDImpute) (Stacklies et al. 2007; Steinfath et al. 2008). Univariate analysis includes three commonly used methods—fold-change analysis, *t*-tests, and volcano plots. The *t*-test attempts to determine whether the means of two groups are distinct. With *t*-value, *P*-value can be calculated which can be used to determine whether the distinction is statistically significant or not. The volcano plots compare the size of the fold change to the statistical significance level (Xia et al. 2009). Directions of maximum variance can be determined by principal component analysis (PCA) and partial least squares discriminant analysis (PLS-DA). PCA is an unsupervised method aiming to find the directions of maximum variance in a data set (X) without referring to the class labels (Y), and PLS-DA is a supervised method that uses multiple linear regression technique to find the direction of maximum covariance between a data set (X) and the class membership (Y). In both PCA and PLS-DA, the original variables are summarized into much fewer variables using their weighted averages called scores. Diversity analysis can be performed by estimating the alpha diversity, which provides a summary statistic of a single population, or beta diversity, which gives organismal composition between populations. Chao1 (Chao 1984), abundance-based coverage estimator (ACE) (Chao et al. 1993), and Jackknife (Heltshe and Forrester 1983) measure alpha diversity, species richness, and evenness (species distribution) expected within a single population. These results in collector's or rarefaction curves (Colwell and Coddington 1994). Alpha diversity is often quantified by the Shannon Index (Shannon 1948) or by Simpson Index (Simpson 1949). Beta diversity can be measured by simple taxa overlap or quantified by the Bray-Curtis dissimilarity (Bray and Curtis 1957) or UniFrac (Lozupone and Knight 2005). Two major approaches of clustering analysis include Hierarchical clustering and partitional clustering. Hierarchical, which is also called as agglomerative clustering, begins with each sample considered as separate cluster and then proceeds to combine them until all samples belong to one cluster. The result of hierarchical clustering is usually presented as a dendrogram or as a heat map, which displays the actual data values using color gradients. Clustering methods include average linkage, complete linkage, single linkage, and Ward's linkage. A dissimilarity measure includes Euclidean distance, Pearson's correlation, and Spearman's rank correlation. On the other hand, partitional clustering attempts to directly decompose the data set into a user-specified number of disjoint clusters. This uses methods like k-means clustering and self-organizing map (SOM). k-Means clustering create k clusters such that the sum of squares from points to the assigned cluster centers' is minimized. SOM is an unsupervised neural network based around the concept of a grid of interconnected nodes, each of which contains a model.

Demands for new statistical methods to support emerging trends in metagenomics applications have resulted in more efficient implementations and better data visualization to lodge the tremendous increase in data analysis workloads. Web-based server with its user-friendly interface, comprehensive data processing options, wide array of statistical methods, and extensive data visualization and analysis support are playing key role. Servers like GEPAS (Herrero et al. 2003) and CARMAweb (Rainer et al. 2006), MG-RAST (Meyer et al. 2008), MEGAN (Huson et al. 2007), QIIME (Caporaso et al. 2010), Mothur (Schloss et al. 2009), and MetaboAnalyst (Xia et al. 2015) are few worth mentioning. Table 12.1 summarizes some the commonly used tools in microbiome analysis and their internet resources.

12.8 Analysis of Human Microbiome

Since birth, continuous exposure to microbial challenges has shaped the human microbiome and whose perturbation affects both human health and disease (Segal and Blaser 2014). In recent years, the knowledge about composition, distribution, and variation of bacteria in the human body has dramatically increased. Besides external factors like air, food, and environment, routine activity, habit, and physiology create selective pressure of each organism. In order to understand the influence of human microbiome, several studies have assessed the microbial compositions in different locations like stool, nasal, skin, vaginal, and oral of health and unhealthy individuals (Kraal et al. 2014). Thus, determining the extent of the variability of the human microbiome is therefore crucial for understanding the microbiology, genetics, and ecology of the microbiome. Besides that, it is useful for practical issues in designing experiments and interpretation of clinical studies (Zhou et al. 2014).

Study demonstrating the feasibility of using the composition of the gut microbiome to detect the presence of precancerous and cancerous lesions (Zackular et al. 2014), ethnic relation to significant differences in the vaginal microbiome (Fettweis et al. 2014), and discovery closely related oligotypes, differing sometimes by as little as a single nucleotide, showing dramatic different distributions among oral sites and among individuals (Eren et al. 2014), a less robustly interrogated placental microbiome by Aagaard et al. (2014), altered interactions between intestinal microbes, and the mucosal immune system resulting in inflammatory bowel disease (IBD) (Kostic et al. 2014) have taken us to the next level of understanding the human microbiome. Other studies like understanding of the etiology and pathogenesis of reflux disorders and esophageal adenocarcinoma (Yang et al. 2014) and altered microbiome on pulmonary responses (Segal and Blaser 2014) will be definitely be critical and open door for future investigations.

Table 12.1 Selected tools and their resources for microbiome analysis

Software	Brief description	URLs
Preprocessing		
FASTX-Toolkit	A collection of command line tools for short-read FASTA/FASTQ files preprocessing	hannonlab.cshl.edu/fastx_toolkit
FastQC	A quality-controlled tool for high-throughput sequence data	www.bioinformatics.babraham.ac.uk/projects/fastqc
SolexaQA	Calculates sequence quality statistics and creates visual representations of data quality for second-generation sequencing data	http://solexaqa.sourceforge.net/
Lucy 2	Raw DNA sequence trimming and visualization tool based on the command-line tool Lucy1	http://www.complex.iastate.edu/download/Lucy2/index.html
CutAdapt	Removal of adapter sequences from high-throughput sequencing data	https://code.google.com/p/cutadapt/
NGS QC Toolkit	Perl-based stand-alone program package for the quality control (QC)	www.nipgr.res.in/ngsqctoolkit.html
Trimmomatic	Employed in trimming tasks for illumina paired-end and single ended data	http://www.usadellab.org/cms/?page=trimmomatic
ngsShoRT	Commonly used preprocessing algorithms in PERL	research.bioinformatics.udel.edu/genomics/ngsShoRT/
QC-Chain	A fast, accurate, and holistic NGS data quality-controlled method	http://www.computationalbioenergy.org/qc-chain.html
Meta-QC-Chain	A tool that combines multiple QC functions like identifying potential errors, quality trimming filters for poor sequencing quality bases and reads, and contamination screening that identifies higher eukaryotic species, which are considered as contamination for metagenomic data	http://computationalbioenergy.org/meta-qc-chain.html
PathoQC	A streamlined toolkit for preprocessing next-generation sequencing data	http://sourceforge.net/projects/PathoScope/
PRINSEQ	Provides summary statistics of FASTA (and QUAL) or FASTQ files	http://prinseq.sourceforge.net/
Denoising		
AmpliconNoise	A collection of programs for the removal of noise from 454 sequenced PCR amplicons	https://code.google.com/p/ampliconnoise/

(continued)

Table 12.1 (continued)

Software	Brief description	URLs
DADA	Algorithm for fast and accurate removal errors from PCR-amplified sequence data	http://sites.google.com/site/dadadenoiser
Acacia	Error corrector for pyrosequenced amplicon reads	http://sourceforge.net/projects/acaciaerrorcorr
Chimera detection		
UCHIME	Detects very low-divergent chimeras with a reference database	http://drive5.com/usearch/manual/uchime_algo.html
ChimeraSlayer	A chimeric sequence detection utility, compatible with near-full-length Sanger sequences and shorter 454-FLX sequences	http://microbiomeutil.sourceforge.net/
DECIPHER	Chimeric sequence detection utility developed using the R statistical programming language	http://decipher.cee.wisc.edu
Reference-based assembly		
Newbler (Roche)	Assembling sequence data generated by the 454 GS-series of pyrosequencing platforms sold by 454 Life Sciences, a Roche Diagnostics company	http://swes.cals.arizona.edu/maier_lab/kartchner/documentation/index.php/home/docs/newbler
MIRA 4	A multi-pass DNA sequence data assembler/mapper for whole-genome and EST/RNASeq projects	http://mira-assembler.sourceforge.net/docs/DefinitiveGuideToMIRA.html
AMOS	A consortium committed to the development of open-source whole-genome assembly software	http://amos.sourceforge.net/wiki/index.php/AMOS
MetAMOS	An integrated assembly and analysis pipeline for metagenomic data	http://www.cbcb.umd.edu/software/metamos
Bowtie 2	Ultrafast and memory-efficient tool for aligning sequencing reads to long reference sequences	http://bowtie-bio.sourceforge.net/bowtie2/index.shtml
De novo assembly		
EULER	A suite of programs for correcting errors in short reads (454 and Illumina) and assembling them	http://euler-assembler.ucsd.edu/
Velvet	de Bruijn graph-based single-genome assembler for short reads	https://www.ebi.ac.uk/~zerbino/velvet/
SOAPdenovo	The program is specially designed to assemble Illumina GA short reads for the human-sized genomes	http://soap.genomics.org.cn/soapdenovo.html
Abyss	A de novo, parallel, paired-end sequence assembler that is designed for short reads	http://www.bcgsc.ca/platform/bioinfo/software/abyss
MetaVelvet	Modified and extended de Bruijn graph-based single-genome	http://metavelvet.dna.bio.keio.ac.jp/

(continued)

Table 12.1 (continued)

Software	Brief description	URLs
	assembler, Velvet, for de novo metagenomic assembly	
MetaVelvet-SL	An extended Velvet assembler for detecting chimeric nodes by using supervised machine learning	metavelvet.dna.bio.keio.ac.jp/
Meta-IDBA	An iterative de Bruijn graph de novo short-read assembler specially designed for de novo metagenomic assembly	http://i.cs.hku.hk/~alse/hkubrg/projects/metaidba/
Genovo	A tool for de novo metagenomic assembly and handle reads with length>1000	http://cs.stanford.edu/group/genovo/
Trinity	Assembles transcript sequences from Illumina RNAseq data	https://github.com/trinityrnaseq/trinityrnaseq/wiki
Binning tools		
TETRA	To calculate how well tetranucleotide usage patterns in DNA sequences correlate	http://www.megx.net/tetra/index.html
PhylopythiaS	Taxonomic assignment of metagenome sequences among from three different models	http://phylopythias.cs.uni-duesseldorf.de/index.php?phase=wait
TACOA	Predicting the taxonomic origin of genomic fragments from metagenomic data sets by combining the advantages of the k-NN approach with a smoothing kernel function	http://www.cebitec.uni-bielefeld.de/index.php/2-uncategorised/99-tacoa?highlight=WyJ0YWNvYSJd
ESOM	A suite of programs to perform data mining tasks like clustering, visualization, and classification	http://databionic-esom.sourceforge.net/
ClaMS	A sequence composition-based classifier for metagenomic sequences	http://clams.jgi-psf.org/
MetaPhyler	Taxonomic classifier for metagenomic shotgun reads	http://metaphyler.cbcb.umd.edu/
SOrt-ITEMS	A similarity-based binning method	http://metagenomics.atc.tcs.com/binning/SOrt-ITEMS/
PhymmBL	Hybrid classifier tool which combines analysis from both Phymm and BLAST and produces even higher accuracy	http://www.cbcb.umd.edu/software/phymm/
MetaCluster	Binning and annotating short paired-end reads	http://i.cs.hku.hk/~alse/MetaCluster/
OTU clustering		
UCLUST	An algorithm that divides a set of sequences into clusters	http://www.drive5.com/usearch

(continued)

Table 12.1 (continued)

Software	Brief description	URLs
CD-HIT-OTU	Fast and accurate in identifying true OTUs and produces much fewer spurious OTUs	http://weizhong-lab.ucsd.edu/cd-hit-otu
TBC	Algorithm for defining operational taxonomic units (OTUs) without multiple sequence alignment	http://www.ezbiocloud.net/sw/tbc
16S databases		
RDP	A database that provides quality-controlled, aligned, and annotated bacterial and archaeal 16S rRNA sequences, fungal 28S rRNA sequences, and a suite of analysis tools to the scientific community	http://rdp.cme.msu.edu/index.jsp
SILVA	A comprehensive online resource for quality-checked and aligned ribosomal RNA sequence data	http://www.arb-silva.de
GreenGenes	A collection of tools for choosing phylogenetically specific probes, interpreting microarray results, and aligning/annotating novel sequences	http://greengenes.lbl.gov
EzTaxon	A database that covers uncultured species often found in microbial ecological studies	http://www.ezbiocloud.net/eztaxon
ITS database		
UNITE	A platform for sequence-borne identification of ectomycorrhizal asco- and basidiomycetes	http://unite.ut.ee
Sub-cellular localization		
CoBaltDB	Predicting prokaryotic protein localizations	http://www.umr6026.univ-rennes1.fr/english/home/research/basic/software/cobalten
PSLpred	To predict the subcellular location for Gram-negative bacteria proteins	http://www.imtech.res.in/raghava/pslpred/
CELLO	Predicting subcellular localization of proteins for Gram-negative bacteria by support vector machines based on n-peptide compositions	http://cello.life.nctu.edu.tw/
PSORT-B	To predict the subcellular location for Gram-positive or Gram-negative bacterial proteins	
Functional annotation databases		
BLAST nr	Basic Local Alignment Search Tool against nonredundant database	http://blast.ncbi.nlm.nih.gov/Blast.cgi
SWISSPROT	Manually annotated and reviewed section of the UniProt Knowledgebase (UniProtKB)	http://www.uniprot.org/

(continued)

Table 12.1 (continued)

Software	Brief description	URLs
KEGG	Kyoto Encyclopedia of Genes and Genomes	http://www.genome.jp/kegg/
SEED	A resource that provide consistent and accurate genome annotations across thousands of genomes	http://pubseed.theseed.org/
EggNOG	A database of orthologous groups and functional annotation	http://eggnogdb.embl.de/#/app/home
COG/KOG	EuKaryotic Orthologous Groups (KOG) is a eukaryote-specific version of the Clusters of Orthologous Groups (COG) tool for identifying ortholog and paralog proteins	http://genome.jgi.doe.gov/Tutorial/tutorial/kog.html
PFAM	Collection of protein families, each represented by multiple sequence alignments and hidden Markov models (HMMs)	http://pfam.xfam.org/
TIGRFAMs	A resource consisting of curated multiple sequence alignments, hidden Markov models (HMMs) for protein sequence classification, and associated information designed to support automated annotation of (mostly prokaryotic) proteins	http://www.jcvi.org/cgi-bin/tigrfams/index.cgi
MetaBioMe	A web resource to find novel homologues for known commercially useful enzymes (CUEs) in metagenomic data sets and completed bacterial genomes	http://metasystems.riken.jp/metabiome/
TSdb	The transporter substrate database (TSdb)—a central repository of formatted substrate information of transporters as well as their annotation	http://tsdb.cbi.pku.edu.cn/
TCDB	Functional and phylogenetic classification of membrane transport proteins	http://www.tcdb.org/
CAZy	A specialist database dedicated to the display and analysis of genomic, structural, and biochemical information on carbohydrate-active enzymes (CAZymes)	http://www.cazy.org/
dbCAN	A database for carbohydrate-active enzymes	http://csbl.bmb.uga.edu/dbCAN/
Annotation of metagenomics sequences		
MetaGeneMark	For gene prediction in metagenomes	http://exon.gatech.edu/meta_gmhmmp.cgi
MetaGeneAnnotator	A gene-finding program for prokaryote and phage	http://metagene.nig.ac.jp/

(continued)

Table 12.1 (continued)

Software	Brief description	URLs
Prodigal	A gene-finding program developed at Oak Ridge National Laboratory and the University of Tennessee	http://prodigal.ornl.gov/
Orphelia	A metagenomic ORF finding tool for the environmental DNA sequences with unknown phylogenetic origin	http://orphelia.gobics.de/
FragGeneScan	Software for predicting prokaryotic genes in incomplete assemblies or complete genomes	http://sourceforge.net/projects/fraggenescan/
PILER-CR	Software for finding CRISPR repeats	http://www.drive5.com/pilercr/
tRNAscan-SE	A web server for predicting tRNAs	http://lowelab.ucsc.edu/tRNAscan-SE/
WebMGA	A web server for rapid metagenomic data analysis using fast and effective algorithms	http://weizhong-lab.ucsd.edu/metagenomic-analysis/
METAREP	An open-source tool to view, query, browse, and compare metagenomics annotation profiles from short reads or assemblies	http://jcvi.org/metarep/
STAMP	A software package for analyzing taxonomic or metabolic profiles	http://kiwi.cs.dal.ca/Software/STAMP
CoMet	A web server for fast comparative functional profiling of metagenomes	http://comet.gobics.de/
RAMMCAP	Analysis and comparison of very large metagenomes with fast clustering and functional annotation	http://weizhong-lab.ucsd.edu/rammcap/cgi-bin/rammcap.cgi
Analytical pipelines for 16S		
CARMA	Software pipeline for characterizing the taxonomic composition and genetic diversity of short-read metagenomes	http://www.cebitec.uni-bielefeld.de/index.php/2-uncategorised/47-carma?highlight=WyJjYXJtYSJd
IMG/M	Integrated Microbial Genomes with Microbiome	http://img.jgi.doe.gov/m/doc/background.html
MG-RAST	An automated analysis platform for metagenomes	http://metagenomics.anl.gov/
Mothur	An open-source software for microbial ecology community analysis	http://www.mothur.org
QIIME	An open-source bioinformatics pipeline for performing microbiome analysis from raw DNA sequencing data	http://qiime.org
SILVAngs	A data analysis service for ribosomal RNA gene (rDNA) amplicon	https://www.arb-silva.de/ngs/

(continued)

Table 12.1 (continued)

Software	Brief description	URLs
	reads from high-throughput sequencing	
MEGAN	Tool for studying the taxonomic content from short-read metagenomes	http://ab.inf.uni-tuebingen.de/software/megan5/
WATERS	From 16S rDNA contigs to biological interpretation and analysis	http://code.google.com/p/waters16s
RDPipeline	For processing large rRNA sequence libraries (single-strand and paired-end reads) obtained through high-throughput sequencing technology	https://pyro.cme.msu.edu
VAMPS	A collection of tools for visualization and analyze data for microbial population structures and distributions	http://vamps.mbl.edu
Genboree	A web-based platform for multi-omic research and data analysis using the latest bioinformatics tools	http://genboree.org
SnoWMan	Pipeline for analysis of microbiome data	https://snowman.genome.tugraz.at/snowman

12.9 Conclusion

Human microbiota includes microorganisms living on the surface and inside the body. They are important for the host's health. These are highly dynamic and can be influenced by a number of factors such as age, diet, and physiology. Studies have shown that most of the human adult microbiota lives in the gut and follows specific microbial signatures but with high intraindividual variability over time. Any alterations of the human gut microbiome can play a role in disease development. Thus, exploring microbiome could make themselves as potent target for diagnostic and therapeutic applications. Since early microbial studies were bases on the direct cultivation and isolation of microbes, clinical applications posed several limitations especially growth conditions. Studies have shown that not all microbes are currently uncultivable. Methods to study cultivable organisms are also not suitable for the study of entire microbiome. Metagenomics helped in the direct genetic analysis of genomes contained within an environmental sample without the need for cultivating. Metagenomic studies using NGS-based methods can be approached by amplifying 16S rRNA genes using specific primers or through whole-genome shotgun sequencing. 16S sequences identified can be used to describe their community relative abundance and/or their phylogenetic relationships by clustering into operational taxonomic units (OTUs) using databases of previously annotated sequences. In whole-genome shotgun sequencing approach, where random primers are used for amplifying all microbial genes, the relative abundances of genes and pathways can be determined by comparing the sequences to functional databases.

Next-generation sequencing (NGS) technologies not only increased the throughput of bases sequenced/run but also reduced sequencing costs. This had a major impact on the field of metagenomics where a specific microbiome can be qualitatively and quantitatively characterized in depth without the selection bias and constraints associated with cultivation methods. Continuous advancements in sequencing technologies have not only allowed address more complex habitats but also have imposed growing demands on bioinformatic data post-processing. Analyzing the huge amount of data by these technologies has become the bottleneck especially in case of larger metagenome projects. From assembly to analysis, bioinformatic post-processing requires dedicated data integration pipelines, some of which have yet to be developed.

Acknowledgments The authors wish to acknowledge the Department of Biotechnology (DBT), Govt. of India, New Delhi for the financial support in the form of State Biotech Hub (BT/04/NE/2009) and Bioinformatics Infrastructure Facility (BT/BI/12/060/2012 (NERBIF-MUA)). KSI acknowledge the financial assistance provided by DST-SERB, New Delhi through young scientist scheme (YSS/2014/000657).

References

Aagaard K, Ma J, Antony KM, Ganu R, Petrosino J, Versalovic J. The placenta harbors a unique microbiome. Sci Transl Med. 2014;6(237):237–65.

Abubucker S, Segata N, Goll J, Schubert A, Izard J, Cantarel B, Rodriguez-Mueller B, Zucker J, Thiagarajan M, Henrissat B, White O, Kelley S, Methe B, Schloss P, Gevers D, Mitreva M, Huttenhower C. Metabolic reconstruction for metagenomic data and its application to the human microbiome. PLoS Comput Biol. 2012;8:e1002358.

Afiahayati SK, Sakakibara Y. MetaVelvet-SL: an extension of the Velvet assembler to a de novo metagenomic assembler utilizing supervised learning. DNA Res. 2014;22(1):69–77.

Altschul S, Madden T, Schäffer A, Zhang J, Zhang Z, Miller W, Lipman D. Gapped BLAST and PSI-BLAST: a new generation of protein database search programs. Nucleic Acids Res. 1997;25(17):3389–402.

Ashburner M, Ball CA, Blake JA, Botstein D, Butler H, Cherry JM, Davis AP, Dolinski K, Dwight SS, Eppig JT, Harris MA, Hill DP, Issel-Tarver L, Kasarskis A, Lewis S, Matese JC, Richardson JE, Ringwald M, Rubin GM, Sherlock G. Gene ontology: tool for the unification of biology. The Gene Ontology Consortium. Nat Genet. 2000;25(1):25–9.

Ashelford K, Chuzhanova N, Fry J, Jones A, Weightman A. At least 1 in 20 16S rRNA sequence records currently held in public repositories is estimated to contain substantial anomalies. Appl Environ Microbiol. 2005;71:7724–36.

Bäckhed F, Ley R, Sonnenburg J, Peterson D, Gordon J. Host-bacterial mutualism in the human intestine. Science. 2005;307(5717):1915–20.

Bakker M, Tu Z, Bradeen J, Kinkel L. Implications of pyrosequencing error correction for biological data interpretation. PLoS One. 2012;7(8):e44357.

Balzer S, Malde K, Grohme M, Jonassen I. Filtering duplicate reads from 454 pyrosequencing data. Bioinformatics. 2013;29(7):830–6.

Bao H, Guo H, Wang J, Zhou R, Lu X, Shi S. MapView: visualization of short reads alignment on a desktop computer. Bioinformatics. 2009;25(12):1554–5.

Bikel S, Valdez-Lara A, Cornejo-Granados F, Rico K, Canizales-Quinteros S, Soberón X, Del Pozo-Yauner L, Ochoa-Leyva A. Combining metagenomics, metatranscriptomics and viromics to explore novel microbial interactions: towards a systems-level understanding of human microbiome. Comput Struct Biotechnol J. 2015;13:390–401.

Birol I, Jackman SD, Nielsen CB, Qian JQ, Varhol R, Stazyk G, Morin RD, Zhao Y, Hirst M, Schein JE, Horsman DE, Connors JM, Gascoyne RD, Marra MA, Jones SJ. De novo transcriptome assembly with ABySS. Bioinformatics. 2009;25(21):2872–7.

Bragg L, Stone G, Imelfort M, Hugenholtz P, Tyson G. Fast, accurate error-correction of amplicon pyrosequences using Acacia. Nat Methods. 2012;9(5):425–6.

Bray J, Curtis J. An ordination of upland forest communities of southern Wisconsin. Ecol Monogr. 1957;27:325–49.

Buttigieg P, Hankeln W, Kostadinov I, Kottmann R, Yilmaz P, Duhaime M, Glöckner F. Ecogenomic perspectives on domains of unknown function: correlation-based exploration of marine metagenomes. PLoS One. 2013;8(3):e50869.

Caporaso J, Kuczynski J, Stombaugh J, Bittinger K, Bushman F, Costello E, Fierer N, Peña A, Goodrich J, Gordon J, Huttley G, Kelley S, Knights D, Koenig J, Ley R, Lozupone C, McDonald D, Muegge B, Pirrung M, Reeder J, Sevinsky J, Turnbaugh P, Walters W, Widmann J, Yatsunenko T, Zaneveld J, Knight R. QIIME allows analysis of high-throughput community sequencing data. Nat Methods. 2010;7(5):335–6.

Caporaso J, Lauber C, Costello E, Berg-Lyons D, Gonzalez A, Stombaugh J, Knights D, Gajer P, Ravel J, Fierer N, Gordon J, Knight R. Moving pictures of the human microbiome. Genome Biol. 2011;12(5):R50.

Caspi R, Altman T, Dreher K, Fulcher C, Subhraveti P, Keseler I, Kothari A, Krummenacker M, Latendresse M, Mueller L, Ong Q, Paley S, Pujar A, Shearer A, Travers M, Weerasinghe D, Zhang P, Karp P. The MetaCyc database of metabolic pathways and enzymes and the BioCyc collection of pathway/genome databases. Nucleic Acids Res. 2012;40:D742–53.

Chao A. Nonparametric estimation of the number of classes in a population. Scand J Stat. 1984;11:265–70.

Chao A, Ma M-C, Yang M. Stopping rules and estimation for recapture debugging with unequal failure rates. Biometrika. 1993;80:93–201.

Chevreux B, Pfisterer T, Drescher B, Driesel A, Müller W, Wetter T, Suhai S. Using the miraEST assembler for reliable and automated mRNA transcript assembly and SNP detection in sequenced ESTs. Genome Res. 2004;14(6):1147–59.

Chistoserdova L. Recent progress and new challenges in metagenomics for biotechnology. Biotechnol Lett. 2010;32:1351–9.

Cho I, Blaser M. The Human Microbiome: at the interface of health and disease. Nat Rev Genet. 2012;13(4):260–70.

Cline J, Braman J, Hogrefe H. PCR fidelity of pfu DNA polymerase and other thermostable DNA polymerases. Nucleic Acids Res. 1996;24:3546–51.

Cole J, Chai B, Farris R, Wang Q, Kulam-Syed-Mohideen A, McGarrell D, Bandela A, Cardenas E, Garrity G, Tiedje J. The ribosomal database project (RDP-II): introducing myRDP space and quality controlled public data. Nucleic Acids Res. 2007;35(Database issue):D169–72.

Cole J, Wang Q, Cardenas E, Fish J, Chai B, Farris R, Kulam-Syed-Mohideen A, McGarrell D, Marsh T, Garrity G, Tiedje J. The Ribosomal Database Project: improved alignments and new tools for rRNA analysis. Nucleic Acids Res. 2009;37:D141–5.

Colwell R, Coddington J. Estimating terrestrial biodiversity through extrapolation. Philos Trans R Soc Lond B. 1994;345:101–18.

DeSantis T, Hugenholtz P, Larsen N, Rojas M, Brodie E, Keller K, Huber T, Dalevi D, Hu P, Andersen G. Greengenes, a chimera-checked 16S rRNA gene database and workbench compatible with ARB. Appl Environ Microbiol. 2006;72(7):5069–72.

Edgar R. Search and clustering orders of magnitude faster than BLAST. Bioinformatics. 2010;26 (19):2460–1.

Edgar RC, Haas BJ, Clemente JC, Quince C, Knight R. UCHIME improves sensitivity and speed of chimera detection. Bioinformatics. 2011;27:2194–200.

Eren AM, Borisy GG, Huse SM, Mark Welch JL. Oligotyping analysis of the human oral microbiome. Proc Natl Acad Sci U S A. 2014;111(28):E2875–84.

Ewing B, Hillier L, Wendl MC, Green P. Base-calling of automated sequencer traces using phred. I. Accuracy assessment. Genome Res. 1998;8(3):175–85.

Fettweis JM, Brooks JP, Serrano MG, Sheth NU, Girerd PH, Edwards DJ, Strauss 3rd JF, Jefferson KK, Buck GA. Differences in vaginal microbiome in African American women versus women of European ancestry. Microbiology. 2014;160(Pt 10):2272–82.

Finn RD, Tate J, Mistry J, Coggill PC, Sammut SJ, Hotz HR, Ceric G, Forslund K, Eddy SR, Sonnhammer EL, Bateman A. The Pfam protein families database. Nucleic Acids Res. 2008;36 (Database issue):D281–8.

Finn RD, Bateman A, Clements J, Coggill P, Eberhardt RY, Eddy SR, Heger A, Hetherington K, Holm L, Mistry J, Sonnhammer EL, Tate J, Punta M. Pfam: the protein families database. Nucleic Acids Res. 2013;42:D222–30.

Franzosa EA, Morgan XC, Segata N, Waldron L, Reyes J, Earl AM, Giannoukos G, Boylan MR, Ciulla D, Gevers D, Izard J, Garrett WS, Chan AT, Huttenhower C. Relating the metatranscriptome and metagenome of the human gut. Proc Natl Acad Sci U S A. 2014;111 (22):E2329–38.

Gaspar JM, Thomas WK. Assessing the consequences of denoising marker-based metagenomic data. PLoS One. 2013;8(3):e60458.

Ghaffari N, Sanchez-Flores A, Doan R, Garcia-Orozco KD, Chen PL, Ochoa-Leyva A, Lopez-Zavala AA, Carrasco JS, Hong C, Brieba LG, Rudiño-Piñera E, Blood PD, Sawyer JE, Johnson CD, Dindot SV, Sotelo-Mundo RR, Criscitiello MF. Novel transcriptome assembly and improved annotation of the whiteleg shrimp (*Litopenaeus vannamei*), a dominant crustacean in global seafood mariculture. Sci Rep. 2014;4:7081.

Goll J, Rusch DB, Tanenbaum DM, Thiagarajan M, Li K, Methé BA, Yooseph S. METAREP: JCVI metagenomics reports—an open source tool for high-performance comparative metagenomics. Bioinformatics. 2010;26(20):2631–2.

Gosalbes MJ, Durbán A, Pignatelli M, Abellan JJ, Jiménez-Hernández N, Pérez-Cobas AE, Latorre A, Moya A. Metatranscriptomic Approach to Analyze the Functional Human Gut Microbiota. PLoS One. 2011;6(3):e17447.

Griffiths-Jones S, Moxon S, Marshall M, Khanna A, Eddy SR, Bateman A. Rfam: annotating non-coding RNAs in complete genomes. Nucleic Acids Res. 2005;33(Database issue):D121–4.

Haas BJ, Gevers D, Earl AM, Feldgarden M, Ward DV, Giannoukos G, Ciulla D, Tabbaa D, Highlander SK, Sodergren E, Methe B, DeSantis TZ, Petrosino JF, Knight R, Birren BW. Chimeric 16S rRNA sequence formation and detection in Sanger and 454-pyrosequenced PCR amplicons. Genome Res. 2011;21(3):494–504.

Haas BJ, Papanicolaou A, Yassour M, Grabherr M, Blood PD, Bowden J, Couger MB, Eccles D, Li B, Lieber M, Macmanes MD, Ott M, Orvis J, Pochet N, Strozzi F, Weeks N, Westerman R, William T, Dewey CN, Henschel R, Leduc RD, Friedman N, Regev A. De novo transcript sequence reconstruction from RNA-seq using the Trinity platform for reference generation and analysis. Nat Protoc. 2013;8(8):1494–512.

Handelsman J, Rondon MR, Brady SF, Clardy J, Goodman RM. Molecular biological access to the chemistry of unknown soil microbes: a new frontier for natural products. Chem Biol. 1998;5 (10):R245–9.

Hansen M, Tolker-Nielsen T, Givskov M, Molin S. Biased 16S rDNA PCR amplification caused by interference from DNA flanking the template region. FEMS Microbiol Ecol. 1998;26:141–9.

Harrington ED, Singh AH, Doerks T, Letunic I, von Mering C, Jensen LJ, Raes J, Bork P. Quantitative assessment of protein function prediction from metagenomics shotgun sequences. Proc Natl Acad Sci U S A. 2007;104(35):13913–8.

Haynes M, Rohwer F. Metagenomics of the Human Body Springer. New: York; 2011.

Heltshe J, Forrester N. Estimating species richness using the jackknife procedure. Biometrics. 1983;39:1–11.

Herrero J, Al-Shahrour F, Diaz-Uriarte R, Mateos A, Vaquerizas JM, Santoyo J, Dopazo J. GEPAS: A web-based resource for microarray gene expression data analysis. Nucleic Acids Res. 2003;31(13):3461–7.

Hoff KJ, Lingner T, Meinicke P, Tech M. Orphelia: predicting genes in metagenomic sequencing reads. Nucleic Acids Res. 2009;37(Web Server issue):W101–5.

Huang W, Marth G. EagleView: a genome assembly viewer for next-generation sequencing technologies. Genome Res. 2008;18(9):1538–43.

Huson DH, Auch AF, Qi J, Schuster SC. MEGAN analysis of metagenomic data. Genome Res. 2007;17(3):377–86.

Iyer S, Bouzek H, Deng W, Larsen B, Casey E, Mullins JI. Quality score based identification and correction of pyrosequencing errors. PLoS One. 2013;8(9):e73015.

Jorth P, Turner KH, Gumus P, Nizam N, Buduneli N, Whiteley M. Metatranscriptomics of the human oral microbiome during health and disease. MBio. 2014;5(2):e01012–4.

Kanehisa M, Goto S. KEGG: kyoto encyclopedia of genes and genomes. Nucleic Acids Res. 2000;28(1):27–30.

Keegan KP, Trimble WL, Wilkening J, Wilke A, Harrison T, D'Souza M, Meyer F. A platform-independent method for detecting errors in metagenomic sequencing data: DRISEE. PLoS Comput Biol. 2012;8(6):e1002541.

Kelley DR, Liu B, Delcher AL, Pop M, Salzberg SL. Gene prediction with Glimmer for metagenomic sequences augmented by classification and clustering. Nucleic Acids Res. 2011;40(1):e9.

Kielbasa SM, Wan R, Sato K, Horton P, Frith MC. Adaptive seeds tame genomic sequence comparison. Genome Res. 2011;21(3):487–93.

Knudsen BS, Kim HL, Erho N, Shin H, Alshalalfa M, Lam LL, Tenggara I, Chadwich K, Van Der Kwast T, Fleshner N, Davicioni E, Carroll PR, Cooperberg MR, Chan JM, Simko JP. Application of a clinical whole-transcriptome assay for staging and prognosis of prostate cancer diagnosed in needle core biopsy specimens. J Mol Diagn. 2016; pii: S1525–1578(16) 00051–9. doi:10.1016/j.jmoldx.2015.12.006.

Koenig JE, Spor A, Scalfone N, Fricker AD, Stombaugh J, Knight R, Angenent LT, Ley RE. Succession of microbial consortia in the developing infant gut microbiome. Proc Natl Acad Sci U S A. 2010;108(Suppl 1):4578–85.

Koljalg U, Nilsson RH, Abarenkov K, Tedersoo L, Taylor AF, Bahram M, Bates ST, Bruns TD, Bengtsson-Palme J, Callaghan TM, Douglas B, Drenkhan T, Eberhardt U, Duenas M, Grebenc T, Griffith GW, Hartmann M, Kirk PM, Kohout P, Larsson E, Lindahl BD, Lucking R, Martin MP, Matheny PB, Nguyen NH, Niskanen T, Oja J, Peay KG, Peintner U, Peterson M, Poldmaa K, Saag L, Saar I, Schussler A, Scott JA, Senes C, Smith ME, Suija A, Taylor DL, Telleria MT, Weiss M, Larsson KH. Towards a unified paradigm for sequence-based identification of fungi. Mol Ecol. 2013;22(21):5271–7.

Kostic AD, Xavier RJ, Gevers D. The microbiome in inflammatory bowel disease: current status and the future ahead. Gastroenterology. 2014;146(6):1489–99.

Kraal L, Abubucker S, Kota K, Fischbach MA, Mitreva M. The prevalence of species and strains in the human microbiome: a resource for experimental efforts. PLoS One. 2014;9(5):e97279.

Kunin V, Copeland A, Lapidus A, Mavromatis K, Hugenholtz P. A bioinformatician's guide to metagenomics. Microbiol Mol Biol Rev. 2008;72(4):557–78. , Table of Contents

Kyrpides NC, Ouzounis CA. Whole-genome sequence annotation: 'going wrong with confidence'. Mol Microbiol. 1999;32(4):886–7.

Laehnemann D, Borkhardt A, McHardy AC (2015) Denoising DNA deep sequencing data-high-throughput sequencing errors and their correction. Brief Bioinform

Lai B, Ding R, Li Y, Duan L, Zhu H. A de novo metagenomic assembly program for shotgun DNA reads. Bioinformatics. 2012;28(11):1455–62.

Lampe JW. The Human Microbiome Project: getting to the guts of the matter in cancer epidemiology. Cancer Epidemiol Biomark Prev. 2008;17(10):2523–4.

Lane DJ, Pace B, Olsen GJ, Stahl DA, Sogin ML, Pace NR. Rapid determination of 16S ribosomal RNA sequences for phylogenetic analyses. Proc Natl Acad Sci U S A. 1985;82(20):6955–9.

Laserson J, Jojic V, Koller D. Genovo: de novo assembly for metagenomes. J Comput Biol. 2011;18(3):429–43.

Li R, Li Y, Kristiansen K, Wang J. SOAP: short oligonucleotide alignment program. Bioinformatics. 2008;24(5):713–4.

Li R, Yu C, Li Y, Lam TW, Yiu SM, Kristiansen K, Wang J. SOAP2: an improved ultrafast tool for short read alignment. Bioinformatics. 2009;25(15):1966–7.

Liu Y, Guo J, Hu G, Zhu H. Gene prediction in metagenomic fragments based on the SVM algorithm. BMC Bioinf. 2013;14(Suppl 5):S12.

Looft T, Johnson TA, Allen HK, Bayles DO, Alt DP, Stedtfeld RD, Sul WJ, Stedtfeld TM, Chai B, Cole JR, Hashsham SA, Tiedje JM, Stanton TB. In-feed antibiotic effects on the swine intestinal microbiome. Proc Natl Acad Sci U S A. 2012;109(5):1691–6.

Lowe TM, Eddy SR. tRNAscan-SE: a program for improved detection of transfer RNA genes in genomic sequence. Nucleic Acids Res. 1997;25(5):955–64.

Lozupone C, Knight R. UniFrac: a new phylogenetic method for comparing microbial communities. Appl Environ Microbiol. 2005;71(12):8228–35.

Luria N, Sela N, Yaari M, Feygenberg O, Kobiler I, Lers A, Prusky D. De-novo assembly of mango fruit peel transcriptome reveals mechanisms of mango response to hot water treatment. BMC Genomics. 2014;15:957.

Maccallum I, Przybylski D, Gnerre S, Burton J, Shlyakhter I, Gnirke A, Malek J, McKernan K, Ranade S, Shea TP, Williams L, Young S, Nusbaum C, Jaffe DB. ALLPATHS 2: small genomes assembled accurately and with high continuity from short paired reads. Genome Biol. 2009;10(10):R103.

Macke TJ, Ecker DJ, Gutell RR, Gautheret D, Case DA, Sampath R. RNAMotif, an RNA secondary structure definition and search algorithm. Nucleic Acids Res. 2001;29(22):4724–35.

Marcotte EM, Pellegrini M, Ng HL, Rice DW, Yeates TO, Eisenberg D. Detecting protein function and protein-protein interactions from genome sequences. Science. 1999a;285 (5428):751–3.

Marcotte EM, Pellegrini M, Thompson MJ, Yeates TO, Eisenberg D. A combined algorithm for genome-wide prediction of protein function. Nature. 1999b;402(6757):83–6.

Margulies M, Egholm M, Altman WE, Attiya S, Bader JS, Bemben LA, Berka J, Braverman MS, Chen YJ, Chen Z, Dewell SB, Du L, Fierro JM, Gomes XV, Godwin BC, He W, Helgesen S, Ho CH, Irzyk GP, Jando SC, Alenquer ML, Jarvie TP, Jirage KB, Kim JB, Knight JR, Lanza JR, Leamon JH, Lefkowitz SM, Lei M, Li J, Lohman KL, Lu H, Makhijani VB, McDade KE, McKenna MP, Myers EW, Nickerson E, Nobile JR, Plant R, Puc BP, Ronan MT, Roth GT, Sarkis GJ, Simons JF, Simpson JW, Srinivasan M, Tartaro KR, Tomasz A, Vogt KA, Volkmer GA, Wang SH, Wang Y, Weiner MP, Yu P, Begley RF, Rothberg JM. Genome sequencing in microfabricated high-density picolitre reactors. Nature. 2005;437(7057):376–80.

Markowitz VM, Ivanova N, Palaniappan K, Szeto E, Korzeniewski F, Lykidis A, Anderson I, Mavromatis K, Kunin V, Garcia Martin H, Dubchak I, Hugenholtz P, Kyrpides NC. An experimental metagenome data management and analysis system. Bioinformatics. 2006;22 (14):e359–67.

Markowitz VM, Ivanova NN, Szeto E, Palaniappan K, Chu K, Dalevi D, Chen IM, Grechkin Y, Dubchak I, Anderson I, Lykidis A, Mavromatis K, Hugenholtz P, Kyrpides NC. IMG/M: a data management and analysis system for metagenomes. Nucleic Acids Res. 2008;36:D534–8.

Mavromatis K, Ivanova N, Barry K, Shapiro H, Goltsman E, McHardy AC, Rigoutsos I, Salamov A, Korzeniewski F, Land M, Lapidus A, Grigoriev I, Richardson P, Hugenholtz P, Kyrpides NC. Use of simulated data sets to evaluate the fidelity of metagenomic processing methods. Nat Methods. 2007;4(6):495–500.

Meyer F, Paarmann D, D'Souza M, Olson R, Glass EM, Kubal M, Paczian T, Rodriguez A, Stevens R, Wilke A, Wilkening J, Edwards RA. The metagenomics RAST server – a public resource for the automatic phylogenetic and functional analysis of metagenomes. BMC Bioinf. 2008;9:386.

Miller DN, Bryant JE, Madsen EL, Ghiorse WC. Evaluation and optimization of DNA extraction and purification procedures for soil and sediment samples. Appl Environ Microbiol. 1999;65 (11):4715–24.

Miller JR, Koren S, Sutton G. Assembly algorithms for next-generation sequencing data. Genomics. 2010;95(6):315–27.

Milne I, Bayer M, Cardle L, Shaw P, Stephen G, Wright F, Marshall D. Tablet – next generation sequence assembly visualization. Bioinformatics. 2009;26(3):401–2.

Morgan XC, Tickle TL, Sokol H, Gevers D, Devaney KL, Ward DV, Reyes JA, Shah SA, LeLeiko N, Snapper SB, Bousvaros A, Korzenik J, Sands BE, Xavier RJ, Huttenhower C. Dysfunction of the intestinal microbiome in inflammatory bowel disease and treatment. Genome Biol. 2012;13(9):R79.

Nacke H, Engelhaupt M, Brady S, Fischer C, Tautzt J, Daniel R. Identification and characterization of novel cellulolytic and hemicellulolytic genes and enzymes derived from German grassland soil metagenomes. Biotechnol Lett. 2011;34(4):663–75.

Namiki T, Hachiya T, Tanaka H, Sakakibara Y. MetaVelvet: an extension of Velvet assembler to de novo metagenome assembly from short sequence reads. Nucleic Acids Res. 2012;40(20): e155.

Nawrocki EP, Eddy SR. Computational identification of functional RNA homologs in metagenomic data. RNA Biol. 2013;10(7):1170–9.

Ness RW, Siol M, Barrett SC. De novo sequence assembly and characterization of the floral transcriptome in cross- and self-fertilizing plants. BMC Genomics. 2011;12:298. [936]

Nilakanta H, Drews KL, Firrell S, Foulkes MA, Jablonski KA. A review of software for analyzing molecular sequences. BMC Res Note. 2014;7:830.

Noguchi H, Park J, Takagi T. MetaGene: prokaryotic gene finding from environmental genome shotgun sequences. Nucleic Acids Res. 2006;34(19):5623–30.

Oliver KM, Degnan PH, Hunter MS, Moran NA. Bacteriophages encode factors required for protection in a symbiotic mutualism. Science. 2009;325(5943):992–4.

Oulas A, Pavloudi C, Polymenakou P, Pavlopoulos GA, Papanikolaou N, Kotoulas G, Arvanitidis C, Iliopoulos I. Metagenomics: tools and insights for analyzing next-generation sequencing data derived from biodiversity studies. Bioinf Biol Insight. 2015;9:75–88.

Overbeek R, Fonstein M, D'Souza M, Pusch GD, Maltsev N. The use of gene clusters to infer functional coupling. Proc Natl Acad Sci U S A. 1999;96(6):2896–901.

Overbeek R, Begley T, Butler RM, Choudhuri JV, Chuang HY, Cohoon M, de Crecy-Lagard V, Diaz N, Disz T, Edwards R, Fonstein M, Frank ED, Gerdes S, Glass EM, Goesmann A, Hanson A, Iwata-Reuyl D, Jensen R, Jamshidi N, Krause L, Kubal M, Larsen N, Linke B, McHardy AC, Meyer F, Neuweger H, Olsen G, Olson R, Osterman A, Portnoy V, Pusch GD, Rodionov DA, Ruckert C, Steiner J, Stevens R, Thiele I, Vassieva O, Ye Y, Zagnitko O, Vonstein V. The subsystems approach to genome annotation and its use in the project to annotate 1000 genomes. Nucleic Acids Res. 2005;33(17):5691–702.

Pellegrini M, Marcotte EM, Thompson MJ, Eisenberg D, Yeates TO. Assigning protein functions by comparative genome analysis: protein phylogenetic profiles. Proc Natl Acad Sci U S A. 1999;96(8):4285–8.

Peng Y, Leung HC, Yiu SM, Chin FY. Meta-IDBA: a de Novo assembler for metagenomic data. Bioinformatics. 2011;27(13):i94–101.

Peng Y, Leung HC, Yiu SM, Chin FY. IDBA-UD: a de novo assembler for single-cell and metagenomic sequencing data with highly uneven depth. Bioinformatics. 2012;28(11):1420–8.

Peterson J, Garges S, Giovanni M, McInnes P, Wang L, Schloss JA, Bonazzi V, McEwen JE, Wetterstrand KA, Deal C, Baker CC, Di Francesco V, Howcroft TK, Karp RW, Lunsford RD, Wellington CR, Belachew T, Wright M, Giblin C, David H, Mills M, Salomon R, Mullins C,

Akolkar B, Begg L, Davis C, Grandison L, Humble M, Khalsa J, Little AR, Peavy H, Pontzer C, Portnoy M, Sayre MH, Starke-Reed P, Zakhari S, Read J, Watson B, Guyer M. The NIH Human Microbiome Project. Genome Res. 2009;19(12):2317–23.

Pevzner PA, Tang H, Waterman MS. An Eulerian path approach to DNA fragment assembly. Proc Natl Acad Sci U S A. 2001;98(17):9748–53.

Pevzner PA, Tang H, Tesler G. De novo repeat classification and fragment assembly. Genome Res. 2004;14(9):1786–96.

Pop M. Genome assembly reborn: recent computational challenges. Brief Bioinform. 2009;10 (4):354–66.

Powell S, Szklarczyk D, Trachana K, Roth A, Kuhn M, Muller J, Arnold R, Rattei T, Letunic I, Doerks T, Jensen LJ, von Mering C, Bork P. eggNOG v3.0: orthologous groups covering 1133 organisms at 41 different taxonomic ranges. Nucleic Acids Res. 2011;40(Database issue): D284–9.

Pruesse E, Quast C, Knittel K, Fuchs BM, Ludwig W, Peplies J, Glockner FO. SILVA: a comprehensive online resource for quality checked and aligned ribosomal RNA sequence data compatible with ARB. Nucleic Acids Res. 2007;35(21):7188–96.

Pruitt KD, Tatusova T, Brown GR, Maglott DR. NCBI reference sequences (RefSeq): current status, new features and genome annotation policy. Nucleic Acids Res. 2011;40(Database issue):D130–5.

Punta M, Coggill PC, Eberhardt RY, Mistry J, Tate J, Boursnell C, Pang N, Forslund K, Ceric G, Clements J, Heger A, Holm L, Sonnhammer EL, Eddy SR, Bateman A, Finn RD. The Pfam protein families database. Nucleic Acids Res. 2011;40(Database issue):D290–301.

Qin J, Li R, Raes J, Arumugam M, Burgdorf KS, Manichanh C, Nielsen T, Pons N, Levenez F, Yamada T, Mende DR, Li J, Xu J, Li S, Li D, Cao J, Wang B, Liang H, Zheng H, Xie Y, Tap J, Lepage P, Bertalan M, Batto JM, Hansen T, Le Paslier D, Linneberg A, Nielsen HB, Pelletier E, Renault P, Sicheritz-Ponten T, Turner K, Zhu H, Yu C, Jian M, Zhou Y, Li Y, Zhang X, Qin N, Yang H, Wang J, Brunak S, Dore J, Guarner F, Kristiansen K, Pedersen O, Parkhill J, Weissenbach J, Bork P, Ehrlich SD. A human gut microbial gene catalogue established by metagenomic sequencing. Nature. 2010;464(7285):59–65.

Quast C, Pruesse E, Yilmaz P, Gerken J, Schweer T, Yarza P, Peplies J, Glockner FO. The SILVA ribosomal RNA gene database project: improved data processing and web-based tools. Nucleic Acids Res. 2012;41(Database issue):D590–6.

Quince C, Lanzen A, Davenport RJ, Turnbaugh PJ. Removing noise from pyrosequenced amplicons. BMC Bioinf. 2011;12:38.

Rainer J, Sanchez-Cabo F, Stocker G, Sturn A, Trajanoski Z. CARMAweb: comprehensive R- and bioconductor-based web service for microarray data analysis. Nucleic Acids Res. 2006;34 (Web Server issue):W498–503.

Reeder J, Knight R. Rapidly denoising pyrosequencing amplicon reads by exploiting rank-abundance distributions. Nat Methods. 2010;7(9):668–9.

Rho M, Tang H, Ye Y. FragGeneScan: predicting genes in short and error-prone reads. Nucleic Acids Res. 2010;38(20):e191.

Rice P, Longden I, Bleasby A. EMBOSS: the European Molecular Biology Open Software Suite. Trends Genet. 2000;16(6):276–7.

Robertson G, Schein J, Chiu R, Corbett R, Field M, Jackman SD, Mungall K, Lee S, Okada HM, Qian JQ, Griffith M, Raymond A, Thiessen N, Cezard T, Butterfield YS, Newsome R, Chan SK, She R, Varhol R, Kamoh B, Prabhu AL, Tam A, Zhao Y, Moore RA, Hirst M, Marra MA, Jones SJ, Hoodless PA, Birol I. De novo assembly and analysis of RNA-seq data. Nat Methods. 2010;7(11):909–12.

Schloss PD, Handelsman J. Introducing DOTUR, a computer program for defining operational taxonomic units and estimating species richness. Appl Environ Microbiol. 2005;71(3):1501–6.

Schloss PD, Handelsman J. Introducing SONS, a tool for operational taxonomic unit-based comparisons of microbial community memberships and structures. Appl Environ Microbiol. 2006a;72(10):6773–9.

Schloss PD, Handelsman J. Introducing TreeClimber, a test to compare microbial community structures. Appl Environ Microbiol. 2006b;72(4):2379–84.

Schloss PD, Westcott SL, Ryabin T, Hall JR, Hartmann M, Hollister EB, Lesniewski RA, Oakley BB, Parks DH, Robinson CJ, Sahl JW, Stres B, Thallinger GG, Van Horn DJ, Weber CF. Introducing mothur: open-source, platform-independent, community-supported software for describing and comparing microbial communities. Appl Environ Microbiol. 2009;75 (23):7537–41.

Schloss PD, Gevers D, Westcott SL. Reducing the effects of PCR amplification and sequencing artifacts on 16S rRNA-based studies. PLoS One. 2011;6(12):e27310.

Scholz MB, Lo CC, Chain PS. Next generation sequencing and bioinformatic bottlenecks: the current state of metagenomic data analysis. Curr Opin Biotechnol. 2011;23(1):9–15.

Schultz J, Milpetz F, Bork P, Ponting CP. SMART, a simple modular architecture research tool: identification of signaling domains. Proc Natl Acad Sci U S A. 1998;95(11):5857–64.

Schulz MH, Zerbino DR, Vingron M, Birney E. Oases: robust de novo RNA-seq assembly across the dynamic range of expression levels. Bioinformatics. 2012;28(8):1086–92.

Segal LN, Blaser MJ. A brave new world: the lung microbiota in an era of change. Ann Am Thorac Soc. 2014;11(Suppl 1):S21–7.

Segata N, Izard J, Waldron L, Gevers D, Miropolsky L, Garrett WS, Huttenhower C. Metagenomic biomarker discovery and explanation. Genome Biol. 2011;12(6):R60.

Selengut JD, Haft DH, Davidsen T, Ganapathy A, Gwinn-Giglio M, Nelson WC, Richter AR, White O. TIGRFAMs and Genome Properties: tools for the assignment of molecular function and biological process in prokaryotic genomes. Nucleic Acids Res. 2007;35(Database issue): D260–4.

Shannon C. A mathematical theory of communication. Bell Syst Tech J. 1948;27:379–423. , 623–656

Shi CY, Yang H, Wei CL, Yu O, Zhang ZZ, Jiang CJ, Sun J, Li YY, Chen Q, Xia T, Wan XC. Deep sequencing of the *Camellia sinensis* transcriptome revealed candidate genes for major metabolic pathways of tea-specific compounds. BMC Genomics. 2011;12:131.

Simon C, Daniel R. Metagenomic analyses: past and future trends. Appl Environ Microbiol. 2010;77(4):1153–61.

Simpson E. Measurement of diversity. Nature. 1949;163:688.

Simpson JT, Wong K, Jackman SD, Schein JE, Jones SJ, Birol I. ABySS: a parallel assembler for short read sequence data. Genome Res. 2009;19(6):1117–23.

Stacklies W, Redestig H, Scholz M, Walther D, Selbig J. pcaMethods – a bioconductor package providing PCA methods for incomplete data. Bioinformatics. 2007;23(9):1164–7.

Steinfath M, Groth D, Lisec J, Selbig J. Metabolite profile analysis: from raw data to regression and classification. Physiol Plant. 2008;132(2):150–61.

Tatusov RL, Koonin EV, Lipman DJ. A genomic perspective on protein families. Science. 1997;278(5338):631–7.

Thomas T, Gilbert J, Meyer F. Metagenomics – a guide from sampling to data analysis. Microb Info Exp. 2012;2(1):3.

Treangen TJ, Koren S, Sommer DD, Liu B, Astrovskaya I, Ondov B, Darling AE, Phillippy AM, Pop M. MetAMOS: a modular and open source metagenomic assembly and analysis pipeline. Genome Biol. 2013;14(1):R2.

Trimble WL, Keegan KP, D'Souza M, Wilke A, Wilkening J, Gilbert J, Meyer F. Short-read reading-frame predictors are not created equal: sequence error causes loss of signal. BMC Bioinf. 2012;13:183.

Tringe SG, Hugenholtz P. A renaissance for the pioneering 16S rRNA gene. Curr Opin Microbiol. 2008;11(5):442–6.

van den Berg RA, Hoefsloot HC, Westerhuis JA, Smilde AK, van der Werf MJ. Centering, scaling, and transformations: improving the biological information content of metabolomics data. BMC Genomics. 2006;7:142.

Virgin HW, Wherry EJ, Ahmed R. Redefining chronic viral infection. Cell. 2009;138(1):30–50.

Wooley JC, Ye Y. Metagenomics: facts and artifacts, and computational challenges. J Comput Sci Technol. 2009;25(1):71–81.

Xia J, Psychogios N, Young N, Wishart DS. MetaboAnalyst: a web server for metabolomic data analysis and interpretation. Nucleic Acids Res. 2009;37(Web Server issue):W652–60.

Xia J, Sinelnikov IV, Han B, Wishart DS. MetaboAnalyst 3.0 – making metabolomics more meaningful. Nucleic Acids Res. 2015;43(W1):W251–7.

Yang L, Chaudhary N, Baghdadi J, Pei Z. Microbiome in reflux disorders and esophageal adenocarcinoma. Cancer J. 2014;20(3):207–10.

Yok NG, Rosen GL. Combining gene prediction methods to improve metagenomic gene annotation. BMC Bioinf. 2011;12:20.

Zackular JP, Rogers MA, Ruffin MT, Schloss PD. The human gut microbiome as a screening tool for colorectal cancer. Cancer Prev Res (Phila). 2014;7(11):1112–21.

Zerbino DR, Birney E. Velvet: algorithms for de novo short read assembly using de Bruijn graphs. Genome Res. 2008;18(5):821–9.

Zhao Y, Tang H, Ye Y. RAPSearch2: a fast and memory-efficient protein similarity search tool for next-generation sequencing data. Bioinformatics. 2011;28(1):125–6.

Zhou Y, Mihindukulasuriya KA, Gao H, La Rosa PS, Wylie KM, Martin JC, Kota K, Shannon WD, Mitreva M, Sodergren E, Weinstock GM. Exploration of bacterial community classes in major human habitats. Genome Biol. 2014;15(5):R66.

Zhu W, Lomsadze A, Borodovsky M. Ab initio gene identification in metagenomic sequences. Nucleic Acids Res. 2010;38(12):e132.

Chapter 13
Pharmacogenomics: Clinical Perspective, Strategies, and Challenges

Dev Bukhsh Singh

Abstract Pharmacogenomics (PGx) defines the genetic basis of variability among individuals in response to drugs. It is an emerging discipline of medical science and is now a challenging and applied area of medical research. Several factors influence the efficacy and toxicity of drugs such as environmental factors, age, weight, gender, liver and kidney function, and applied drug therapy. Another crucial factor that influences the drug response of a patient is the genetic makeup of the patient. Polymorphism affects the drug efficacy, bioavailability, and toxicity. Human Genome Project (HGP) has provided a foundation for PGx study by identifying genes related to a disease. PGx knowledge derived from genetic profiling and associated drug response must be translated into clinical applications. A drug label contains information about PGx biomarker and drug related to a therapeutic area and also provides specific information for safe and effective medication based on a biomarker. PGx drugs have improved therapeutic response and also avoid events of adverse drug reactions (ADRs). There are some important ethical, social justice, and economic issues related to PGx which create hurdles in the drug development via PGx. The objective of this chapter is to discuss the basic principle of PGx and its application and also to put forward the ethical, social, technological, and economic challenges in the way of PGx. In spite of many challenges, it is expected that PGx may offer significant promises toward the goal of personalized medicine in the future.

Keywords PGx • Polymorphism • Drug development • Clinical trial • Biomarker • Personalized medicine

---

D.B. Singh (✉)
Department of Biotechnology, Institute of Biosciences and Biotechnology,
Chhatrapati Shahu Ji Maharaj University, Kanpur 208024, Uttar Pradesh, India
e-mail: answer.dev@gmail.com

© Shanghai Jiao Tong University Press, Shanghai and Springer Science+Business Media Dordrecht 2017

D.-Q. Wei et al. (eds.), *Translational Bioinformatics and Its Application*,
Translational Medicine Research, DOI 10.1007/978-94-024-1045-7_13

13.1 Introduction

PGx is the study of how genes affect drug response within a person or population. This is an emerging field which combines pharmacology and genomics to find a safe and effective treatment for a disease based on the genetic makeup of an individual. The long-term goal of PGx is to help doctors in selecting the drug and dosage best suited for each patient, based on patient's gene, an environment, lifestyle, and other characteristics. The major objective of PGx is to identify all the genetic and epigenetic differences that cause phenotypic variations in patient's response to drug therapy (Hess et al. 2015). Most of the drugs that are currently available are not genome specific and also do not have the similar response to each individual. These drugs generate three possibilities after treatment: positive response, adverse reactions, and no response to a population (Fig. 13.1). PGx promises a dramatic improvement in drug safety and efficacy. HGP played a very important role in learning the significance of inherited differences in genes on individual's drug response to medication. Patient's response to a drug depends on pharmacokinetics and pharmacodynamics. Pharmacokinetic effects are due to differences in absorption, distribution, metabolism, or excretion of the drug. The inappropriate concentration of drugs and metabolites can result in toxicity. In contrast, pharmacodynamics defines the efficacy of drugs among individuals despite the presence of an effective concentration of drug at the site of action (Wispelwey 2005).

The goal of PGx is to develop genetic-based strategies that will optimize the therapeutic outcomes. PGx uses the differences in genetic makeup to find an effective treatment for a particular individual and also avoids the chances of ADR. PGx approach has been used for cancer, anticonvulsant, anti-infective, cardiovascular, opioid, proton pump inhibitor, and psychotropic drugs, as well as other types of therapies. The use of PGx is quite limited, but new approaches are

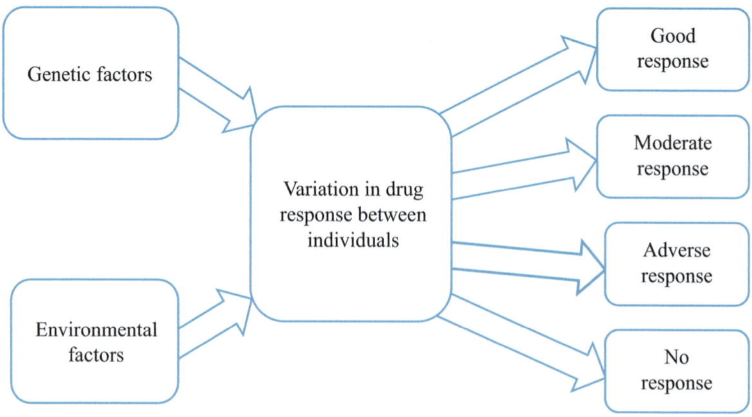

Fig. 13.1 Responses of a drug in traditional treatment

under study in clinical trials. In the future, PGx will offer a potential and effective medication to a wide range of health problems, including cancer, AIDS, Alzheimer's disease, and other fatal diseases (Genetic Home Reference n.d.). ADRs lead to a large number of injuries and deaths every year. PGx tests decrease the possibility of ADR. It also reduces the need for trial and error treatment to find the best therapy. Codeine is an effective painkiller, and it acts after its conversion to morphine, which needs to be detoxified and excreted. For example, codeine may have a toxic effect because of the high amount of morphine produced and/or impairment of excretion. Individual genetic differences, as well as prescribed drugs, affect the related metabolic pathways with clinical implications. Adverse drug events result in high cost and experimenting with treatments is expensive. Thus, PGx test reduces the overall cost for patients and physicians significantly.

Genetic basis of drug response enables us to understand the most critical aspects of drug action, improves drug safety, and makes it easier to prescribe the right dose for each person. PGx supports the drug development process which could be achieved through a more rationalized, safer, and less expensive clinical trial process. Drugs that suit to an individual with a particular genetic profile could be marketed only for those with that genetic makeup, while drugs previously in use could be recommended to the patient or limited population for whom they are safer. Differences between individuals can affect drug absorption, metabolism, toxicity, or activity. Therefore, while one treatment may work well for one individual, the same may cause adverse effects to other individuals (Kitzmiller et al. 2011). Currently, the majority of drug prescriptions are based on clinical factors such as patient's age, weight, sex, and liver and kidney function. For a small subset of drugs, scientists have identified the genetic variations that affect people's response to a drug. The Food and Drug Administration (FDA) of the United States includes the PGx information such as dosage guidelines, possible side effects, or difference in effectiveness for people with certain genomic variations for more than 150 medications (U.S. FDA n.d.).

PGx utilizes the variations in genes for proteins or enzyme that affects the response of a drug. Such proteins include a number of liver enzymes that convert drugs into their active or inactive forms. Even small variations in the genetic composition of these enzymes can have a big impact on a drug's safety or effectiveness. A gene may exist in many forms/versions, many of which vary by only a single difference in their DNA sequence or some may have larger changes. Most of these genetic variants do not influence the drug responses. Some patients may have much more copies of a gene. Those with extra copies of this gene manufacture an overabundance of enzyme molecules and show a different response to treatment with a drug (NIH, NIGMS n.d.). Pharmaceutical companies are using the PGx knowledge to develop and market drugs for patients with specific genetic profile. PGx also raises a lot of ethical issues. There is a need to protect informed consent and confidentiality and to promote justice and equity both nationally and globally. There is a need of a public policy related to PGx for the betterment of individual and society. The potential benefits of PGx should not be underestimated even from an ethical point of view.

13.2 Genetic Polymorphism

Studies have shown that about 85% of human diversity at short tandem repeat (STR) and restriction fragment length polymorphism (RFLP) autosomal loci is due to differences between individuals of the same population, whereas differences between populations of the same continent account for 5–10% (Romualdi et al. 2002). A study based on more than 350 microsatellites from a global sample of humans showed that individuals could be grouped according to their continental origin (Serre and Pääbo 2004). Results indicated that the pattern seen is one of the gradients of allele frequencies that extend over the entire world and also disqualify the assumption that major genetic discontinuities exist between different continents or races. Human genetic variation is based on patterns of gene flow and genetic drift (Jorde and Wooding 2004). Therefore, ancestry or racial study may prove useful in the biomedical testing, but the results directly associated with disease-related genetic variation will be more accurate and beneficial. CYP2D6 allele frequency varies among racial groups. In European Caucasians and their descendants, the functional group of alleles is predominant, with a frequency of 71% (Bradford 2002). The alleles which encode for no or reduced functioning will affect the activity of the CYP2D6-mediated drug. Therefore, allele-related studies are necessary to assure the optimal dosing recommendations.

HGP has provided a foundation for PGx study by identifying genes related to a disease. Genetic information derived from genomic research must be translated into clinical applications for the welfare of society. Most differences in drug response among individuals are not caused by mutation of a single gene but by the altered function of genes. Variations in absorption, distribution, metabolism, and excretion (ADME) genes and the genes associated with drug targets may result in the absence of protein or the production of a protein with altered or no activity. These variations decide overall metabolism of the drug and the therapeutic index of the drug, as well as the activity of its metabolites (Nadine and Theresa 2008). The clinical association of a genetic variation can be related with a disease. Cyclooxygenases are the key enzymes in several physiopathological processes. Genes coding for these enzymes (*PTGS1* and *PTGS2*) are highly variable, and variations in these genes cause the risk of developing several diseases and ADR. Major variations in the *PTGS1* and *PTGS2* genes, allele frequencies, functional consequences, and population genetics have been analyzed (Agúndez et al. 2015). The most salient clinical associations of *PTGS* gene variations are related to colorectal cancer and stroke. Genes responsible for a clinical outcome can be identified by correlating variability in genotype with phenotypic differences.

The study of genetic variants associated with cancer or any disease helps in early detection of disease and also opens the way for personalized cancer therapy. In recent years, the GWAS studies have provided information about many genetic variants in cancer. For example, the presence of genetic variants rs1447295 and rs6983267 on 8q24 contributes to prostate cancer in Europeans (Yeager et al. 2007). These variants provide a useful biomarker for diagnosis and therapeutic categorization. The systematic cataloging of genetic variants may provide the information of pathways controlling the cellular activities in cancer (Cho 2010). Genetic variants can be more clinically useful if they are combined with the family history records of the disease. New diagnostic test and therapeutic strategies can be developed on the basis of biochemical targets to control the disease. Rare genetic variants are too rare to be identified by GWAS and they have large effect on disease risk (Cirulli and Goldstein 2010). The 1000 Genomes Project has identified many genetic variants at lower frequencies. Some examples of polymorphism and its effect on metabolism and therapeutic role have been discussed here.

13.2.1 Polymorphism in miRNA

MicroRNAs (miRNAs) are small, single-stranded, 19–21 nucleotide long nonprotein-coding RNA molecules. miRNAs act as negative regulators of gene expression through binding to their target mRNAs and consequently lead to mRNA cleavage or translational repression (Bartel 2004). The miRNAs regulate the expression of roughly 10–30 % of all human genes, including the genes related to cell differentiation, proliferation, and apoptosis (Berezikov et al. 2005). miRNA may contribute to cancer development with changes in the miRNA's properties and/or maturation process. A study was performed to validate the potential association between the four common SNPs (miR-196a2C.T, rs11614913; miR-146aG.C, rs2910164; miR-499A.G, rs3746444; miR-149C.T, rs2292832) and the risk for developing cancer (He et al. 2012). The results of this study indicated that the rs11614913TT genotype is significantly associated with a decreased risk for colorectal cancer and lung cancer. The rs2910164C allele is associated with decreased risk for cervical cancer, esophageal cancer, prostate cancer, and hepatocellular carcinoma. SNPs in miRNA may prevent the pathogenesis of some cancers, and some SNPs may also increase risk for cancer.

13.2.2 Urate Transporter 1 (URAT1) Polymorphisms

Genetic variation is routinely seen in all drug targets. African populations show more genetic variation than Asian (Gurdasani et al. 2015). In most cases, these polymorphisms do not alter the encoded amino acid and probably have no functional effect. Approximately 75% of drug targets sequenced have at least one amino acid changing genetic variant and >35% having more than three variants (McHale 2008). Sequence polymorphism reflects the variability in chemical target interactions, but the real effect can only be tested in vitro or in clinical trials. In a polymorphism-related study, the effect of urate transporter 1 (URAT1) polymorphisms in the hypertensive patients with hyperuricemia and the uricosuric action of losartan therapy were explored. Results suggest that URAT1 rs3825016 and rs1529909 polymorphism affects the uricosuric action of losartan (Sun et al. 2015). Polymorphism affects the drug efficacy, bioavailability, and toxicity.

13.2.3 Opioid Receptor Polymorphism

Knowledge of polymorphism is important as a specific type of polymorphism is responsible for a particular characteristic which might not be exhibited by other types. The pharmacologic actions of opioids are due to their interaction with the opioid receptors (G-protein-coupled receptors) located in the brain and spinal cord (Feng et al. 2012). Three subtypes of opioid receptors are mu-opioid receptors, kappa-opioid receptors, and delta-opioid receptors. The mu-opioid receptor is the primary site of action of opioid analgesics including morphine, fentanyl, and methadone. More than hundred polymorphisms are reported for the human mu-opioid peptide receptor gene. These polymorphisms are associated with both agonistic and antagonistic opioid effects. Studies have shown that *Gpr88*, *Ttr*, *Gh*, and *Tac1* mRNAs were altered in mice exposed to chronic stress (Ubaldi et al. 2015). These transcripts represent a biomarker and therapeutic targets for diagnosis and can also help in treatment of chronic stress-associated disease in humans.

13.2.4 The HapMap Project

The International HapMap Project was initiated in the year 2002 (International HapMap Project n.d.). The goal of this project was to map the common patterns of DNA sequence variation in the human genome. An international consortium was involved in the mapping of these patterns across the genome by determining the genotypes of sequence variants, their frequencies, and the degree of association between them. The HapMap guides the discovery of sequence variants that affect common disease. The HapMap facilitates the development of diagnostic tools and also helps in drug target selection for therapeutic intervention. The HapMap enhances our understanding of the hereditary factors involved in health and disease. The International HapMap Project has much in common with the Human Genome Project. The Human Genome Project covered the sequencing of the entire genome, including the 99.9% of the genome where all human beings are identical in genetic makeup (International HapMap Consortium 2003). The HapMap project characterizes the common patterns of DNA sequence within the 0.1% where individuals differ from each other.

13.2.5 The 1000 Genomes Project

The data from 1000 Genomes Project is publically available through the 1000 Genomes Project website and dbSNP. This project was completed between 2008 and 2015 and provides the largest public catalog of human variation and genotype data (IGSR: The International Genome Sample Resource n.d.). The goal of this project was to find most genetic variants with frequencies of at least 1% in the population. The International Genome Sample Resource (IGSR) was set up to ensure the future usability and accessibility of data from the 1000 Genomes Project. The goal of IGSR is to expand the data collection to include new populations and ensure the future usability of the 1000 Genomes reference data (IGSR n.d.).

13.2.6 PGx Biomarkers and Drug Labeling

PGx plays an important role in identifying responders and nonresponders to a drug, avoiding adverse events, and prescribing drug dose. Biomarkers include genetic or somatic gene variants, changes in expression level, functional irregularities, and chromosomal abnormalities. Drug labeling provides information about genomic biomarkers and can describe (1) drug exposure and clinical response variability, (2) risk of ADR, (3) genotype-specific dosing, (4) mechanism of drug action, and (5) polymorphic drug target and disposition genes. FDA-approved drugs with PGx information in their labeling are listed in Table 13.1. The labeling for the products includes specific information for safe and effective medication based on biomarker information.

EGFR has been approved as a biomarker in the therapeutic area of oncology (lung cancer). The efficacy of epidermal growth factor receptor-tyrosine kinase inhibitors (EGFR-TKIs) is superior to that of cytotoxic chemotherapy in advanced non-small cell lung cancer (NSCLC) patients (Zhang et al. 2014). The efficacy of EGFR-TKIs (gefitinib, erlotinib, and afatinib) differs between exon 19 deletion and exon 21 L858R mutations. The L858R mutation (exon 21) results in an amino acid substitution at position 858 in EGFR, from a leucine (L) to an arginine (R) (My Cancer Genome n.d.). Patients with EGFR-mutated tumors display a longer progression-free survival (PFS) on EGFR-TKI therapy. It has been reported that patients with EGFR exon 19 deletions were associated with longer PFS compared with L858 mutation at exon 21 (Zhang et al. 2014). The investigators should consider the sensitive EGFR mutation as an important factor in clinical studies regarding target therapy. After a demonstration of a genetic association with response phenotype, there is the need to validating the biomarker for a diagnostic test.

Cytochrome P450 (CYP) enzyme polymorphisms are a determining factor in a patient's ability to respond to different drugs (Lynch and Price 2007). CYP enzymes metabolize the drugs within the endoplasmic reticulum of liver cells,

Table 13.1 List of some PGx biomarkers in drug labeling

Drug	Therapeutic area	Biomarker	Referenced subgroup	Labeling section
Abacavir	Infectious diseases	HLA-B	HLA-B*5701 allele carriers	Boxed warning, contraindications, warnings and precautions
Afatinib	Oncology	EGFR	EGFR exon 19 deletion or exon 21 substitution (L858R) positive	Indications and usage, dosage and administration, adverse reactions, clinical pharmacology
Aripiprazole	Psychiatry	CYP2D6	CYP2D6 poor metabolizers	Dosage and administration, clinical pharmacology
Busulfan	Oncology	BCR-ABL1	Philadelphia chromosome negative	Clinical studies
Carvedilol	Cardiology	CYP2D6	CYP2D6 poor metabolizers	Drug interactions, clinical pharmacology
Clobazam	Neurology	CYP2C19	CYP2C19 poor metabolizers	Dosage and administration, use in specific populations, clinical pharmacology
Glipizide	Endocrinology	G6PD	G6PD deficient	Precautions
Celecoxib	Rheumatology	CYP2C9	CYP2C9 poor metabolizers	Dosage and administration, use in specific populations, clinical pharmacology
Chlorpropamide	Endocrinology	G6PD	G6PD deficient	Precautions
Cisplatin	Oncology	TPMT	TPMT intermediate or poor metabolizers	Clinical pharmacology, warning, precautions
Diazepam	Psychiatry	CYP2C19	CYP2C19 poor metabolizers	Clinical pharmacology
Mafenide	Infectious diseases	G6PD	G6PD deficient	Warnings, adverse reactions
Omeprazole	Gastroenterology	CYP2C19	CYP2C9 poor metabolizers	Drug interactions

U.S. FDA (n.d.)

and waste is excreted through urine. Most phase I metabolizing enzymes belong to CYP family (CYP3A4, CYP3A5, CYP2D6, CYP1A1, CYP1B1, and CYP2E1) and convert a wide range of substrates into more water-soluble form (Yiannakopoulou 2015). The polymorphisms of transporter enzymes, such as the OATP family of transporters, have also been linked to differences in the pharmacokinetics of drug absorption. For example, a single nucleotide polymorphism (SNP) in the SLCO1B1 gene, which encodes the OATP1B1 enzyme, leads to impaired absorption of statins (Kalliokoski and Niemi 2009). Recent studies have highlighted the role of

CYP2C19 polymorphism for the action of clopidogrel, whereas the CYP2C9 polymorphism has a role in anticoagulant treatment (Sim et al. 2013). Furthermore, the analgesic and side effect of codeine is related with CYP2D6 polymorphism, and CYP2D6 genotype influences the breast cancer recurrence during tamoxifen treatment. In another study, the effect of polymorphisms in *CYP2C9* and *CYP2C8* and gender on the pharmacokinetics of the enantiomeric (R, S) forms of ibuprofen was studied (Ochoa et al. 2015). The *CYP2C9* polymorphisms and gender affect the pharmacokinetics of *S*-ibuprofen and *R*-ibuprofen. *CYP2C8* polymorphisms do not have a significant role on the pharmacokinetics of ibuprofen. Polymorphism affects drug metabolism and is also responsible for the diverse response of the same drug to different individuals. An incidence of drug-induced toxicity also depends on polymorphism.

13.2.7 Effect of Age, Sex, and Other Factors on Drug Response

Many other factors such as age, sex, smoking or alcohol intake, intake of multiple drugs, past history of ADR, presence of other diseases, pregnancy, breastfeeding, kidney problem, and the liver function also affect the drug response (Alomar 2014). Understanding of these factors on drug response enables healthcare professionals to prescribe the most appropriate medication for a particular patient. Both very young and very old individuals are more vulnerable to ADR than other age groups. Because of all age-related changes, many drugs stay much longer in the body of old-age person than younger person's body and increase the risk of side effects (Klotz 2009). Genetic, hormonal, and physiological differences between male and female affect the prevalence, incidence, and severity of diseases and responses to therapy (Soldin et al. 2011). In an age- and sex-based study of cortisol plasma level in normal control and Alzheimer's diseases (AD), a significant difference in cortisol plasma levels between female AD patients and age-matched female controls and between female and male AD patients has been reported (Leblhuber et al. 1993). Important pharmacokinetic and pharmacodynamic changes occur with advancing age. Pharmacokinetic changes include a reduction in renal and hepatic clearance and an increase in the level of lipid-soluble drugs, whereas pharmacodynamic changes involve altered sensitivity to several drugs such as anticoagulant, cardiovascular, and psychotropic drugs (Mangoni and Jackson 2004).

13.2.8 Herb-Drug and Drug-Drug Interactions

Environmental chemicals, coadministered drugs, dietary constituents, tobacco smoking, and alcohol intake are known to induce or inhibit drug-metabolizing enzymes and drug transporters (Ma and Lu 2011). The factors alter drug efficacy, induce drug-drug and drug-chemical interactions, and result in drug side effects. Herbs in combination with therapeutic drugs result in herb-drug interactions. Herb-drug interactions may lead to serious clinical consequences. *Ginkgo biloba* (ginkgo) causes bleeding when combined with warfarin or aspirin, raises the blood pressure when combined with a thiazide diuretic, and may cause coma when combined with trazodone in patients (Hu et al. 2005). Herbs should be labeled to alert patients when used in combination with a drug. A drug-drug interaction (DDI) involves pharmacokinetic or pharmacodynamic mechanisms. Adverse drug reactions may occur due to DDIs, and health service providers are often unaware of the ADR of certain drug combinations (Magro et al. 2012). DDIs can lead to ADR, particularly in cancer patients, because of polypharmacy and age-related organ dysfunction (Chan et al. 2009). The number of clinically relevant DDIs is probably low. In most cases, DDIs may be responsible for a substantial number of hospital admissions (Becker et al. 2005). Specifically, pharmacists should have good knowledge of combinations of drugs that may cause serious DDIs. The pharmacokinetics and pharmacodynamics of many drugs are well known, but the role of coadministered herbs has not been well explored due to complex components of herbal products (Zuo et al. 2015). The pharmacokinetics and pharmacodynamics of drug-drug and herb-drug interactions cannot be ignored. The safety of coadministration of herbs together with drug should be kept in mind. These interactions need to be addressed by conducting the high-quality scientific research.

13.3 PGx Testing and Drug Discovery Process

This section describes PGx studies performed on patients with a particular disease and presents the major outcome of these studies (Fig. 13.2). The genome-wide association study (GWAS) is extensively used to analyze hundreds of thousands of SNP by high-throughput genotyping. In addition to the candidate gene approach, the GWAS approach is utilized to investigate the determinants of antidepressant response to therapy (Lin and Lane 2015). PGx has shown less impact on human health than initially expected. One reason for this is that many diseases' and patients' response to the drug treatments is affected by both genetic and environmental factors. Pure genomics should also consider the role of environmental elements (Everett 2015). Pharmacometabolomics describes the role of both genetic and environmental influences on physiology. It is concerned with the study of drug effects through the analysis of predose, biofluid, and metabolite profiles. Polymorphisms that are clinically relevant show population-specific allele frequencies. Fifteen polymorphisms from 12 genes have been assessed in 81 Peruvian and 95 Mexican individuals (Marsh et al. 2015). Six polymorphism frequencies differed significantly between these two populations.

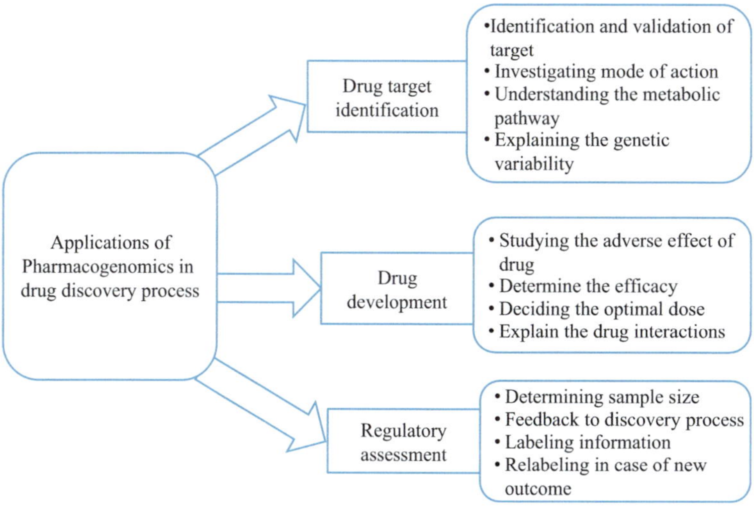

Fig. 13.2 Applications of pharmacogenomics across the drug discovery process

13.3.1 Adverse Drug Reactions and Hypersensitivity

Interindividual genetic differences are important causes of ADRs and lack of drug response. The majority of phase I and phase II drug-metabolizing enzymes are polymorphic and responsible for varying drug response (Ingelman-Sundberg and Rodriguez-Antona 2005). GWAS related to drug response and genes encoding drug-metabolizing enzymes have extracted knowledgeable information for variation in drug response and drug metabolism. For example, PGx markers in the HLA-coding genes are associated with drug hypersensitivity of multiple drugs. The HLA-B*5801 allele was significantly associated with the risk of severe cutaneous ADRs (cADRs) in the Han Chinese, Korean, Thai, Japanese, and European populations (Jarjour et al. 2015). All SNPs identified in GWAS of common variants are also located in or nearby HLA-B*5801. Five specific HLA alleles that predict drug-induced hypersensitivity reactions (HSR) were tagged by seven SNPs (He et al. 2015). It was concluded that SNP tagging is a “real-time” approach to identify the specific HLA alleles associated with drug-induced hypersensitivity across diverse racial groups. The influence of SNPs has been studied on efficacy and safety of calcineurin inhibitors upon heart transplantation (Sánchez-Lázaro et al. 2015). A panel of 36 SNPs was correlated with a series of clinical parameters. Such types of studies can identify the patients at increased risk of clinical complications.

13.3.2 Lung Adenocarcinoma

There is need to expand the scope of geographic data in PGx. Lung adenocarcinoma is the most common form of lung cancer, and it begins in the tissues that lie near the outer parts of the lung. The impact of the genetic polymorphisms on the therapeutic efficacy of pemetrexed in lung adenocarcinoma patients has been investigated. The genotyping of 51 polymorphisms of 13 genes in 243 lung adenocarcinoma patients treated with pemetrexed was performed (Woo et al. 2015). Twelve polymorphisms in six genes were found statistically significant in univariate analysis. Finally, two polymorphisms (ATIC and GGH genes) were associated with therapeutic efficacy in multivariate analysis. Genetic polymorphisms have been identified for many enzymes, drug receptors, and transporters that are significant in clinical pharmacology. These polymorphisms can cause alterations in the amount, structure, binding, and/or function of these proteins and also affect the drug interaction with the target.

13.3.3 Breast Cancer

Cancer has become a great threat and challenge to public health. Breast cancer accounts for 23% of the total cancer burden and 14% of cancer deaths worldwide (Jemal et al. 2011). There is need of new diagnostic markers for the early detection and prevention of breast cancer. Many studies have shown that the pathogenesis of various tumors, including breast cancer, occurs due to suppression of apoptosis (Wang et al. 2012). The effects of Fas and FasL polymorphisms on breast cancer risk have been studied among the Chinese population. The Fas−1377GA, Fas−1377AA, Fas−670AG, Fa−670GG, and FasL−844TC genotypes have been associated with a lower risk of breast cancer (Xu et al. 2014). The genotype Fas−1377G/−670A was associated with an increased risk of breast cancer. This study also revealed that the Fas−1377GA/AA (−670AG/GG) and FasL−844CC or TC/TT genotypes were associated with a decreased risk of breast cancer. This study indicates that Fas polymorphisms may affect the breast cancer risk by regulating the soluble Fas concentration.

Tamoxifen is used for the treatment of breast cancer. Tamoxifen is not effective in all estrogen receptor (ER)-positive breast cancer patients and has side effects. CYP2D6 is an important enzyme responsible for the production of endoxifen, a potent tamoxifen metabolite (De Souza and Olopade 2011). Studies have shown that genetic variation reduces CYP2D6 enzyme activity and results in poor clinical outcome when treated with tamoxifen (Zembutsu 2015). Dose-adjustment study of tamoxifen based on *CYP2D6* genotypes suggests that dose adjustment is beneficial for the patients carrying reduced or null allele of *CYP2D6* to maintain the effective endoxifen level.

13.3.4 Acute Myeloid Leukemia

Variation in terms of efficacy and toxic side effects exists among acute myeloid leukemia (AML) patients on chemotherapy with cytarabine (Ara-C). Differentially expressed genes between Ara-C-sensitive and Ara-C-resistant samples were identified by global gene expression profiling (Abraham et al. 2015). Variations in Ara-C cytotoxicity were seen among samples from AML patients and categorized into sensitive, intermediately sensitive, and resistant groups, based on IC_{50} values. Ara-C resistance index could be a potential biomarker for AML treatment outcome and toxicity.

13.3.5 Tyrosine Kinase Inhibitors in Cancer Therapy

PGx informations are being widely used for drug discovery process and are already used in clinical practice for the treatment of many diseases. *EGFR* family of receptor tyrosine kinase regulates many metabolic, developmental, and physiological processes. In tumor cells, the tyrosine kinase activity of EGFR is dysregulated by various oncogenic mechanisms, including EGFR gene mutation and overexpression and increased gene copy number. Many mutations in the kinase domain of the *EGFR* gene provide sensitivity to tyrosine kinase inhibitors (TKIs). Most of these patients acquired resistance to EGFR inhibitors after treatment. EGFR-TKI resistance mechanisms include amplification and mutation in MET, resulting in tumor cell growth (Pérez-Ramírez et al. 2015). Therefore, MET is considered as an attractive target for anticancer therapy. MET promotes cell proliferation, scattering, invasion, survival, and angiogenesis. Because of the important role of MET in cancer development and progression, it has been recommended as potential target for cancer therapy.

In chronic myeloid leukemia, the bone marrow produces too many white blood cells. These cells crowd the bone marrow and interfere with the normal blood cell production. In an interesting case, a patient bearing a T315I-mutant chronic myeloid leukemia resistant to nilotinib was successfully treated with two cycles of omacetaxine and then with dasatinib (Venton et al. 2015). This study has suggested that eradication of the T315I mutation could be achieved without third-generation tyrosine kinase inhibitors. In the future, it is expected that other genetic markers of drug response for a disease will further improve the efficacy and safety of therapies. In vitro human cell line models may be used for PGx studies to know the clinical response and to identify mechanisms associated with variation in drug response.

13.3.6 Human Cell Line Models and Pharmacogenomics

In vitro human cell line models are used for cancer pharmacogenomics to predict clinical response, to generate a pharmacogenomic hypothesis, and to identify mechanisms associated with variation in drug response. Among cell line model systems, Epstein-Barr virus-transformed lymphoblastoid cell lines (LCLs) have been used to test the effect of genetic variation on drug efficacy and toxicity (Niu and Wang 2015). In the future, patient-specific inducible pluripotent stem cells could improve the predictive validity. The human LCLs comprise a useful model system for identifying genetic variants associated with pharmacologic phenotypes. Many GWAS for drug-induced phenotypes have been tested in LCLs, often incorporating gene expression data (Wheeler and Dolan 2012).

The large-scale genome-wide studies in both human and model systems have allowed us to understand how cell-based models help in finding an association between clinically relevant genetic and drug response (Cox et al. 2012). A genome-wide cell-based model was used to evaluate genetic variants for their contribution to cellular sensitivity to tamoxifen. This model has included multidimensional datasets, including genome-wide genotype, gene expression, and endoxifen-induced cellular growth inhibition in lymphoblastoid LCLs (Weng et al. 2013a). Genome-wide findings were further evaluated in NCI60 cancer cell lines. Furthermore, SNPs that were associated with tamoxifen-induced toxicities in breast cancer patients were identified. The cell-based models are very useful in genome-wide identification of pharmacogenomic markers.

13.4 Clinical Perspective and Implications

PGx has many advantages over traditional treatment options. The availability of low-cost genotyping methods can make PGx drugs cost-effective and affordable to poor people. Clinical and economic status should be identified under which a PGx test might be a cost-effective option for patients (Shabaruddin et al. 2015). It is considered that PGx tests are cost saving and better in improving human health than no-testing approach for the cure of a disease. Pharmacists should advise clinicians and patients on matters related to the implementation of PGx. The genetic variants evaluated in PGx include SNPs, nucleotide insertion, deletion, copy number variation, tandem repeat, and chromosomal translocation. In addition, gene expression is also commonly studied in PGx for relevancy in tumorigenesis and chemotherapy response. In PGx, drug prescription is purely based on the knowledge derived from association study between genetic profile and drug response. Basic steps and principles in the PGx approach of treatment are represented in Fig. 13.3. However, the current status of PGx in pharmacy colleges is poor and fails to produce pharmacists with the required knowledge or practical training in this discipline (Rao et al. 2015). More than 135 medications in the United States describe PGx

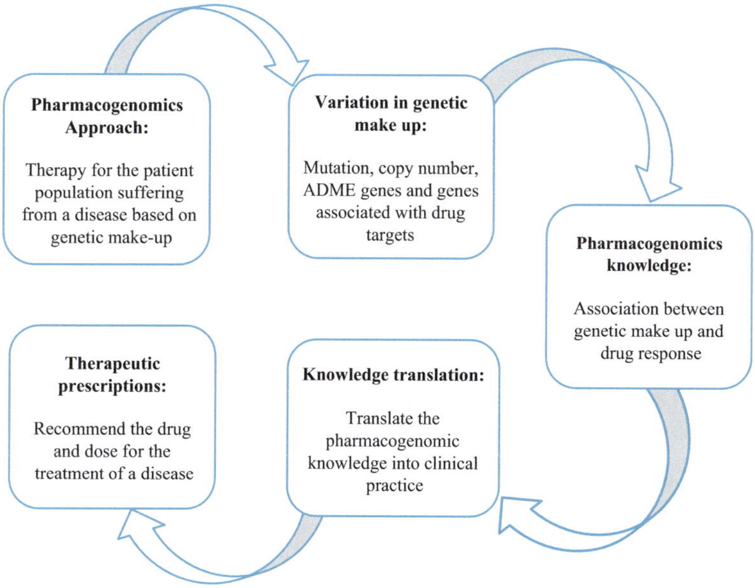

Fig. 13.3 Basic principles of PGx approach for medication

informations related to drug response or drug safety on their package inserts. Pharmacists offering PGx testing services receive billing questions specific to the laboratory tests (O'Connor et al. 2015). Pharmacists must be able to discuss these issues with the concerned patients. The goal of a pharmacist must always be to provide a high quality of result and patient care. Some testing companies offer extensive literature resources to help guide prescribers for suggesting medications.

Researchers are trying to establish an association between the response of a drug and genotype of patients suffering from that disease. Most extensive genetic studies such as GWAS, rare variant exome sequencing, copy number variant analysis, and allele-related analysis can provide an answer to the altered response of the same drug for the different individual (Cirulli and Goldstein 2010). The causal variants in single-gene disorders are necessary and sufficient to impart large effects (Marian and Belmont 2011). Knowledge of association between the genetic makeup of patients and drug response must be translated into clinical practice. If genetic variation controls the risk of drug-induced side effects, then it is recommended to identify the variants and translate them into a highly sensitive PGx test.

Understanding of autoimmune diseases such as rheumatoid arthritis, systemic lupus erythematosus, and psoriasis has improved considerably over the last decades, and several PGx studies of these diseases have been carried out (Gregersen and Olsson 2009). But the clinical applications still need to be improved. A striking failure of modern medicine is an ADR which leads to death and illness in the developed world with a high healthcare cost. For autoimmune disease, several DNA-based tests are in practice to improve drug selection and dose optimization and reduce the risk of toxicity. The "GATC" project was a nationwide project established in Canada to identify novel markers of severe ADRs in children (Ross et al. 2007). The goal of this project was to identify ADR patients, collect their DNA samples, and apply genomic studies to identify ADR-associated genetic markers. Different individuals may have a diverse response to the same drug, in terms of efficacy and toxicity. ADRs cause about 6% of all hospital admissions and account for up to 9% of hospitalization costs. Drug-induced skin injury (DISI) has been reported as the most common ADR, ranging from maculopapular eruptions to severe adverse cutaneous drug reactions (SCARs) with mortality of up to 40% (Borroni 2015). Specific genetic polymorphisms present susceptibility to different types of DISI. The cases of SCARs are now less frequent, with a low rate of morbidity and mortality.

Abnormal genes related to cancer may be inherited or acquired. Genetic changes that occur because of mutation in tumor suppressor genes, mismatch of repair genes, and mutations in oncogenes alter the cell signaling pathways and other cellular functions. Tumor-associated somatic mutations are used to identify cellular signaling pathways in tumors. Somatic mutations serve as genomic predictors of tumor response and represent a new target for drug development. A deep sequencing of 145 genes in colorectal and non-small cell lung cancers reported somatic mutations in 98% and 83% of tumors, respectively. More than half (52.5%) of colorectal cancers and 72% of non-small cell lung cancers contained at least one mutation that was associated with a specific chemotherapy approach (Lipson et al. 2012). Clinical PGx applies deep sequencing to reveal the mechanism of sensitivity or resistance to drug therapy. Drugs used in traditional cancer therapy destroy both malignant and healthy cells.

PGx drugs target the specific molecules of a pathway that is related to the division, growth, and spreading of cancer cells. One example of personalized cancer treatment is trastuzumab (Herceptin), a recombinant monoclonal antibody used for the treatment of breast cancer. Herceptin targets the human epidermal growth factor receptor 2 gene (HER2) on the tumor cell surface and induces cell-mediated cytotoxicity against the tumor cells (American Nurse Today n.d.). Leukemia is the most common cancer affecting children, accounting for 25–35% of childhood malignancies worldwide with acute lymphoblastic leukemia comprising 80% of leukemia cases. In certain leukemia patients, treatment fails due to drug resistance that is why acute lymphoblastic leukemia is the leading cause of cancer-related death in children (Ansari and Krajinovic 2007). Many advances have been made in the field of anticancer therapy. Currently, the US FDA is providing the package inserts of approximately 30 anticancer agents to include PGx information (Weng et al. 2013b). FDA recommendation and potential action needed vary among drugs. Scientific values of PGx knowledge should be used for improving therapeutic efficacy and reducing side effects. There are significant limitations to PGx discovery for anticancer therapies, because of unavailability of enough patients for both discovery and validation purpose. A clinical study is a time-consuming process, and outcomes of the clinical trial can then be used for PGx discovery. The same is true for a validation of result, which requires enough discovery and replication studies in the literature.

Breast cancer is the most frequently diagnosed cancer in women. In breast cancer, somatic mutations in only three genes are observed with a greater than 10% incidence across primary breast cancer (Stjepanovic and Bedard 2015). Breast cancer treatment is based on the identification of expression of estrogen receptor or protein overexpression of HER2/ERBB2. HER2 amplification is tested for clinical practices related to breast cancer, as HER2-targeted therapies are approved. Genomic alternations in different types of cancer diseases can be reviewed, and genotype-based treatments can be a common practice in the future. Efforts should be made in the direction to translate the PGx knowledge to clinical application. The clozapine is used for the treatment of resistant schizophrenia patients. Agranulocytosis, an ADR, was reported in 0.8% of clozapine-treated patients, and this adverse event was not associated with dose (Verbelen and Lewis 2015). Later on, PGx evidence has established an association between HLA regions with agranulocytosis in clozapine patients, but this knowledge has not been translated into clinical practice yet.

The CYP2D6 polymorphisms have an impact on the clearance and response to a series of cardiovascular drugs. Clinical studies indicate the relationships between the CYP2D6 genotype and concentrations of drugs perphenazine, zuclopenthixol, risperidone, and haloperidol. CYP2D6 is used as an independent predictor of the outcome of tamoxifen treatment in breast cancer. Genotype testing for CYP2D6 is not customarily performed in clinical practice, and there is uncertainty regarding genotype-phenotype, gene concentration, and gene-dose relationships (Zhou 2009). Further, prospective studies on the clinical impact of CYP2D6 are required. Genetic polymorphisms of CYP, and the presence of the human leukocyte antigen (HLA)-B*1502 allele, influence drug disposition and/or response in patients (Ma et al. 2012). Pharmacokinetic and pharmacodynamic variability can be explained by polymorphism of genotype. However, conflicting evidence exists in some cases. The effect of CYP2D6 polymorphisms on codeine efficacy and toxicity is not well studied. The CYP2D6 genotyping tests are available, but its clinical utility is limited.

Lack of sufficient resources, lack of knowledge provider, and ethical, legal, and social issues are major limitations and challenges in the implementation of PGx testing for clinical application. Understanding of the technologies and their application is limited among practitioners (Collins et al. 2016). Pretreatment genetic testing is very useful in preventing ADR in cardiovascular, cancer, HIV, and many other diseases. Patients should be encouraged for genetic testing-based treatment for cancer and other diseases, and such types of testing centers should be distributed globally. Genetic testing is far from being realized because of low specificity and sensitivity and a low incidence of an ADR or the high cost of genotyping for a disease.

13.4.1 Metabolomics and Pharmacology

Metabolomics is the study of metabolome (small molecules) present in the cells, tissues, and body fluids. Metabolic status of a person provides the close representation of the health status that is not obvious from gene expression analysis (Beger et al. 2016). The metabolic status reflects the effect of gene expression, environmental factors, diets, and the gut microbiome. For researchers in the field of clinical pharmacology, metabolomics offers a systems biology approach to understand genotype-phenotype associations, disease signatures, severity and subclass, and variability in drug response (James 2013). Clinicians measure only a small part of information contained in the metabolome to assess disease status. In the future, the narrow range of chemical analysis in medical community will be replaced by the more comprehensive metabolic signatures (Kaddurah-Daouk et al. 2015). Metabolic signatures are expected to more accurately describe specific disease and their progression and also help in differential diagnosis of disease and healthy status. The phenotypic outcome of complex interactions between genotype, diet, drug therapy, environmental exposure, and gut microflora can be investigated at the molecular level to see the overall drug response (Huang et al. 2015). Metabolic phenotyping provides an insight into disease pathophysiology and mechanisms of drug response and also predicts the risk of toxicity.

Pharmacometabolomics defines the efficacy, toxicity, or other outcomes of a drug based on a mathematical model of a preintervention metabolite signatures. Pharmacometabolomics complements genomic, transcriptomic, proteomic, and epigenomic "systems biology" approaches to drug development by taking into account the interindividual variation in drug response (Burt and Dhillon 2013). Metabolomics provides the useful prognostic indicator to complement other personalized biomarker related to genomics, transcriptomics, and proteomics because endogenous metabolites or small molecules are more closer and directly interacts with the components affecting the human health. Before utilizing biomarkers in drug development, a candidate omics-based test should be clearly defined and validated using a two-step process: (i) discovery and (ii) evaluation of clinical utility and use (Burt and Nandal 2016). Metabolomic data can be integrated with genomic results to get some novel insight into mechanisms of variation in drug response. Many scientific advances have been made to detect, identify, and quantify the large numbers of metabolites. These advances have enabled us to study hundreds or thousands of metabolites and millions of genomic variants in a single cell (Neavin et al. 2016). It is now possible to analyze the large datasets generated by omics studies to understand molecular basis of variation in disease risk and drug response.

13.4.2 CPIC and DPWG in Clinical Implementation of PGx

Dosing guideline takes into consideration patient's genotype and has been published by Clinical Pharmacogenetics Implementation Consortium (CPIC), Dutch Pharmacogenetics Working Group (DPWG), or other organizations. One barrier to clinical implementation of pharmacogenetics is the lack of freely available, clinical practice guidelines (PharmGKB n.d.). CPIC provides guidelines that enable the translation of genetic test results for prescribing specific drugs. The guidelines are focused on genes or around drugs. CPIC guidelines are peer reviewed, published, and posted to PharmGKB with supplemental information/data and updates. CPIC's goal is to address barriers to the implementation of pharmacogenetic tests into clinical practice (CPIC n.d.). DPWG was established in 2005 by the Royal Dutch Pharmacists Association. The DPWG is multidisciplinary and includes clinical pharmacists, physicians, clinical pharmacologists, clinical chemists, epidemiologists, and toxicologists (PharmGKB, n.d.). The objective of the DPWG is to develop pharmacogenetic-based therapeutic (dose) recommendations and assist the prescribers and pharmacist by recommending drug prescription. DPWG has evaluated therapeutic dose recommendations for tamoxifen based on CYP2D6 genotypes (Swen et al. 2011). For PM and IM genotypes, aromatase inhibitors have been recommended for postmenopausal women due to the risk of breast cancer with tamoxifen. For IM genotypes, the recommendation is to avoid the use of a CYP2D6 inhibitor.

13.5 Therapeutic Advances in Pharmacogenomics

Progress has been achieved in the pharmacogenomics of SCAR, warfarin, and antiplatelet therapy, and a summary of developments has been represented. In recent years, many genetic polymorphisms were reported as contributing to ADR. A recent study in Japan found 1010 ADRs in 3459 adult patients, and of these, 1.6%, 4.9%, and 33% were fatal, life-threatening, and serious, respectively (Morimoto et al. 2011). The ability to predict ADR-related issues would prevent drug administration to high-risk patients. However, genetic markers were studied for several ADRs, especially for SCARs and drug-induced liver injury (DILI). As for SCARs, associations of alleles HLA-B*15:02 or HLA-A*31:01 and HLA-B*58:01 were reported for carbamazepine- and allopurinol-related Stevens-Johnson syndrome and toxic epidermal necrolysis, respectively (Kaniwa and Saito 2013). Several HLA alleles also demonstrate drug-specific associations with DILI, such as HLA-A*33:03 for ticlopidine, HLA-B*57:01 for flucloxacillin, and HLA-DQA1*02:01 for lapatinib.

13.5.1 Warfarin Therapy

Therapeutic advances have been achieved in pharmacogenomics of warfarin. Warfarin is a commonly used oral anticoagulant for the prevention of thromboembolism in patients with deep vein thrombosis, atrial fibrillation, or prosthetic heart valve replacement (Hirsh et al. 2001). Warfarin exerts its anticoagulation effect by blocking the vitamin K regeneration cycle. Potential lethal side effects of warfarin therapy have been found, and efforts were made to reduce the ADR. Efforts were focused on developing dosing algorithms using clinical variables to predict warfarin dose (Ageno et al. 2000). However, this approach was not very efficient due to the lack of effectiveness of the programs. Warfarin maintenance was found to be associated with polymorphisms in cytochrome P450 2C9 and vitamin K epoxide reductase subunit 1. With the identifications of associated genetic factors, efforts have been made on developing dosing algorithms incorporating both clinical and genetic variables (Lee and Klein 2013).

13.5.2 Antiplatelet Therapy

Antiplatelet drugs are used in the prevention of thrombotic events associated with cardiovascular disease. The adenosine diphosphate (ADP) receptor inhibitors are a subclass of antiplatelet medications, which include clopidogrel, prasugrel, ticagrelor, and ticlopidine. Clopidogrel is one of the most commonly prescribed medications for the patients with acute coronary syndrome (ACS) and in patients undergoing percutaneous coronary intervention (PCI) (Kushner et al. 2009). Recent evidence supports a role of loss-of-function (LOF) variants in CYP2C19 as a determinant of clopidogrel response. Patients who carry LOF variants do not metabolize clopidogrel, a prodrug, into its active form resulting in decreased inhibition of platelet function and a higher risk of cardiovascular events (Perry and Shuldiner 2013). CYP2C19 LOF variants have been demonstrated to be clinically significant determinants of poor outcomes in ACS/PCI patients. With the addition of new antiplatelet therapy, the promise of translating these pharmacogenetic insights into more effective individualized antiplatelet therapy has excited the hope for future of personalized medicine.

13.5.3 Type 2 Diabetes

Genetic variants associated with incretin-based therapeutic approach for type 2 diabetes were also determined. Incretin-based therapies are used to treat patients with type 2 diabetes. Incretin effect enhancers include GLP-1 receptor agonists and dipeptidyl peptidase-4 (DPP4) inhibitors. Gliptins act by increasing endogenous incretin levels. GLP-1 receptor (GLP-1R) and GIP receptor (GIPR) are their indirect drug targets (Tkáč and Gotthardová 2016). Several genetic variants were predicted to be involved in the physiology of incretin secretion. Only two gene variants TCF7L2 rs7903146 C>T and CTRB1/2 rs7202877 T>G minor allele carriers were associated with a smaller reduction in HbA1c after gliptin treatment (Javorský et al. 2016). HbA1c is a form of hemoglobin that is bound to glucose and indicates how well diabetes is controlled. These clinical observations could be helpful to identify patients with lower or higher response to gliptin inhibitor.

13.5.4 Cancer Therapy

Mapping of the human genome has been a boon for cancer therapy. Both somatic and germline genome provide some insight into the decision-making of cancer treatment (Hertz and McLeod 2013). The somatic genome is involved in predicting tumor behavior. Germline genome assists in determining drug exposure and toxicity. These somatic and germline informations will be very helpful in personalized therapy for cancer patients. Several new chemotherapeutic agents are available for the treatment of colorectal cancer, and it has increased the decision complexity in treatment planning. Treatment decision-making should be guided by predictive and prognostic markers. Most cytotoxic drugs induce DNA damage; the DNA damage repair pathways hold potential for yielding, predicting, and prognostic biomarkers (Kap et al. 2016). The involvement of the nucleotide excision repair pathway in the efficacy of chemotherapeutic agents should be validated for the treatment of colorectal cancer. Vincristine induces distinct death programs in primary acute lymphoblastic leukemia (ALL) cells depending on cell-cycle phase (Kothari et al. 2016). Vincristine is an important component of ALL treatment that can cause neurotoxicity. Recently, a GWAS study reported a SNP, involved in vincristine pharmacodynamics, with neurotoxicity during later phases of therapy. The strongest associations with neurotoxicity were observed for two SNPs in ABCC2, and the genotypes rs3740066 GG and rs12826 GG were associated with increased neurotoxicity (Lopez-Lopez et al. 2016). Polymorphisms in ABCC2 could be novel markers for vincristine-related neurotoxicity in pediatric ALL in early phases. These results indicate that polymorphisms in pharmacokinetic genes are associated with drug toxicity. The level of vincristine transporters or metabolizers could be used as predictors of vincristine-related neurotoxicity in ALL patients.

13.5.5 Invasive Aspergillosis

Many genetic polymorphisms have been reported that are known to alter CYP enzymes and drug receptors, drug targets, and transporters. These genetic variants can greatly influence pharmacokinetics, dose requirement and response, and therapeutic outcomes. The clinical applications of these findings can significantly improve drug efficacy and safety. For example, invasive aspergillosis (IA) is one of the leading causes of morbidity and mortality in hematological patients (Kimura 2016). Voriconazole is used for initial therapy for IA. Individuals who carry the CYP2C19*17 gain-of-function allele were shown lower voriconazole exposure and are therefore at risk of failing IA therapy. However, there are limited data to establish a predicted relationship between voriconazole dosage and CYP2C19 metabolic capacity. A pediatric CYP2C19 rapid metabolizer (i.e., CYP2C19*1/*17) requires a voriconazole dose of 14 mg/kg twice daily (usual dose from 7 to 9 mg/kg twice daily) (Hicks et al. 2016). CYP2C19 genotype could be utilized to optimize voriconazole dose and this may be a cost-effective to improve IA therapy.

13.5.6 New Drug Labels for Clopidogrel and Warfarin

The common, complex diseases have environmental and multiple genetic influences. Therefore, drugs targeting a specific mutation can be highly successful in cancer, but we could not expect same success for chronic disease treatments. However, genes identified through GWAS and other studies provide the important protein targets. Substantial advances in the understanding of the genetic determinants of drug response have been reported, and most frequent use of pharmacogenetic data to guide drug therapy decisions can be seen in the future. Implementation of CYP2C19 genotyping for clopidogrel treatment in patients undergoing PCI is also occurring with increasing frequency, and centers Scripps Health, Vanderbilt, University of Florida, and University of North Carolina have adopted this approach (Pulley et al. 2012; Johnson et al. 2012) For instance, understanding the relationship between genetics and drug metabolism causes to issue a new drug label. In the case of a clopidogrel, new findings demonstrated that patients with genetic variants of CYP2C19 may not effectively convert the drug to its active form. After that, FDA issued a new label warning in 2010. Similarly, new labels were issued for warfarin based on genetic findings (Lesko 2008). Changes in drug labeling are likely to continue as more genetic findings are disclosed from studies on approved drugs.

13.5.7 Collaborative Efforts to Achieve the Goal of PGx

New applications and processes are needed to integrate emerging pharmacogenomic data into clinical practice. Current barriers, concerns, system limitations, and requisite infrastructure need to be addressed to achieve the true goal of pharmacogenomics. In 2010, the Pharmacy e-Health Information Technology (HIT) Collaborative was formed by nine national nonprofit organizations (Reiss and American Pharmacists Association 2011). The goal of HIT collaborative is to ensure the pharmacist's role of providing patient care services and medication. The Evaluation of Genomic Applications in Practice and Prevention (EGAPP) working group has noticed that there is insufficient evidence to recommend for or against genetic testing in case of the most genetic test (Khoury et al. 2010). The challenge with such cases is that pharmacist still makes some decisions in the lack of proper evidence. In this context Veenstra and colleagues proposed a three-tiered approach, focusing on how to deal with cases with insufficient evidence for or against testing (Veenstra et al. 2010).

The ACCE (analytical validity, clinical validity, clinical utility, and associated ethical, legal, and social implications (ELSI)) model project, sponsored by the Centers for Disease Control and Prevention (CDC), has recommended the evaluation of pharmacogenomic biomarker tests (Berg 2009). Analytical validity determines how well diagnostic test measures what it is intended to measure, regardless of whether it is an expression pattern, a mutation, or a protein (Lam 2013). Clinical validity measures the ability of the test to differentiate between responders and nonresponders or to identify ADR. The clinical utility measures the ability of the test result to determine the outcome of clinical testing. Obviously, any biomarker with validation and FDA approval could enhance test implementation and utilization in the clinical settings. It is interesting to note that a large amount of PGx information has been generated, but most of the findings have not yet been applied in clinical testing and treatment. The clinical application of PGx is slow, and some physicians do not know how to interpret and apply the clinical findings in patient care (Ventola 2013).

13.6 PGx Resources

PGx discovery is based on two approaches: the candidate gene approach and GWAS. The candidate gene study focuses on genes involved in transport, drug metabolism, and targeting pathways. On the other hand, GWAS considers all genes and noncoding sequences of the human genome, assuming that all genetic elements have equal chances to affect the response of a drug. GWAS approaches came into existence after the completion of the HGP in 2000. Genome-wide studies have become more popular due to the public availability of human genomic information and low cost of sequencing.

13.6.1 PharmGKB

The Pharmacogenomics Knowledge Base (PharmGKB) is a comprehensive online resource that provides knowledge about the impact of genetic variation on drug response (Whirl-Carrillo et al. 2012). Informations retrieved from PharmGKB are very useful for clinical implementation and interpretation of clinical results. It provides a well-known PGx association between drug and gene on the basis of dosing guidelines, drug label, clinical annotation, variant annotation, pharmacology, mechanism of action, and related pathway. It also provides the details of genes that are associated with a drug based on variant annotation, literature reviews, pathway, and information retrieved from DrugBank. PharmGKB annotates drug labels containing pharmacogenetic information approved by FDA, European Medicines Agency (EMA), the Pharmaceuticals and Medical Devices Agency, Japan (PMDA), and Health Canada (Santé Canada) (HCSC) and provides a brief summary of the PGx in the label (PharmGKB, n.d.). National Institute of Health (NIH)-funded scientists have studied the effect of genes on medications relevant to a wide range of conditions, including asthma, depression, cancer, and heart disease. The research findings are collected in PharmGKB.

13.6.2 PGxOne™

PGxOne™ is a proprietary clinical PGx test that provides relevant medical and clinical data and its interpretation for the treatment of patients. PGxOne™ results indicate dosing recommendations for 76 drugs. All 76 drugs are directly influenced by the 13 PGx genes (drug metabolism) covered by PGxOne™ testing. Copy number variations (CNVs) are a major source of genetic variation within an individual's genome. PGxOne™ is capable of detecting and providing information about CNVs in the CYP2D6 gene. Importantly, CYP2D6 is responsible for metabolizing approximately 25% of drugs on the market, and CYP2D6 CNVs impact the metabolism of 50% of these drugs (Ingelman-Sundberg et al. 2007). Features of PGxOne™ test (1) screens all well-established PGx genes in a single, cost-effective test; (2) detects multiple types of variations, including substitutions, insertions/ deletions, and copy number variations using next-generation amplicon sequencing technology; (3) delivers results quickly via intuitive, clinically relevant, medically actionable report; and (4) provides lifetime utility of data, decreasing the need for future testing (GENEWIZ PGxOne™ n.d.).

13.6.3 Biobank

Biobank is a collection of human tissue samples or blood and medical information about donors, which are stored for long periods of time and are used for research studies. Donors voluntarily decide to give a blood or tissue sample or information about themselves for free. Imagine if every person offers and shares their health information with biobanks, then there will be a vast amount of health and clinical data, which could be used in health study for decision-making (Genetic Alliance Registry and Biobank n.d.). Biobank has the advantages of being considered in cell-, tissue-, blood-, or DNA-related studies as minimal risk research since there is no harm to individual if their sample is examined (Biobank n.d.). Disease-related biobanks were established initially with the goal of personalized medicine. In the majority of clinical trials, the samples and blood are stored for future genetic analysis. The UK Biobank is a major health resource, with the objective of improving the health status, diagnosis, and treatment of a wide range of serious diseases – including cancer, heart diseases, stroke, diabetes, arthritis, and osteoporosis (Budimir et al. 2011). Scientist uses data from the questionnaire, physical measures, and biological samples to undertake studies to improve the health of future generations. In due course, it may be possible to find out more information about the use of the resource and follow their results.

13.7 Challenges in Implementation of PGx

13.7.1 Informed Consent and Confidentiality

Drugs that are tailored to individual genomes may require extensive genetic information of the participant in the clinical trial. Data collected from clinical trials could be stored and utilized for future research. Hence, these genetic samples could provide other information about the subject that could be unrelated to the intended study but yet might prove useful for other genetic research (Singh 2003). The patient should have high motivation to participate in the clinical trial, where the subject and patient are the same person. Therefore, the guarantee of informed consent, for all participants, is essential for the current as well as the future study. Patient, whose genetic information has been collected for clinical diagnosis, may not be interested in disclosing his/her health status and wants to keep it confidential. Sometimes, family members show their interest to know the status of inherited diseases in the patient's genetic profile. Employers or health insurers may also desire to access genetic profile of the person. Indeed, the fear of losing a job and health insurance discourage the people's interest to participate in genetic disease-related study (Nass et al. 2009). To promote the participation of patients in health and clinical study, the informations related to donor must be confidential and anonymous.

13.7.2 Technical and Educational Status

The success of PGx testing depends on the accuracy of the genomic information. The accuracy rate of hundred percent is impractical to expect in sequencing. Now, the question is how to address the sequencing errors produced. Fundamental of all sequencing method is DNA amplification. DNA amplification is well known for introducing errors and these technical errors are impossible to avoid. During the data analysis, variation due to sequencing errors might be incorporated as a "natural" characteristic of the cell (Bavarva et al. 2015). The next challenge would be a question of how close our analyzed sequence is to the real sequence. NGS technologies and their data analysis methods must be standardized for accurate interpretation of biological problems. Auxiliary labels can be used as a tool to promote patient's awareness about PGx testing. Auxiliary labels highlight the informations related to the use and risk of drugs (Haga and Moaddeb 2015). This approach motivates the patients to consult with health provider for PGx testing. It is necessary to educate patients about PGx testing using new educational strategies. Pharmacist, physician, and other health service providers should increase and update their knowledge about PGx testing to effectively respond to patient's inquiries.

13.7.3 Economic Status, Justice, and Equity

The cost of a drug developed by PGx approach will be high due to increased research in order to identify genetic profile, develop genetic tests, and conduct clinical trials. It is also believed that few pharmaceutical companies will show their interest in developing personalized medicine due to the high cost of drug development and limited availability of market. Pharmaceutical industries show their interest in developing drugs against diseases with largest market value. An orphan population has a genotype leading to a condition for which no effective treatment is available, and pharmaceutical companies are also not attracted toward these diseases due to low market potential. Therefore, incentives must be given to pharmaceutical industries to promote drug development for rare genotypes or less common diseases (Sharma et al. 2010). One challenging issue for PGx is to develop the effective therapy for those who do not show the response to a drug (nonresponders) or difficult to treat.

There is also a disconnection between the funding agencies and the prioritization of PGx research, in terms of financial commitment, clinical trial infrastructure, and ability to adopt new strategies. Now, the question is whether national health insurance or private medical insurance companies are willing to pay the cost of drug therapy or the tests needed to prescribe them. It is also a challenge for the governments to allocate grant for drug therapy in healthcare budget. Due to high cost, PGx drugs will flow primarily into developed countries where most of the individuals can afford them. Public policy must be altered to encourage drug development via PGx, by promoting researcher, pharmaceutical companies, and market. Justice and equity are other important issues in PGx. The idea behind this is that every person should enjoy equal access to medical treatment irrespective of the virtue of race, origin, or economic status. But these inequalities in access to healthcare exist worldwide. Beyond that, it is important to address whether and how justice related to healthcare can be well served between developed and poor countries.

The most controversial issue is whether and how to integrate category of race into drug devolvement. It is discouraging that many members of the racial/ethnic community do not show their interest in participating clinical drug trials. It is noticed that a racial community receives a lower quality of healthcare than others (Nsiah-Jefferson 2003). In recent years, few drugs got approval for use by a particular racial group. Such injustice and difference are in practice because biomedical research may be biased in favor of a particular race.

13.7.4 Recommendations for Implementation of PGx

Ossorio and Duster 2005 (2005) argue that "While attempting to provide medical benefit, or market products, scientists and the pharmaceutical industry may reinvigorate the very notions of biological difference that may have resulted in racially disparate treatment and racially disparate health." Justice demands that benefits of personalized medicine must be available to individuals of all racial and socioeconomic status. The policy maker must keep in view the racial, social, or economic disparity that exists in healthcare system. Major points suggested by Peterson-Iyer (2008) for consideration in policy of PGx are (1) informed consent for the use of genetic samples, (2) improvements to subject/patient confidentiality, (3) increased post-marketing surveillance, (4) increased incentives for the development of orphan drugs, (5) revision of patent law to encourage the "rescuc" of drugs, (6) subsidies to ensure that the less wealthy have fair access, (7) approval of gene-specific drugs over race-specific drugs, (8) inclusion of racial/ethnic minority groups in drug research, and (9) incentives for pharmaceutical companies to invest in and provide drug to developing countries. There are many technical, financial, and ethical hurdles in clinical implementation of PGx. In spite of these hurdles, we should start this journey with a patience and high motivation to achieve the goal.

13.8 Future Perspective

Human genome sequencing and advances in techniques that correlate specific genetic variations to diseases have played an important role in developing more effective therapy against disease. Currently, the most studied genetic variant is SNP due to low cost and high accuracy of SNP genotyping. The location and allele frequencies of genome-wide SNPs in human can be retrieved from SNP database (dbSNP) of NCBI. The future of PGx is very wide. Biotechnology industries have provided very advanced technologies for sequencing, genome annotation, expression analysis, and pharmacology. A major drawback in the study of PGx is the common occurrence of false-positive association between polymorphisms and the investigated outcome. Identification of biologically relevant polymorphism can trigger the application of PGx. Next-generation sequencing has created a plethora of analytical and biological consideration. Scientists are very optimistic that single-cell sequencing technology will better characterize cancer and other complex diseases. The genome sequencing contributes a large amount of data but with limited insight into therapies. It has become necessary to elucidate the clinical implications of available data as well as to define the guidelines for the clinical application of PGx data. PGx knowledge is not fully utilized in clinical practice due to lack of in-depth understanding of PGx principles among the healthcare professionals. Recent PGx studies have paid attention over mitochondrial genome along with nuclear genome, because of its role in metabolism, cell cycling, cellular differentiation, and signaling. The high rate of polymorphisms in mitochondrial genome further highlights the significance of studying genetic variants in mitochondria.

There is a need to promote the field of PGx globally and aware the peoples with merits and demerits related to this field. Future advancements in PGx technologies might be able to make firm and cost-effective recommendations for drug therapy. Pharmaceutical companies are considering the importance of conducting PGx research in the early stages of drug development so that the derived knowledge could be utilized for new drug approval to avoid the risk of rejection or delayed approval. Most importantly, efforts are needed to translate the scientific outcome of PGx study into clinical practice. Patients living in urban areas are educated and aware of benefits of PGx. Efforts should be made to improve and upgrade the current status of PGx and also to implement the potential of PGx. Many PGx biomarkers corresponding to a therapeutic agent have been evaluated and more are in the process of study. These biomarkers have shown to improve the status of medication with reduced toxicity and high efficacy, which could subsequently lower the overall healthcare cost. Clinical feasibility of implementing PGx tests is dependent on medical service providers and practitioners. Patients are optimistic about the potential of PGx tests, but cost and testing time frame are barriers in the implementation of PGx.

Now, the question is the accessibility of PGx testing to common and poor people. It is surprising to know that few pharmaceutical industries are indeed in favor of the race-based medicine. FDA has approved a heart failure drug (BiDil), for its use by a self-identified racial group African-Americans, although there was no solid evidence that BiDil would be ineffective for the rest of the population. Therefore, race-based drug development practices should be discouraged to avoid racial discrimination. There are many technical- and policy-related issues associated with the wide-scale implementation of PGx. These issues must be resolved to cater the benefits of PGx equally and globally. Clinical application and cost-effectiveness cannot be the only criteria for determining the relative value of pharmacogenomics for drug therapy. Rather, it should be aimed to supplement the best practice strategies to achieve optimal drug therapy.

References

Abraham A, Varatharajan S, Karathedath S, Philip C, Lakshmi KM, Jayavelu AK, Mohanan E, Janet NB, Srivastava VM, Shaji RV, Zhang W, Abraham A, Viswabandya A, George B, Chandy M, Srivastava A, Mathews V, Balasubramanian P. RNA expression of genes involved in cytarabine metabolism and transport predicts cytarabine response in acute myeloid leukemia. Pharmacogenomics. 2015;16:877–90.

Ageno W, Johnson J, Nowacki B, Turpie AG. A computer generated induction system for hospitalized patients starting on oral anticoagulant therapy. Thromb Haemost. 2000;83:849–52.

Agúndez JA, Blanca M, Cornejo-García JA, García-Martín E. Pharmacogenomics of cyclooxygenases. Pharmacogenomics. 2015;16:501–22.

Alomar MJ. Factors affecting the development of adverse drug reactions (Review article). Saudi Pharm J. 2014;22:83–94.

American Nurse Today. The role of pharmacogenomics in cancer. n.d. http://www.americannursetoday.com/the-role-of-pharmacogenomics-in-cancer. Accessed 12 Sept 2015

Ansari M, Krajinovic M. Pharmacogenomics in cancer treatment defining genetic bases for inter-individual differences in responses to chemotherapy. Curr Opin Pediatr. 2007;19:15–22.

Bartel DP. MicroRNAs: genomics, biogenesis, mechanism, and function. Cell. 2004;116:281–97.

Bavarva JH, Bavarva MJ, Karunasena E. Next in line in next-generation sequencing: are we there yet? Pharmacogenomics. 2015;16:1–4.

Becker ML, Kallewaard M, Caspers PW, Schalekamp T, Stricker BH. Potential determinants of drug-drug interaction associated dispensing in community pharmacies. Drug Saf. 2005;28:371–8.

Beger RD, Dunn W, Schmidt MA, Gross SS, Kirwan JA, Cascante M, Brennan L, Wishart DS, Oresic M, Hankemeier T, Broadhurst DI, Lane AN, Suhre K, Kastenmüller G, Sumner SJ, Thiele I, Fiehn O, Kaddurah-Daouk R, for "Precision Medicine and Pharmacometabolomics Task Group"-Metabolomics Society Initiative. Metabolomics enables precision medicine: "A white paper, community perspective". Metabolomics. 2016;12:149.

Berezikov E, Guryev V, van de Belt J, Wienholds E, Plasterk RH, Cuppen E. Phylogenetic shadowing and computational identification of human microRNA genes. Cell. 2005;120:21–4.

Berg AO. The CDC's EGAPP initiative: evaluating the clinical evidence for genetic tests. Am Fam Physician. 2009;80:1218.

Biobank. About UK Biobank. n.d. http://www.ukbiobank.ac.uk/about-biobank-uk. Accessed 25 July 2015.

Borroni RG. Role of dermatology in pharmacogenomics: drug-induced skin injury. Pharmacogenomics. 2015;16:401–12.

Bradford LD. CYP2D6 allele frequency in European Caucasians, Asians, Africans and their descendants. Pharmacogenomics. 2002;3:229–43.

Budimir D, Polasek O, Marusić A, Kolcić I, Zemunik T, Boraska V, Jeroncić A, Boban M, Campbell H, Rudan I. Ethical aspects of human biobanks: a systematic review. Croat Med J. 2011;52:262–79.

Burt T, Dhillon S. Pharmacogenomics in early-phase clinical development. Pharmacogenomics. 2013;14:1085–97.

Burt T, Nandal S. Pharmacometabolomics in early-phase clinical development. Clin Transl Sci. 2016;9:128–38.

Chan A, Tan SH, Wong CM, Yap KY, Ko Y. Clinically significant drug-drug interactions between oral anticancer agents and nonanticancer agents: a Delphi survey of oncology pharmacists. Clin Ther. 2009;2:2379–86.

Cho WC. Recent progress in genetic variants associated with cancer and their implications in diagnostics development. Expert Rev Mol Diagn. 2010;10:699–703.

Cirulli ET, Goldstein DB. Uncovering the roles of rare variants in common disease through whole-genome sequencing. Nat Rev Genet. 2010;11:415–25.

Collins SL, Carr DF, Pirmohamed M. Advances in the pharmacogenomics of adverse drug reactions. Drug Saf. 2016;39:15–27.

Cox NJ, Gamazon ER, Wheeler HE, Dolan ME. Clinical translation of cell-based pharmacogenomic discovery. Clin Pharmacol Ther. 2012;92:425–7.

CPIC. Clinical Pharmacogenetics Implementation Consortium. n.d. https://cpicpgx.org. Accessed 03 May 2016.

De Souza JA, Olopade OI. CYP2D6 genotyping and tamoxifen: an unfinished story in the quest for personalized medicine. Semin Oncol. 2011;38:263–73.

Everett JR. Pharmacometabonomics in humans: a new tool for personalized medicine. Pharmacogenomics. 2015;16:737–54.

Feng Y, He X, Yang Y, Chao D, Lazarus LH, Xia Y. Current research on opioid receptor function. Curr Drug Targets. 2012;13:230–46.

Genetic Alliance Registry and Biobank. Overview. n.d. http://biobank.org/biobanks. Accessed 28 July 2015.

Genetic Home Reference What is Pharmacogenomics? n.d. http://ghr.nlm.nih.gov/handbook/genomicresearch/pharmacogenomics. Accessed 28 July 2015.

GENEWIZ PGxOne™. Comprehensive pharmacogenomics test. n.d. http://www.genewiz.com/public/PGxOne-pharmacogenomics-test.aspx. Accessed 20 July 2015.

Gregersen PK, Olsson LM. Recent advances in the genetics of autoimmune disease. Annu Rev Immunol. 2009;27:363–91.

Gurdasani D, Carstensen T, Tekola-Ayele F, Pagani L, Tachmazidou I, Hatzikotoulas K, Karthikeyan S, Iles L, Pollard MO, Choudhury A, Ritchie GR, Xue Y, Asimit J, Nsubuga RN, Young EH, Pomilla C, Kivinen K, Rockett K, Kamali A, Doumatey AP, Asiki G, Seeley J, Sisay-Joof F, Jallow M, Tollman S, Mekonnen E, Ekong R, Oljira T, Bradman N, Bojang K, Ramsay M, Adeyemo A, Bekele E, Motala A, Norris SA, Pirie F, Kaleebu P, Kwiatkowski D, Tyler-Smith C, Rotimi C, Zeggini E, Sandhu MS. The African genome variation project shapes medical genetics in Africa. Nature. 2015;517:327–32.

Haga SB, Moaddeb J. Potential use of auxiliary labels to promote patient awareness of pharmacogenetic testing. Pharmacogenomics. 2015;16:299–301.

He B, Pan Y, Cho WC, Xu Y, Gu L, Nie Z, Chen L, Song G, Gao T, Li R, Wang S. The association between four genetic variants in microRNAs (rs11614913, rs2910164, rs3746444, rs2292832) and cancer risk: evidence from published studies. PLoS One. 2012;7(11):e49032.

He Y, Hoskins JM, Clark S, Campbell NH, Wagner K, Motsinger-Reif AA, McLeod HL. Accuracy of SNPs to predict risk of HLA alleles associated with drug-induced hypersensitivity events across racial groups. Pharmacogenomics. 2015;16:817–24.

Hertz DL, McLeod HL. Use of pharmacogenetics for predicting cancer prognosis and treatment exposure, response and toxicity. J Hum Genet. 2013;58:346–52.

Hess GP, Fonseca E, Scott R, Fagerness J. Pharmacogenomic and pharmacogenetic-guided therapy as a tool in precision medicine: current state and factors impacting acceptance by stakeholders. Genet Res (Camb). 2015;97:e13. doi:10.1017/S0016672315000099.

Hicks JK, Gonzalez BE, Zembillas AS, Kusick K, Murthy S, Raja S, Gordon SM, Hanna R. Invasive Aspergillus infection requiring lobectomy in a CYP2C19 rapid metabolizer with subtherapeutic voriconazole concentrations. Pharmacogenomics. 2016;17:663–7.

Hirsh J, Dalen J, Anderson DR, Poller L, Bussey H, Ansell J, Deykin D. Oral anticoagulants: mechanism of action, clinical effectiveness, and optimal therapeutic range. Chest. 2001;119:8S–21S.

Hu Z, Yang X, Ho PC, Chan SY, Heng PW, Chan E, Duan W, Koh HL, Zhou S. Herb-drug interactions: a literature review. Drugs. 2005;65:1239–82.

Huang Q, Aa J, Jia H, Xin X, Tao C, Liu L, Zou B, Song Q, Shi J, Cao B, Yong Y, Wang G, Zhou G. A pharmacometabonomic approach to predicting metabolic phenotypes and pharmacokinetic parameters of atorvastatin in healthy volunteers. J Proteome Res. 2015;14:3970–81.

IGSR. Supporting the 1000 genomes data. n.d. http://www.internationalgenome.org. Accessed 01 May 2016.

IGSR: The International Genome Sample Resource. n.d. http://www.1000genomes.org/about. Accessed 02 May 2016.
Ingelman-Sundberg M, Rodriguez-Antona C. Pharmacogenetics of drug metabolizing enzymes: implications for a safer and more effective drug therapy. Philos Trans R Soc Lond Ser B Biol Sci. 2005;360:1563–70.
Ingelman-Sundberg M, Sim SC, Gomez A, Rodriguez-Antona C. Influence of cytochrome P450 polymorphisms on drug therapies: pharmacogenetic, pharmacoepigenetic and clinical aspects. Pharmacol Ther. 2007;116:496–526.
International HapMap Consortium. The international HapMap project. Nature. 2003;426:789–96.
International HapMap Project. n.d. http://www.hapmap.org. Accessed 02 May 2016.
James LP. Metabolomics: integration of a new "omics" with clinical pharmacology. Clin Pharmacol Ther. 2013;94:547–51.
Jarjour S, Barrette M, Normand V, Rouleau JL, Dubé MP, de Denus S. Genetic markers associated with cutaneous adverse drug reactions to allopurinol: a systematic review. Pharmacogenomics. 2015;16:755–67.
Javorský M, Gotthardová I, Klimčáková L, Kvapil M, Židzik J, Schroner Z, Doubravová P, Gaľa I, Dravecká I, Tkáč I. A missense variant in GLP1R gene is associated with the glycemic response to treatment with gliptins. Diabetes Obes Metab. 2016;18:941. doi:10.1111/dom.12682.
Jemal A, Bray F, Center MM, Ferlay J, Ward E, Forman D. Global cancer statistics. CA Cancer J Clin. 2011;61:69–90.
Johnson JA, Burkley BM, Langaee TY, Clare-Salzler MJ, Klein TE, Altman RB. Implementing personalized medicine: development of a cost-effective custom pharmacogenetics genotyping array. Clin Pharmacol Ther. 2012;92:437–9.
Jorde LB, Wooding SP. Genetic variation, classification and 'race'. Nat Genet. 2004;36:S28–33.
Kaddurah-Daouk R, Weinshilboum R, Pharmacometabolomics Research Network. Metabolomic signatures for drug response phenotypes-pharmacometabolomics enables precision medicine. Clin Pharmacol Ther. 2015;98:71–5.
Kalliokoski A, Niemi M. Impact of OATP transporters on pharmacokinetics. Br J Pharmacol. 2009;158:693–705.
Kaniwa N, Saito Y. Pharmacogenomics of severe cutaneous adverse reactions and drug-induced liver injury. J Hum Genet. 2013;58:317–26.
Kap EJ, Popanda O, Chang-Claude J. Nucleotide excision repair and response and survival to chemotherapy in colorectal cancer patients. Pharmacogenomics. 2016;17:755–94.
Khoury MJ, Coates RJ, Evans JP. Evidence-based classification of recommendations on use of genomic tests in clinical practice: dealing with insufficient evidence. Genitourin Med. 2010;12:680–3.
Kimura S. Invasive Aspergillosis in Hematological Patients. Med Mycol J. 2016;57:77–88.
Kitzmiller JP, Groen DK, Phelps MA, Sadee W. Pharmacogenomic testing: relevance in medical practice: why drugs work in some patients but not in others. Cleve Clin J Med. 2011;78:243–57.
Klotz U. Pharmacokinetics and drug metabolism in the elderly. Drug Metab Rev. 2009;41:67–76.
Kothari A, Hittelman WN, Chambers TC. Cell cycle-dependent mechanisms underlie Vincristine-induced death of primary acute lymphoblastic leukemia cells. Cancer Res. 2016;76:3553. doi:10.1158/0008-5472.CAN-15-2104.
Kushner FG, Hand M, Smith Jr SC, King 3rd SB, Anderson JL, Antman EM, Bailey SR, Bates ER, Blankenship JC, Casey Jr DE, Green LA, Hochman JS, Jacobs AK, Krumholz HM, Morrison DA, Ornato JP, Pearle DL, Peterson ED, Sloan MA, Whitlow PL, Williams DO. Focused updates: ACC/AHA guidelines for the management of patients with ST-elevation myocardial infarction (updating the 2004 guideline and 2007 focused update) and ACC/AHA/SCAI guidelines on percutaneous coronary intervention (updating the 2005 guideline and 2007 focused update) a report of the American College of Cardiology Foundation/American Heart Association Task Force on Practice Guidelines. J Am Coll Cardiol. 2009;54:2205–41.

Lam YW. Scientific challenges and implementation barriers to translation of pharmacogenomics in clinical practice. ISRN Pharmacol. 2013;2013:1. doi:10.1155/2013/641089.

Leblhuber F, Neubauer C, Peichl M, Reisecker F, Steinparz FX, Windhager E, Dienstl E. Age and sex differences of dehydroepiandrosterone sulfate (DHEAS) and cortisol (CRT) plasma levels in normal controls and Alzheimer's disease (AD). Psychopharmacology. 1993;111:23–6.

Lee MT, Klein TE. Pharmacogenetics of warfarin: challenges and opportunities. J Hum Genet. 2013;58:334–8.

Lesko LJ. The critical path of warfarin dosing: finding an optimal dosing strategy using pharmacogenetics. Clin Pharmacol Ther. 2008;84:301–3.

Lin E, Lane HY. Genome-wide association studies in pharmacogenomics of antidepressants. Pharmacogenomics. 2015;16:555–66.

Lipson D, Capelletti M, Yelensky R, Otto G, Parker A, Jarosz M, Curran JA, Balasubramanian S, Bloom T, Brennan KW, Donahue A, Downing SR, Frampton GM, Garcia L, Juhn F, Mitchell KC, White E, White J, Zwirko Z, Peretz T, Nechushtan H, Soussan-Gutman L, Kim J, Sasaki H, Kim HR, Park SI, Ercan D, Sheehan CE, Ross JS, Cronin MT, Jänne PA, Stephens PJ. Identification of new ALK and RET gene fusions from colorectal and lung cancer biopsies. Nat Med. 2012;18:382–4.

Lopez-Lopez E, Gutierrez-Camino A, Astigarraga I, Navajas A, Echebarria-Barona A, Garcia-Miguel P, Garcia de Andoin N, Lobo C, Guerra-Merino I, Martin-Guerrero I, Garcia-Orad A. Vincristine pharmacokinetics pathway and neurotoxicity during early phases of treatment in pediatric acute lymphoblastic leukemia. Pharmacogenomics. 2016;17:731–41.

Lynch T, Price A. The effect of cytochrome P450 metabolism on drug response, interactions, and adverse effects. Am Fam Physician. 2007;76:391–6.

Ma Q, Lu AY. Pharmacogenetics, pharmacogenomics, and individualized medicine. Pharmacol Rev. 2011;63:437–59.

Ma JD, Lee KC, Kuo GM. Clinical application of pharmacogenomics. J Pharm Pract. 2012;25:417–27.

Magro L, Moretti U, Leone R. Epidemiology and characteristics of adverse drug reactions caused by drug-drug interactions. Expert Opin Drug Saf. 2012;11:83–94.

Mangoni AA, Jackson SH. Age-related changes in pharmacokinetics and pharmacodynamics: basic principles and practical applications. Br J Clin Pharmacol. 2004;57:6–14.

Marian AJ, Belmont J. Strategic approaches to unraveling genetic causes of cardiovascular diseases. Circ Res. 2011;108:1252–69.

Marsh S, King CR, Van Booven DJ, Revollo JY, Gilman RH, McLeod HL. Pharmacogenomic assessment of Mexican and Peruvian populations. Pharmacogenomics. 2015;16:441–8.

McHale D. Applications of pharmacogenomics in drug discovery. In: Cohen N, editor. Pharmacogenomics and personalized medicine. Totowa: Humana Press; 2008. p. 73–87.

Morimoto T, Sakuma M, Matsui K, Kuramoto N, Toshiro J, Murakami J, Fukui T, Saito M, Hiraide A, Bates DW. Incidence of adverse drug events and medication errors in Japan: the JADE study. J Gen Intern Med. 2011;26:148–53.

My Cancer Genome. EGFR c.2573 T>G (L858R) mutation in non-small cell lung cancer. n.d. https://www.mycancergenome.org/content/disease/lung-cancer/egfr/5. Accessed 03 May 2016.

Nadine C, Theresa F. Challenges, opportunities, and evolving landscapes in pharmacogenomics and personalized medicine. In: Cohen N, editor. Pharmacogenomics and personalized medicine. Totowa: Humana Press; 2008. p. 1–26.

Nass SJ, Levit LA, Gostin LO. The value and importance of health information privacy. In: Nass SJ, Levit LA, Gostin LO, editors. Beyond the HIPAA privacy rule: enhancing privacy, improving health through research. Washington DC: National Academies Press (US); 2009.

Neavin D, Kaddurah-Daouk R, Weinshilboum R. Pharmacometabolomics informs pharmacogenomics. Metabolomics. 2016;12:121.

NIH, NIGMS. Pharmacogenomics fact sheet. n.d. http://www.nigms.nih.gov/education/pages/factsheet-pharmacogenomics.aspx. Accessed 20 July 2015.

Niu N, Wang L. In vitro human cell line models to predict clinical response to anticancer drugs. Pharmacogenomics. 2015;16:273–85.

Nsiah-Jefferson L. Pharmacogenomics: considerations for communities of color. In: Rothstein MA, editor. Pharmacogenomics social, ethical, and clinical dimensions. Hoboken: Wiley-Liss; 2003. p 26790.

O'Connor SK, Michaels N, Ferreri S. Expansion of pharmacogenomics into the community pharmacy: billing considerations. Pharmacogenomics. 2015;16:175–80.

Ochoa D, Prieto-Pérez R, Román M, Talegón M, Rivas A, Galicia I, Abad-Santos F, Cabaleiro T. Effect of gender and CYP2C9 and CYP2C8 polymorphisms on the pharmacokinetics of ibuprofen enantiomers. Pharmacogenomics. 2015;16:939–48.

Ossorio P, Duster T. Race and genetics: controversies in biomedical, behavioral, and forensic sciences. Am Psychol. 2005;60:11528.

Pérez-Ramírez C, Cañadas-Garre M, Jiménez-Varo E, Faus-Dáder MJ, Calleja-Hernández MÁ. MET: a new promising biomarker in non-small-cell lung carcinoma. Pharmacogenomics. 2015;16:631–47.

Perry CG, Shuldiner AR. Pharmacogenomics of anti-platelet therapy: how much evidence is enough for clinical implementation? J Hum Genet. 2013;58:339–45.

Peterson-Iyer K. Pharmacogenomics, ethics, and public policy. Kennedy Inst Ethics J. 2008;18:1.

PharmGKB. CPIC: Clinical Pharmacogenetics Implementation Consortium. n.d.-a. https://www.pharmgkb.org/page/cpic. Accessed 03 May 2016.

PharmGKB. DPWG: Dutch Pharmacogenetics Working Group. n.d.-b. https://www.pharmgkb.org/page/dpwg. Accessed 03 May 2016.

PharmGKB. Drug labels. n.d.-c. https://www.pharmgkb.org/view/drug-labels.do. Accessed 21 July 2015.

Pulley JM, Denny JC, Peterson JF, Bernard GR, Vnencak-Jones CL, Ramirez AH, Delaney JT, Bowton E, Brothers K, Johnson K, Crawford DC, Schildcrout J, Masys DR, Dilks HH, Wilke RA, Clayton EW, Shultz E, Laposata M, McPherson J, Jirjis JN, Roden DM. Operational implementation of prospective genotyping for personalized medicine: the design of the Vanderbilt PREDICT project. Clin Pharmacol Ther. 2012;92:87–95.

Rao US, Mayhew SL, Rao PS. Strategies for implementation of an effective pharmacogenomics program in pharmacy education. Pharmacogenomics. 2015;16:905–11.

Reiss SM, American Pharmacists Association. Integrating pharmacogenomics into pharmacy practice via medication therapy management. J Am Pharm Assoc (2003). 2011;51:e64–74.

Romualdi C, Balding D, Nasidze IS, Risch G, Robichaux M, Sherry ST, Stoneking M, Batzer MA, Barbujani G. Patterns of human diversity, within and among continents, inferred from biallelic DNA polymorphisms. Genome Res. 2002;12:602–12.

Ross CJ, Katzov H, Carleton B, Hayden MR. Pharmacogenomics and its implications for autoimmune disease. J Autoimmun. 2007;28:122–8.

Sánchez-Lázaro I, Herrero MJ, Jordán-De Luna C, Bosó V, Almenar L, Rojas L, Martínez-Dolz L, Megías-Vericat JE, Sendra L, Miguel A, Poveda JL, Aliño SF. Association of SNPs with the efficacy and safety of immunosuppressant therapy after heart transplantation. Pharmacogenomics. 2015;16:971–9.

Serre D, Pääbo S. Evidence for gradients of human genetic diversity within and among continents. Genome Res. 2004;14:1679–85.

Shabaruddin FH, Fleeman ND, Payne K. Economic evaluations of personalized medicine: existing challenges and current developments. Pharmgenomics Pers Med. 2015;8:115–26.

Sharma A, Jacob A, Tandon M, Kumar D. Orphan drug: development trends and strategies. J Pharm Bioallied Sci. 2010;2:290–9.

Sim SC, Kacevska M, Ingelman-Sundberg M. Pharmacogenomics of drug-metabolizing enzymes: a recent update on clinical implications and endogenous effects. Pharmacogenomics J. 2013;13:1–11.

Singh D. Ethical issues of pharmacogenetics must be addressed, says Nuffield Council. BMJ. 2003;327:701.

Soldin OP, Chung SH, Mattison DR. Sex differences in drug disposition. J Biomed Biotechnol. 2011;2011:187103. doi:10.1155/2011/187103.

Stjepanovic N, Bedard PL. Elucidating the genomic landscape of breast cancer: how will this affect treatment? Pharmacogenomics. 2015;16:569–72.

Sun H, Qu Q, Qu J, Lou XY, Peng Y, Zeng Y, Wang G. URAT1 gene polymorphisms influence uricosuric action of losartan in hypertensive patients with hyperuricemia. Pharmacogenomics. 2015;16:855–63.

Swen JJ, Nijenhuis M, de Boer A, Grandia L, Maitland-van der Zee AH, Mulder H, Rongen GA, van Schaik RH, Schalekamp T, Touw DJ, van der Weide J, Wilffert B, Deneer VH, Guchelaar HJ. Pharmacogenetics: from bench to byte--an update of guidelines. Clin Pharmacol Ther. 2011;89:662–73.

Tkáč I, Gotthardová I. Pharmacogenetic aspects of the treatment of Type 2 diabetes with the incretin effect enhancers. Pharmacogenomics. 2016;17:795–804.

U.S. FDA. Table of pharmacogenomic biomarkers in drug labeling. n.d. http://www.fda.gov/drugs/scienceresearch/researchareas/pharmacogenetics/ucm083378.htm. Accessed 21 July 2015.

Ubaldi M, Ricciardelli E, Pasqualini L, Sannino G, Soverchia L, Ruggeri B, Falcinelli S, Renzi A, Ludka C, Ciccocioppo R, Hardiman G. Biomarkers of hippocampal gene expression in a mouse restraint chronic stress model. Pharmacogenomics. 2015;16:471–82.

Veenstra DL, Roth JA, Garrison Jr LP, Ramsey SD, Burke W. A formal risk benefit framework for genomic tests: facilitating the appropriate translation of genomics into clinical practice. Genitourin Med. 2010;12:686–93.

Ventola CL. The role of pharmacogenomic biomarkers in predicting and improving drug response: part 2: challenges impeding clinical implementation. PT. 2013;38:624–7.

Venton G, Colle J, Mercier C, Fanciullino R, Ciccolini J, Ivanov V, Suchon P, Sebahoun G, Beaufils N, Gabert J, Hadjaj D, Costello R. Eradication of T315I mutation in chronic myeloid leukemia without third-generation tyrosine kinase inhibitor: a case report. Pharmacogenomics. 2015;16:677–9.

Verbelen M, Lewis CM. How close are we to a pharmacogenomic test for clozapine-induced agranulocytosis? Pharmacogenomics. 2015;16:915–7.

Wang W, Zheng Z, YuW LH, Cui B, Cao F. Polymorphisms of the FAS and FASL genes and risk of breast cancer. Oncol Lett. 2012;3:625–8.

Weng L, Ziliak D, Im HK, Gamazon ER, Philips S, Nguyen AT, Desta Z, Skaar TC, Consortium on Breast Cancer Pharmacogenomics (COBRA), Flockhart DA, Huang RS. Genome-wide discovery of genetic variants affecting tamoxifen sensitivity and their clinical and functional validation. Ann Oncol. 2013a;24:1867–73.

Weng L, Zhang L, Peng Y, Huang RS. Pharmacogenetics and pharmacogenomics: a bridge to individualized cancer therapy. Pharmacogenomics. 2013b;14:315–24.

Wheeler HE, Dolan ME. Lymphoblastoid cell lines in pharmacogenomic discovery and clinical translation. Pharmacogenomics. 2012;13:55–70.

Whirl-Carrillo M, McDonagh EM, Hebert JM, Gong L, Sangkuhl K, Thorn CF, Altman RB, Klein TE. Pharmacogenomics knowledge for personalized medicine. Clin Pharmacol Ther. 2012;92:414–7.

Wispelwey B. Clinical implications of pharmacokinetics and pharmacodynamics of fluoroquinolones. Clin Infect Dis. 2005;41(Suppl 2):S127–35.

Woo HI, Kim JA, Jung HA, Kim KK, Lee JY, Sun JM, Ahn JS, Park K, Lee SY, Ahn MJ. Correlation of genetic polymorphisms with clinical outcomes in pemetrexed-treated advanced lung adenocarcinoma patients. Pharmacogenomics. 2015;16:383–91.

Xu Y, Deng Q, He B, et al. The diplotype Fas-1377A/−670G as a genetic marker to predict a lower risk of breast cancer in Chinese women. Tumour Biol. 2014;35:9147–61.

Yeager M, Orr N, Hayes RB, Jacobs KB, Kraft P, Wacholder S, Minichiello MJ, Fearnhead P, Yu K, Chatterjee N, Wang Z, Welch R, Staats BJ, Calle EE, Feigelson HS, Thun MJ, Rodriguez C, Albanes D, Virtamo J, Weinstein S, Schumacher FR, Giovannucci E, Willett

WC, Cancel-Tassin G, Cussenot O, Valeri A, Andriole GL, Gelmann EP, Tucker M, Gerhard DS, Fraumeni Jr JF, Hoover R, Hunter DJ, Chanock SJ, Thomas G. Genome-wide association study of prostate cancer identifies a second risk locus at 8q24. Nat Genet. 2007;39:645–9.

Yiannakopoulou E. Pharmacogenomics and opioid analgesics: clinical implications. Int J Genet. 2015;2015:1. doi:10.1155/2015/368979.

Zembutsu H. Pharmacogenomics toward personalized tamoxifen therapy for breast cancer. Pharmacogenomics. 2015;16:287–96.

Zhang Y, Sheng J, Kang S, Fang W, Yan Y, Hu Z, Hong S, Wu X, Qin T, Liang W, Zhang L. Patients with exon 19 deletion were associated with longer progression-free survival compared to those with L858R mutation after first-line EGFR-TKIs for advanced non-small cell lung cancer: a meta-analysis. PLoS One. 2014;9(9):e107161. doi:10.1371/journal.pone.0107161.

Zhou SF. Polymorphism of human cytochrome P450 2D6 and its clinical significance: part II. Clin Pharmacokinet. 2009;48:761–804.

Zuo Z, Huang M, Kanfer I, Chow MS, Cho WC. Herb-drug interactions: systematic review, mechanisms, and therapies. Evid Based Complement Alternat Med. 2015;2015:1. doi:10.1155/2015/239150.

Part IV
Biostatistics, Bioinformatics, and System Biology Approaches to Complex Diseases

Chapter 14
Computational Network Approaches and Their Applications for Complex Diseases

Ankita Shukla and Tiratha Raj Singh

Abstract Network biology has been widely used for the interaction studies and analysis in modern era. Studies associated with biological networks, their modeling, analysis, and visualization are imperative to the biological world. The advancements in network biology have helped us better understand the biomolecular complexities which in earlier times were difficult to study in vivo. High-throughput technologies have revolutionized the genomic-sequencing procedures for the generation of tremendous data that need to be analyzed and interpreted rigorously. However it has still been difficult to construe biological networks completely due to the complexity of the interactions that exist between its components. Great efforts have been made to disclose the maximum possible interactions that are significant to maintain the potential mechanisms with the help of network biology. With improvement in the analysis process, network biology has thought to be playing a key role in understanding the complex biological behavior of the networks. In this chapter we will cover the basics of network biology and its expansion in various disciplines and its implementations in complex diseases and disorders, current resources, and tools available for studying diverse forms of pathways (transcriptional regulation, protein–protein interaction, signal transduction, and metabolism). This chapter therefore deals with the core of network biology, its role in various disease studies, and the advancements introduced so far in this field.

Keywords Interactome • Cancers • Autism • Network motif • Cardiovascular • Aging • Diabetes

A. Shukla • T.R. Singh (✉)
Department of Biotechnology and Bioinformatics, Jaypee University of Information Technology (JUIT), Waknaghat, Solan, India
e-mail: tiratharaj.singh@juit.ac.in

© Shanghai Jiao Tong University Press, Shanghai and Springer Science+Business Media Dordrecht 2017

D.-Q. Wei et al. (eds.), *Translational Bioinformatics and Its Application*, Translational Medicine Research, DOI 10.1007/978-94-024-1045-7_14

14.1 Introduction

Network biology is an amalgamation of systems biology, graph theory, and computational and statistical analysis techniques wherein the topology of the graphs represents the molecular interaction (Barabási and Oltvai 2004). Network biology deals with biological complexities of the cellular components that comprise macromolecules (genes, RNA and DNA) and metabolites. Most biological characteristics arise from complex interactions between these cellular components. There are diverse forms of biological networks which include metabolic networks, cell signaling networks, kinase–substrate networks, gene regulatory networks, protein–protein interaction networks, epistasis interaction networks, disease–gene interaction networks, and drug interaction networks (Ma'ayan 2011; Zhu et al. 2007; Winterbach et al. 2013). In network biology, a whole system is summarized in the form of nodes and edges, wherein nodes represent the biological component and edges represent interactions between them. These interaction studies can prove to be helpful to unlock the mystery behind interrelatedness of biological pathways. The use of standardized and efficient approaches is the first step in knowledge casting to provide unbiased maps of functional interactions. The next challenge in network biology is to examine the resulting interactome and to excavate global or local network properties. These graph properties can be used to improve understanding of biological processes that are crucial for organism's livelihood. Advancement in network biology offers a novel conceptual scaffold that will possibly revolutionize our view of biology and disease pathologies (Barabási and Oltvai 2004).

It has been seen from various studies that behavior of the cell is not the result of functioning of single pathway but of multiple ones that work together. So the rule of unity can also be seen in biological environment wherein if there is a problem with one pathway, it will not directly affect the system as the interrelated pathways try to compensate for it. Therefore, a key challenge today is to understand the structure and the dynamics of the complex intercellular web of interactions that contribute to the structure and functioning of a living cell. Numerous experimental techniques have been developed to date for identifying the physical interactions as well as the state of the cell at any point of time. It includes protein chips, yeast two-hybrid screens for physical interactions, and high-throughput screening for determining the status of the cell (Srinivasa et al. 2014). On the other hand, network biology proposes a quantifiable description of the networks that helps researchers characterize diverse biological systems.

To understand the network topology, one needs to know the properties related to the nodes and edges that should be considered while performing network analysis. Properties of nodes include (a) connectivity degree, the number of links each node possesses; (b) node betweenness centrality, the number of the shortest paths among all the shortest paths between all possible pairs of nodes; (c) closeness centrality, the average shortest path from one node to all other nodes; and (d) eigenvector centrality, a more sophisticated centrality measure that assesses the closeness to highly connected nodes. Properties of edges include (a) betweenness centrality, the number of the shortest paths among all possible shortest paths between all pairs of nodes; (b) types of relationship, for example, edges may represent activating or inhibiting relationships between a pair of nodes; and (c) edge directionality, the upstream and downstream nodes connected by a particular link.

Properties of network include:

(a) Connectivity distribution: quantitative links between nodes and edges
(b) Characteristic path length: the average shortest path between all pairs of nodes
(c) Clustering coefficient: local density of interactions measured by the connectivity of neighbors for each node averaged over the entire network
(d) Grid coefficient: extends the clustering coefficient by only looking at first neighbors to also examine second neighbors
(e) Network diameter: represents the longest of the shortest paths
(f) Assortativity: assesses whether nodes prefer to attach to other nodes on the basis of common nodal properties (Barabási and Oltvai 2004; Ma'ayan 2011).

Along with the abovesaid properties, there are two significant characteristics that are found in the network topology; one is network motifs which are recurring patterns composed of few nodes and edges. It has been observed that these network motifs appear in the regulatory networks much more frequently than in random or shuffled networks. There are also subcategories that underlie motifs including autoregulatory motifs, feedback loops, feedforward loops, bifans, diamonds, 3 and 4 chains, and other types of cycles that directly influence system's overall dynamics (Kim et al. 2011). These biologically significant subgraphs are vital to uncover the structural aspect of the complex network. Although network motifs provide insight to the functional properties, its detection is quite challenging (Masoudi-Nejad et al. 2012). Another characteristic of networks is their modularity, representing the modules or network clusters. These modules are dense regions of connectivity and are separated by low connectivity regions. Nearest neighbors clustering, Markov clustering, and betweenness centrality-based clustering which comes under the category of unsupervised clustering algorithms are used to identify the modules in networks (Ma'ayan 2011). Modularity is often used for detecting community structure in networks which have significant functionality at local levels. In the subcategory of modules come party hubs, nodes that interact with several proteins in one cellular compartment at a specific time, and date hubs, proteins that can be found in many places inside the cell and interact with diverse partners at different times (Chang et al. 2013). It has been observed from various studies that most of the biological molecular regulatory networks are scale-free, meaning that their degree distribution, i.e., distribution of edges per node, fits a power law. The overall functioning and homeostasis are being maintained by the scale-free architectures which make the network robust to the random failures.

Networks can be directed or undirected depending on the nature of the interactions. The directed network represents the interaction between any two nodes with a well-defined direction. These directionalities can be an inference of the material flow from a substrate to a product in a metabolic reaction or the information flow from a transcription factor to the gene. However, in case of the undirected networks, the links do not show any assigned direction. Network motifs identified within directed or undirected networks are called graphlets. One of the examples of graphlets is those found in protein–protein interaction networks (Przulj et al. 2004). The fundamental method in understanding the biological networks is the network visualization that helps scientists in uncovering important properties of the underlying biochemical processes. The computational methods are proved helpful for analysis purpose, but the major disadvantage is in data peculiarity that makes the network interpretation difficult due to the complexity of the relationships.

As mentioned earlier, complex diseases are generally caused not from the malfunction of individual molecules but from the interplay of a group of correlated molecules or a network (Schadt 2009). Molecular biomarkers are widely employed today as they are helpful to discriminate normal vs. disease samples. However, there is a serious problem regarding their usage, i.e., they suffer from low coverage along with the high false-positive/false-negative rates and further limit their clinical applications. The limitation in traditional concept of biomarkers has been now conquered with the modern concept of network biomarkers (also called module biomarkers), and they achieve better performance because of the involvement of diverse interactions of the molecules. Networks are considered more robust to characterize the disease conditions than individual molecules. One drawback allied both to the molecular biomarkers and the network biomarkers is that they can only differentiate disease and normal conditions but cannot reliably identify predisease conditions, therefore lacking the ability for early diagnosis.

Regarding the abovementioned condition, a new concept of dynamical network biomarkers (DNBs) has been developed based on nonlinear dynamical theory and complex network theory. One of the advantages of the DNB is its ability to distinguish a predisease state from normal and disease states for even a small number of samples, providing great potential to achieve authentic early diagnosis for the complex diseases. Network biomarkers offer quantifiable and stable forms to characterize biomedical phenotypes or diseases in contrast to individual molecular biomarkers, which has inspired the development of systems medicine in the network level (Liu et al. 2012; Kitano 2002; Aryee et al. 2013). Unlike molecular biomarkers and network biomarkers, a DNB does not always contain a group of fixed members even for the same disease but might have different molecules depending on individual variations that can be identified by high-throughput data. As compared to the edge biomarkers (or network biomarkers), which exploit correlation or association information between molecules (expected to uncover better biomarkers relating genotypes to phenotypes), DNB explores dynamical information of data together with network information. An unavoidable problem for both edge biomarkers and network biomarkers is the requirement of multiple samples in the predicting step. Thus, DNB is used for detecting the predisease condition and therefore provides the early signal of a disease. DNB method is relatively easy to implement as it can be achieved with a smaller number of samples and is a model-free method (Liu et al. 2014). Therefore, network biomarkers can exploit network information to unravel mechanisms of disease initiation and progression and thus improve the accuracy of diagnosis and prognosis.

14.2 Importance of Network Biology Toward Disease Prevention and Cure

The study of networks has emerged in diverse disciplines for analyzing complex relationships. The analysis of biological networks with respect to the human diseases has led to the discovery of field known as network medicine. Network biology has revolutionized the disease interrogation by uncovering many complex linkages that reflect perturbations in the biological networks. The functional interdependencies play major roles in maintaining the potential mechanisms which if not worked properly could lead to a disease condition. It has been found that disease is rarely a consequence of abnormality in single gene; the majority is due to the irregularities in multiple linked genes. The advent of network medicine leads to the emergence of a variety of tools which provide platforms to explore not only the molecular complexity of a particular disease but also the molecular relationships that exist between distinct phenotypes (Barabási et al. 2011). Advances in this direction are vital for the detection of new disease genes and for revealing the biological significance of disease-associated mutations. Since network also influences functioning of the other related networks, this interconnectivity implies that the impact of a specific genetic abnormality is not only limited to the activity of the gene product that holds it but can extend along the links of the network. These anomalies therefore alter the activity of gene products that otherwise had no defects initially. Hence, identifying phenotypic impact of a defect is not solely dependent on the known function of the mutated gene but also on the functions of components with which the gene and its products interact.

From the field of network medicine, it was found that it is the essential genes that are encoding hubs and not the diseased ones (Barabási et al. 2011). This statement is justified in terms of evolutionary perception because if we assume that mutations disrupt hubs, the absence of hubs will create so many disruptions that the host may not survive long enough to evolve and reproduce. Thus, only mutations that impair functioning lie at the periphery, accounting for the numerous disease conditions (Park et al. 2008). To determine the network-based position of disease genes, we need to understand three distinct phenomena which comprise (a) topological module, represents a locally dense neighborhood in a network, (b) functional module, represents the aggregation of nodes of related function, and (c) disease module, represents a group of network components that together contribute to a cellular function and their disruption results in a particular disease phenotype (Vidal et al. 2011). These three concepts are interrelated given that the cellular components that form a topological module have closely related functions and thus correspond to a functional module, and a disease is an outcome of the breakdown in a particular functional module.

Network-based approaches to human diseases can have numerous biological and clinical applications. It therefore provides a better understanding of the implications of cellular interconnectedness on disease progression which offers better targets for drug development by the identification of disease genes and pathways. These advances will possibly reshape clinical practice, through the discovery of better and more accurate biomarkers for better disease classification, paving the way to personalized treatments and therapies.

14.3 Network-Based Computational Approaches from Network Biology Available for Complex Diseases

In recent years, network-based approaches emerged as powerful tools for studying complex diseases. Computational biologists are working continuously in utilizing various approaches for understanding implicated pathways in complex diseases. This leads to the expansion of diverse algorithmic approaches to expose a range of facet of network biology. Today there is a dire need to enhance this field so as to tackle the most challenging diseases that fall under the category of complex one (named so as currently there is no effective therapy available to treat them). Many diseases fall in this category including cancer, autism, diabetes, obesity, Alzheimer's disease (AD), and cardiovascular diseases (CVDs). Genetic unbalancing is the root cause in complex diseases along with internal and external perturbations (Cho et al. 2012; Mitchell 2012). Primary difficulties to deal with complex diseases are that each of them might be caused by different genetic conditions. In addition, if a disease is caused by a combinatorial effect of many mutations, the individual effects of each mutation might be small and therefore hard to discover. According to the study, autism is considered to be one of the most inbred complex disorders, but its principal genetic causes are still largely unknown (Diaz-Beltran et al. 2013). Rare genetic disparity and its heterogeneity aid in the emergence of the complex disease (Kristiansson et al. 2008). Therefore, in case of the complex diseases, researchers are gradually focusing more on groups of related genes, referred to as modules or subnetworks. Typically, these topologies hold information like whether a given molecule acts as an activator or inhibitor (Mitra et al. 2013). An important advantage of working with modules rather than individual gene is that it is often easier to predict the function of a module than the function of a gene.

It is important to keep in mind that there are some disadvantages pertaining to the modules; those identified from high-throughput techniques are noisy, containing both false-negative and false-positive edges (Cho et al. 2012). Also, many times they skip information about the nature of an interaction. Not only the experimental ones but the computationally identified network modules also lack a mechanistic explanation of pathway activities. Therefore, selection of data and associated methods of analysis has to be chosen carefully. Network biology enlightens the diverse ways to deal with complex forms of the disease.

Aging Aging is the most prominent factor allied to the more complex forms of diseases, such as cancer, diabetes, CVDs, and neurodegenerative disorders (Cevenini et al. 2010). Aging phenotypes are coupled to the large and complex networks where cross talk occurs between assorted components. The main challenge in post-genomic aging research will be the dissection and analysis of the complex gene regulatory networks involved in aging processes. Structure and behavior studies are helpful in deducing the phenotypic effects of the responses that take place inside the cell. Since it is hard to infer logic of genetic networks experimentally, the union of new experiments and computational modeling techniques has been pursued currently.

Cardiovascular Diseases CVD covers a wide variety of disorders which influence different parts of cardiovascular system and includes coronary diseases, carotid diseases, peripheral arterial diseases, and aneurysms (Sarajlić and Pržulj 2014). There are also other forms of CVDs that are Mendelian disorders resulting from a mutation of the single gene. Genome-wide association studies have revealed that cardiovascular diseases, like the majority of complex diseases, have amazing complex genetic architecture, and they actually do not possess any major genes. Researchers have tried to investigate whether basic topological information such as connectivity of the nodes can be interrelated with biological properties which underlie CVD onset and progression. Module-based approaches are applied to determine functional modules related to the disease and to discover new associations between genes and disease. Various types of network biology approaches have been applied so far for CVD like in analysis done by Diez et al. who had created a combined gene association and correlation network (Diez et al. 2010). Rende et al. used topological features of PPI networks in search of genes common to CVDs and other diseases by identifying functional modules of genes (Rende et al. 2011). Approaches based on the biomolecular interaction networks provide better insight into network topology of the disease and thus could help researchers discover novel CVD genes and pathways.

Autism Autism is an early onset complex neurodevelopment disorder manifested in a broad phenotypic range (Diaz-Beltran et al. 2013). Although it is recognized as a highly heritable disorder, it is still uncertain whether the genetic variations are due to few common variants or because of many rare ones. Multifactorial nature of the complex disease makes use of systems biology perspective to embrace the complexity of the biological processes and the vast variety of molecular interactions that take place. Results showed that more than half of the published autism genes have been also allied to related neurological disorders (Wall et al. 2009). These findings provide evidence of molecular overlapping and indicate that these disorders might share molecular mechanisms which will perhaps help us understand the etiology of this complex disorder. The main objective of exploring the autism network is that the researchers will be able to locate those genes that cause the disorder and pave the way toward faster diagnosis as well as treatment.

Diabetes About 90% of the diabetic population is affected with type 2 diabetes which poses serious health issues to the society (Bergholdt et al. 2007). Elevated blood glucose level is the primary marker which occurs as a consequence of declined insulin activity. The adverse form of the disease could lead to the cardiovascular, renal, neurological, and organ complications. To date, a key challenge has been to identify the biological processes or signaling pathways that play significant roles in the disorder. There are various system-level studies done for diabetes like the one contributed by Davis et al. who found the loci contributing to diabetes-related traits along with the candidate genes with variation in gene expression (Davis et al. 2012). Integrating high-throughput microarray studies, with protein–protein interaction networks, seems to give the benefit in elucidating the underlying biological processes associated with chronic diabetic conditions. Therefore, there is a need to place more emphasis on the network biology methods to envisage the alteration caused by the summation of disordered pathways.

Cancers Cancer is caused by deleterious mutations leading to the abnormal functioning of a complex network. In cancer, dysregulation of multiple pathways, which govern fundamental cell processes, leads to different consequences like cell death, proliferation, differentiation, and migration. Like in case of other complex diseases, interrelatedness of biological interactions affects multiple cellular functioning. A major challenge is to find how diverse genetic mutations could help researchers build actionable understanding of this multivariate dysregulation. Therefore, the availability of diverse forms of networks which are amenable for computational analysis offers successful application of bioinformatics and systems biology methods for analysis of high-throughput data in cancer research. However, the key challenge is how significant advances can be made by applying computational modeling approaches to expose the pathways most critically involved in tumor formation and succession (Shukla et al. 2015, 2016; Sehgal et al. 2015).

Alzheimer's Disease Alzheimer's disease also abbreviated as AD is another category of complex disease of the central nervous system that occurs as a result of abnormal increase in levels of beta-amyloid (Aβ) and hyperphosphorylation of the tau protein (Mondragón-Rodríguez et al. 2012). Although AD is the most common form of dementia, its pathogenesis is still not well understood. Network modeling offers a unique opportunity for better understanding of AD by combining the current knowledge with a quantitative framework. The use of network biology in elucidating AD markers has been widely reported by various researchers like in one study performed by Ray et al. where severity across brain regions was examined by topological analysis of gene co-expression (Ray and Zhang 2010).

14.4 Available Databases/Resources/Computational Tools/Servers for the Network Analysis

One of the challenges in network analysis is data visualization. Here we present several useful databases and software tools that exist for network analysis. First we introduce the databases that we categorized in four groups: transcriptional regulation pathways, protein–protein interaction pathways, signal transduction pathways, and metabolic pathways (Table 14.1). Thereafter we present network biology tools for network analysis and visualization (Table 14.2). With the advancement of techniques and availability of the data, a myriad of such resources and tools have been developed as shown in Fig. 14.1. We have compiled the list based upon accuracy and applications.

14.5 Current Status of These Tools and Their Future Enhancements

We have discussed a wide variety of tools that encompass features essential for network visualization. As observed from various studies, three major challenges are faced while performing data visualization, i.e., large amount of data, heterogeneous data integration, and the representation of multiple linkages between nodes with heterogeneous biological meaning (Pavlopoulos et al. 2008). Each visualization tool differs from others in terms of specific features they possess and therefore tackles the aforementioned challenges in its own way. Regarding the heterogeneity tools, Ondex, Pivot, or Medusa offers some possible solutions. However, Medusa and other tools that can handle multi-edged networks are also used when working with systems biology data such as in case of highly interlinked nodes. Cytoscape or BioLayout Express3D tools have good resolution and scaling features which further augment the visualization process. Pajek is ideal for pattern recognition and for studying the properties like density, centrality, and frequency of nodes. Osprey is suitable for comparative biological analysis. The tools presented in this chapter have a wide range of applicability in the network-related problems.

Besides a wide range of applicability of the tools, there are some drawbacks associated with them. Firstly, the majority of tools can handle datasets only up to a certain limit. As the size of datasets increases rapidly, there is a need for new generations of visualization tools that can withstand this problem. Secondly, there is a scaling problem posting a challenge to this field. Although many algorithms have been developed, they are still not able to deal with the layout problem and follow a heuristic approach instead of exhaustive ones. Therefore, there is an urgent need to develop fast and more efficient algorithms for speedy analysis of large-scale networks. One possible way to evade this problem is the parallel processing that makes use of powerful machines which greatly speed up the process of visualization and hence reduce the computational load. In addition, layout can be extended

Table 14.1 Tools and resources for the analysis of transcriptional regulation pathway, protein–protein interaction pathways, signal transduction pathways, and metabolic pathways

Transcriptional regulation pathway		
PAZAR	http://www.pazar.info/	A public database of transcription factor and regulatory sequence annotation
RegulonDB	regulondb.ccg.unam.mx	A reference database of *Escherichia coli* K-12 offering curated knowledge of the regulatory network and operon organization
TRANSFAC	http://transfac.gbf.de/	A manually curated database of eukaryotic transcription factors, their genomic-binding sites and DNA-binding profiles
TRRUST	http://www.grnpedia.org/trrust/	A reference database of human transcriptional regulatory interactions
YTRP (Yeast Transcriptional Regulatory Pathway) Database	http://cosbi3.ee.ncku.edu.tw/YTRP/	A repository for yeast transcriptional regulatory pathways
Protein–protein interaction pathways		
BIND/BOND (Biomolecular Interaction Network Database)	http://bind.ca	Archives biomolecular interaction, complex and pathway information
BioGRID (Biological General Repository for Interaction Datasets)	http://thebiogrid.org/	A repository for set of physical and genetic interactions that include interactions, chemical associations, and post-translational modifications (PTM)
CYGD (Comprehensive Yeast Genome Database)	http://mips.gsf.de/genre/proj/yeast/index.jsp	Present information on the molecular structure and functional network of the entirely sequenced, well-studied model eukaryote, the budding yeast *Saccharomyces cerevisiae*
DIP (Database of Interacting Proteins)	http://dip.doembi.ucla.edu/	Catalogs experimentally determined interactions between proteins
HPRD (Human Protein Reference Database)	http://www.hprd.org/	A web-based resource for protein–protein interactions, posttranslational modifications, enzyme–substrate relationships, and disease associations
MINT (Molecular INTeraction) Database	http://mint.bio.uniroma2.it/mint/Welcome.do	A public repository for protein–protein interactions (PPI)
STRING (Search Tool for the Retrieval of Interacting Genes/Proteins)	http://string-db.org/	A database of known and predicted protein interactions
Signal transduction pathways		
BBID (Biological Biochemical Image Database)	http://bbid.grc.nia.nih.gov/	A searchable database of images of putative biological pathways, macromolecular structures, gene families, and cellular relationships

(continued)

Table 14.1 (continued)

CSNDB (Cell Signaling Networks Database)	http://geo.nihs.go.jp/csndb/	A data and knowledge base for signaling pathways of human cells. It compiles the information on biological molecules, sequences, structures, functions, and biological reactions which transfer the cellular signals
SPAD (Signaling PAthway Database)	http://www.grt.kyushu-u.ac.jp/spad/	An integrated database for genetic information and signal transduction systems
TransPath	http://transpath.gbf.de/	A database system about gene regulatory networks that combines encyclopedic information on signal transduction with tools for visualization and analysis
Metabolic pathways		
BioCyc	http://www.biocyc.org/	The BioCyc database collection is a set of more than 7600 pathway/genome databases (PGDBs) describing many sequenced genomes
BIOPATH (Biochemical Pathways) Database	http://www.mol-net.de/databases/biopath.html	A database of biochemical pathways that provides access to metabolic transformations and cellular regulations
ECMDB (E. coli Metabolome database)	http://ecmdb.ca/	An expertly curated database containing extensive metabolomic data and metabolic pathway diagrams about *Escherichia coli* (strain K12, MG1655)
EMP (Enzymes and Meta bolic Pathways) Database	http://emp.mcs.anl.gov/	A database on the biochemistry of some 1800 different organisms
GMD (Golm Metabolome Database)	http://gmd.mpimp-golm.mpg.de/	Facilitates the search for and dissemination of reference mass spectra from biologically active metabolites quantified using gas chromatography (GC) coupled to mass spectrometry (MS)
HMDB (Human Metabolome Database)	http://www.hmdb.ca/	A freely available electronic database containing detailed information about small molecule metabolites found in the human body
KEGG (Kyoto Encyclopedia of Genes and Genomes)	http://www.genome.ad.jp/kegg/	A collection of databases dealing with genomes, biological pathways, diseases, drugs, and chemical substances
Metacyc	http://metacyc.org/	A curated database of experimentally elucidated metabolic pathways from all domains of life
MANET (Molecular Ancestry Network)	http://www.manet.illinois.edu/index.php	A database tracing the evolution of protein architecture in metabolic networks
PathCase (Pathways Database System)	http://nashua.case.edu/pathwaysweb/	Store, query, and visualize metabolic pathways, in addition to their specialized tasks

(continued)

Table 14.1 (continued)

PMN (Plant Metabolic Network)	http://www.plantcyc.org/	A broad network of plant metabolic pathway databases that contain curated information from the literature and computational analyses about the genes, enzymes, compounds, reactions, and pathways involved in primary and secondary metabolism in plants
UniPathway	http://www.unipathway.org/	A resource for the exploration of metabolic pathways

The left most column has been arranged alphabetically

Table 14.2 Important and popular tools and resources for the analysis of biological networks

Network biology tools and resources		
Arena3D	http://arena3d.org/	Use multilayered graphs to visualize biological networks. In such a way, heterogeneous data will be distinguished between each other
BioLayout Express3D	http://www.biolayout.org/	Offers different analytical approaches to microarray data analysis
BioTapestry	http://www.biotapestry.org/	A tool to visualize the dynamic properties of gene regulatory networks
CellDesigner	http://www.celldesigner.org/	A structured diagram editor for drawing gene regulatory and biochemical networks
CellML	https://www.cellml.org/	An XML-based markup language for describing mathematical models
COPASI	http://copasi.org/	COPASI is a software application for simulation and analysis of biochemical networks and their dynamics
CSB.DB	http://csbdb.mpimp-golm.mpg.de/csbdb/dbcor/cor.html	A comprehensive systems biology database
Cytoscape	http://www.cytoscape.org/	Incorporates statistical analysis of the network, and it makes it easy to cluster or detect highly interconnected regions
EAWAG-BBD	http://eawag-bbd.ethz.ch/	Contains information on microbial biocatalytic reactions and biodegradation pathways for primarily xenobiotic, chemical compounds
E-Cell	http://www.e-cell.org/	Develops general technologies and theoretical supports for computational biology with the grand aim to make precise whole cell simulation at the molecular level possible
FANMOD	http://theinf1.informatik.uni-jena.de/~wernicke/motifs/index.html	A tool for fast network motif detection
Genes2Networks	http://actin.pharm.mssm.edu/genes2networks/	Powerful web-based software that can help experimental biologists to interpret lists of genes and proteins such as those commonly produced through genomic and proteomic experiments, as well as lists of genes and proteins associated with disease processes

(continued)

Table 14.2 (continued)

Network biology tools and resources		
GEPHI	https://gephi.org/	Interactive visualization and exploration platform for all kinds of networks and complex systems, dynamic and hierarchical graphs
Igraph	http://igraph.org/	A collection of network analysis tools with the emphasis on efficiency, portability, and ease of use
JWS Online	http://omictools.com/jws-online-tool	A systems biology tool for the construction, modification, and simulation of kinetic models and for the storage of curated models
Medusa	http://coot.embl.de/medusa/	Medusa is optimized for protein–protein interaction data as taken from STRING or protein–chemical and chemical–chemical interactions as taken from STITCH
Ondex	http://ondex.sourceforge.net/	Ondex main strength is the ability to combine heterogeneous data types into one network. It is suitable for text mining and sequence and data integration analysis
Osprey	http://biodata.mshri.on.ca/osprey/servlet/Index	The ability to incorporate new interactions into an already existing network
Pajek	http://pajek.imfm.si/doku.php?id=pajek	Main strength is the variety of layout algorithms which greatly facilitate exploration and pattern identification within networks
PATIKA (Acquisition)	http://www.patika.org/	Integrated software environment designed to provide researchers a complete solution for modeling and analyzing cellular processes
PIVOT	http://acgt.cs.tau.ac.il/pivot/	Best suited for visualizing protein–protein interactions and identifying relationships between them
ProViz	http://cbi.labri.fr/eng/proviz.htm	Performs protein–protein interaction and their analysis using arbitrary properties, like for example annotations or taxonomic identifier
Reactome	http://www.reactome.org/	A free, open-source, curated, and peer-reviewed pathway database
VisANT	http://visant.bu.edu	An online visualization and analysis tool for biological interaction data

The left most column has been arranged alphabetically

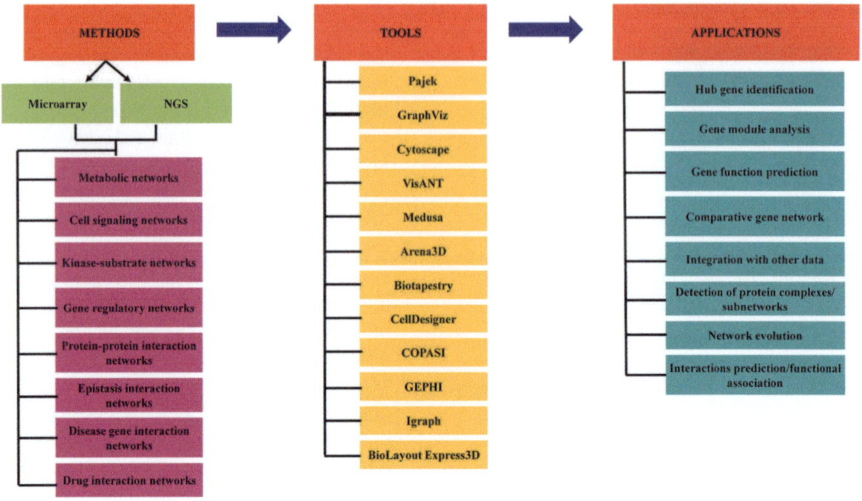

Fig. 14.1 Flowchart depicting the methods, tools, and applications of network biology

by adding a third dimension which would allow a clearer visualization of structures and strongly facilitate a better navigation within the network. Currently, most of the network tools only generate static snapshots of the interactions and provide no methods to visualize a time series data (Suderman and Hallett 2007). By introducing time series data, the process of network visualization would thus achieve a more complete picture of complex and highly dynamic biological systems. It is highly expected that this will provide breakthroughs in the pathway analysis process or the observation of interaction at different time points of cell cycles.

Systems memory is another important issue that should be taken care of while performing computational visualization and analysis. The limited functionalities of existing visualization tools make it necessary to constantly switch between different applications to complete different levels of analysis. Frequent information and data sharing between different tools has become possible with the availability of standard file formats. The visualization tools designed by taking care of aforementioned functionalities would greatly simplify large-scale research in molecular biology and would significantly cut down time and effort spent on data processing and analysis (Pavlopoulos et al. 2008).

14.6 Conclusion

Network biology is the revolution in the field of life science as it provides information not only on the significant interactions but also on the functionalities allied to them. The network complexities can be studied with the help of a variety of methods available in network biology, and the most significant of them is the module-based approach. Modularity focuses on the local significant regions that share high-functional relationships comparative to the rest of the network. These methods not only uncover the complex biological mysteries but also help investigators understand various disease networks or their interactions and hence provide the potential therapeutic targets. Network biology has accelerated the biomolecular analysis process by offering different tools that have diverse applications depending on the number of features embed in them. Network-based methods also have several limitations including the lack of mechanistic explanations. Despite the limitations, network analysis has been applied successfully to understand the complexities of many disease states. It is anticipated that with rapid advancements, network biology will serve as an excellent research complement to annotate biomolecules at a system level and will also help in the generation of more biologically meaningful information.

References

Aryee MJ, Liu W, Engelmann JC, Nuhn P, Gurel M, Haffner MC. DNA methylation alterations exhibit intraindividual stability and interindividual heterogeneity in prostate cancer metastases. Sci Transl Med. 2013;5(169):169ra10.

Barabási AL, Oltvai ZN. Network biology: understanding the cell's functional organization. Nat Rev Genet. 2004;5(2):101–13.

Barabási AL, Gulbahce N, Loscalzo J. Network medicine: a network-based approach to human disease. Nat Rev Genet. 2011;12(1):56–68.

Bergholdt R, Størling ZM, Lage K, Karlberg EO, Olason PI, Aalund M. Integrative analysis for finding genes and networks involved in diabetes and other complex diseases. Genome Biol. 2007;8(11):R253.

Cevenini E, Bellavista E, Tieri P, Castellani G, Lescai F, Francesconi M. Systems biology and longevity: an emerging approach to identify innovative anti-aging targets and strategies. Curr Pharm Des. 2010;16(7):802–13.

Chang X, Xu T, Li Y, Wang K. Dynamic modular architecture of protein-protein interaction networks beyond the dichotomy of 'date' and 'party' hubs. Sci Rep. 2013;3:1691.

Cho DY, Kim YA, Przytycka TM. Chapter 5: network biology approach to complex diseases. PLoS Comput Biol. 2012;8(12):e1002820.

Davis RC, van Nas A, Castellani LW, Zhao Y, Zhou Z, Wen P. Systems genetics of susceptibility to obesity-induced diabetes in mice. Physiol Genomics. 2012;44(1):1–13.

Diaz-Beltran L, Cano C, Wall DP, Esteban FJ. Systems biology as a comparative approach to understand complex gene expression in neurological diseases. Behav Sci (Basel). 2013;3 (2):253–72.

Diez D, Wheelock AM, Goto S, Haeggström JZ, Paulsson-Berne G, Hansson GK. The use of network analyses for elucidating mechanisms in cardiovascular disease. Mol BioSyst. 2010;6 (2):289–304.

Kim W, Li M, Wang J, Pan Y. Biological network motif detection and evaluation. BMC Syst Biol. 2011;5(Suppl 3):S5.

Kitano H. Systems biology: a brief overview. Science. 2002;295(5560):1662–4.

Kristiansson K, Naukkarinen J, Peltonen L. Isolated populations and complex disease gene identification. Genome Biol. 2008;9(8):109.

Liu ZP, Wang Y, Zhang XS, Chen L. Network-based analysis of complex diseases. IET Syst Biol. 2012;6(1):22–33.

Liu R, Wang X, Aihara K, Chen L. Early diagnosis of complex diseases by molecular biomarkers, network biomarkers, and dynamical network biomarkers. Med Res Rev. 2014;34(3):455–78.

Ma'ayan A. Introduction to network analysis in systems biology. Sci Signal. 2011;4(190):tr5.

Masoudi-Nejad A, Schreiber F, Kashani ZR. Building blocks of biological networks: a review on major network motif discovery algorithms. IET Syst Biol. 2012;6(5):164–74.

Mitchell KJ. What is complex about complex disorders? Genome Biol. 2012;13(1):237.

Mitra K, Carvunis AR, Ramesh SK, Ideker T. Integrative approaches for finding modular structure in biological networks. Nat Rev Genet. 2013;14(10):719–32.

Mondragón-Rodríguez S, Perry G, Zhu X, Boehm J. Amyloid beta and tau proteins as therapeutic targets for Alzheimer's disease treatment: rethinking the current strategy. Int J Alzheimers Dis. 2012;2012:630182.

Park D, Park J, Park SG, Park T, Choi SS. Analysis of human disease genes in the context of gene essentiality. Genomics. 2008;92(6):414–8.

Pavlopoulos GA, Wegener AL, Schneider R. A survey of visualization tools for biological network analysis. BioData Min. 2008;1:12.

Przulj N, Corneil DG, Jurisica I. Modeling interactome: scale-free or geometric? Bioinformatics. 2004;20(18):3508–15.

Ray M, Zhang W. Analysis of Alzheimer's disease severity across brain regions by topological analysis of gene co-expression networks. BMC Syst Biol. 2010;4:136.

Rende D, Baysal N, Kirdar B. A novel integrative network approach to understand the interplay between cardiovascular disease and other complex disorders. Mol BioSyst. 2011;7(7):2205–19.

Sarajlić A, Pržulj N. Survey of network-based approaches to research of cardiovascular diseases. Biomed Res Int. 2014;2014:527029.

Schadt EE. Molecular networks as sensors and drivers of common human diseases. Nature. 2009;461(7261):218–23.

Sehgal M, Gupta R, Moussa A, Singh TR. An integrative approach for mapping differentially expressed genes and network components using novel parameters to elucidate key regulatory genes in colorectal cancer. PLoS One. 2015;10(7):e0133901.

Shukla A, Sehgal M, Singh TR. Hydroxymethylation and its potential implication in DNA repair system: a review and future perspectives. Gene. 2015;564(2):109–18.

Shukla A, Moussa A, Singh TR. DREMECELS: a curated database for base excision and mismatch repair mechanisms associated human malignancies. PLoS One. 2016;11(6):e0157031.

Srinivasa RV, Srinivas K, Sujini NG, Kumar SNG. Protein-protein interaction detection: methods and analysis. Int J Proteomics. 2014;2014:12. Article ID 147648.

Suderman M, Hallett M. Tools for visually exploring biological networks. Bioinformatics. 2007;23(20):2651–9.

Vidal M, Cusick ME, Barabási AL. Interactome networks and human disease. Cell. 2011;144(6):986–98.

Wall DP, Esteban FJ, Deluca TF, Huyck M, Monaghan T, Velez de Mendizabal N. Comparative analysis of neurological disorders focuses genome-wide search for autism genes. Genomics. 2009;93(2):120–9.

Winterbach W, Van Mieghem P, Reinders M, Wang H, de Ridder D. Topology of molecular interaction networks. BMC Syst Biol. 2013;7:90. doi:10.1186/1752-0509-7-90.

Zhu X, Gerstein M, Snyder M. Getting connected: analysis and principles of biological networks. Genes Dev. 2007;21(9):1010–24.

Chapter 15
Bioinformatics Applications in Clinical Microbiology

Chao Zhang, Shunfu Xu, and Dong Xu

Abstract The human body is believed to house over 100 trillion microbes. These microbial communities have a tremendously influential impact on their human hosts. Although increasing evidence indicated a key role for the specific microbial species in carcinogenesis, such as *Helicobacter pylori* (*H. pylori*), Epstein-Barr virus, *Human papillomavirus*, and *Hepatitis C virus*, the underlying roles of human microbiome in cancers are still unclear. Using the bioinformatics algorithms and tools to integrate the microbiological data and clinical data could be very helpful to better understand the mechanisms of diseases. During the past decade, we have kept working on microbiome research and utilized bioinformatics methods to discover host-pathogen interactions, relationships between microbiome dynamics and diseases, and correlations between bacterial sequence variation and clinical outcomes. In this chapter, we use *H. pylori* as an example to demonstrate the procedure of related data integration, virulence classification, and prognosis model construction.

Keywords Microbiome • *Helicobacter pylori* • CagA • Gastric cancer • Bioinformatics • SVM

15.1 Introduction

As the most abundant domain of all living organisms on earth, bacteria are estimated to have more than five nonillion (10^{30}) individuals worldwide (Whitman et al. 1998), and these small single-cell organisms can be found everywhere. They

---

C. Zhang
Department of Medicine and Institute for Computational Biomedicine, Weill Cornell Medical College, New York, NY, USA

S. Xu
Department of Gastroenterology, Nanjing Medical University, Nanjing, Jiangsu, China

D. Xu (✉)
Department of Computer Science & C.S. Bond Life Science Center, University of Missouri, Columbia, MO, USA
e-mail: XuDong@missouri.edu

© Shanghai Jiao Tong University Press, Shanghai and Springer Science+Business Media Dordrecht 2017

D.-Q. Wei et al. (eds.), *Translational Bioinformatics and Its Application*, Translational Medicine Research, DOI 10.1007/978-94-024-1045-7_15

are playing very important roles in our life, and we actually benefit from the microorganisms in many cases, e.g., food production, human health (Turnbaugh et al. 2009), environmental biotechnologies (Dinsdale et al. 2008), and chemical industry (Lorenz and Eck 2005). On the other hand, pathogenic bacteria are one of the most serious threats to human life. For example, tuberculosis, the most common fatal bacterial disease, kills about two million people every year (Andries et al. 2005).

In the past, analysis of microbial communities was a complicated task due to their high diversity and inaccessibility via culturing. The emerging next-generation sequencing technologies provide a potential way for doing this analysis on a routine basis (Petrosino et al. 2009). The Human Microbiome Project (Turnbaugh et al. 2007) undoubtedly provides new insight into many aspects of complex microbial communities, such as metabolic capabilities of microorganisms, coevolution of bacteria and host, interactions among microbial cells, and so on (Medini et al. 2008). Meanwhile, the unprecedented amount of genome data also poses major challenges for computational analysis, which is an essential tool for microbial genomics. In fact, computational methods for massive genomic sequence analysis have become a bottleneck of microbial genomics. In our previous study, we reviewed the major computational methods on metagenomic/genomic analyses and the future computational challenges on general microbial identification (Zhang et al. 2012a, 2015), and we will focus on bioinformatics applications in clinical microbiology in this chapter.

Immediately after birth, humans undergo a lifelong process of colonization by foreign microorganisms. Although we benefit from some host-bacterial associations, bacterial pathogens have long been known to play important roles in the development of many diseases (Hacker et al. 2003) including cancer (Ullman and Itzkowitz 2011). The host-bacteria interactions include many complicated mechanisms, and discovering associations between bacteria and diseases in a clinical setting is even more challenging. Due to the explosion of metagenomic/genomic data, DNA sequence-based identification and classification are becoming more and more important in exploring microbial diversity in clinical research. For example, *Bradyrhizobium enterica* was discovered in cord colitis syndrome with shotgun DNA sequencing of biopsy specimens (Bhatt et al. 2013). Recently we also found that the *Helicobacter pylori* (*H. pylori*) infection can change the gastric microbiome according to whole genome sequencing (WGS) on endoscopic biopsy. WGS gives a much more accurate identification on *H. pylori* infection than traditional methods, such as ELISA test and C-13 breath test. Besides *H. pylori* infection identification, we also spent much effort on discovering the molecular mechanisms that underlie different gastroduodenal diseases caused by *H. pylori* infection.

In this chapter, we use *H. pylori* as the example to describe how we utilized computational methods to discover the relationships between *H. pylori* virulence factor and diseases and built a potential model for clinical diagnosis or prognosis. At first, we collected and curated the data from public databases, and then through studying the distribution and polymorphism of EPIYA motif in CagA sequences, we attempted to better understand the function of EPIYA motif, especially the role of EPIYA motif during the interaction process between *H. pylori* and hosts. We also constructed a computational model to access gastric cancer risk by using detected important residues in CagA intervening sequences.

15.2 Public Data Collection and Curation

H. pylori is a Gram-negative helix-shaped bacterium inhabiting the human stomach for possibly more than thousands of years. By far as one of the oldest known human pathogens, it infects more than half of the world's population (Suerbaum and Michetti 2002). *H. pylori* has shown a strong correlation with all gastroduodenal diseases, including duodenal ulcers (Covacci et al. 1993), gastric ulcers (Ernst and Gold 2000), and chronic gastritis, especially being an important risk factor for developing gastric cancer (Uemura et al. 2001). *H. pylori* is becoming more and more important not only for gastroenterologists and pathologists but also for phylogenists who use it as the evidence to study human's origin and migration (Linz et al. 2007).

As one of the most important model bacteria, the data of *H. pylori* have been increasing dramatically in recent years. As of January 2014, 399 genome-sequencing projects are almost complete or "in progress." 37,304 nucleotide sequences, 65,684 protein sequences, 61 primers, and 9953 publications were collected from several major databases, e.g., NCBI databases (http://www.ncbi.nlm.nih.gov), EBI databases (http://www.ebi.ac.uk), DDBJ (http://www.ddbj.nig.ac.jp), and PDB (http://www.pdb.org). We searched the above databases with the keywords "Helicobacter pylori" and "H. pylori" and then verified all results based on the taxonomy information. References were collected from PubMed (http://www.ncbi.nlm.nih.gov/pubmed).

As we know, geographical diversity and disease diversity are two most significant features and hottest topics on *H. pylori* research. Without these types of information, the sequences of *H. pylori* are not very useful for studying the underline mechanisms of *H. pylori* causing gastric diseases. Based on our research experience, collecting *H. pylori* data from various sources is laborious and difficult, and currently no database/website can provide the corresponding accurate information, and collecting comprehensive information of *H. pylori* specifically for a particular country or disease is even more time-consuming. We manually curated the information not only based on the records from the above major databases but also by reviewing related literature.

15.3 Data Deposit

Based on the information we collected and curated, a web-based database, HPbase (www.hpbase.org) has been built for providing a one-stop shop for *H. pylori* data from multiple sources together with multiple embedded search/analysis tools for querying the database. This website is not only for depositing collected public data but also for providing curated information, new data generated by users, and other features derived from original experimental data. By continuously accumulating and updating the data, we anticipate that HPbase will serve as an important resource for studying *H. pylori* and gastroduodenal diseases.

15.3.1 Implementation

The web interface is constructed using PHP, CSS, and the JavaScript jQuery framework for a flexible user interaction with the system. The HPbase database is implemented through a MySQL relational database as the backend data storage system. A Java-based tool was developed to periodically synchronize data with major sources, and it is also used to import related manually curated diseases and geographical information into the MySQL database.

15.3.2 Other Information

Besides the basic information we collected from other sources, we generated sequence profiles for all 65,684 protein sequences by running PSIBLAST (Altschul et al. 1997) (2007 release version) three rounds against nonredundant (NR) database (as of 2013) with the e-value cutoff of 0.001, and then we predicted secondary structures by using PSIPRED (McGuffin et al. 2000) with the sequence profiles generated above. We also predicted 3D structures for most of proteins, including all major ones, e.g., CagA and VacA, by using our in-house software MUFOLD (Zhang et al. 2010), which integrates whole and partial template information to cover both template-based and ab initio predictions in the same package. The predicted secondary and tertiary structural information could help users to better understand the interaction between human proteins and *H. pylori* proteins.

15.3.3 Browsing Data

Users can search *H. pylori* data by different entries, such as GI number, accession number, strain ID, keywords, disease type, geographical information, and so on

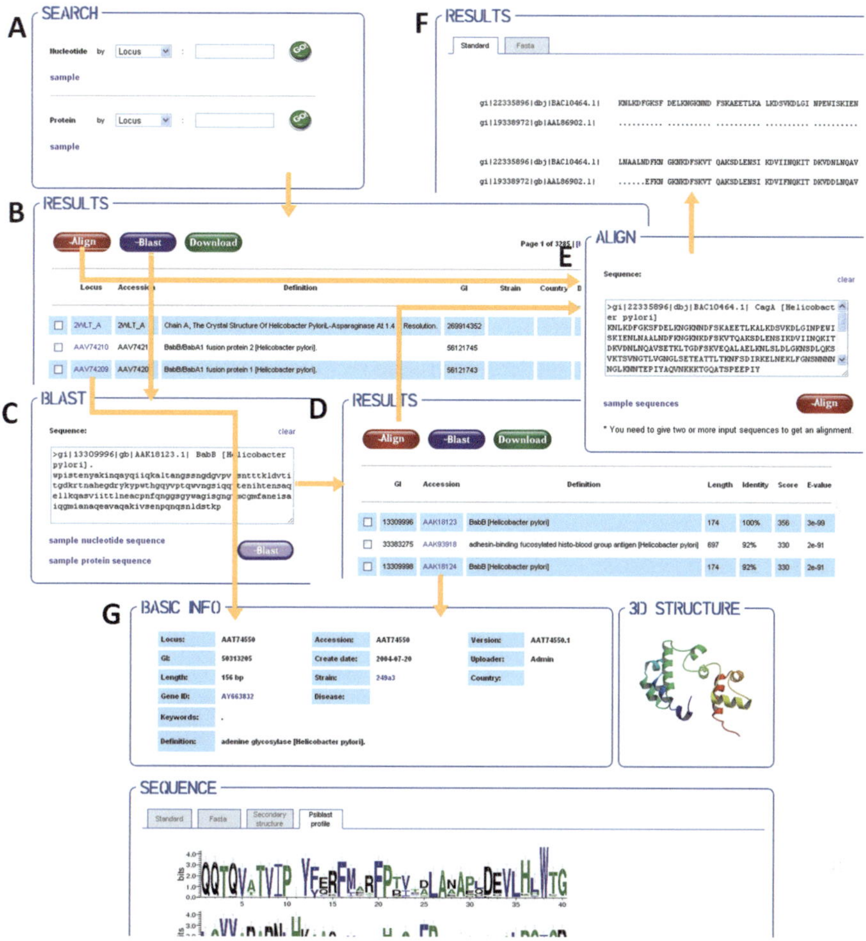

Fig. 15.1 Screenshots and basic workflow of HPbase. (**a**) Dialog for entering search criteria. (**b**) Table for displaying searching results. (**c**) Input dialog for BLAST. (**d**) Table for displaying BLAST results. (**e**) Input dialog for sequence alignment. (**f**) Dialog for displaying alignments between different sequences. (**g**) Page for displaying detail information of one particular protein/DNA sequence

(Fig. 15.1a). After submission of the search condition, the results will be displayed in a new page as a list of records with some brief information, including GI number, accession number, definition, strain ID, keywords, corresponding diseases, and geographical information. As shown in Fig. 15.1b, users can simply pick all sequence records or part of them in search results to download in the FASTA or Genbank/GenPept format for further analyses provided by HPbase, e.g., BLAST and multiple sequence alignment (MSA) as in Fig. 15.1c and e. Furthermore, users can also navigate detailed information of any particular nucleotide/protein record in the result page by clicking on the corresponding "locus" link, and it will redirect to the nucleotide/protein detail page as shown in Fig. 15.1g. It not only provides the brief information as other major databases do but also includes manually curated information, e.g., disease-related and geographical information and computational information, e.g., PSIBLAST sequence profiles, secondary structures, and 3D structures for proteins. In this page, the sequence will be displayed with several formats. PSIBLAST sequence profile is represented as a sequence logo, and it is generated by using the WebLogo (Crooks et al. 2004) for the top 100 alignments of the last PSIBLAST round with no gap in the query sequence. Protein secondary structures are colored with the FASTA format. Jmol (http://www.jmol.org) is used as a viewer for displaying protein 3D structures. Users are also encouraged to add their own comments to each nucleotide/protein record and use the reference voting function to improve the correlation between each sequence record and its references, which could be helpful for others to better understand *H. pylori*.

15.3.4 Other Tools

Some further functions have also been embedded into the HPbase website to improve the power of search and data analysis. As shown in Fig. 15.1c, a BLAST utility was integrated as one useful feature, and two different BLAST programs have been included, e.g., BLASTn and BLASTp. By selecting gene entries from search results or uploading a protein/nucleotide sequence, users can retrieve identical or similar nucleotides/peptides in the database through BLAST according to user-defined parameters, which can be freely chosen including E-value, number of alignments, mutation matrix, and so on. As shown in Fig. 15.1d, a typical result page contains the collected information including GI number, accession, definition, length of sequence, E-value, identity, score, and alignment. Users can download records and further execute BLAST for database search or MUSCLE (Edgar 2004) for MSA by selecting records of their own interest from the BLAST results. Users can also upload their own sequences to perform multiple sequence alignment. In addition, the entire sequence data can be downloaded directly in the FASTA or Genbank/GenPept format. Users can also download data for one particular "strain," "disease," or "country." Some statistical analysis of the most important virulence factor – CagA from our previous work (Zhang et al. 2012b) – is also included in the website, including the relations between CagA sequence subtypes and diseases, the geographical diversity of CagA sequences, and the geographical diversity of different diseases.

15.4 Computational Model for CagA

15.4.1 Motivation

As one of the most important virulence markers of *H. pylori*, the cytotoxin-associated gene A (CagA) has been revealed to be related to the gastric disease occurrence. *H. pylori* strains carrying the CagA gene increase the risk factor of gastroduodenal diseases by threefold over CagA-negative strains (Blaser et al. 1995). CagA contains 1142–1320 amino acids, and at the C-terminal region, it has a variable region in which various short sequences (EPIYA motif) repeat 1–7 times. After colonizing on the surface of the gastric epithelium, *H. pylori* translocated into the gastric epithelial cell through type IV secretion system. Once injected into the host cell, CagA could localize to the plasma membrane. Src family tyrosine kinases can phosphorylate CagA on the specific tyrosine residues of a five-amino-acid (EPIYA) motif (Odenbreit et al. 2000). Then tyrosine-phosphorylated CagA binds specifically to SHP-2 tyrosine phosphatase (Higashi et al. 2002) to activate a phosphorylase, which causes the cascade effect that interferes with the signal transduction pathway of the host cell, leading to a restructuring of the host cell cytoskeleton and formation of hummingbird phenotype (Argent et al. 2004). At the same time through activating mitogen-activated protein kinase (MAPK), extracellular signal-regulated kinase (ERK) (Fu et al. 2009), and focal adhesion kinase (FAK), CagA also can cause cell dissociation and infiltrative tumor growth (Amieva et al. 2003).

CagA protein carries two unique features. One is the geographical diversity. There are some different intervening sequences between those EPIYA motifs. One copy of EPIYA plus intervening sequence is identified as an EPIYA segment. Four unique types of EPIYA segments have been found in CagA, defined as EPIYA-A, EPIYA-B, EPIYA-C, and EPIYA-D (Higashi et al. 2002). Among them, EPIYA-D motif only can be found from the East Asian subtype, and for the CagA from Western countries, EPIYA-D is replaced by EPIYA-C. EPIYA-D has stronger phosphorylation motif binding activity which leads to greater morphological changes than what the EPIYA-C motif can cause in infected cells (Higashi et al. 2002). And it explains the higher incidence of gastric cancer in East Asian countries (Jones et al. 2009).

Another feature of CagA is the variation in the number of EPIYA motif copies. Many studies attempted to reveal the relations between number of EPIYA motif repeats and clinical diseases (Lai et al. 2003). Although increasing of number of EPIYA motif copies will affect biological activities, due to the sample size limitation and geographic limitations of studies, none of the studies can draw a statistically significant conclusion about the relation.

Aside from the number of the EPIYA motif repeats, the sequence difference of strains in variable regions could also cause a significant difference of virulence, which might relate to the different pathogenic abilities of *H. pylori* (Naito et al. 2006). We speculate not only the number of EPIYA motif repeats, but also polymorphism of CagA sequences will affect the virulence of *H. pylori* and then cause the different diseases. In this study, we focused on identifying the informative residues, quantifying information of these selected residues, and then using it to design a classifier that can predict whether a new sequence belongs to the cancer group or the noncancer group. This method not only sheds light on the relations between CagA sequences and gastric diseases but also may provide a potentially useful tool for gastric cancer diagnosis or prognosis.

15.4.2 Data Preprocessing

According to our collected data, 535 strains of *H. pylori* CagA protein with disease information will be used for this study. Among them, 287 strains belong to the East Asian subgroup, and the rest 248 are Western strains. In the East Asian subtype group, 47 out of 287 strains are from gastric cancer patients, and the rest are from other diseases. In the Western subtype group, there are 37 strains from the gastric cancer patients, and the remainders are from other diseases or the normal controls, including 24 strains from volunteers whose health (disease) status was unknown. Due to the significant difference between two subgroups, the East Asian subtype and the Western subtype were treated as two independent groups and analyzed within each group individually.

CagA sequences of each subtype were put into the corresponding disease groups, and then the multiple sequence alignments were applied for each group individually by using Clustal X version 2.0.3 (Larkin et al. 2007). Based on the aligned sequences, for each column of multiple alignments, we computed the background entropy B_i and the combinatorial entropy C_i based on the disease groups for each column i as follow:

$$C_i = \sum_k \ln \frac{N_k!}{\prod\limits_{\alpha=1\ldots20} N_{\alpha,i,k}!}$$

$$B_i = \sum_k \ln \frac{N_k!}{\prod\limits_{\alpha=1\ldots20} \widetilde{N}_{\alpha,i,k}!}$$

$$\widetilde{N}_{\alpha,i,k} = N_k N_{\alpha,i}/N$$

where N_k represents the number of sequences in group k, $N_{\alpha,i,k}$ indicates the number of residues of type α in the column i of group k, $N_{\alpha,i}$ is the number of residues of type α in the column i, and N represents the total number of aligned sequences. Then the entropy difference between the combinatorial entropy and the background entropy was calculated as feature values:

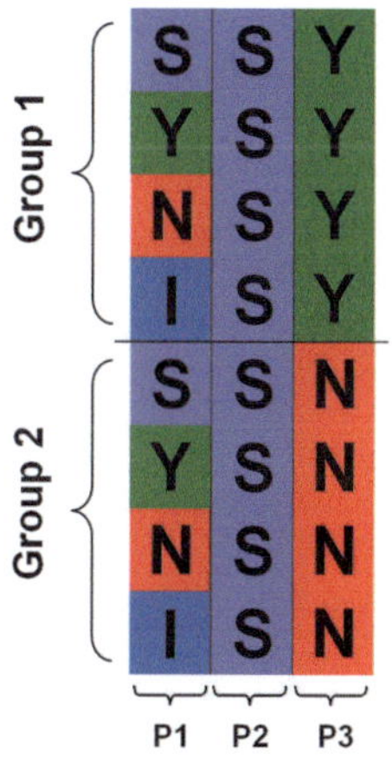

Fig. 15.2 An example to present different cases for the entropy calculation

$$\Delta E = C_i - B_i$$

Figure 15.2 illustrates the entropy concept using three extreme cases for a given column of residues from sequence alignment. Case P1 is the so-called randomly distributed or uniformly distributed, and the amino acids are "randomly distributed" over all groups and without significantly conserved pattern. Case P2 represents a "globally conserved" pattern, and all the amino acids are all almost same across different groups. In "locally conserved" case P3, some specific amino acids are only conserved in particular groups, and different groups have different conserved pattern.

According to the calculation results of the entropy difference for the above three cases, the combinatorial entropy is $C_i = 0$ for both "globally conserved" and "locally conserved" cases. For "randomly or uniformly distributed" case, C_i gets the maximum value. "Conserved" and "randomly distributed" cases can be distinguished based on the value of combinatorial entropy, but it won't help pick "locally conserved" case from all "conserved" cases. Then we look at the value of background entropy, B_i gets the maximum value, 0 and medium value for the "randomly and uniformly distributed" case, "globally conserved" case, and "locally conserved" case, respectively. Finally, "locally conserved" case could be selected based on the differences between combinatorial entropy and the background entropy. The value of differences for the above three cases are $\Delta E_1 = 0$, $\Delta E_2 = 0$, and ΔE_3 gets the minimum value.

15.4.3 Modeling

The training/identification procedure has been implemented based on the workflow shown as follows (Fig. 15.3):

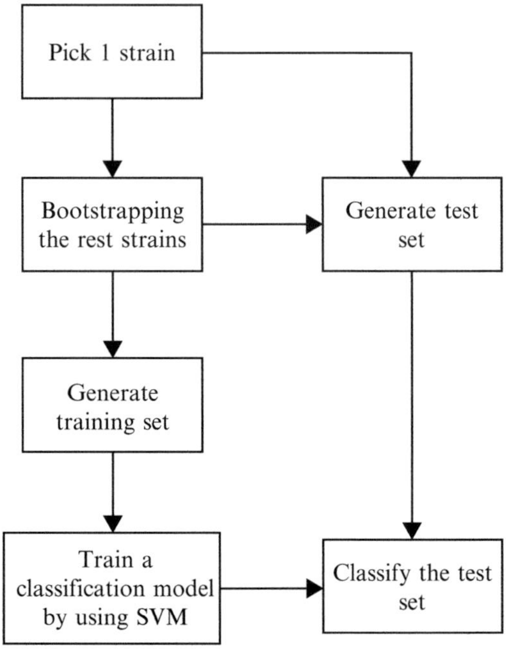

Fig. 15.3 Workflow of classification/prediction procedure for one specific CagA sequence

- Select one strain as the test strain.
- Apply a bootstrap procedure to the rest of the strains to get the training strains.
- Calculate the feature entropy for the test strain based on training strains and save it as the test data.
- Calculate the feature entropy for each strain in the training strain set based on training strains and save them as the training data.
- Generate classification model by using the training data.
- Classify the test data according to the classification model.
- Repeat this procedure five times, and then calculate the average as the final result.

A bootstrapping procedure was applied to avoid the classification bias, since the extremely unbalanced number of cases from different disease group. Usually gastric cancer cases will be much less frequent than other diseases, such as ulcer or gastritis. So we used all samples from noncancer group, and stains from the cancer group were continuously drawn on a random basis until getting the same number of samples as noncancer group. We also repeated this process five times to generate five independent training sets for each test strain, and the final decision is based on the average of five independent classification results. Due to the same reason, traditional n-fold cross validation won't fit our data. Then a leave-one-out (LOO) cross validation procedure was performed. This is not only an assessment of the classifier performance on training/test data but also an estimate of prediction power for novel cases.

SVM^{Light} package V6.02 (http://svmlight.joachims.org/) (Joachims 1999) has been employed as the classifier, and radial basis function (RBF) has been chosen as kernel function. Two parameters were tuned to obtain the optimal F-vlaue by using grid search with above-generated training data. The feature values of each test stain were then fed into the optimized model to get the classification decision. Overall classification performances were evaluated by using the following measurements accuracy (Acc), sensitivity (*Sn*), specificity (*Sp*), Matthews correlation coefficient (*MCC*), and F-value:

$$\text{Accuracy} = \frac{TP + TN}{TP + FP + TN + FN}$$

$$Sp = \frac{TN}{FP + TN}$$

$$Sn = \frac{TP}{TP + FN}$$

$$MCC = \frac{TP \times TN - FP \times FN}{\sqrt{(TP + FP)(TP + FN)(TN + FP)(TN + FN)}}$$

$$F = \frac{2(Sp \times Sn)}{Sp + Sn}$$

where a true positive (*TP*) is a cancer-related sequence classified as such, while a false positive (*FP*) is a noncancer-related sequence classified as cancer related, a false negative (*FN*) is a cancer-related sequence classified as noncancer related, and a true negative (*TN*) is a noncancer-related sequence classified as noncancer related.

15.4.4 Comparison with Other Methods

Two popular identification methods, BLAST (Altschul et al. 1990) and HMMER (Eddy 1998), were selected as the representative methods for comparison. We applied the same evaluation procedure and measurements to above two tools as our method, such as LOO cross validation. The default parameters have been used for both BLAST and HMMER. Comparing the results for three methods, our method achieved 76% and 71% classification accuracy for Western and East Asian subtypes, respectively, which performed significantly better than the rest of the two methods (Table 15.1).

Table 15.1 Classification performances of different methods

Subtype	No. of cancer cases	No. of noncancer cases	Method	Sn	Sp	Accuracy	F-value	MCC
Western	37	211	Entropy	0.86	0.74	0.76	0.80	0.45
			BLAST	0.22	0.77	0.69	0.34	−0.01
			HMMER	0.94	0.005	0.14	0.009	−0.16
East Asian	47	240	Entropy	0.74	0.71	0.71	0.73	0.35
			BLAST	0.17	0.75	0.65	0.28	−0.07
			HMMER	1	0.003	0.19	0.05	0.06

15.4.5 Discussion

It was found that CagA multimerizes in mammalian cells (Ren et al. 2006). This multimerization is independent to the tyrosine phosphorylation, but it is related to the "FPLxRxxxVxDLSKVG" motif, which is named CM motif following EPIYA-C motif. The CM motif plays an important role in CagA-positive *H. pylori*-mediated gastric pathogenesis, since the multimerization is a prerequisite for the CagA-SHP-2 signaling complex and subsequent deregulation of SHP-2. With multiple CM motifs, *H. pylori* strains are much likely associated with severe gastroduodenal diseases (Lu et al. 2008), but this observation cannot explain why different gastroduodenal diseases can be developed with the exact same number of CM motifs. Our study detected two residues in the CM motif, which might lead to the change of multimerization, thus changing the virulence of CagA. This is in consistent with a previous discovery (Sicinschi et al. 2010) that the sequence difference between the East Asian CM and the Western CM determines the binding affinity between CagA and SHP-2.

However, we also found that there is no simple relation between any single residue and cancer occurrence, and hence, it is not possible to just use one single residue to be the marker for identifying cancer. We speculate that one special combination of all or partial important residues could have a high correlation with one particular disease. The classification result strongly supports our hypothesis, i.e., the information of the selected residues in intervening regions can be used to classify the relation between CagA sequences and gastric cancer, although the difference between the profiles of cancer and noncancer groups is not very strong.

15.5 Summary

We described the procedures for collecting, curating, and depositing public data into a web-based database. With a user-friendly interface, those data could be easily downloaded, browsed, and searched by different entries. Some computational information (PSIBLAST sequence profile, protein secondary structures, and 3D structures) have also been integrated into the database. This database is not only useful for our research but also could benefit the *H. pylori* and gastroduodenal disease research community.

Based on the curated CagA data, an entropy-based calculation was used to detect key residues of CagA intervening sequences as the gastric cancer biomarker. For each residue, both combinatorial entropy and background entropy were calculated, and the entropy difference was used as the criterion for feature residue selection. The feature values were then fed into SVM with the RBF kernel, and two parameters were tuned to obtain the optimal F-value by using a grid search. Two other popular sequence classification methods, the BLAST and HMMER, were also applied to the same data for comparison. Our study indicates that small variations of amino acids in those important residues might lead to the virulence variance of CagA strains resulting in different gastroduodenal diseases. This study provides not only a useful tool to predict the correlation between the novel CagA strain and diseases but also a general new framework for detecting biological sequence biomarkers in population studies.

References

Altschul SF, Gish W, Miller W, Myers EW, Lipman DJ. Basic local alignment search tool. J Mol Biol. 1990;215(3):403–10. doi:10.1016/S0022-2836(05)80360-2.

Altschul SF, Madden TL, Schaffer AA, Zhang J, Zhang Z, Miller W, Lipman DJ. Gapped BLAST and PSI-BLAST: a new generation of protein database search programs. Nucleic Acids Res. 1997;25(17):3389–402.

Amieva MR, Vogelmann R, Covacci A, Tompkins LS, Nelson WJ, Falkow S. Disruption of the epithelial apical-junctional complex by Helicobacter pylori CagA. Science. 2003;300(5624):1430–4. doi:10.1126/science.1081919.

Andries K, Verhasselt P, Guillemont J, Gohlmann HW, Neefs JM, Winkler H, Van Gestel J, Timmerman P, Zhu M, Lee E, Williams P, de Chaffoy D, Huitric E, Hoffner S, Cambau E, Truffot-Pernot C, Lounis N, Jarlier V. A diarylquinoline drug active on the ATP synthase of Mycobacterium tuberculosis. Science. 2005;307(5707):223–7. doi:10.1126/science.1106753.

Argent RH, Kidd M, Owen RJ, Thomas RJ, Limb MC, Atherton JC. Determinants and consequences of different levels of CagA phosphorylation for clinical isolates of Helicobacter pylori. Gastroenterology. 2004;127(2):514–23.

Bhatt AS, Freeman SS, Herrera AF, Pedamallu CS, Gevers D, Duke F, Jung J, Michaud M, Walker BJ, Young S, Earl AM, Kostic AD, Ojesina AI, Hasserjian R, Ballen KK, Chen YB, Hobbs G, Antin JH, Soiffer RJ, Baden LR, Garrett WS, Hornick JL, Marty FM, Meyerson M. Sequence-based discovery of Bradyrhizobium enterica in cord colitis syndrome. N Engl J Med. 2013;369(6):517–28. doi:10.1056/NEJMoa1211115.

Blaser MJ, Perez-Perez GI, Kleanthous H, Cover TL, Peek RM, Chyou PH, Stemmermann GN, Nomura A. Infection with Helicobacter pylori strains possessing cagA is associated with an increased risk of developing adenocarcinoma of the stomach. Cancer Res. 1995;55(10):2111–5.

Covacci A, Censini S, Bugnoli M, Petracca R, Burroni D, Macchia G, Massone A, Papini E, Xiang Z, Figura N, et al. Molecular characterization of the 128-kDa immunodominant antigen of Helicobacter pylori associated with cytotoxicity and duodenal ulcer. Proc Natl Acad Sci U S A. 1993;90(12):5791–5.

Crooks GE, Hon G, Chandonia JM, Brenner SE. WebLogo: a sequence logo generator. Genome Res. 2004;14(6):1188–90. doi:10.1101/gr.849004.

Dinsdale EA, Pantos O, Smriga S, Edwards RA, Angly F, Wegley L, Hatay M, Hall D, Brown E, Haynes M, Krause L, Sala E, Sandin SA, Thurber RV, Willis BL, Azam F, Knowlton N, Rohwer F. Microbial ecology of four coral atolls in the Northern Line Islands. PLoS One. 2008;3(2):e1584. doi:10.1371/journal.pone.0001584.

Eddy SR. Profile hidden Markov models. Bioinformatics. 1998;14(9):755–763. doi:btb114 [pii]

Edgar RC. MUSCLE: multiple sequence alignment with high accuracy and high throughput. Nucleic Acids Res. 2004;32(5):1792–7. doi:10.1093/nar/gkh340.

Ernst PB, Gold BD. The disease spectrum of Helicobacter pylori: the immunopathogenesis of gastroduodenal ulcer and gastric cancer. Annu Rev Microbiol. 2000;54:615–40. doi:10.1146/annurev.micro.54.1.615.

Fu H, Hu Z, Wen J, Wang K, Liu Y. TGF-beta promotes invasion and metastasis of gastric cancer cells by increasing fascin1 expression via ERK and JNK signal pathways. Acta Biochim Biophys Sin. 2009;41(8):648–56.

Hacker J, Hentschel U, Dobrindt U. Prokaryotic chromosomes and disease. Science. 2003;301 (5634):790–3. doi:10.1126/science.1086802.

Higashi H, Tsutsumi R, Fujita A, Yamazaki S, Asaka M, Azuma T, Hatakeyama M. Biological activity of the Helicobacter pylori virulence factor CagA is determined by variation in the tyrosine phosphorylation sites. Proc Natl Acad Sci U S A. 2002;99(22):14428–33. doi:10.1073/pnas.222375399.

Joachims T. Making large-scale support vector machine learning practical. In: Schölkopf B, editors. Advances in kernel methods: support vector machines. Cambridge, MA: MIT Press; 1999. doi:citeulike-article-id:227265.

Jones KR, Joo YM, Jang S, Yoo YJ, Lee HS, Chung IS, Olsen CH, Whitmire JM, Merrell DS, Cha JH. Polymorphism in the CagA EPIYA motif impacts development of gastric cancer. J Clin Microbiol. 2009;47(4):959–68. doi:10.1128/JCM.02330-08.

Lai YP, Yang JC, Lin TZ, Wang JT, Lin JT. CagA tyrosine phosphorylation in gastric epithelial cells caused by Helicobacter pylori in patients with gastric adenocarcinoma. Helicobacter. 2003;8(3):235–43.

Larkin MA, Blackshields G, Brown NP, Chenna R, McGettigan PA, McWilliam H, Valentin F, Wallace IM, Wilm A, Lopez R, Thompson JD, Gibson TJ, Higgins DG. Clustal W and Clustal X version 2.0. Bioinformatics. 2007;23(21):2947–8. doi:10.1093/bioinformatics/btm404.

Linz B, Balloux F, Moodley Y, Manica A, Liu H, Roumagnac P, Falush D, Stamer C, Prugnolle F, van der Merwe SW, Yamaoka Y, Graham DY, Perez-Trallero E, Wadstrom T, Suerbaum S, Achtman M. An African origin for the intimate association between humans and Helicobacter pylori. Nature. 2007;445(7130):915–8. doi:10.1038/nature05562.

Lorenz P, Eck J. Metagenomics and industrial applications. Nat Rev Microbiol. 2005;3(6):510–6. doi:10.1038/nrmicro1161.

Lu HS, Saito Y, Umeda M, Murata-Kamiya N, Zhang HM, Higashi H, Hatakeyama M. Structural and functional diversity in the PAR1b/MARK2-binding region of Helicobacter pylori CagA. Cancer Sci. 2008;99(10):2004–11. doi:10.1111/j.1349-7006.2008.00950.x.

McGuffin LJ, Bryson K, Jones DT. The PSIPRED protein structure prediction server. Bioinformatics. 2000;16(4):404–5.

Medini D, Serruto D, Parkhill J, Relman DA, Donati C, Moxon R, Falkow S, Rappuoli R. Microbiology in the post-genomic era. Nat Rev Microbiol. 2008;6(6):419–30. doi:10.1038/nrmicro1901.

Naito M, Yamazaki T, Tsutsumi R, Higashi H, Onoe K, Yamazaki S, Azuma T, Hatakeyama M. Influence of EPIYA-repeat polymorphism on the phosphorylation-dependent biological activity of Helicobacter pylori CagA. Gastroenterology. 2006;130(4):1181–90. doi:10.1053/j.gastro.2005.12.038.

Odenbreit S, Puls J, Sedlmaier B, Gerland E, Fischer W, Haas R. Translocation of Helicobacter pylori CagA into gastric epithelial cells by type IV secretion. Science. 2000;287 (5457):1497–500.

Petrosino JF, Highlander S, Luna RA, Gibbs RA, Versalovic J. Metagenomic pyrosequencing and microbial identification. Clin Chem. 2009;55(5):856–66. doi:10.1373/clinchem.2008.107565.

Ren S, Higashi H, Lu H, Azuma T, Hatakeyama M. Structural basis and functional consequence of Helicobacter pylori CagA multimerization in cells. J Biol Chem. 2006;281(43):32344–52. doi:10.1074/jbc.M606172200.

Sicinschi LA, Correa P, Peek RM, Camargo MC, Piazuelo MB, Romero-Gallo J, Hobbs SS, Krishna U, Delgado A, Mera R, Bravo LE, Schneider BG. CagA C-terminal variations in Helicobacter pylori strains from Colombian patients with gastric precancerous lesions. Clin Microbiol Infect. 2010;16(4):369–78. doi:10.1111/j.1469-0691.2009.02811.x.

Suerbaum S, Michetti P. Helicobacter pylori infection. N Engl J Med. 2002;347(15):1175–86. doi:10.1056/NEJMra020542.

Turnbaugh PJ, Ley RE, Hamady M, Fraser-Liggett CM, Knight R, Gordon JI. The human microbiome project. Nature. 2007;449(7164):804–10. doi:10.1038/nature06244.

Turnbaugh PJ, Hamady M, Yatsunenko T, Cantarel BL, Duncan A, Ley RE, Sogin ML, Jones WJ, Roe BA, Affourtit JP, Egholm M, Henrissat B, Heath AC, Knight R, Gordon JI. A core gut microbiome in obese and lean twins. Nature. 2009;457(7228):480–4. doi:10.1038/nature07540.

Uemura N, Okamoto S, Yamamoto S, Matsumura N, Yamaguchi S, Yamakido M, Taniyama K, Sasaki N, Schlemper RJ. Helicobacter pylori infection and the development of gastric cancer. N Engl J Med. 2001;345(11):784–9. doi:10.1056/NEJMoa001999.

Ullman TA, Itzkowitz SH. Intestinal inflammation and cancer. Gastroenterology. 2011;140 (6):1807–16. doi:10.1053/j.gastro.2011.01.057.

Whitman WB, Coleman DC, Wiebe WJ. Prokaryotes: the unseen majority. Proc Natl Acad Sci U S A. 1998;95(12):6578–83.

Zhang J, Wang Q, Barz B, He Z, Kosztin I, Shang Y, Xu D. MUFOLD: a new solution for protein 3D structure prediction. Proteins. 2010;78(5):1137–52. doi:10.1002/prot.22634.

Zhang C, Zheng G, Xu S-F, Xu D. Computational challenges in characterization of bacteria and bacteria-host interactions based on genomic data. J Comput Sci Technol. 2012a;27(2):225–39. doi:10.1007/s11390-012-1219-y.

Zhang C, Xu S, Xu D. Risk assessment of gastric cancer caused by Helicobacter pylori using CagA sequence markers. PLoS One. 2012b;7(5):e36844. doi:10.1371/journal.pone.0036844.

Zhang C, Cleveland K, Schnoll-Sussman F, McClure B, Bigg M, Thakkar P, Schultz N, Shah MA, Betel D. Identification of low abundance microbiome in clinical samples using whole genome sequencing. Genome Biol. 2015;16:265. doi:10.1186/s13059-015-0821-z.

Part V
Bioimaging and Other Applications of Informatics Techniques in Translational Medicine

Chapter 16
Artificial Intelligence and Automatic Image Interpretation in Modern Medicine

Costin Teodor Streba, Mihaela Ionescu, Cristin Constantin Vere, and Ion Rogoveanu

Abstract The need for intelligent computerized systems in medicine has increased over recent years; artificial neural networks (ANN) have become increasingly popular for various classification tasks during diagnosis procedures.

One of the most dynamic branches of modern diagnosis is medical imaging, with the emergence of various new methods for finding disease, especially cancer. Quantifying and interpreting the results represents a constant challenge for the already overburdened physician.

Gastroenterology is a dynamic field of medicine with many recent advances in imaging methods for cancer diagnosis. Computerized systems have gained significant traction in the last few years, with promising future prospects of reducing diagnosis time and workload on the performing physicians.

We describe here two image analysis systems that take advantage of the latest accomplishments in ANN and computerized decision making. The first system describes a computerized diagnosis system that takes into account a blend of both clinical and biological set of parameters, combining them with advanced image analysis of contrast-enhanced ultrasound imagery in an attempt to diagnose and classify primary liver malignancies.

The second part of the chapter is dedicated to another advanced imaging method in gastroenterology – wireless videocapsule endoscopy. The combination of different novel image analysis techniques described here greatly reduces interpretation

---

CT Streba and M Ionescu have equally contributed to preparing the manuscript and share first authorship.

C.T. Streba • C.C. Vere (✉) • I. Rogoveanu
Research Center of Gastroenterology and Hepatology of Craiova, Craiova, Romania

Department of Bioinformatics, University of Medicine and Pharmacy of Craiova, Petru Rares St. No. 2-4, Craiova, Romania
e-mail: cc.vere.umf@gmail.com

M. Ionescu
Department of Gastroenterology, University of Medicine and Pharmacy of Craiova, Petru Rares St. No. 2-4, Craiova, Romania

© Shanghai Jiao Tong University Press, Shanghai and Springer Science+Business Media Dordrecht 2017

D.-Q. Wei et al. (eds.), *Translational Bioinformatics and Its Application*, Translational Medicine Research, DOI 10.1007/978-94-024-1045-7_16

times of an extensive investigation and helps doctors in the decision-making process.

Keywords Hepatocellular carcinoma • Artificial neural network • Contrast-enhanced ultrasound • Wireless capsule endoscopy • Computer-aided diagnosis

16.1 Introduction

Computer-aided diagnosis (CAD) systems have gained a reputation for providing integrative solutions for the diagnosis of several types of malignancies (Lisboa and Taktak 2006; Grossi et al. 2007; Cucchetti et al. 2010), with many applications in gastroenterology and tumour pathology associated with the digestive tract. The use of artificial neural networks (ANN) or other adaptive, machine-based learning systems, can substantially improve the accuracy of any quantitative-based image analysis method (Cucchetti et al. 2010; Chiu et al. 2009; Markaki et al. 2009). Current approaches lack integration with clinical and laboratory data, thus not providing an integrative system designed for actual medical use (Markaki et al. 2009; Saftoiu et al. 2008; 2012).

Early correct diagnosis and appropriate staging of liver malignancies are of utmost importance for patient survival, as curative surgical interventions have narrow indications and are extremely specific to certain types of tumours (El-Serag and Rudolph 2007; El-Serag 2011; European Association for the Study of the Liver and European Organisation for Research and Treatment of Cancer 2012; Bruix and Sherman 2011). Hepatocellular carcinoma (HCC) currently ranks third in terms of mortality worldwide and fifth in incidence with almost 750,000 new cases diagnosed each year while being second in mortality among digestive cancers (Ferlay et al. 2010). Differential diagnosis is of the highest importance for the patient, as correct identification of a tumour as being HCC in an earlier stage drastically improves survival. The diagnostic criteria for HCC, the most common primary liver malignancy worldwide, relay primarily on imagistic methods (European Association for the Study of the Liver and European Organisation for Research and Treatment of Cancer 2012; Bruix and Sherman 2011). With the widespread use in recent years of imaging techniques, the differential diagnosis of a newly discovered liver mass became less invasive for the patient.

Current accepted guidelines worldwide are those proposed by the American Association for the Study of Liver Disease (AASLD, revised in 2010) (European Association for the Study of the Liver and European Organisation for Research and Treatment of Cancer 2012) and those of the association between the European Society for the Study of Liver and the European Organization for Research and Treatment of Cancer (EASL-EORTC, recently revised in April 2012) (Bruix and Sherman 2011). The introduction of new generations of contrast agents marked the adoption of a generally accepted radiological hallmark for positive HCC diagnosis, namely, contrast uptake in the arterial phase followed by washout in the venous/late phase. The American proposed guidelines stipulate that only one imaging technique with contrast uptake, either computer tomography (CT) or magnetic resonance imaging (MRI), showing the radiological hallmark, is sufficient for positive diagnosis of tumours between 1 and 2 cm in diameter (this being the optimum tumour size for curative surgery) (European Association for the Study of the Liver and European Organisation for Research and Treatment of Cancer 2012). Their European counterparts, however, are more cautious in applying a single imaging method and recommend two coincidental techniques in suboptimal settings due to technical limitations (Bruix and Sherman 2011).

16.2 Dynamic Interpretation of Contrast-Enhanced Ultrasonography (CEUS) and Endoscopic Ultrasound (EUS) Recordings

Second-generation contrast agents use gas bubbles between 2 and 7 microns in diameter which resonate under the ultrasound (US) probe, the amplified signal being registered by the US machine through the same probe (Dietrich 2004; Rettenbacher 2007). While the radiological hallmark can be clearly identified in this technique, the contrast microbubbles are bound to the intravascular space, as opposed to iodinated contrast CT or gadolinium-based MR imaging in which the contrast agents are rapidly cleared from the bloodstream into the surrounding parenchymal space (Bruix and Sherman 2011; Albrecht et al. 2004; Lencioni et al. 2008a; Rimola et al. 2009; Colli et al. 2005). According to some studies, intrahepatic cholangiocarcinoma (ICC) or even some highly vascularized liver metastases can display uptake patterns similar to HCC during contrast-enhanced ultrasound (CEUS) or gadolinium-based MRI, thus providing an important source of error (Rimola et al. 2009).

16.2.1 Time-Intensity Curve Analysis – Basic Principles

Currently, efforts are being made to overcome the inherited issues with CEUS investigations and diminish the rate of false interpretations (Ignee et al. 2010; Goertz et al. 2010; Huang-Wei et al. 2006; Salvatore et al. 2012; Jiang et al. 2010). The use of time-intensity curves (TICs) in the interpretation of CEUS movies seems a feasible method of increasing the specificity of this investigation (Huang-Wei et al. 2006; Salvatore et al. 2012). The method relies on plotting and comparing on a timescale the median intensities of two user-defined areas, one within the suspected tumour and one in a parenchymal area with no major vessels, thus producing two curves which depict contrast uptake during CEUS (Huang-Wei et al. 2006; Salvatore et al. 2012). The usual graphical representation shows contrast uptake in the first 30–60 s, followed by tumour washout in portal and venous phases, when the maximum intensity is similar to those of the parenchymal-selected area of interest. While TIC quantification does add more precision to the investigation, several user-dependent and technique-dependent limitations have been identified, such as different depths of the analysed tumour/parenchyma areas of interest, moving artefacts due to patient breathing or the interference of large blood vessels in the selected areas of interest (Ignee et al. 2010).

Automated quantitative image analysis techniques have been introduced in medical practice for a number of years (Jiang et al. 2010; Zhang et al. 2009; Guo et al. 2009), gaining significant importance in various fields of micro- or macroscopic assessment. Automated or semiautomated image segmentation with feature extraction and overtime comparative quantification of independent elements has been employed with various grades of success in interpreting medical imagistic data (Jiang et al. 2010; Zhang et al. 2009; Guo et al. 2009; Mittal et al. 2011; Zhang et al. 2008; Kondo et al. 2009; Verma et al. 2009). However, current image analysis methods employed with US or CEUS lack clinical applications and are not sufficient in order to provide effective aid to the clinician (Mittal et al. 2011; Zhang et al. 2008; Kondo et al. 2009; Verma et al. 2009).

A complete ultrasound evaluation of the liver can be performed using conventional ultrasound methods (the “B-mode” US), this step being mandatory before injecting the contrast agent. The contrast enhancement method is then used to investigate at a low mechanical index. This ensures adequate cancellation of tissue signal, leaving visible only the major vascular structures and some anatomical landmarks such as the diaphragm. All US devices have a “dual view” which puts contrast enhancement and conventional B-mode image side by side, in order to facilitate tumour localization. This also helps the clinician when accurately demarcating edges of a tumour. Full examination of the liver takes about 4–5 min, only the first 120 s usually being of importance for a proper TIC analysis.

Bolus models of arterial peak enhancement described contain the three stages of vascular dissemination (an arterial phase, the portal venous phase and late phase corresponding to bubble exhaustion). An inherent advantage of CEUS is being able to assess patterns of contrast agent filling in real time with a higher resolution than CT or MRI, avoiding the need to predefine benchmarks or performance time tracking of contrast bolus. Due to different pharmacokinetics of contrast agent used in CEUS, the vascular pattern is highlighted in this investigation differently compared with CT or MRI imaging, where the agent is eliminated more rapidly in the extracellular space (Singh et al. 2007).

The usual traceable parameters are the maximum value of the intensity (IMAX), represented by the average intensities at each pixel of a region of interest (ROI); mean transit time (MTT), which approximates the time needed by the contrast agent to stop being represented as a significant increase in light intensity in the ROI; the area under the curve (AUC), which provides a parallel interpretation of the impact of changes in intensity; perfusion index (PI), which is an appreciation of the vasculature in the area of interest; rise time (RT) which is a measurement of the rate of increase of intensity until the upper threshold; and time to peak (TTP) which

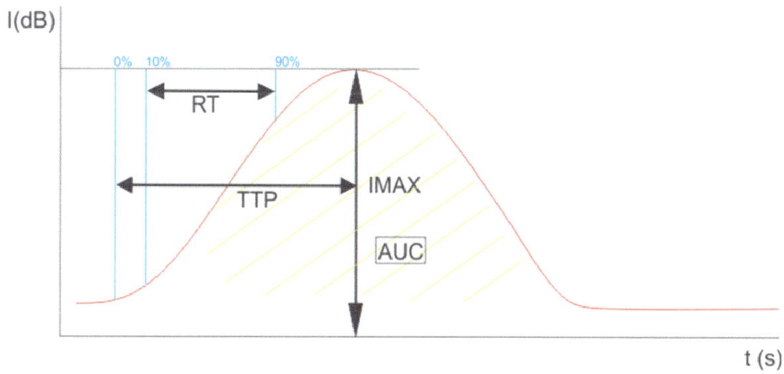

Fig. 16.1 Schematic representation of the TIC parameters, graphed as a function of intensity against the time that the US transducer records the image

is the time taken by the contrast agent to reach peak intensity, from the point of zero intensity (Fig. 16.1).

The importance of contrast agents to enhance the rate of diagnosis of liver tumours has been demonstrated in several studies (Caturelli et al. 2004; Forner et al. 2008; Bolondi et al. 2005; Lencioni et al. 2008b; Youk et al. 2003; Kim et al. 2000; Catalano et al. 2005; Leen et al. 2006). For example, a comparative analysis using ROC curves showed an improvement of up to 91% of the specificity of diagnosis and an increase in the number of positive diagnoses of up to 75% from a maximum of 48% for the conventional ultrasound (Kim et al. 2005).

A recent study published by the team led by Adre Ignea (Ignee et al. 2010) assessed using TICs obtained from analysis of CEUS movies, identifying possible sources of error and developing a set of general rules for this investigation. Thus, it is important to choose the region for comparison to normal parenchyma at the same depth in the tumour parenchyma examined, but no more than 4–6 cm from the US probe. They also showed that there are no restrictions in regard to the shape of selected regions of normal liver parenchyma.

All recordings should previously be calibrated, with adjusted signal for light intensity, colour and saturation channels. Recent studies using intensity-time curves showed the great importance of proper standardization of analysed recordings (Ignee et al. 2010; Goertz et al. 2010). The usual protocol that is generally employed is focused on the selection of representative areas for focal liver formation, excluding large blood vessels and areas of necrosis which could influence the results. Also, when choosing areas of interest surrounding parenchyma, the user should avoid large blood vessels, which could influence the results by artificially elevated values. Moreover, a recent study (Salvatore et al. 2012) concludes that analysis of only ten frames chosen by different combinations of the three distinct phases of CEUS recording gives the same result in the interpretation of best-fit TICs. Currently, it is considered that the frames of interest must be selected from a minimum interval of 100–180 s.

In our experience, parameters resulting from the analysis of TICs show a low degree of variability among the studied cases. In addition to these, we introduced statistical comparison of values string for each phase of CEUS recording (Gheonea et al. 2013; Streba et al. 2012a). The degree of correlation between the two averages can be quantified by the computer, the observations a human operator can make concerning the appearance of the two curves thus being objectified by the artificial diagnosis system. Considering this approach, even longer recordings can be employed in the analysis.

Outliers resulting from artefacts due to a short interposition of a blood vessel (resulting in artificially increased intensity for the frame) or deeper breathing movement (which can lead to a repositioning of an area of necrosis within the selected region) should be excluded by plotting the curve on a "best-fit" model, thus compensating with values following a theoretical distribution.

Hepatocellular carcinomas show higher enhancement in the tumour ROI compared to parenchyma, similarly to hypervascular metastases and hepatic haemangiomas. Differentiating these formations can be achieved mainly based on the appearance of the terminal phase where the correlation of the two curves, resulting in the appearance of overlapping routes, coincides with the existence of the phenomenon of washout. Primitive malign tumours showed partial washout, the approximate paths of the two ROIs being slightly offset during the late portal phase, as hypervascular metastasis showed a complete and premature washout phenomenon. The signal strength in these areas decreases more markedly than in the area of parenchyma, therefore resulting in statistically significant differences between the two rows of values. For haemangiomas, the particular aspect is the lack of convergence of the two rows of values, following almost parallel routes during these phases. It is also possible to quantify the centripetal contrast load, with a characteristic appearance in the arterial phase. This can be recorded as a slope of the curve at an angle less than the surge in cases with HCC or hypervascular metastases. A comparative study (Huang-Wei et al. 2006) conducted on a group of patients with diverse liver tumour pathology revealed significant differences in these parameters between hepatocellular carcinomas and haemangiomas or hypervascular liver metastases. This study did not identify significant differences between the maximum intensities reached in cases with HCC and hypervascular metastases.

Hypovascular metastases have a negative peak, with the maximum intensity value actually lower than the parenchymal ROI. Literature lacks conclusive studies on the appearance of time-intensity curves in this type of metastatic tumours; however, it is logical that large areas of necrosis are avascular, thus resulting in minimal contrast uptake (Goertz et al. 2010; Huang-Wei et al. 2006).

In cases of focal steatosis, the TICs of the two selected ROIs show parallel contrast loading, determined by the lack of blood vessels alteration in these cases. When referring to malignant tumours, there is an intense process of neogenesis with predominantly arterial vasculature, leading to contrast uptake more pronounced during the first phase of contrast dispersion. The washout phenomenon is due to the lack of blood supply to the venous vasculature; therefore contrast uptake during this step remains low. In cases with steatosis, veins remain approximately unchanged,

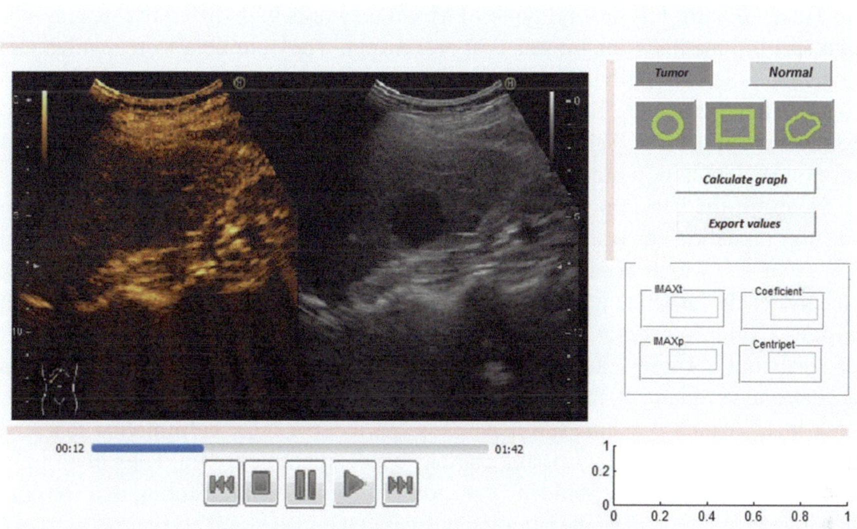

Fig. 16.2 Example of the interface designed to select ROIs and plot TICs with automatic extraction of imaging parameters for later analysis in the ANN system

also missing the "unpaired arteries" found in other pathologies, thus resulting in parallel TICs.

Complex analysis performed on thc TIC parameters resulting from CEUS investigation proves extremely important in enhancing the diagnostic capabilities of the ANN, as the final diagnostic rate is above 97% in the case of complex neural networks (Gheonea et al. 2013; Streba et al. 2011, 2012a, b, c). An example of an interface designed for plotting TICs and extracting relevant parameters for an ANN system is shown in Fig. 16.2.

16.3 Artificial Neural Networks – Basic Principles and Common Applications in Medicine

Computer-aided diagnosis systems may offer new possibilities in the diagnosis and staging of tumours. Artificial neural networks (ANNs) are the result of research in medical informatics, representing a form of artificial intelligence. Neural networks are computer language translation of the principles of functioning of the human central nervous system. Basic components of the brain, namely, neurons and neuronal synapses, are thus reproduced by dedicated software applications, creating networks that have the ability to retrieve and process information to solve complex problems, with the added ability to learn and take calculated decisions (Chiu et al. 2009).

There are currently several types of neural networks with different topographies and specific methods for information processing. They are able to recognize and generalize the correlations and rules of different databases, which then apply them to solve new problems. They are also able to recognize patterns of any type and to formulate hypotheses, being particularly useful in solving problems that are commonly encountered in clinical practice (Kondo et al. 2009).

Neurons, the basic unit of these networks, are organized in layers accessible to the user ("visible") or without direct access ("hidden layers"), being interconnected via synapses, which ensure the transmission of information between each layer. Briefly, each variable introduced in the system is received by a neuron, which assigns a value – "weight" – quantifying its importance and a "bias" – a coefficient of "disbelief". These values can be both positive and negative, thus ensuring an accurate assessment of the variables. All received information cross-links to the next layer, which contains fewer neurons but receive more information than the first, and thus made the logical associations between problem data. They in turn can communicate with other "hidden" decision layers in the end resulting in a solution or a logical assessment of the situation presented (Chiu et al. 2009; Jiang et al. 2010; Kondo et al. 2009).

To be able to produce more accurate results, neural networks must be "trained", either by a human operator (supervised training) or by intrinsic systems (self/unsupervised learning networks). Redundant data is eliminated through these procedures, networks being able to properly assess each variable introduced to make the right decision at the end of a logic cycle (Chiu et al. 2009; Markaki et al. 2009; Jiang et al. 2010).

There are several architectures of neural networks useful in processing medical data, the "feedforward" type being one of the most widely used. In such networks, the neurons in each layer are connected only with the subsequent layers. These connections are unidirectional, information being transmitted by input neurons to the hidden decision layer and further to the output layer, where a conclusion is formulated (Verma et al. 2009). The preferred learning algorithm is typically a type of supervised learning with successive propagation (back-propagation supervised learning), in which the coefficients of each neuron are changed after each successive iteration (the weight and bias of each variable).

A typical example of such a network is the *multilayer perceptron* (MLP) network comprising a nonlinear transfer function in the hidden layer (Fig. 16.3). MLPs are capable of associating learning patterns to outcomes in non-homogeneous data sets, thus being particularly useful in medical applications that contain imaging data, with or without numerical data obtained from laboratory investigations and medical history (Guo et al. 2009; Verma et al. 2009).

Radial neural networks (radial basis function, RBF) contain three neural layers with vertical nonlinear transmission and linear transmission within the hidden decision layer. These networks are extremely flexible in topology and size, being suited to solve a variety of problems, in particular the analysis of three-dimensional shapes or processing of quantities of data that refer to time series (Goodband et al. 2008).

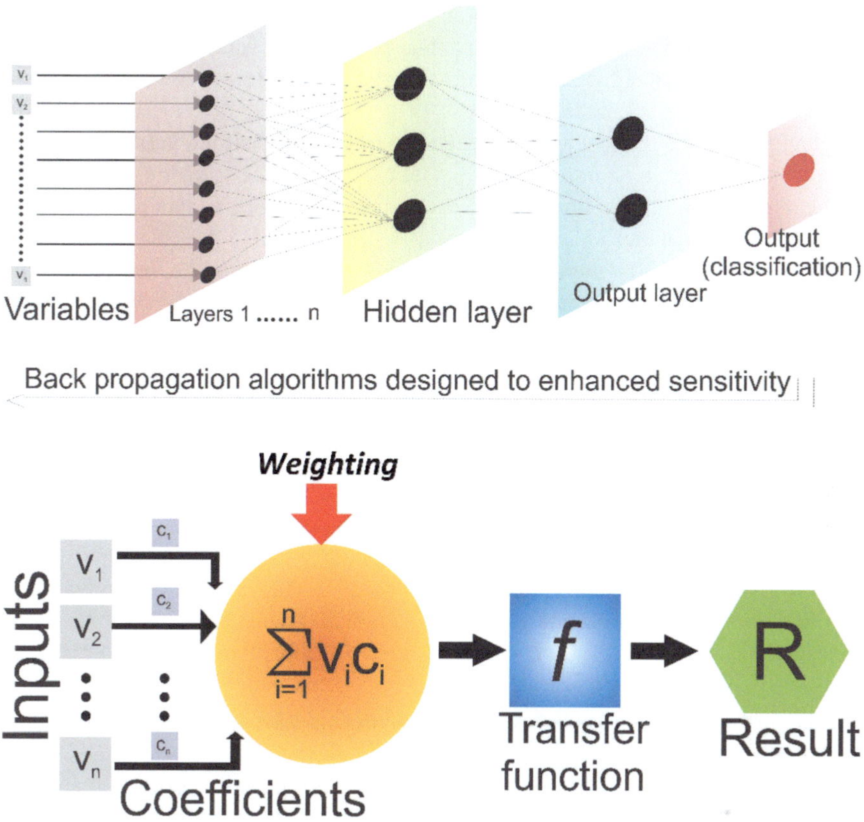

Fig. 16.3 Schematic representation of a typical MLP network and of a neuron

In *recurrent networks*, information can travel in both directions by introducing loops that propagates the values from hidden decision layers back to the input layer. Their status changes continuously until a point of equilibrium is reached, where they remain until a radical change occurs with input data, when a new balance must be found. These have the potential to become very powerful tools to process large amounts of data. They are primarily used to reproduce the associative memory, which helps when the user has a partial original data set and the network needs to fill certain parameters in the learning cycle. They are particularly useful in medical practice, especially in characterizing and determining the degree of malignancy of tumour formations, when a result should be given with regard to partial or similar models that were identified during the learning phase (Markaki et al. 2009; Guo et al. 2009; Verma et al. 2009; Goodband et al. 2008).

Choosing the optimum neural network architecture is very important in every medical task, as it must be able to intervene in both image analysis and when complex patient data has to be processed during the investigation of a new case.

Training artificial intelligence system ensures the production of results with a high degree of confidence for the clinician, thus proving the versatility of the system which could be used by health professionals with different levels of training regarding the complex pathology.

Having both the clinical and biological set of parameters for a sufficiently large number of patients, a neural network model can attempt classification and staging and can determine the optimum therapeutic indication for patients with tumours, especially malignancies.

Neural networks that are used to classify tumours are usually a variation of the feedforward back propagation, optimal for solving classification problems. Networks have a number of input neurons equal to the number of parameters used, a hidden layer and an output neuron (which provides the outcome). Each input parameter introduced in numerical form (either predefined values or binary variable 0/1) is assigned a weight based on the network decision, in order to calculate their importance in the final outcome. It is usually indicated not to set strict rules for the weights, as their values will be determined based on the training set. Transfer functions are the equivalent of human synapses, as they allow the connection between neural layers.

Neural networks are now modern systems available in terms of technology, flexible enough and particularly dynamic, which recommends them as semi-independent diagnostic tools. They are currently worldwide preferred to classical tracking methods and statistical modelling for population groups with complex diseases, such as those with malignant neoplastic disease (Lisboa and Taktak 2006; Cucchetti et al. 2010; Chiu et al. 2009; Mittal et al. 2011).

From a given patient lot, some cases are used to train the network, while the rest of the data is involved in the validation of results. Data entered is stored in a dedicated database for easy retrieval. Iteratively modifying the weights of the input data in the global algorithm during the neural network training, after a certain time they converge to a solution that provides a logical correlation between the input data and output. In the training phase, the network becomes able to generalize the relationships between the input data and output classes, based on the training set that included only some of the possible combinations of inputs and outputs. The training set is usually chosen so as to give the network the ability to identify one or more characteristics of the input data corresponding to a specific output data (Fig. 16.4).

16.3.1 Testing of an ANN in Classifying Liver Tumours

Although the concept is not new, neural network applied in medicine existing since the mid-1990s (Lisboa and Taktak 2006; Cucchetti et al. 2010), advancing modern diagnostic techniques has provided new opportunities for the networks which today no longer rely only on clinical data, anamnesis and simple laboratory tests, but also integrate imaging parameters or genetics and cellular biology methods (Lisboa and

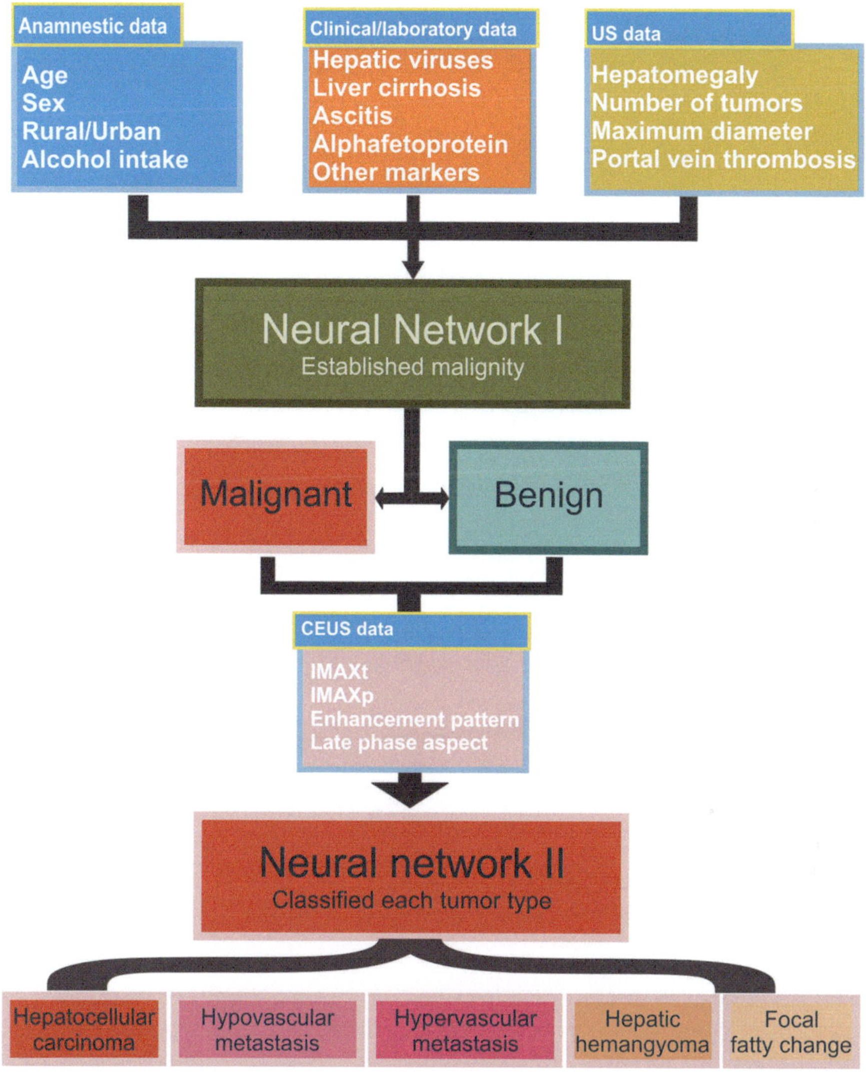

Fig. 16.4 Schematic representation of the cascading ANNs and annexed subsystems for different diagnostic, staging and prognostic modules employed in the management of liver tumours

Taktak 2006; Chiu et al. 2009). The system designed in this study belongs to this second category – that of the complex systems diagnosis, being able to take over, interpret and sort complex imaging parameters. They come from the innovative application of state-of-the-art methods such qualitative and quantitative analysis of time-intensity curves resulting from CEUS.

We chose a representative number of variables for the studied pathology, which contained patient history data, clinical, laboratory and imaging investigations as

Table 16.1 Variables entered into the ANN network

Variable	Type	Special remarks
Personal data		
Sex	Binary	Male sex is more predisposed to develop HCC
Age	Predefined intervals	10–30, 30–60, >60 years
Background	Binary	Urban/rural
Risk factors		
Alcohol consumption	Binary	Abuse/normal
Hepatitis viruses	Variable	B+C/B/C
Significant patient history data		
Liver cirrhosis	Binary	Yes/no
Tumoural markers		
AFP	Binary	1 = above limit, 0 = normal
Other tumour markers	Binary	1 = above, 0 = normal
Standard US tumour parameters		
Hepatomegaly	Binary	Da/Nu
Size (mm)	Value	Cu valoare prognostică
Number	Value	Cu valoare prognostică
Malignant PVT	Binary	
CEUS parameters		
IMAX	Value	These parameters entered in the complex analysis for determining the type of focal liver lesion
RT	Value	
TTP	Value	

Some of these parameters also played a role in prognosis
AFP alpha fetoprotein (marker for HCC), *TVP* portal vein thrombosis

well as TIC-derived specific data (described above), which have been distributed to separate subsystems in the complete neural network. The most important variables can be found in Table 16.1.

All complex parameters for the neural network training were stored in a dedicated database. This step allows the future use of this data for better training of the ANN. To ensure continuity we chose a standardized format for entering patients using additional fields for storing additional information that were not subject to the interpretation of the neural network.

Data obtained through TIC analysis is fed directly into the database, and the ANN makes use of it in its interpretation. Data can also be recorded on standard forms and later fed into the database. Dynamic input fields can be organized according to the immediate needs of training the neural network, changed during the evolution of a study. The interface has been structured so that the options are limited by subsequent logical steps – fields that are not subject to certain immediate

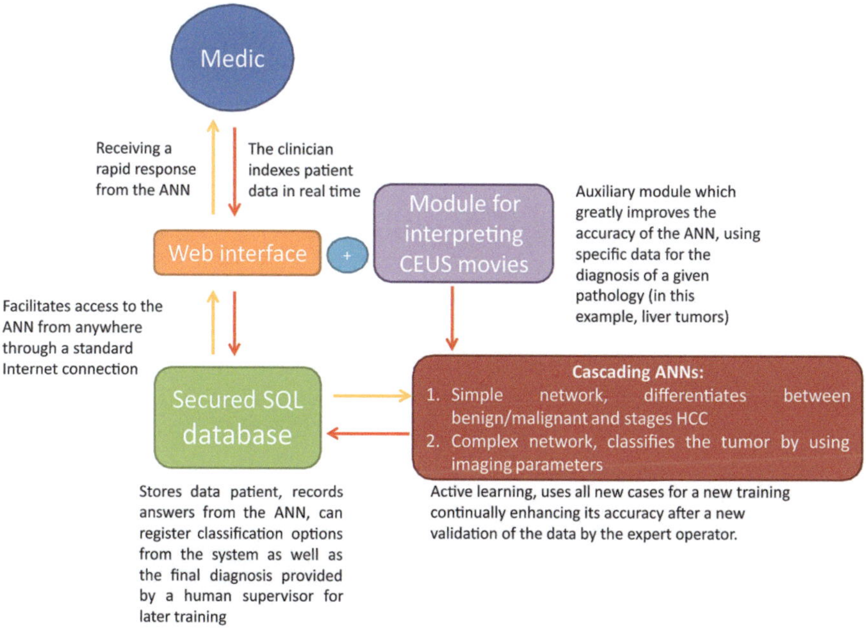

Fig. 16.5 Schematic representation of the computerized diagnostic system based on ANNs and the TIC analysis integrated module

interaction (e.g. if a patient does not have liver cirrhosis, the user cannot enter data on the Child-Pugh score) are inactive for the user. This method of data validation eliminates errors inherent during data recording.

In a final stage of development a web interface accessible from within an Internet browser can be added. It realizes the communication between the doctor and the database which takes the inputs and outputs of the neural network. Between the database and the ANN, there is a two-way communication, the results of newly classified tumours are stored, and after a subsequent validation by the clinician, the results can enter the ANN, thereby increasing diagnostic sensitivity considerably through another training cycle. The major advantage of using a web interface for controlling neural network is the ability to use the ANN diagnostic system at any time and from any geographic location where Internet access is available for entering all or part of the data (Fig. 16.5). The peculiarity of this system, which distinguishes neural networks from any other statistical method, consists in particular plasticity, the ability to learn actively or passively from the new problems which are offered.

For staging of HCC cases, we used the diagnostic algorithm proposed by the Barcelona group study of the liver (Barcelona Clinic Liver Cancer, BCLC), which was translated into a format easier to apply the computer system (Fig. 16.6). If the liver tumour was determined by the ANN to be HCC, the diagnostic system

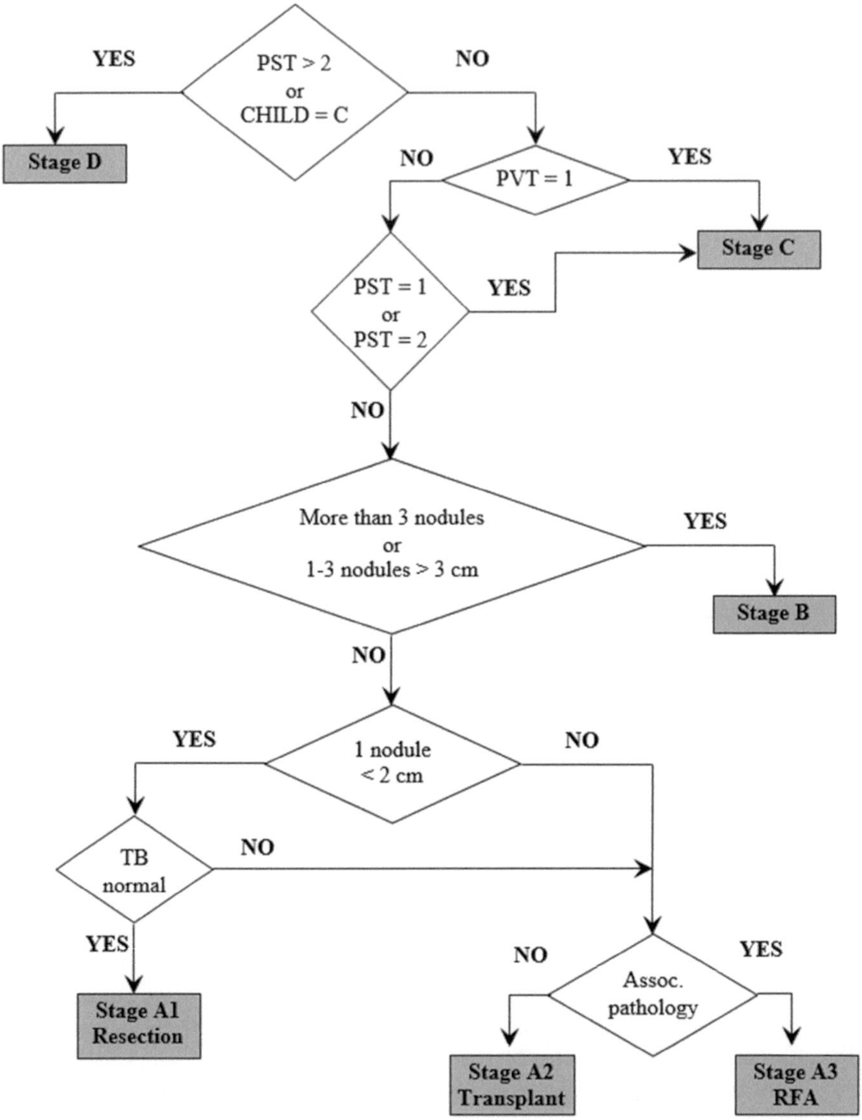

Fig. 16.6 Algorithm used in the computerized model to stage HCC cases. It follows the principles set by the BCLC set of diagnostic and prognostic criteria (European Association for the Study of the Liver and European Organisation for Research and Treatment of Cancer 2012)

proposes a possible classification according to these criteria, and a complete indication is provided to the clinician.

The final step was to design, train and test a neural network ensemble able to receive an extensive panoply of clinical and paraclinical parameters. They offered a diagnosis of certainty, prognosis and staging of HCC, completed by the complex imaging investigations described above. Moreover, computer-aided diagnosis system was able to perform a differential diagnosis for five distinct classes of focal liver tumours identifiable by imaging: hepatocellular carcinoma, hypo- and hypervascular metastases, hepatic haemangiomas and focal steatosis.

A recent study conducted by Cucchetti (Cucchetti et al. 2010) and the HCC Bologna study group examined the capability of a neural network based on clinical and paraclinical parameters used in pretransplant or liver resection for noninvasive assessment of the degree of differentiation of the tumour and microscopic vascular invasion compared with classic logistic regression used in current practice for such assessment. The study demonstrated the absolute superiority of the neural network to the simple statistical techniques, having predictive values of 0.94 and 0.92, respectively, comparing to only 0.85 for logistic regression. Parameters introduced into the network resulted from noninvasive techniques, being represented by the serum level of alpha-fetoprotein, the number, volume and diameter of tumour formation (Cucchetti et al. 2010). In the described ANN model, clinical and paraclinical parameters are diversified, the intention being to diagnose the underlying disease and not to assess the morphological characteristics of tumour. It is interesting to note that, with available pathological data for the set of cases used in training, the network developed here would have been able to provide the same results. In addition, the imaging parameters resulting from the CEUS investigation provide additional information and can help to more accurately establish the stage of the tumour.

The system is constantly evolving by its very nature; any new case presented may undergo external validation of the clinician, thus adding a new set of training. This ensures safety while increasing diagnostic accuracy, eliminating sources of error. It is also noted that the complex system that encompasses imaging parameters exhibits significantly improved capabilities compared to a simple network based only on anamnesis data, clinical and simple imaging (sensitivity and specificity of 97.3 and 97.4%, respectively, compared with 93.2 and 89.7%, respectively). The system is also extremely tolerant and avoids overfitting, as the learning algorithm does not allow strict rules for the chosen weights.

16.4 Automated Computer-Aided Interpretation of Wireless Capsule Endoscopy Recordings

16.4.1 Introduction to Wireless Capsule Endoscopy

Medicine has lately become a domain in which the latest technology is turning out to be not only necessary, but more and more compulsory, starting from the first contact with the patient and case management up to various examination techniques, necessary tests, interventional therapies and post-treatment monitoring. Computers have become deductive and intellectual instruments that represent an integral part in medical system structure, being involved in almost all phases of the medical act.

Wireless capsule endoscopy (WCE) represents a modern investigation technique of the small bowel, being comfortable for the patient, with little constraints and rare complications. This procedure is the result of combining the most recent discoveries in medical and engineering fields, where the latest progress in technology finds its place in medical explorations, leading to a high-tech investigation of the small bowel. WCE delivers a set of over 50,000 images that enable a visual analysis of the intestinal mucosa, thus allowing the identification of possible lesions present at patient's small bowel level.

The rich informational content of this imaging technique represents both an advantage, by emphasizing the interior of the intestine, and a disadvantage, due to the prolonged analysis and interpretation period (for more than 50,000 frames), even for an experienced physician. The minimum time interval necessary for the examination is 3–4 h that adds to the 8–9 h required for the procedure itself, leading thus to a total of more than 12 h dedicated to a single patient. Another disadvantage in the image analysis of the WCE images is represented by the fact that a lesion may be very small compared to the intestinal region present in the image, or it may be incompletely captured, thus being difficult to detect and analysed. In the same time, a lesion may appear in a very small number of successive frames, and it may be overseen by the examiner physician, especially if there are also distraction elements that interfere within the analysis dedicated period.

These disadvantages represent a strong motivation for engineers to approach software applications specific for computer-aided diagnosis, with support and assistance in potential lesion detection, by automatic analysis of the images acquired by WCE and the enhancement of particular elements present in these images. Acknowledging the various types of aspects and the multitude of potential lesions present in the set of frames captured by the WCE, the automatic detection techniques must be designed, developed and applied, depending on each lesion's characteristics as well as on its differences from the normal small bowel mucosa. Thus, in order to help the physician, it is necessary on one hand to have a unique application, which comprises multiple stages of lesion detection, and, on the other hand, to have an intelligent system that is able to analyse the entire set of images, to identify frames that have a different aspect from the normal mucosa and to classify potential lesions, offering as output a set of images corresponding only to lesions, together with their classification and the time of appearance.

The aim of our efforts is the reduction in the overall analysis time necessary for the physician in the interpretation of the images provided by the WCE, as well as improving the accuracy of intestinal lesion automatic detection, offering computer-aided support in diagnosis.

Currently, WCE examination represents the "golden standard" in assessing lesions present in the interior of the digestive tract, in the same time offering a set of images that are difficult to obtain with other techniques of exploration. It represents an essential and important method in the exploration of the entire digestive tract, thus confirming one more time the symbiosis between medicine and technology in the field of gastroenterology by enhancing the exploration with software applications that offer assistance to gastroenterologists, in the assessment and evaluation of images provided by the WCE. In the same time, this technique offers decision support in image processing in a shorter period of time and with better performances.

16.4.2 Brief History of the Wireless Capsule Endoscopy

Three decades ago, an idea of a medical procedure that offers information and images from the interior of the digestive tract first appeared. It aimed in principal the exploration of the small bowel, which is a more difficult segment to explore with other investigation techniques. The first experiments took place later, when technologic progress in this field had made possible a series of trials that led to the development and improvement of this technique.

In 1996, Swain Paul conducted a series of experiments achieving the first image transmission from inside a pig's stomach (Gay et al. 2004). The concept of endoscopic capsule has materialized after collaboration with Ing. Gavriel Iddan, PhD., from Israel, who is considered the inventor of this miniature device (Meron 2000). The development of this new exploration technique was first announced in 2000 in Nature; in the same time, there were conducted the first studies on animals (Iddan et al. 2000).

The first videocapsule was launched in 2001 by Given Imaging Ltd. (PillCam SB), together with the first clinical trials conducted on human subjects (Kornbluth et al. 2004; Mackiewicz 2011). The device was approved by FDA (American Food and Drug Administration) at the beginning of August 2001, being recommended as a supplementary method of exploring the small bowel, next to other endoscopic and radiological techniques, but not necessary a replacement for these. In 2003, WCE examination was directly recommended for the visualization of the intestinal mucosa, being used as a technique for detecting small bowel lesions. The next decade was very active, with over 100,000 examinations conducted and almost 1000 papers published (Meron 2000).

The first videocapsule produced by Olympus (EndoCapsule) first appeared in October 2005. Its characteristics were similar to those of PillCam, having an acquisition rate of two frames per second and the same resolution. It had a higher quality provided by its acquisition system – CCD (charge-coupled device) instead of CMOS (complementary metal oxide semiconductor). EndoCapsule had also an automatic system to control brightness (ABC, automatic brightness control) adapted from classical endoscopic equipment, as well as the possibility to view the captured images in real time (Mackiewicz 2011).

In 2007, IntroMedic introduced a new videocapsule (MicroCam). It is shorter than the other existing videocapsules, having only 24 mm and a 150° capture angle. The lifespan of its batteries is about 11 h, in which period it can record around 120,000 images with a 320 $\times$ 320 pixel resolution, at an acquisition rate of three frames per second. This videocapsule is superior to other types by the quantity of data it can gather in a same time period, thus increasing the chances to record images of all lesions. On the other hand, the necessary time to analyse all the recorded frames is increased.

OMOM videocapsule was developed in 2008 (approved in France and the USA), bringing something new in this field – data transmission in both ways – allowing the control of image acquisition rate and allowing to adjust brightness (Mackiewicz 2011). The lack of speed and motion direction control increases the risk to modify the rate of acquisition and brightness, in the detriment of best level and conditions to capture images.

Regardless of videocapsule producer, one of the main disadvantages of this investigation technique is the absence of a self-propulsion system, which would allow speed and motion direction control. There are several studies that suggest different solutions, based upon magnetic fields (Gao et al. 2010; Kosa et al. 2008), or on other technologies (Lenaerts and Puers 2006; Quirini et al. 2007), but nothing has materialized yet.

16.4.3 Wireless Capsule Endoscopy Procedure

WCE investigation is a modern noninvasive technique of diagnosis, which allows a complete exploration of the small bowel without patient sedation. The WCE investigation system consists of a capsule which will be swallowed by the patient, a set of sensors attached on patient's abdomen or chest, a small digital recorder attached on a belt that is carried by the patient and a working station (which has evolved in time from a classic desktop to laptops and tablets). This working station has a software application which can take recorded images and store them for further analysis (Vere et al. 2009, 2012a, b).

The videocapsule has an image acquisition rate of 2–3 frames per second (depending on producer), and after activation it transmits about 50,000 images, during the lifespan of its batteries, which can vary from 8 to 11 h (Hadithi et al. 2006). At the beginning of the examination, the videocapsule is unsealed thus becoming activated and is swallowed by the patient with a glass of water. It has a diameter of 11 mm and a length which varies from 26 to 27.9 mm resembling a normal tablet. Once activated, it starts recording images, thus the examination must begin immediately. Normal peristaltic movement of the small bowel propels the videocapsule that records images from its interior. It is naturally eliminated after 1–7 days from ingestion.

The WCE examination technique can be resumed in some important stages: activation and swallowing of the videocapsule, acquisition of a set of frames from the interior of the digestive tract (Fig. 16.7), transmission of recorded images to the external working station, image processing and analysis, in order to detect potential small bowel lesions. In present, WCE system is the most common technique of small bowel exploration, being the only method almost without discomfort for the

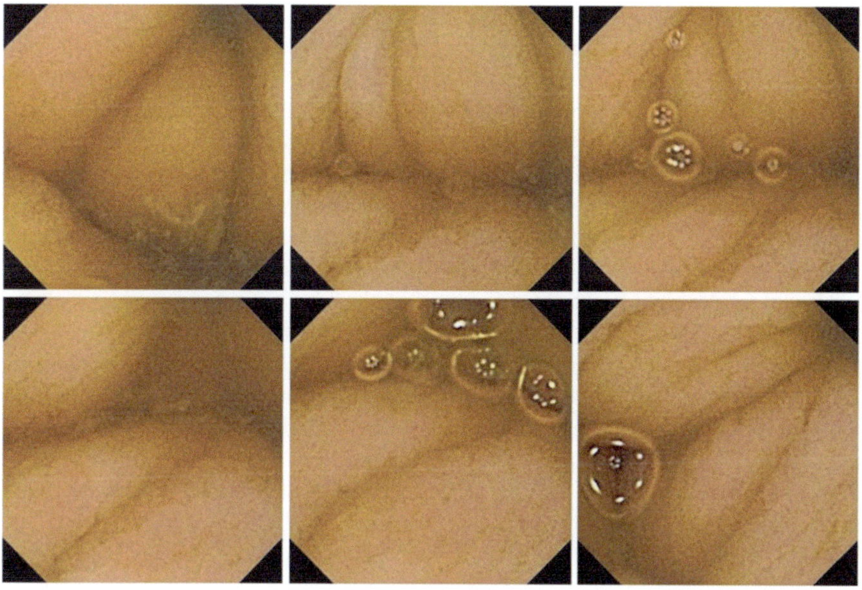

Fig. 16.7 Six successive frames (that present only normal intestinal mucosa, without pathology), acquired during wireless capsule endoscopy

patient, noninvasive and with rare complications. Having an acquisition rate of two or three frames per second that offers a high amount of information, over 50,000 images, it has become the "golden standard" in assessing potential lesions of the small bowel, allowing access to a set of images otherwise difficult to obtain using other exploration techniques.

16.4.4 *Software Applications Used for Automatic Analysis of WCE Results*

Any exploration technique, imagistic or not, is analysed from both patient and physician perspectives. From a patient's point of view, WCE examination is preferable, being comfortable, without anaesthesia and requires only a previous preparation of the small bowel, followed by the ingestion of the videocapsule. Patient's comfort is incomparable relative to classic endoscopic techniques for the extremities of the digestive tract (gastroscopy and colonoscopy), which have a higher degree of discomfort during the investigation. For physicians, WCE examination offers a high amount of information, providing clear images from the interior of the small bowel, thus facilitating the visual analysis of the intestinal mucosa and the detection of potential lesions.

From the medical point of view, the aim of WCE examination, regardless of the digestive segment it was designed to explore, is the identification and classification of lesions present in the captured images acquired by it during its voyage in the interior of the digestive tract. This requires attention and concentration for about 3–4 h from an experienced examiner. In order to reduce the time period necessary for the analysis of the images recorded by the WCE, there have been developed a series of software applications that assist the physician and reduce the amount of time required to examine the whole set of frames. These applications work by automatically analysing the images, removing unnecessary frames or frames with irrelevant content and highlighting sequences showing potential intestinal lesions.

Another advantage of the computer-assisted image analysis of frames recorded by the WCE is the improvement of clinical diagnosis. Thus, the software becomes an expert system able to establish patient's diagnosis (Mackiewicz 2011). Nonetheless, the physician has the final word in confirming or infirming the lesions indicated by the software application. Beside the long period of time required for image analysis, the speed which the frames succeed can be sometimes problematic especially for lesions captured in a reduced number of frames or viewed incomplete.

From a human point of view, the speed at which WCE captured frames succeed is relatively constant, and it can be improved only until a certain point (achieved after continuous exercise). On the other hand, the speed at which software applications can process images depends only upon the processing power of the computer which performs the analysis. Current processors can perform an impressive number or operations per second – over 12 digits – thus exceeding any human ability of analysis. These applications not only reduce the amount of time necessary to process the images generated by the WCE, but in the same time, they amplify WCE's accuracy. If, in general, a human examiner has a tendency to rapidly overlook certain frames or to perform a superficial analysis of some images, a software application assures objectivity in analysing each image assuring the same steps processing and interpretation.

16.4.5 Important Elements in WCE Software Analysis

16.4.5.1 WCE Illumination

According to its technical characteristics, the wireless capsule is equipped with six white light LEDs, which provide the necessary illumination for image acquisition. This normally ensures the visibility of the field corresponding to the CCD camera, characterized by a wide angle of 145° and a depth between 0 and 20 mm (Olympus 2007). All sources based on LEDs have an efficient light capacity, determined by the ratio between the illuminance and the associated power. The illuminance mainly specifies the light quantity expressed in lumens (lm), measured independently from the light distribution direction. The illumination of the digestive tract is

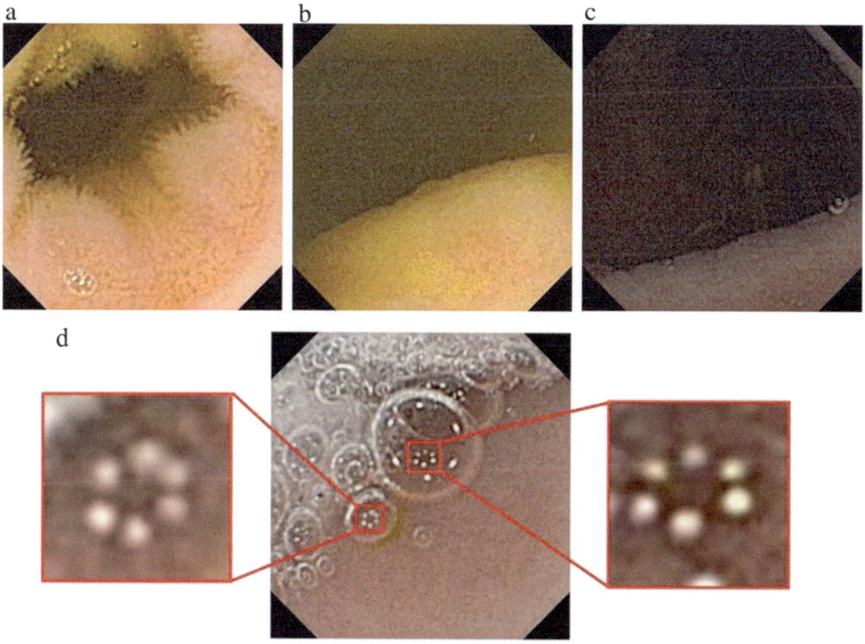

Fig. 16.8 (**a–c**) Different illumination degrees in WCE frames. (**d**) Reflection of the six LEDs within WCE images

thus dependent on the uniform distribution of the illuminance upon the intestinal mucosa surface, being mainly influenced by the fact that the tissue is not flat and, instead, it presents plicae of different dimensions. Also, the capsule is not always parallel with the tract central axis; thus the projection angles differ in accordance with its position (Mang et al. 2008).

The luminance quantifies the light appearance of an object or a surface. The entire section of the digestive tract located near the capsule absorbs a part of the illuminance, reflecting the residual part (Filip et al. 2011). Thus, the two main features that characterize the intestinal mucosa from a visual point of view – colour and texture – define the quantity of absorbed light and the quantity of reflected light. The human examiner apprehends the brightness of an image relative to the quantity of reflected light, expressed by the illuminance. A particular case of images acquired by WCE presents relative straight sections of the digestive tract, with a lumen wide enough to let the light be dispersed, allowing only a small quantity of light to be reflected back by the intestinal tissue (Fig. 16.8a–c). Even in these conditions, the illumination must be powerful enough to acquire in optimum conditions at least a section of 10 mm (in case there is no solid tissue to reflect back the emitted light).

Generally, the spatial distribution of light produced by the WCE's LED sources is not distributed in a uniform manner in the foreground, which leads to an image acquired with a lower quality relative to the capacity of the acquisition system. Due to the physical characteristics of the digestive tract, the luminance varies quantitatively from one section to another. Thus, the LEDs are circularly disposed around the image acquisition system, in order to improve the degree of uniformity of the light produced. Moreover, this disposition should compensate the drawback of the LEDs' size, which should be small enough relative to the size of the capsule (in order to maintain its global size as small as possible), but in the same time they should offer a proper light that ensures the acquisition of good quality images, useful for the examiner (Aihara et al. 2011).

In case certain solid elements from the front end of the capsule present a smooth and glare surface (like air bubbles), the quantity of absorbed light is minimal; thus the reflected light offer a visual aspect easy to identify, being similar to another light source. In fact, these elements placed right in front of the capsule reflect exactly the six LEDs circularly positioned (Fig. 16.8d).

These reflections may influence the lesions' automatic detection, being misidentified as potential lesions (potential ulcerations that are characterized by circular shapes, bright colours, undefined edges and normal mucosa in its neighbourhood). In the same time, they might be useful in the automatic detection process, by allowing a differential diagnostic between air bubbles that reflect the LEDs and polyps that also have a round shape, but do not present such reflections. They may be eliminated from the frames, either alone or together with the air bubbles that contain them.

16.4.5.2 Artefacts

There are cases when the physical characteristics of the small bowel, as well as other different elements (artefacts) found inside it, are misconsidered as mucosa anomalies or even lesions (especially polyps, due to the round shape of intestinal bubbles). Debris and bubbles are the most common elements present in the images acquired through WCE, influencing the identification of potential lesions, as they do not reflect relevant data, but they affect the global features of the analysed images. The intestinal mucosa presents constant intensity, colour and texture, while the areas containing artefacts present sudden changes in contrast, due to well-defined edges, shadows, LEDs' reflections for bubbles but also colour and texture for debris.

Lumen, debris and bubble identification from the images acquired through WCE represents an important phase that leads to a set of images free of artefacts, reducing thus the risk of indicating them as potential lesions through the use of automatic detection applications.

For each artefact potentially present within the images acquired by WCE, its associated general characteristics were determined. Lumen detection was best realized based on colour features – being characterized by darker shades relative to the rest of the informational content of all frames – varying in the interval brown to dark brown. The colour corresponding to the lumen is progressively changed from the surrounding normal mucosa, thus the edge detection is not really effective unless the luminance of all areas in the frame is relatively uniform. The texture has a lower weight in lumen automatic detection.

Debris present a similar behaviour, being mostly identified based on their specific colour (relatively well delimited in the colour palette encountered within WCE images), as well as its texture. The edges have sometimes a lower weight in their automatic identification, especially when the passage towards the intestinal mucosa is progressive; thus the content of debris is not very compact.

Air bubbles are best represented by their well-delimited margins, the best results in their segmentation being obtained after applying edge detection techniques. In some cases, favoured also by their close position relative to the wireless capsule or in an optimum angle of illumination, their contour turns out to be thick, leading to double edge detection; this represents a unique feature and ensures the differential diagnosis relative to intestinal polyps. On the other hand, their interior is transparent, and it allows the visualization of intestinal mucosa placed behind them even if its aspect is rather unclear. From a colour point of view, they do not present a constant palette, which gives this feature a lower weight in the automatic detection process. Their texture has similar weight, being useful only for characterizing the interior generated by double edges. Overall, the correct identification of air bubbles is performed especially through an optimum combination of detected edges and the interior with normal mucosa aspect that adds to the pattern composed by the reflection of the six LEDs.

Since there is no unique feature that uniquely identifies artefacts, the solution implies the creation of a feature combination, texture, colour, edges and specific elements, according to the weight of each feature in the automatic detection of a certain artefact. Also, in some cases, one should take into account the invariance of attributes to light variation (Hansen and Gegenfurtner 2006; Saarela and Landy 2007). For a better visual analysis, the segments of the original image were superimposed over the original images, for eventual comparisons and an objective evaluation of the results (Fig. 16.9).

The process of artefacts removal is not simple, due to the potential unstable state of the image afterwards. All artefacts include colour and the associated shades in their detection process, more or less accurate. This process involved an analysis at pixel level; thus the removal of pixels with colours corresponding to their palettes may lead to "holes" in the images and isolated regions that lead to a loss of coherence within the WCE frame. This incoherent state may be obtained by removing the bubble contours (as well as the reflections of LEDs).

Moreover, the separation area between sections in original images and the areas created following the removal of artefacts will not be clearly emphasized, which will lead to an increase of uncertainty in the following stages of lesion detection. A potential solution for this problem is represented by the erosion morphological operator, when the original image section between two or more regions is small enough to be completely absorbed within the artefact section and thus subsequently removed, without risking losing relevant information. But erosion also is applied at pixel level, which does not always lead to proper expected results.

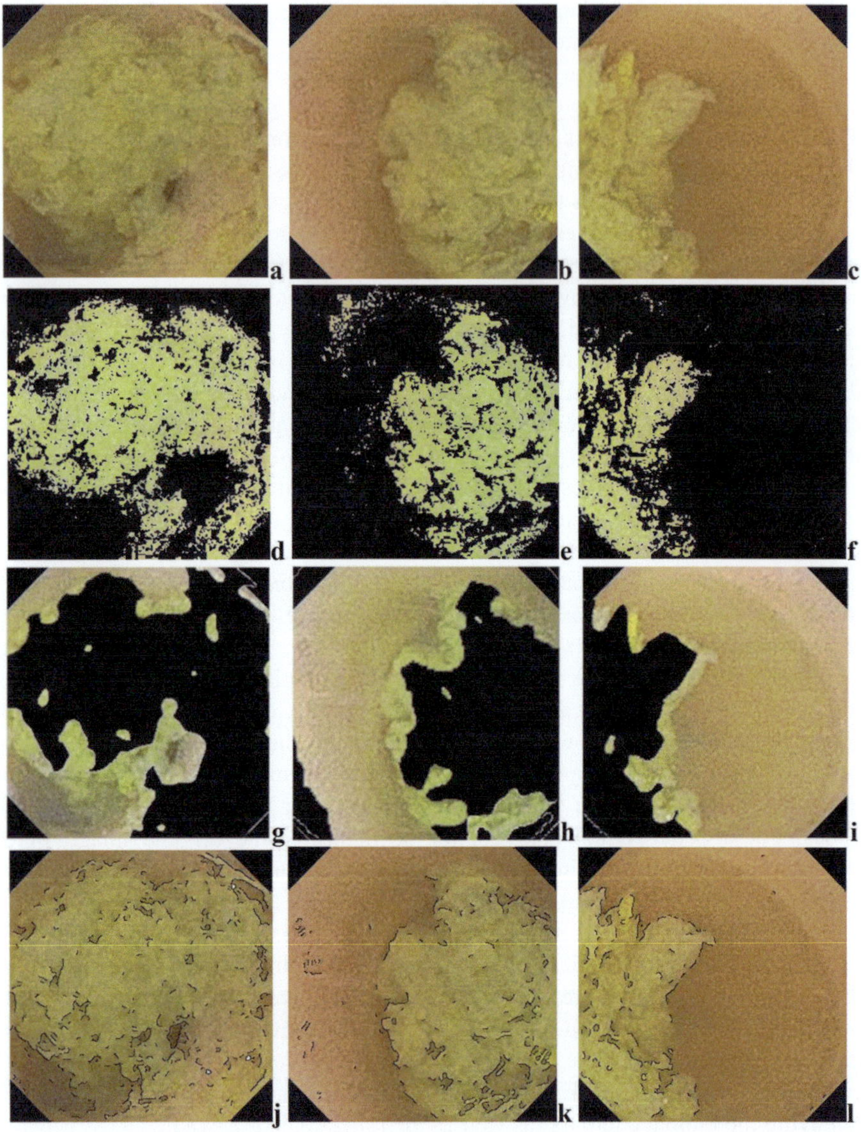

Fig. 16.9 (**a–c**) Three WCE frames presenting intestinal debris; (**d–f**) segmentation based on colour; (**g–i**) debris texture identification using Gabor filters; (**j–l**) edge detection based on Canny edge detector (with $\sigma = 1.4$)

Due to these reasons, artefacts removal must be performed in a controlled manner, taking into account the maintenance of a clear region of the original image, with a good delimitation from the removed areas, in order to ensure a good analysis and detection of potential intestinal lesions. Thus, the region specific to artefacts must be removed based on general characteristics previously identified, but applied to the same area in the original image. This was achieved by overimposing regions with features specific to artefacts and separating those regions that correspond with a high probability to these artefacts. Colour, feature and edge detection were identified in three different images that were subsequently applied as marks upon the original image. The delimitation area was mainly represented by texture, as it presents a reasonable border compared to surrounding regions. Each area obtained following this process was reanalysed, based on specific features, for the detection of a potential artefact and its associated classification.

16.4.5.3 WCE Frames – Similarity Analysis

From the moment of the ingestion, the wireless capsule begins its journey inside the patient's digestive tract, constantly acquiring, every half second, a snapshot of its current location. The content of these frames also reflects the movement of the capsule. A higher movement speed implies a greater variation of the informational content of WCE frames. Vice versa, a lower speed amplifies the degree of similarity between two or more successive frames, all being correlated with the images acquisition rates.

In order to reduce the analysis time of an entire WCE movie, acquired while the batteries were still active, a frame almost identical with the previous one may be eliminated from the activities cycle, considering that it has the same features as the previous one; therefore it does not bring any new information or other benefits for the automatic lesion detection process. Moreover, this phase allows an efficient use of the available resources and the access period, implicitly leading to an optimization in the global processing time.

Regarding the informational content of a frame, two images are similar if they present objects, elements or scenes from the same category, or they represent different perspectives of the same object or scene, but acquired in different acquisition conditions (Tirilly et al. 2010). Thus, image similarity may reflect an object equivalence or a category equivalence.

The main techniques for defining image similarity may be based on the representation method, or the definition of a proper comparison technique between images, in a special representation space (Goldberger et al. 2003; Chen and Chu 2005; Chechik et al. 2010).

Imagistic differences between two successive WCE images are given by different acquisition conditions, generated by the continuous movement of the capsule. Even so, the speed of the acquisition system may compensate the physical movement (combined in the same time with the peristaltic movement of the small bowel), so that a series of pairs of successive frames may reflect a higher degree of similarity, through a higher acquisition rate. This notion expresses from a quantitative point of view, the image equivalence evaluated according to a set of

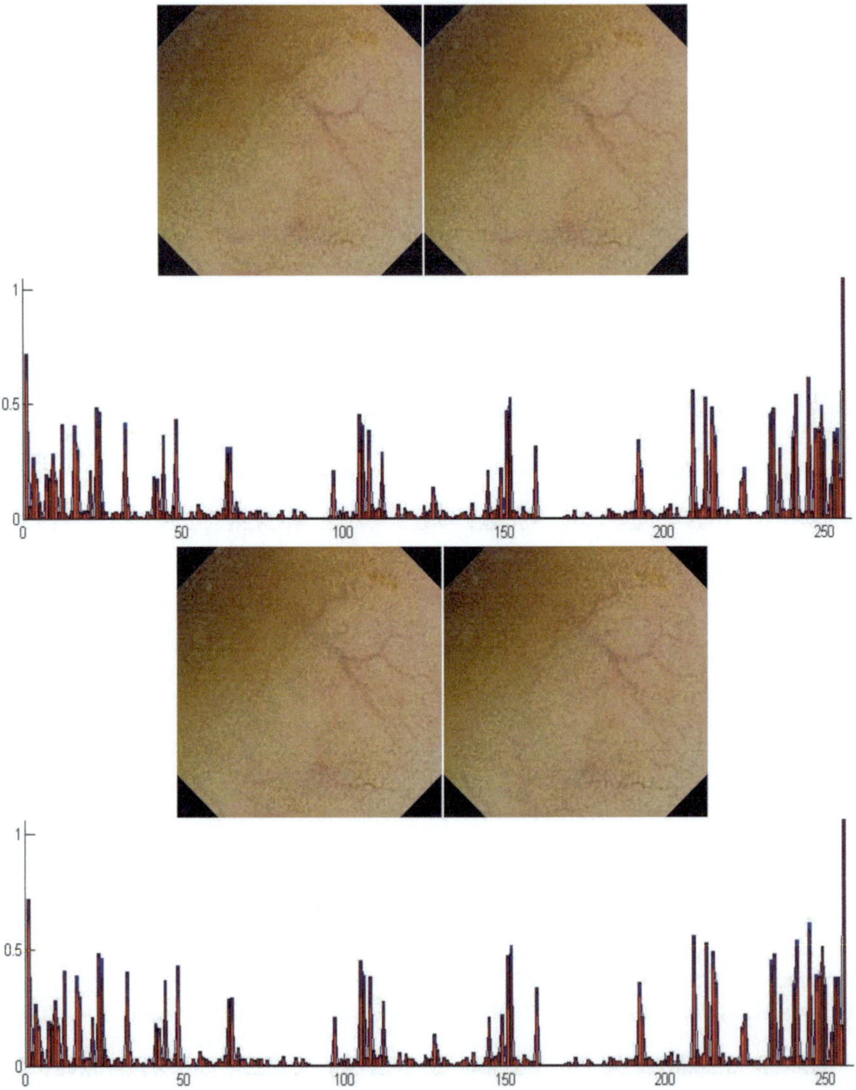

Fig. 16.10 Colour and texture histograms for two pairs of similar WCE frames

predefined criteria. The image similarity is defined according to the necessary data for the evaluation of image descriptors and corresponding measurements.

In a global analysis process of multiple WCE movies, the category equivalence allows the classification of images in groups associated to the sections of the digestive tract (oesophagus, stomach, duodenum, small bowel, colon – if necessary) or associated to specific pathology. On the other hand, the similarity between WCE frames, computed in order to reduce the analysis time, is based on object equivalence – if two successive frames display exactly the same area of the intestinal mucosa, then a single analysis is sufficient to extract the informational content.

The similarity between two WCE images may be computed based on the difference between their histograms – the smaller the difference, the higher the similarity degree. Both colour and texture may be represented as histograms (Fig. 16.10).

Colour histograms represent the distribution of colours within an image, being defined as the number of pixels whose colour corresponds to the elements in the list of shades associated to the image colour space (comprising all colours within an image). Texture histogram requires a method of expression in a quantitative manner. An optimum method is LBP (local binary pattern). For each pixel within the image, the LPB is computed, according to the chosen values for neighbourhood and radius.

In case there is a uniform pattern present, the number of elements in the corresponding category is incremented. Otherwise, the number of elements in the non-uniform category is incremented. Those categories will subsequently represent the basic classes for computing the associated histogram.

Histograms present the following drawback: it is known that similar images have similar histograms. However, in the same time, different images may also have similar histograms. Basically, both images may have the same colour distribution, but content totally different from an informational point of view. In this case, an analysis must be performed regarding the WCE frames and the applicability of histograms for the necessary time reduction in the global analysis process, taking into account that we do not employ a content-based image retrieval approach, but an approach related to an imagistic equivalence:

- WCE images are acquired in the same global context.
- Similarity analysis is only performed for successive images or, at most, within a sequence of 5–6 images (basically, multiple similarity analysis may be executed within a complete sequence).
- All images are acquired with the same camera.
- Images comprise relatively the same informational content that cannot be radically changed from one frame to another (except in a small number of cases, relative to the size of an entire WCE movie).
- In any moment, the next frame will mainly present the same area of the digestive tract, eventually with a different brightness level.

Basically, the purpose of using histograms is not represented by the identification of a similar image or the identification of the most similar image within a database of images; the purpose is to define the degree of equivalence between two frames from the same movie sequence.

According to experiments performed on sets of WCE images, the colour histogram reflected better the differences and similarities between successive frames. For successive similar images, the removal should take place only for images reflecting normal intestinal mucosa or artefacts clearly identified, reducing thus

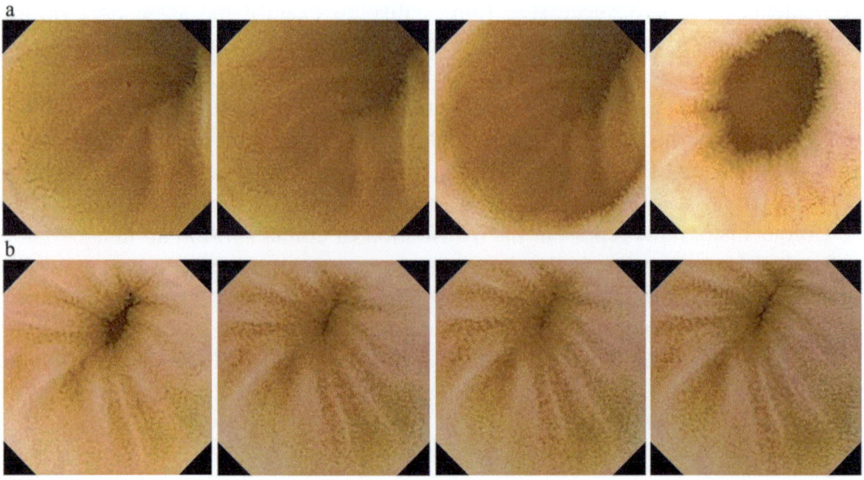

Fig. 16.11 Eight successive WCE images that emphasize the peristaltic motion of the small bowel

the analysis time allocated for the entire WCE movies. For lesion sequences, the similarity analysis is not performed in order to remove those frames from the sequence, but to confirm once more the presence of that lesion in the sequence, which may represent an extra certainty degree regarding the complete identification of appearance/disappearance sequences.

16.4.5.4 Motion Analysis Within WCE Movies

Motion analysis within WCE movies implies not one but two different perspectives, as the wireless capsule moves while it acquires images, but in the same time, the digestive tract is in a continuous motion that represents the reason of the capsule motion. Thus, every frame is influenced by two elements that are simultaneously moving, ensuring implicitly the variability of the acquired images.

Figure 16.11 emphasizes, in a few successive frames, the peristaltic motion of the small bowel, while the capsule is propelled forward.

Block matching algorithms represent one of the most known methods of motion estimation by evaluating the similarity between two images, starting from the premise that a block of pixels has the same translation movement from one frame to the next. The image extremities are not taken into account. The block sizes may vary, according to imposed requirements, but they may not exceed certain limits. Based on a subset of 20 sequences extracted from a set of WCE movies, an optimum size of 35 pixels was defined for the matching block – with a possibility to decrease until 25, in case the important element covers a smaller area within WCE images (Fig. 16.12).

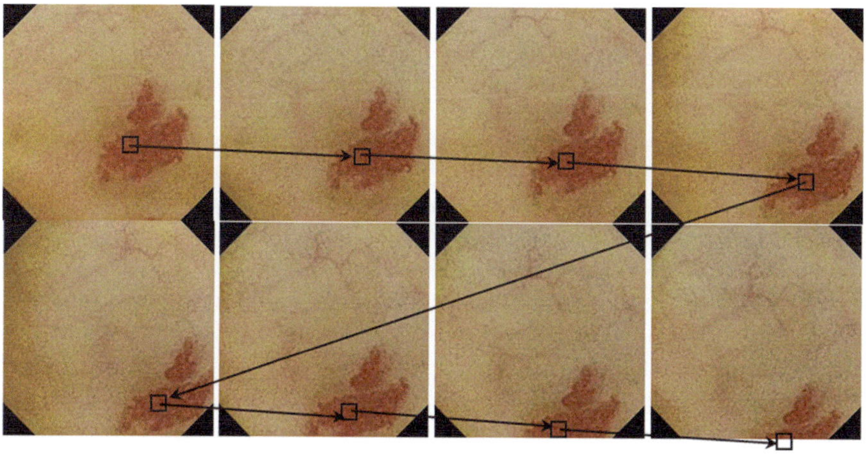

Fig. 16.12 Motion direction of a matching block identified within a telangiectasia lesion

The choice of the matching block depends on the main purpose of building the motion sequence. For sequences of frames with normal intestinal mucosa, the block may represent any area within this image. For sequences containing artefacts or lesions, the block must reflect a region of them.

Building sequences that contain artefacts (and the moments of appearance, disappearance) may be performed in a similar manner, being useful in the correct removal of those artefacts from WCE images.

Motion analysis is useful in constructing sequences of lesions – sets of images that present the same lesion, from different perspectives, from the moment it is first captured and the moment when it is no longer visible in WCE images. The lesion sequence is build based on similarity, the characteristic features useful in the automatic detection and classification processes, and the motion direction of the wireless capsule and the associated movement context. Considering the fact that the capsule does not have its own propelling system, there is no control over the motion itself, which may lead to sudden direction changes from one frame to another. Even so, the acquisition rate compensates most of the times these changes that are not always obvious in the WCE movie.

16.4.5.5 Telangiectasia Lesions

Automatic detection of telangiectasia lesions implies the analysis of WCE frames following the phase of artefacts removal. Compared to artefact analysis, the entire approach is changed, so that the particular identified elements are not removed from the image, but instead they are analysed and classified, in order to increase the accuracy of the diagnosis procedure.

The main criterion used by physicians in the evaluation of telangiectasia lesions when defining a diagnosis is represented by colour. These lesions are characterized by red shades, different from the colour palette of WCE images presenting normal intestinal mucosa, with no pathology.

The majority of studies upon this matter also use colour as main characteristic specific for vascular lesions. Lau et al. have proposed a detection method based on a combination of characteristics at pixel level: R value of every pixel from a WCE image (limited by two fixed thresholds), RGB characteristic triplet, mutual information given by the difference between triplets corresponding to pixels in two different images as well as the associated HSV triplet. Contrast segmentation and threshold values were added to this feature set. The authors explained that the experiments were performed on a set of images presenting occult gastrointestinal bleedings, without offering supplementary statistical data (Lau and Correia 2007; Bourbakis et al. 2005; Karargyris and Bourbakis 2008). Both Lv et al. and Mackiewicz et al. used colour information expressed as histograms, based on the colour distribution within images, the classification being subsequently performed with SVM elements (Mackiewicz et al. 2008; Lv et al. 2011). Following the experiments, both studies obtain sensitivity and specificity values between 94 and 97%, but for relatively small sets of images.

Other studies analysed different colour systems and expressing the colour of vascular lesions through these systems, computing Euler or Euclidean distances, angular vectors and covariance matrices, the image classification process being based on weights or neural networks (Signorelli et al. 2005). Pan et al. extracted a series of colour and texture features that were subsequently classified using a back-propagation neural network, obtaining a sensitivity of 93% and a specificity of 96% (Pan et al. 2009). Later on, they have performed experiments on the same data set, but with a different probabilistic neural network, obtaining a sensitivity of 93.1%, but a lower specificity value of 85.6% (Pan et al. 2010). A higher value of specificity – 97.97% – was obtained by Shah et al. by combining colour characteristics expressed using the HIS system with image region segmentation, in the detriment of sensitivity – 70.96%; the study lot was relatively small, being composed by 100 images (among which only 50 represented active bleeding) (Shah et al. 2007).

Even though the analysis of vascular lesion colour was thoroughly studied, the relation between their shades and the colour palette of the surrounding normal intestinal mucosa and, in general, the context corresponding to these lesions of the small bowel did not benefit from the same attention over the time. By analysing the individual frame acquired through WCE, the uniqueness of the digestive tract aspect of each patient is lost.

The red shades interval corresponding to potential vascular lesions may be determined using a sample set of telangiectasia images, previously identified by an experienced physician (Fig. 16.13). The final objective is to evaluate the colour interval corresponding to telangiectasia lesions, relative to the colour interval corresponding to normal intestinal mucosa.

For a global evaluation of an entire WCE movie, as well as a proper presentation of the results for the examining physicians, in a concise and optimum manner, a

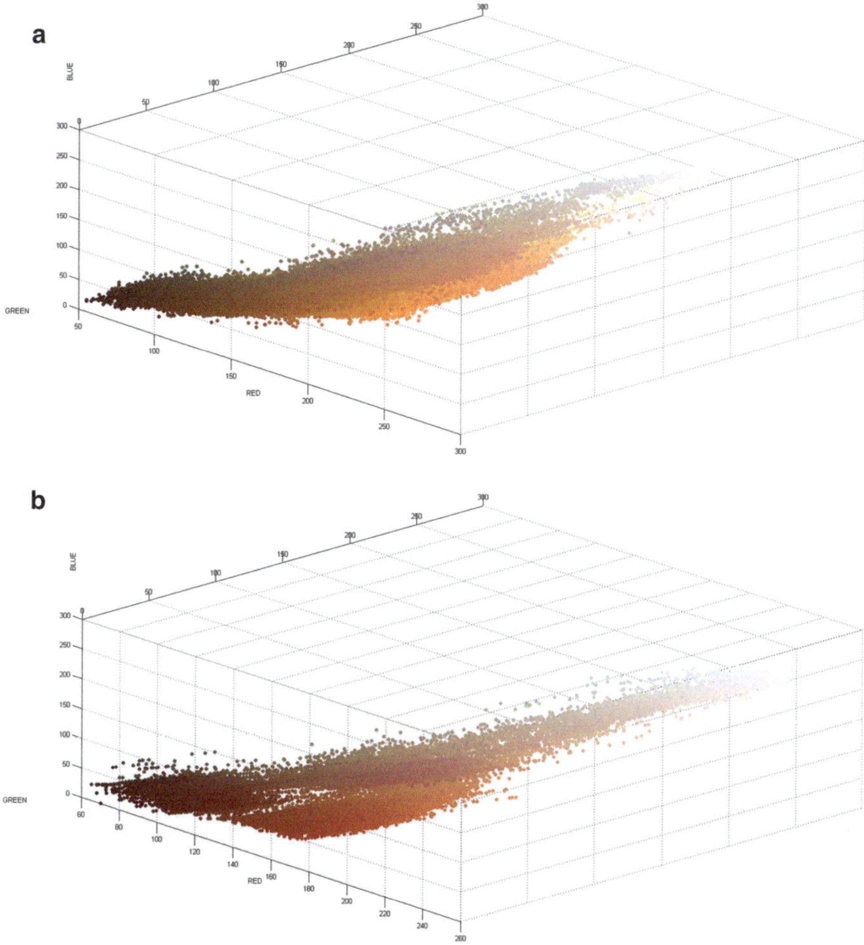

Fig. 16.13 Associated colour interval for: (**a**) telangiectasia lesions and (**b**) normal intestinal mucosa

quantitative analysis of the movie should be performed, expressing colour characteristics and parameters obtained from independent frames.

16.4.5.6 Polyps

The main feature of intestinal sessile polyps is represented by their partial round/oval shape (partial hemisphere), which is visible as a slight contour on the intestinal mucosa (Fig. 16.14). A polyp is a growth of tissue, projecting on the surface of the mucosa. Therefore, its contour is created by the light from the six LEDs of the videocapsule, due to light's reflection on their surface that creates a narrow area of

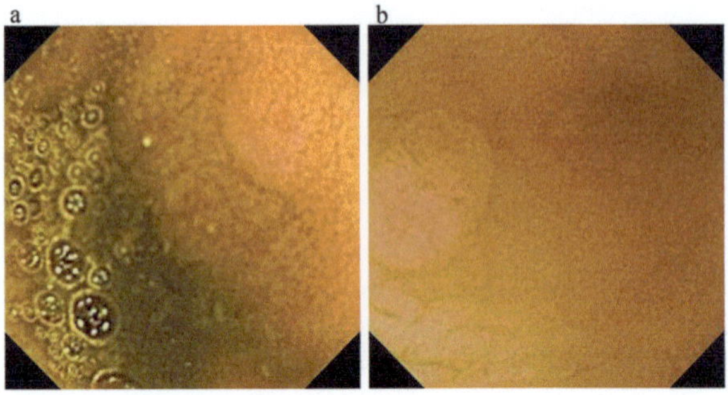

Fig. 16.14 Intestinal polyps present in images acquired through wireless capsule endoscopy

shadow on the intestinal mucosa. By contrast, air bubbles are transparent and allow the visualization of the mucosa through them. Intestinal polyp's structure is dense and does not allow light to cross through.

Depending on a polyp's degree of development and type, the contour can be better highlighted on less of half its circumference (transposed in two dimensions).

The texture of a polyp's surface is not so different from the rest of the intestinal mucosa, which makes their identification difficult within different frames recorded by the WCE, even for experienced physicians. In fact, from an imagistic point of view, polyps are abnormal protuberances of the normal tissue, having similar characteristics with the surrounding mucosa. In some cases, certain polyps present small ulcerations, which give them a different aspect, with a modified texture from normal areas.

According to polyp's size, their apical region can be situated at a smaller distance from the videocapsule's dome and from the intestinal wall, generating a higher reflection of the light emitted by videocapsule's LEDs; thus a brighter area can appear in their central area. This feature is very useful in differential diagnosis process.

The analysis of intestinal polyp's imagistic aspect has revealed as a main feature their specific partial round form. In this case, the main automatic detection method will be focused on contour detection from the original frames provided by the WCE, aiming to identify partial more or less closed contours, which present a curvature specific to intestinal polyps.

Most studies regarding different methods for intestinal polyps detection focus on identifying their curvature. Karargyris and Bourbakis have proposed a synergistic detection method for protuberances inside the small bowel (potential polyps) using a segmentation detector (SUSAN) associated with Gabor log filters and a set of geometric rules (Karargyris and Bourbakis 2011). Studies were conducted upon a set of 50 images, 10 presenting intestinal polyps. Their results showed a sensitivity of 100% and a specificity of 67.5%.

Qian and Meng proposed another technique for the detection of polyp's texture according to LPB (local binary pattern) operator based upon LLE (locally linear embedding), which leads to a detection accuracy at about 97% (Qian and Meng 2011). Other studies were mainly focused on analysing curvatures present in the

WCE image as a main element in polyp identification throughout graphic interpretation of functions, defined in a pixel domain (Brown et al. 2006), or on surface deformation according to physical folds (van Wijk et al. 2006). Other authors obtained good results in intestinal polyp's identification process, by using a series of different characteristics like colour, texture, contour and associated curvature together with SURF (speeded up robust features) descriptors (Zhao et al. 2012; Hwang 2011).

Following a physician's approach in analysing images provided by the WCE, a more appropriate solution in identifying intestinal polyps will take in consideration a combination of their physical characteristics like contour, curvature, colour and texture. Our work begun with a comparative study between three detection methods: Sobel operator, Canny operator and Gabor filter's phase response. Considering that the intestinal mucosa is characterized by the presence of intestinal vilosities, which induce detectable edges at their level, a new preprocessing phase was introduced in order to level the aspect of the intestinal mucosa.

Subsequently, based on the edges detected in images – emphasized by the phase response of Gabor filters – the real contours of the individual elements present in WCE frames were determined (following a process of refinement that removed the fake contour generated by the intestinal mucosa or the noise present in images, induced by the image acquisition system and the wireless transmission towards external recorders). For each independent contour, the associated curvature was determined according to a generated Bezier curve from the points of the contour. If this curve may be fitted in an ellipse with specific characteristics, then the original edge curvature belongs to a polyp (depending on the sizes of the escribed ellipse relative to the sizes of the entire WCE image).

Differential diagnostic may be performed based on texture or brightness that represent extra information useful in establishing a clear opinion.

Acknowledgments The research presented here was partially financed by the following scientific grants: CNCS – UEFISCDI: Partnership project VIP SYSTEM, ID: 2011–3,2-0503, CNCS-UEFISCDI Partnership project, ID: PN-II-PT-PCCA-2013-4-1931 and CNCS-UEFISCDI Partnership project ID: PN-II-PT-PCCA-2013-4-1930.

References

Aihara H, Ikeda K, Tajiri H. Image-enhanced capsule endoscopy based on the diagnosis of vascularity when using a new type of capsule. Gastrointest Endosc. 2011;73(6):1274–9.

Albrecht T, Blomley M, Bolondi L, et al. EFSUMB Study Group. Guidelines for the use of contrast agents in ultrasound. Ultraschall Med. 2004;25:249–56.

Bolondi L, Gaiani S, Celli N, Golfieri R, Grigioni WF, Leoni S, Venturi AM, Piscaglia F. Characterization of small nodules in cirrhosis by assessment of vascularity: the problem of hypovascular hepatocellular carcinoma. Hepatology. 2005;42:27–34.

Bourbakis N, Makrogiannis S, Kavraki D. A neural network-based detection of bleeding in sequences of WCE images. In: IEEE international symposium on bioinformatic and bioengineering, Minneapolis, USA. Oct. 2005. p. 324–7; 2005.

Brown G, Fraser C, Schofield G, Taylor S, Bartram C, Phillips R, Saunders B. Video capsule endoscopy in peutz-jeghers syndrome: a blinded comparison with barium follow-through for detection of small-bowel polyps. Endoscopy. 2006;38(4):385–90.

Bruix J, Sherman M. Management of hepatocellular carcinoma: an update. Hepatology. 2011;53 (3):1020–2.

Catalano O, Nunziata A, Lobianco R, Siani A. Real-time harmonic contrast material-specific US of focal liver lesions. Radiographics. 2005;25:333.

Caturelli E, Solmi L, Anti M, Fusilli S, Roselli P, Andriulli A, Fornari F, Del Vecchio BC, de Sio I. Ultrasound guided fine needle biopsy of early hepatocellular carcinoma complicating liver cirrhosis: a multicentre study. Gut. 2004;53:1356–62.

Chechik G, Sharma V, Shalit U, Bengio S. Large scale online learning of image similarity through ranking. J Mach Learn Res. 2010;11:1109–35.

Chen CC, Chu HT. Similarity measurement between images. in COMPSAC-W'05. In: Proceedings of the 29th annual international conference on computer software and applications conference. p. 41–42; 2005.

Chiu JS, Wang YF, Su YC, Wei LH, Liao JG, Li YC. Artificial neural network to predict skeletal metastasis in patients with prostate cancer. J Med Syst. 2009;33:91–100.

Colli A, Fraquelli M, Casazza G, et al. Characterization of small nodules in cirrhosis by assessment of vascularity: the problem of hypovascular hepatocellular carcinoma. Hepatology. 2005;42:27–34.

Cucchetti A, Piscaglia F, Grigioni AD, et al. Preoperative prediction of hepatocellular carcinoma tumour grade and micro-vascular invasion by means of artificial neural network: a pilot study. J Hepatol. 2010;52(6):880–8.

Dietrich CF. Characterization of focal liver lesions with contrast enhanced ultrasonography. Eur J Radiol. 2004;51S:S9.

El-Serag HB. Hepatocellular Carcinoma. N Engl J Med. 2011;365:1118–27.

El-Serag HB, Rudolph L. Hepatocellular carcinoma: epidemiology and molecular carcinogenesis. Gastroenterology. 2007;132:2557–76.

European Association for the Study of the Liver, European Organisation for Research and Treatment of Cancer. EASL-EORTC clinical practice guidelines: management of hepatocellular carcinoma. J Hepatol. 2012;56:908–43.

Ferlay J, Shin HR, Bray F, Forman D, Mathers C, Parkin DM. GLOBOCAN 2008 v1.2, cancer incidence and mortality worldwide: IARC CancerBase No. 10 [Internet]. Lyon, France: International Agency for Research on Cancer; 2010. http://globocan.iarc.fr. Accessed 10 Apr 2012.

Filip D, Yadid-Pecht O, Andrews CN, Mintchev MP. Design, implementation, and testing of a miniature self-stabilizing capsule endoscope with wireless image transmission capabilities. Int J Inf Technol Knowledge. 2011;5(1):3–24.

Forner A, Vilana R, Ayuso C, Bianchi L, Solé M, Ayuso JR, Boix L, Sala M, Varela M, Llovet JM, Brú C, Bruix J. Diagnosis of hepatic nodules 20 mm or smaller in cirrhosis: prospective validation of the noninvasive diagnostic criteria for hepatocellular carcinoma. Hepatology. 2008;47:97–104.

Gao M, Hu C, Chen Z, Zhang H, Liu S. Design and fabrication of a magnetic propulsion system for self-propelled capsule endoscope. IEEE Trans Biomed Eng. 2010;57(12):2891–902.

Gay G, Delvaux M, Rey J. The role of video capsule endoscopy in the diagnosis of digestive diseases: a review of current possibilities. Endoscopy. 2004;36:913–20.

Gheonea DI, Streba CT, Ciurea T, Saftoiu A. Quantitative low mechanical index contrast-enhanced endoscopic ultrasound for the differential diagnosis of chronic pseudotumoral pancreatitis and pancreatic cancer. BMC Gastroenterol. 2013;13:2.

Goertz RS, Bernatik T, Strobel D, Hahn EG, Haendl T. Software-based quantification of contrast-enhanced ultrasound in focal liver lesions – a feasibility study. Eur J Radiol. 2010;75(2):e22–6.

Goldberger J, Gordon S, Greenspan H. An efficient image similarity measure based on approximations of KL-divergence between two gaussian mixtures. In: Proceedings of ninth IEEE international conference on computer vision. vol 1, p. 487–93; 2003.

Goodband JH, Haas OCL, Mills JA. A comparison of neural network approaches for on-line prediction in IGRT. Med Phys. 2008;35:1113–22.

Grossi E, Mancini A, Buscema M. International experience on the use of artificial neural networks in gastroenterology. Dig Liver Dis. 2007;39(3):278–85.

Guo D, Qiu T, Bian J, Kang W, Zhang L. A computer-aided diagnostic system to discriminate SPIO-enhanced magnetic resonance hepatocellular carcinoma by a neural network classifier. Comput Med Imaging Graph. 2009;33(8):588–92.

Hadithi M, Heine GD, Jacobs MA, van Bodegraven AA, Mulder CJ. A prospective study comparing video capsule endoscopy with double-balloon enteroscopy in patients with obscure gastrointestinal bleeding. Gastroenterology. 2006;131(1):327–9.

Hansen T, Gegenfurtner KR. Higher level chromatic mechanisms for image segmentation. J Vis. 2006;6:239–59.

Huang-Wei C, Bleuzen A, Bourlier P, Roumy J, Bouakaz A, Pourcelot L, Tranquart F. Differential diagnosis of focal nodular hyperplasia with quantitative parametric analysis in contrast-enhanced sonography. Investig Radiol. 2006;41(3):363–8.

Hwang S. Bag-of-visual-words approach based on SURF features to polyp detection in wireless capsule endoscopy videos. In: ISVC'11 Proceedings of the 7th international conference on advances in visual computing. Berlin/Heidelberg: Springer-Verlag. p. 320–7; 2011. Part II.

Iddan G, Meron G, Glu-khovsky A, Swain P. Wireless capsule endoscopy. Nature. 2000;405:725–9.

Ignee A, Jedrejczyk M, Schuessler G, Jakubowski W, Dietrich CF. Quantitative contrast enhanced ultrasound of the liver for time intensity curves-Reliability and potential sources of errors. Eur J Radiol. 2010;73(1):153–8.

Jiang J, Trundle P, Ren J. Medical image analysis with artificial neural networks. Comput Med Imaging Graph. 2010;34:617–31.

Karargyris A, Bourbakis N. A methodology for detecting blood-based abnormalities in wireless capsule endoscopy videos. In: 8th IEEE international conference on bioinformatics and bioengineering. p. 1–6; 2008.

Karargyris A, Bourbakis N. Detection of small bowel polyps and ulcers in wireless capsule endoscopy videos. IEEE Trans Biomed Eng. 2011;58(10):2777–86.

Kim TK, Choi BI, Han JK, Hong HS, Park SH, Moon SG. Hepatic tumours: contrast agent enhancement patterns with pulse-inversion harmonic US. Radiology. 2000;216:411.

Kim SH, Lee JM, Lee JY, Han JK, An SK, Han CJ, Lee KH, Hwang SS, Choi BI. Value of contrast-enhanced sonography for the characterization of focal hepatic lesions in patients with diffuse liver disease: receiver operating characteristic analysis. AJR Am J Roentgenol. 2005;184(4):1077–84.

Kondo C, Kondo T, Ueno J. Three-dimensional medical image analysis of the heart by the revised GMDH-type neural network self-selecting optimum neural network architecture. Artif Life Robot. 2009;14(2):123–8.

Kornbluth A, Legnani P, Lewis BS. Video capsule endoscopy in inflammatory bowel disease: past, present, and future. Inflamm Bowel Dis. 2004;10:278–85.

Kosa G, Jakab P, Jolesz F, Hata N. Swimming capsule endoscope using static and rf magnetic field of mri for propulsion. In: IEEE International Conference Robotics and Automation, ICRA. 2008; p. 2922–7.

Lau PY, Correia PL. Analyzing gastrointestinal tissue images using multiple features. In: 6th Conference on telecommunications Peniche. p. 435–8; 2007.

Leen E, Ceccotti P, Kalogeropoulou C, Angerson WJ, Moug SJ, Horgan PG. Prospective multicenter trial evaluating a novel method of characterizing focal liver lesions using contrast-enhanced sonography. Am J Roentgenol. 2006;186:1551.

Lenaerts B, Puers R. An omnidirectional transcutaneous power link for capsule endoscopy. In: Proceedings of international workshop on wearable and implantable body sensor networks. 2006; p. 46–9.

Lencioni R, Piscaglia F, Bolondi L. Contrast-enhanced ultrasound in the diagnosis of hepatocellular carcinoma. J Hepatol. 2008a;48:848–57.

Lencioni R, Piscaglia F, Bolondi L. Contrast-enhanced ultrasound in the diagnosis of hepatocellular carcinoma. J Hepatol. 2008b;48:848–57.

Lisboa PJ, Taktak AF. The use of artificial neural networks in decision support in cancer: a systematic review. Neural Netw. 2006;19(4):408–15.

Lv G, Yan G, Wang Z. Bleeding detection in wireless capsule endoscopy images based on colour invariants and spatial pyramids using support vector machines. In: 33rd annual international conference of the IEEE EMBS. September 2011; 2011.

Mackiewicz M. Capsule endoscopy – state of the technology and computer vision tools after the first decade. In: Pascu O, Seicean A, editors. New techniques in gastrointestinal endoscopy. Croatia: InTech Open; 2011. p. 103–24.

Mackiewicz M, Fisher M, Jamieson C. Bleeding detection in wireless capsule endoscopy using adaptive colour histogram model and support vector classification. In: Medical imaging image processing, Proceedings of SPIE, 69140R; 2008.

Mang OY, et al. Multiple LEDs luminous system in capsule endoscope. In: Proceedings SPIE the International Society for Optical Engineering. Feb. 6 2008;8(7).

Markaki VE, Asvestas PA, Matsopoulos GK. Application of Kohonen network for automatic point correspondence in 2D medical images. Comput Biol Med. 2009;39:630–5.

Meron GD. The development of a swallowable video-capsule (M2A). Gastrointest Endosc. 2000;52:817–9.

Mittal D, Kumar V, Saxena SC, Khandelwal N, Kalra N. Neural network based focal liver lesion diagnosis using ultrasound images. Comput Med Imaging Graph. 2011;35(4):315–23.

Pan G, Yan G, Song X, Qiu X. BP neural network classification for bleeding detection in wireless capsule endoscopy. J Med Eng Technol. 2009;33(7):575–81.

Pan G, Yan G, Song X, Qiu X. Bleeding detection from wireless capsule endoscopy images using improved euler distance in CIELab. J Shanghai Jiaotong Univ (Science) 2010;15(2):218–23.

Qian Z, Meng MQH. Polyp detection in wireless capsule endoscopy images using novel color texture features. In: 9th World Congress on Intelligent Control and Automation (WCICA). p. 948–52; 2011.

Quirini M, Ill RW, Menciassi A, Dario P. Design of a pill-sized 12-legged endoscopic capsule robot. In: Proceedings of IEEE international conference on robotics and automation. 2007; p. 1856–62.

Rettenbacher T. Focal liver lesions: role of contrast-enhanced ultrasound. Eur J Radiol. 2007;64:173–82.

Rimola J, Forner A, Reig M, Vilana R, de Lope CR, Ayuso C, et al. Cholangiocarcinoma in cirrhosis: absence of contrast washout in delayed phases by magnetic resonance imaging avoids misdiagnosis of hepatocellular carcinoma. Hepatology. 2009;50:791–8.

Saarela TP, Landy MS. Combination of texture and color cues in visual segmentation. Vis Res. 2007;58:59–67.

Săftoiu A, Vilmann P, Gorunescu F, Gheonea DI, Gorunescu M, Ciurea T, Popescu GL, Iordache A, Hassan H, Iordache S. Neural network analysis of dynamic sequences of EUS elastography used for the differential diagnosis of chronic pancreatitis and pancreatic cancer. Gastrointest Endosc. 2008;68(6):1086–94.

Săftoiu A, Vilmann P, Gorunescu F, Janssen J, Hocke M, Larsen M, Iglesias-Garcia J, Arcidiacono P, Will U, Giovannini M, Dietrich CF, Havre R, Gheorghe C, McKay C, Gheonea DI, Ciurea T, European EUS Elastography Multicentric Study Group. Efficacy of an artificial neural network-based approach to endoscopic ultrasound elastography in diagnosis of focal pancreatic masses. Clin Gastroenterol Hepatol. 2012;10(1):84–90.

Salvatore V, Borghi A, Sagrini E, Galassi M, Gianstefani A, Bolondi L, Piscaglia F. Quantification of enhancement of focal liver lesions during contrast-enhanced ultrasound (CEUS). Analysis of ten selected frames is more simple but as reliable as the analysis of the entire loop for most parameters. Eur J Radiol. 2012;81:709–13.

Shah SK, Rajauria PP, Lee J, Celebi ME. Classification of bleeding images in wireless capsule endoscopy using HIS colour domain and region segmentation. In: URI- NE ASEE conference; 2007.

Signorelli C, Villa F, Rondonotti E, Abbiati C, Beccari G, de Franchis R. Sensitivity and specificity of the suspected blood identification system in video capsule enteroscopy. Endoscopy. 2005;37:1170–3.

Singh P, Erickson RA, Mukhopadhyay P, Gopal S, Kiss A, Khan A, Ulf WT. EUS for detection of the hepatocellular carcinoma: results of a prospective study. Gastrointest Endosc. 2007;66:65–273.

Streba CT, Gheonea DI, Sandulescu L, Ciurea T, Saftoiu A, Vere CC, Rogoveanu I. Using contrast-enhanced ultrasonography (CEUS) time-intensity curves (TICs) as classifiers in neural network diagnosis of focal liver lesions. Gastroenterology. 2011;142(5, Suppl 1):S-1004.

Streba CT, Ionescu M, Gheonea D, et al. Using contrast-enhanced ultrasonography time-intensity curves as classifiers in neural network diagnosis of focal liver lesions. World J Gastroenterol. 2012a;18(32):4427–34.

Streba CT, Sandulescu L, Vere CC, Saftoiu A, Gheonea IA, Streba L, Rogoveanu I. Quantitative analysis of dynamic image modalities represents the best classifiers of focal liver lesions in an artificial neural network approach. United European Gastroenterology Week, Amsterdam, Olanda. Gut/Endoscopy. 2012b; 60(Suppl. II):A45.

Streba CT, Sandulescu L, Vere CC, Streba L, Rogoveanu I. Computer aided differentiation model for automatic classification of focal liver lesions based on contrast-enhanced ultrasonography (CEUS) time intensity curve (TIC) analysis. J Hepatol. 2012c;56(Suppl. 2):S296.

Tirilly P, Mu X, Huang C, Xie I, Jeong W, Zhang J. Image similarity as assessed by users: a quantitative study. Proc Am Soc Inf Sci Technol. 2010;49:1–10.

van Wijk C, van Ravesteijn VF, Vos FM, Truyen R, de Vries AH, Stoker J, van Vliet LJ. Detection of protrusions in curved folded surfaces applied to automated polyp detection in CT colonography. Berlin Heidelberg: Springer-Verlag; 2006. p. 471–8. MICCAI 2006. LNCS 4191.

Olympus capsule endoscope system, User Manual ver 1.5, 2007.

Vere CC, Foarfă C, Streba CT, Cazacu S, Pîrvu D, Ciurea T. Videocapsule endoscopy and single balloon enteroscopy: novel diagnostic techniques in small bowel pathology. Rom J Morphol Embryol. 2009;50(3):467–74.

Vere CC, Rogoveanu I, Streba CT, Popescu A, Ciocalteu A, Ciurea T. The role of capsule endoscopy in the detection of small bowel disease. Chirurgia (Bucharest). 2012a;107 (3):352–60.

Vere CC, Streba CT, Rogoveanu I, Georgescu M, Pirvu D, Iordache S, Gheonea DI, Saftoiu A, Ciurea T. The contribution of the video capsule endoscopy in establishing the indication of surgical treatment in the tumoral pathology of the small bowel. Curr Health Sci J. 2012b;2 (38):69–72.

Verma B, McLeod P, Klevansky A. A novel soft cluster neural network for the classification of suspicious areas in digital mammograms. Pattern Recogn. 2009;42(9):1845–52.

Youk JH, Kim CS, Lee JM. Contrast-enhanced agent detection imaging: value in the characterization of focal hepatic lesions. J Ultrasound Med. 2003;22:897.

Zhang X, Fujita H, Qin T, et al. CAD on Liver Using CT and MRI. MIMI LNCS. 2008;4987:367–76.

Zhang X, Kanematsu M, Fujita H, Zhou X, Hara T, Yokoyama R, Hoshi H. Application of an artificial neural network to the computer-aided differentiation of focal liver disease in MR imaging. Radiol Phys Technol. 2009;2(2):175–82.

Zhao Q, Dassopoulos T, Mullin GE, Meng MQH, Kumar R. A decision fusion strategy for polyp detection in capsule endoscopy. Stud Health Technol Inform. 2012;173:559–65.

Chapter 17
Computation in Medicine: Medical Image Analysis and Visualization

Adekunle Micheal Adeshina

Abstract Computation in medicine has recently revolutionized those ideal procedures for translating fundamentally proven mathematical concepts in medical imaging and analysis into relevant routines of algorithms. Modern computational techniques, such as CUDA, a parallel computing platform, enabling direct access to the GPU instruction and parallel processing capability, are currently providing flexibility in the use of high-performance computational approaches. Similarly are the other software optimization procedures that assure low-cost and high-throughput visualization of medical datasets. Without mincing words, significant impact of such hardware and software optimization algorithms in medical image analysis and visualization cannot be overemphasized. In the same vein, acquisition of appropriate clinical datasets plays a great role in the accurate diagnosis of diseases and therapy management. The use of appropriate datasets and suitable image modalities are both important in order to successfully prove the effectiveness of any applied computational approaches in medical image analysis and visualization. Moreover, data reconstruction and representation from 2-D to 3-D usually follow notable mathematical approaches such as Euclidean plane, projective plane, and Cartesian coordinate systems and involve other interactive properties such as rotation, scaling, and translation which are also relying on various renderable concepts of data representation. This chapter documents some of the image procedures for acquiring morphological and functional information of patients with more emphasis on mathematical computations of commonly used techniques, such as X-ray, computed tomography (CT), and magnetic resonance imaging (MRI). Interestingly, a typical framework for medical imaging and visualization has been conceptualized in the course of this documentation. Relevant approaches to medical data representation, restructuring, and modeling procedures such as volume segmentation, classification, shading, gradient computation, interpolation, and resampling are presented along with all the significant processes required before generating informative composition of images. In order to facilitate better

A.M. Adeshina (✉)
Faculty of Computer Science and Information Technology, Universiti Tun Hussein Onn Malaysia, 86400 Parit Raja, Batu Pahat, Johor, Malaysia
e-mail: codedengineer@yahoo.com

© Shanghai Jiao Tong University Press, Shanghai and Springer Science+Business Media Dordrecht 2017

D.-Q. Wei et al. (eds.), *Translational Bioinformatics and Its Application*, Translational Medicine Research, DOI 10.1007/978-94-024-1045-7_17

understanding of some of the concepts introduced in this chapter, real-world examples of CT and MRI datasets in 2-D and in their 3-D correspondence are showcased to depict the significance of the mapped structures in the 2-D.

Keywords CUDA • GPU • Image analysis • Medical imaging • Visualization

17.1 Introduction

Recent evolvement of translational informatics has been a strong driving force for translation of laboratories' data. The term "translation" is seen to involve correlating genotype with phenotype, which often requires dealing with information at all structural levels, ranging from molecules and cells to tissues and organs and from individuals to populations (Chen et al. 2013). A relatively new concept was coined with translational bioinformatics, introducing profound changes which include the identification of conviction biology as an informational science; the application of high-throughput genomic and proteomic platforms for global analyses; the requirement to bring computer science, mathematics, and statistics into biology; the use of model organisms as *Rosetta Stones* for deciphering biological complexity; and also the power of comparative genomics for coming to understanding the logic of life (Hood 2003). Apparently, such conceptual analysis opens up a new dawn in medicine. Translational bioinformatics involves the development and the use of computational methods that can reason over the enormous amounts of life science data being collected and stored for the purpose of creating new tools for medicine (Butte 2008). This field has been identified as a revolutional domain addressing some of the hindering computational challenges in medicine. Translational bioinformatics is seen as an emerging field addressing the computational challenges in biomedical research and the analysis of the vast amount of clinical data generated from it (Butte 2008). Technically, the term "computational" involves certain specific procedures for translating those ideal and fundamentally proven mathematical concepts into routines of algorithms. However, all the accurate diagnosis, surgical treatment, and assessment of response to treatment depend on the ability to see through the affected tissues or organs (Aldrich et al. 2012), and this brings medical image analysis and visualization forward into play in translational bioinformatics, thereby forming both combinatory and an integral part of the revolutionary processes of translational bioinformatics. Those actions requiring the use of scientific mathematics and execution of algorithms in order to attain significant and more precise results in medical analysis are deeply rooted in the word *computation in medicine*.

Computational approaches in medical image analysis have also gain more attention due to the recent overwhelming rate of generation of biomolecular data. This accumulated information explosion is being driven by the development of low-cost, high-throughput experimental technologies in genomics, proteomics, and molecular imaging, among others, tying the anticipated success in the life sciences

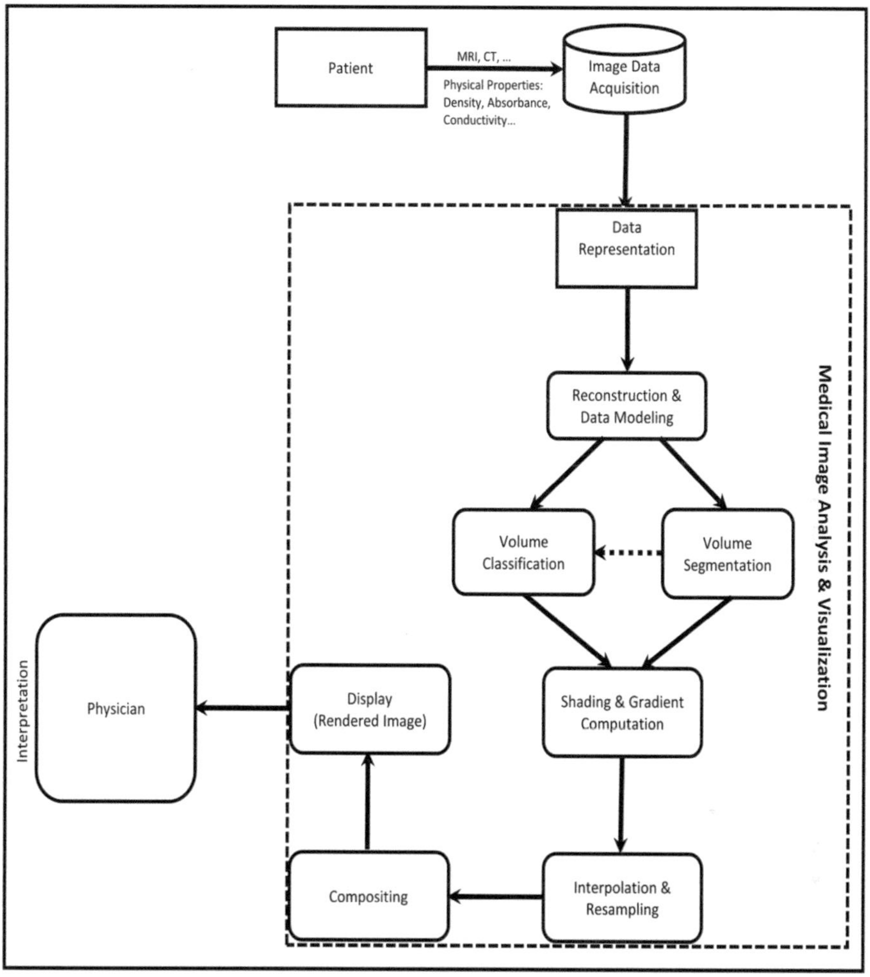

Fig. 17.1 Medical imaging and visualization architecture

to our ability to rationally interpret these large-scale, high-dimensional datasets into clinically understandable and useful information, which in turn requires us to adopt advances in informatics (Chen et al. 2013). Overview of medical imaging and visualization architecture is presented in Fig. 17.1. However, with current alignment of computational medicine with high-performance computation, computer models and efficient software could be leveraged in figuring out, within a considerable interactive speed, how diseases develop and how to thwart it. Invariably, paramedical research and computational approaches seem inseparable. This chapter introduces computation in medicine. Medical image acquisition techniques, their numerical computations, structuring, and data visualization procedures are presented.

17.2 Acquisition of Medical Image Data

In medical diagnosis and disease therapy management, acquisition of medical image data is a crucial process immediately after the diagnosis of the concerned patients. However, in certain circumstances, acquisition of patient images may be considered a priority, overriding the usual medical doctors' preexamination and interpretation of the health situation of patients. Such cases could be in the case of emergency situation either as a result of severe injury especially when handling unconscious patients. Nevertheless, in clinical practices, medical image data could be acquired for diagnosis, therapy planning, intraoperative navigation, or postoperative monitoring (Preim and Bartz 2007). According to Dhawan et al. (2008), medical imaging could be seen as a process of collecting information about a specific physiological structure (an organ or tissue) using a predefined characteristic property that is displayed in the form of an image. Such predefined characteristic property may be physical properties such density, absorbance, or conductivity. Image acquisition technique required in any case largely depends on the intended information from patient medical examination. Image modalities such X-ray, computed tomography (CT), and magnetic resonance imaging (MRI) are useful in extracting "morphological information" from the patient. Other specialized MRI techniques include MR spectroscopy, MR angiography, and MR microscopy. However, in order to obtain physiological or functional information from patients, positron emission tomography (PET) and single-photon emission computerized tomography (SPECT) are appropriate. Diffusion tensor imaging (DTI) also plays significant roles in diagnosis procedures that require measuring of the diffusion of water and in tracking of the brain's nerve fibers, the white matter. Apparently, suitability of image modalities solemnly depends on the required medical examination, and thus, image modalities are seen to be complementary to each other in the medical diagnosis and disease and therapy management procedures (Adeshina et al. 2012). This section briefly discusses the X-ray, computed tomography, and magnetic resonance imaging being the most commonly used image modalities.

17.2.1 X-Ray

In 1895, Wilhelm Conrad Röntgen (or "Roentgen" in anglicized typography) discovered X-ray (Roentgen 1898) as a high-voltage discharge between electrodes in a gas at very low pressure producing a penetrating radiation which causes certain materials to fluoresce visible light. X-ray is a medical imaging technique that utilizes the radiation that is partly transmitted and partly absorbed through irradiated objects. The X-ray photons are a form of electromagnetic radiation produced following the ejection of an inner orbital electron and subsequent transition of atomic orbital electrons from states of high to low energy (Jenkins 2000). X-ray is widely used in projection of images based on absorption and scattering with very high spatial resolution, and it has been seen to be greatly useful in imaging fractured bones, such as a broken arm or wrist, often used by surgeons during therapeutic procedures, such as a coronary angioplasty, to help guide equipment to the area being treated and in highlighting a lung infection, such as pneumonia. Meanwhile, X-rays can only produce 2-D images; it exposes patients to radiation and not suitable for imaging soft tissues. According to Bingham (1998), considering the following assumptions, the inside structures of an object could be investigated:

1. There exists an object with n-dimensional space where n is fixed as 2.
2. $f(x)$ is the X-ray attenuation coefficient at point $X \in \mathbb{R}$ where the attenuation coefficinet depends on the material through which the ray passes. Therefore, f is expected to give information about object.
3. Suppose the object is contained in a ball of radius $\mathcal{R}$ with the center at the origin and that the X-ray attenuation coefficient f is zero outside the object.
4. If the object is x-rayed in a direction $\theta \epsilon S^{n-1}$ from a point $a \epsilon A := S^{n-1}(0, R)$, the attenuation of the X-ray intensity I at each point $a + t\theta$, $t \geq 0$.

According to assumption (iii), we can have Eq. (17.1):

$$\operatorname{supp} f \subset B(0, R) \tag{17.1}$$

and then Eq. (17.2):

$$\Leftrightarrow dI = f(a + t\theta)\, I dt \tag{17.2}$$

By solving this differential equation, we see that the intensity of the X-ray measured by a detector situated behind the object is as Eq. (17.3):

$$I_{\text{meas}} = I_0 \exp\left(\Leftrightarrow \int_0^\infty f\,(a + t\theta)\, dt\right) \tag{17.3}$$

Therefore, we can derive formulae for reconstructing $f(x)$ from the measurements I_{meas} or equivalently from Eq. (17.4):

$$\int_0^\infty f\,(a + t\theta) dt = \ln\left(\frac{I_O}{I_{\text{meas}}}\right) \tag{17.4}$$

The above equation could be seen as cases of having different combinations of $a \epsilon A$ *and* $\theta \in S^{n-1}$.

17.2.2 Computed Tomography

X-ray imaging techniques follow the same scenario of allowing radiation to pass through different parts of the patients' body. Such passage of X-rays is dependent on the amount of X-rays that could be absorbed or exit the body of the patients, which in turn determines the radiation dose of the patient. Computed tomography (CT) is not exceptional to this; however, in CT multiple X-ray images are taken from different directions producing cross-sectional images or "slices" of patients' anatomy. The cross-sectional images could be used in medical diagnosis and disease therapy. CT entails the reconstruction of a function *f* from a finite number of line integrals *f* (Faridani and Ritman 2000). With such understanding, it becomes apparent that the goal of CT is to recover an approximation to $f(x)$ from CT datasets over a finite number of lines.

X-rays from a located source travel and pass through the patient. However, some energy of rays are attenuated, and rays with less energy eventually reach the detector. Rays of CT are able to produce a map of gray values representing a close resemblance of the insides of the patient. This situation can be understood either through a monochromatic beam or polychromatic beam considering the intensity I_{in} at distance x; I_{out}, the intensity at the detector's end; and μ, the attenuation coefficient or absorption coefficient. If we consider a situation whereby the radiation passes through a body with the same property at every point, a homogeneous body, we expect the intensity of radiation passing through the body to decrease exponentially with distance; hence, we have Eq. (17.5):

$$I_{(x)} = I_{in}\exp(-\mu x) \tag{17.5}$$

Therefore, if we differentiate Eq. (17.5),

$$\frac{dI}{dx} = -\mu I \tag{17.6}$$

However, for a nonhomogeneous body where the absorption coefficient varies with distance x,

$$I_{(x)} = I_{in} \exp\left(-\int udx\right) \tag{17.7}$$

Similarly, we can consider a specific interval a, b where a *and* b have values between 0 *and* n in order to get a more specific approximation, thus Eq. (17.8):

$$I_n = I(n) = I_{in}\exp\left(-\int_a^b \mu dx\right) \tag{17.8}$$

If we know I_n, the total absorption, A_t, could be calculated:

$$A_t = \int_a^b \mu dx = -\log\left(\frac{I_n}{I_{in}}\right) = \log\frac{I_{in}}{I_n} \tag{17.9}$$

However, even if we know I_{in} and I_n, we still cannot clearly say the distribution of the material within the interval a, b as being illustrated. Resolution of this was first attempted by Radon (1917). Similarly, the absorption coefficient could be analyzed following a related approach based on a number of assumptions established for computed tomography (Faridani 2003):

1. $f(x)$ = density of the cross-section at $x \in \mathbb{R}^2$
2. L = the line of X-rays
3. $I(x)$ = the intensity of X-rays at $x \in L$

Apparently, in theoretical physics, $I(x)$ decreases proportional to $f(x)$; thereby, we can have Eqs. (17.10) and (17.11):

$$dI/dx = -f(x)I(x) \tag{17.10}$$

$$dI/I = -f(x)dx \tag{17.11}$$

Therefore we can have the measured data calculated as,

$$\text{meas} = \frac{I_{in}}{I_{out}} = e^{RF(L)} \tag{17.12}$$

where the total attenuation along L,

$$RF(L) = \int_{x \in L} f(x)ds \tag{17.13}$$

Radon (1917) referred to the expression in Eq. (17.8) as the total "material" along L. To construct the absorption coefficient μ (x, y) as a function of position using Radon approach, we assume:

1. Projection is a line integral.
2. Projection p $(s, \varnothing)$ at angle $\varnothing$, s is coordinate on detector.

The Radon transform (RT) of a distribution f(x; y) is given by Eq. (17.14):

$$p(s, \varnothing) = \int_{-\infty}^{\infty} \int_{-\infty}^{\infty} f(x, y).\delta(x \cos \varnothing + y \sin \varnothing - s)dxdy \tag{17.14}$$

where δ is the Dirac delta function and x, y, $\varnothing$, and s are respective coordinates. The Radon transform of an off – center point source is a sinusoid; hence, the function $p(s, \varnothing)$ is usually being referred to as a sinogram.

17.2.3 Magnetic Resonance Imaging

Magnetic resonance imaging (MRI) is an application of nuclear magnetic resonance (NMR) which is a subtle quantum mechanical phenomenon that has played a major role in medical imaging revolution over the last 30 years. Hydrogen in water molecules possesses an inherent ability referred to as *spin* which gives it potential to act as magnet. Nuclear magnetic resonance is a phenomenon which occurs when the nuclei of certain atoms are immersed in a static magnetic field and exposed to second oscillating magnetic field (Hornak 1997). The *spin* property in proton makes the nucleus that produces NMR signal. Mathematical description of NMR could be better presented using 2-D Fourier transform, a standard Fourier transformation of two variables $f(x, y)$, wave forms $e^{2\pi i(k_x \ x + k_y y)}$ and k – space $\left(k_x k_y\right)$.For $f \in L^2$ $\left(R^2\right)$, the Fourier transformation of f is presented in Eq. (17.15):

$$F(f)\left(k_x, k_y\right) = \int_{-\infty}^{\infty} \int_{-\infty}^{\infty} f(x, y) e^{-i2\pi\left(k_x x + k_y y\right)} \, dxdy \tag{17.15}$$

In the same vein, we can re-represent Eq. (17.15) to portray a reverse approach to Fourier transformation as in Eq. (17.16):

$$F^{-1}(f)(x,) = \int_{-\infty}^{\infty} \int_{-\infty}^{\infty} f(x, y) e^{-i2\pi\left(k_x x + k_y y\right)} \, dk_x dk_y \tag{17.16}$$

MRI is the most suitable and widely used imaging technique for brain and other soft tissues. It is capable of producing detail image of patients in any plane. MRI is highly flexible to use and it provides better spatial resolution with higher discrimination, making it very relevance in contrasting soft tissue. Moreover, unlike X-ray and CT, MRI has no ionizing radiation.

17.3 Medical Data Visualization

17.3.1 Reconstruction and Data Representation

Projective plane is seen as a geometric structure with extended concept of a plane. However, in ordinary Euclidean plane, unless the line crosses each other and intersects, parallel lines do not intersect. Meanwhile a projective plane with any two lines intersect in one and only one point called vanishing point, a point where parallel lines that are not parallel to the image plane appear to converge, which could be better interpreted with projective plane.

In the development of medical imaging and visualization framework, using homogeneous coordinate principle, a point x, y of $2-\text{D}$ slice in the Euclidean plane is represented in the projective plane $(3-\text{D})$ by adding a third

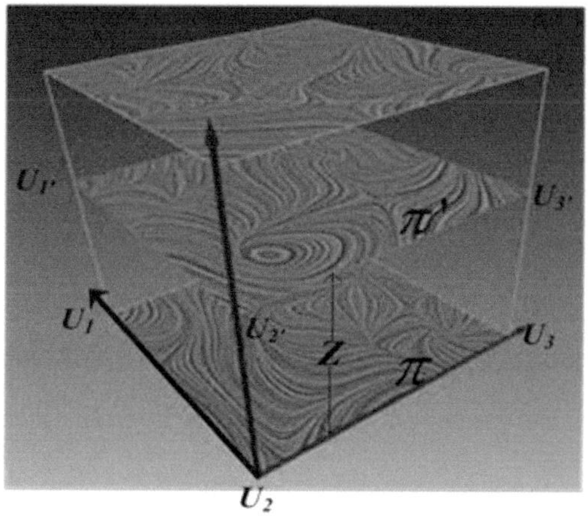

Fig. 17.2 Point estimation of 2-D slices

coordinate 1 at the end, $x, y, 1$. This is based on the fundamental Euclidean theorem which states that a point in an $n-$dimensional Euclidean space is represented as appoint in an $(n+1)$-dimensional projective space. However, overall scaling is not important.

The MRI slices are abstractly represented as stack of images as in Fig. 17.2. It is assumed that there are points U_i with $U_i = 1, 2, 3, 4$ arranged parallel in line with plane π. Since there exist such many slices, we assume the slices are moved up a distance Z as shown in Fig. 17.2. With such moved distance of the slices, there will be a formation of new sets of points $U_{i'}$ with $i' = 1, 2, 3, 4$ leaning on a new plane π'. The first issue to address is estimation of the new points $U_{i'}$ automatically which can be done by estimating directly from the first plane π.

At this point, it can be assumed that U_1 and U_3 are known ; hence, U_2 and U_4 can be estimated and computed by applying intrinsic properties of the vanishing points. Figure 17.2 shows the point estimation of 2-D slices. The vanishing point of the parallel lines leaning on plane π could be computed as in Eq. (17.17):

$$V = (U_1 \times U_2) \times (U_3 \times U_4) \tag{17.17}$$

However, based on projective geometry, which describes the physical characteristics of the virtual camera and the relationships between the images, the projection of a point X_w in the object space to a point U_i in the image space using projective camera is expressed in terms of a direct linear mapping in homogeneous coordinates as in Eq. (17.18):

$$\lambda U_i = PX_w = \begin{bmatrix} P_1 & P_2 & P_3 & P_4 \end{bmatrix} \begin{bmatrix} X \\ Y \\ Z \\ 1 \end{bmatrix} \tag{17.18}$$

where λ is the scale factor due to projective equivalency of $(k_x\,; k_y\,; k) = (x; y; 1)$, P is a 3^*4 camera projection matrix, and P_i is the *ith* column of P.

As earlier discussed, with homogeneous coordinate representation, value 1 in the last row of the vector denotes that the defined point leans on the image plane. However, if the point in the object space leans on ground plane $Z = 0$, hence the linear mapping will change to Eq. (17.19):

$$sU = HX'_w = \begin{bmatrix} P_1 & P_2 & P_4 \end{bmatrix} \begin{bmatrix} X \\ Y \\ 1 \end{bmatrix} \tag{17.19}$$

H is the homography matrix mapping points lying on a plane in the object space across different images; s introduced scaling factor in the mapping equation stems from setting Z *to* 0.

In order to establish relationship between U_i and U'_i, we can restate Eq. (17.18) above as (17.20):

$$\lambda U_i = \begin{bmatrix} P_1 & P_2 & P_4 \end{bmatrix} \begin{bmatrix} X_i \\ Y_i \\ 1 \end{bmatrix} + P_3 Z \tag{17.20}$$

where P_3 corresponds to the vanishing point in the direction of Z axis or the normal of the ground plane.

The main target is to project the lines and points that made up the 2-D slices in 3-D. The Euclidean formula for s line is $ax + ay + c = 0$; this is regarded as nonzero scaling factor, and since the equation is unaffected by scaling, we can however arrive at the following:

$$qX + rY + sZ = 0$$

where q, r, and s are the homogeneous coordinates of points (x, y) in the line:

$$t^T P = P^T t = 0$$

$$t = \begin{bmatrix} q, & r, & s \end{bmatrix}^T \text{ representing the line}$$

$$p = \begin{bmatrix} X & Y & Z \end{bmatrix}^T \text{representing the point}$$

Substituting V_z for P_3 in Eq. (17.20) and combining the result with Eq. (17.19), we have

$$\lambda_i U_i = s_i U_i + V_z Z \tag{17.21}$$

λ_i and s_i are the unknowns from this equation, though they were both defined earlier in Eq. (17.18) and Eq. (17.19). We can estimate the respective values by using Eq. (17.22):

$$\begin{bmatrix} \lambda_i \\ s_i \end{bmatrix} = \left(A_j^{\,T} A_i\right)^{-1} A^T{}_i b_i \tag{17.22}$$

$$A_i = [U_i \mid -U_i] \tag{17.23}$$

$$b_i = V_z\, Z \text{ and } V_z = P_3$$

Since s_i is estimated, we can continue setting different values for Z in order to estimate any other image point along the lines. Hence, we can equally estimate U_2 and U_4 as follows:

$$U_2 = (U_2 \times V_z) \times (U_1 \times V) \tag{17.24}$$

$$U_4 = (U_4 \times V_z) \times (U_3 \times V) \tag{17.25}$$

17.3.1.1 Renderable Representation

Memory system architecture of medical image analysis and visualization framework is typically concerned with making the data available for its optimal architectural use. It is at this stage that series of the original data slices are stacked, shaped, and positioned for flow. Properties such as *rotation*, *scaling*, and *translation* are likewise necessary in the data for better *value distribution*. Coordinate system is greatly useful for the success of data preparation; hence, medical image analysis and visualization framework is usually developed to use the *model*, *world*, *view*, and *display coordinate systems*.

The *model coordinate system* is typically a local *Cartesian coordinate system*. As the name *model* implies, it is the coordinate system in which model is defined. This type of coordinate system is locally defined by the modeler. We can refer to this as an *inherent coordinate system* based on the decision of the person that generates it. The units used in its definition may be *meters*, *inches*, or *feet* and its axis might be arbitrary; these are based on discretion of the modeler.

The *world coordinate system* is the *3-D space* where *actors* are positioned. Unlike model coordinate system, which is a typical local Cartesian coordinate system, world coordinate system is the only standard coordinate system where all actors *locally defined coordinate systems* are converted to. The world coordinate system is the coordinate system where all the actors are *scaled*, *rotated*, and *translated into*. Moreover, the position and orientation of cameras and light are specified in the world coordinate system.

The *view coordinate system* is directly referenced to the *camera*; it represents what is *visible* to the *camera*. It consists of *x*, *y*, *z values*. The *x* and *y* specify location of the *image plane* and it ranges from – 1, 1, while *z* is the *depth coordinate* that represents the distance or ranges from the camera. In order to convert from the *world coordinates* to *view coordinates*, a four by four (4^*4) *coordinate transformation* matrix is applied, introducing the perspective effects of a camera.

The usual way to represent element in three dimensions is through *Cartesian vector* x, y, z. However, in order to project 2 – D image to 3 – D plane, vanishing point must be included in the projection; hence, homogeneous coordinate system is needed. Unlike *cartesian vector* with three (3)elements x, y, z, homogeneous coordinate has four *element vectors* represented as X, Y, Z, W as earlier explained in the previous section. The conversion from Cartesian coordinates to homogeneous coordinates is presented in Eqs. (17.26), (17.27), and (17.28):

$$x = \frac{X}{W} \tag{17.26}$$

$$y = \frac{Y}{W} \tag{17.27}$$

$$z = \frac{Z}{W} \tag{17.28}$$

Four by four (4^*4)matrix is used for the performance of translation, scaling, and rotation through repeated multiplication of matrix. We can create a transformation of matrix that translates a point x, y, z in Cartesian space by vector t_x, t_y, t_z as in Eq. (17.29). Figure 17.3 illustrates translation:

$$T_T = \begin{bmatrix} 1 & 0 & 0 & t_x \\ 0 & 1 & 0 & t_y \\ 0 & 0 & 1 & t_z \\ 0 & 0 & 0 & 1 \end{bmatrix} \tag{17.29}$$

where T_T is the matrix for translation.

The created translated matrix needs to be post-multiplied with homogeneous coordinate X, Y, Z, W. Meanwhile, we have to construct the homogeneous coordinate from the Cartesian coordinate before such multiplication; hence, if we set $W = 1$ representing *finite point*, X, Y, Z will yield $X, Y, Z, 1$. In the same vein, we pre – multiply the current position by the transformation matrix T_T in order to determine the translated point X', Y', Z' for yielding the translated coordinate. Hence, we have Eq. (17.30):

$$\begin{bmatrix} x' \\ y' \\ z' \\ w' \end{bmatrix} = \begin{bmatrix} 1 & 0 & t_x \\ 0 & 1 & t_y \\ 0 & 0 & t_z \\ 0 & 0 & 1 \end{bmatrix} \cdot \begin{bmatrix} x \\ y \\ z \\ 1 \end{bmatrix} \tag{17.30}$$

Using the general pattern of conversion back to Cartesian coordinates as in Eqs. (17.26), (17.27), and (17.28), we have Eqs. (17.31), (17.32), and (17.33):

Fig. 17.3 Translation

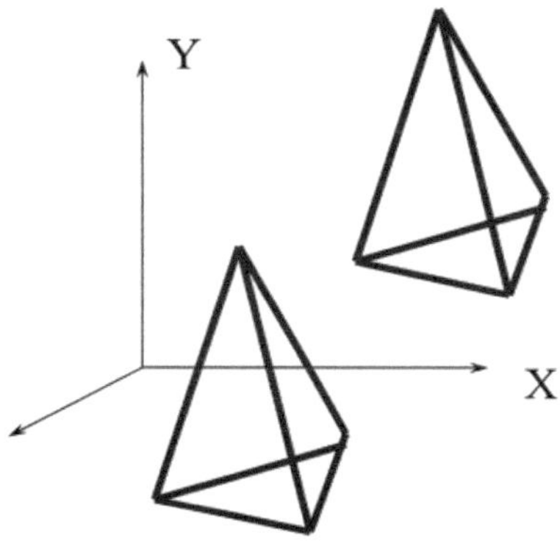

$$x' = x + t_x \tag{17.31}$$

$$y' = y + t_y \tag{17.32}$$

$$z' = z + t_z \tag{17.33}$$

Equations (17.32) and (17.33) are the procedure to translate an object. Similar procedure can be employed for scaling or rotating of an object. Using the transformation matrix as Eq. (17.34) where T_s is the transformation matrix for scaling, s_x, s_y, s_z represent the scale factors along x, y, z axes, respectively. Figure 17.4 illustrates scaling about the origin:

$$T_s = \begin{bmatrix} s_x & 0 & 0 & 0 \\ 0 & s_y & 0 & 0 \\ 0 & 0 & s_z & 0 \\ 0 & 0 & 0 & 1 \end{bmatrix} \tag{17.34}$$

In the same vein, we can do rotation around x, y, *and* z *axes* by angle θ as illustrated in Figs. 17.5, 17.6, and 17.7 to produce T_{R_x}, T_{R_y}, and T_{R_z}, respectively.

$$R_x(\theta): y' = y\cos\theta - z\sin\theta$$

$$z' = y\sin\theta + z\cos\theta$$

$$x' = x$$

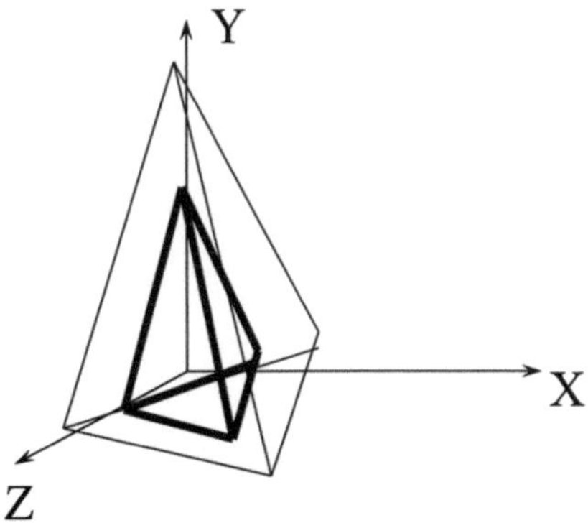

Fig. 17.4 Scaling about the origin

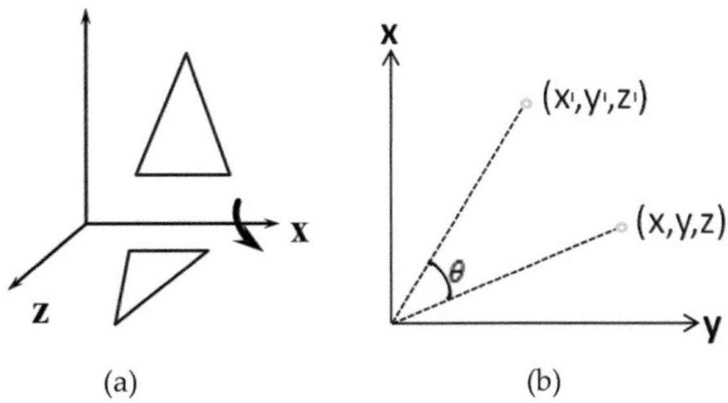

Fig. 17.5 Rotation about x axis

$$T_{R_x} = \begin{bmatrix} 1 & 0 & 0 & 0 \\ 0 & \cos\theta & -\sin\theta & 0 \\ 0 & \sin\theta & \cos\theta & 0 \\ 0 & 0 & 0 & 1 \end{bmatrix} \tag{17.35}$$

Illustration of rotation about y axis is given in Fig. 17.6.

$$R_y(\theta): z' = z\cos\theta - x\sin\theta$$
$$x' = z\sin\theta + x\cos\theta$$
$$\mathrm{y}' = \mathrm{y}$$
$$T_{R_y} = \begin{bmatrix} \cos\theta & 0 & \sin\theta & 0 \\ 0 & 1 & 0 & 0 \\ -\sin\theta & 0 & \cos\theta & 0 \\ 0 & 0 & 0 & 1 \end{bmatrix} \tag{17.36}$$

Similarly, Fig. 17.7 illustrates rotation about z axis producing T_{R_z}.

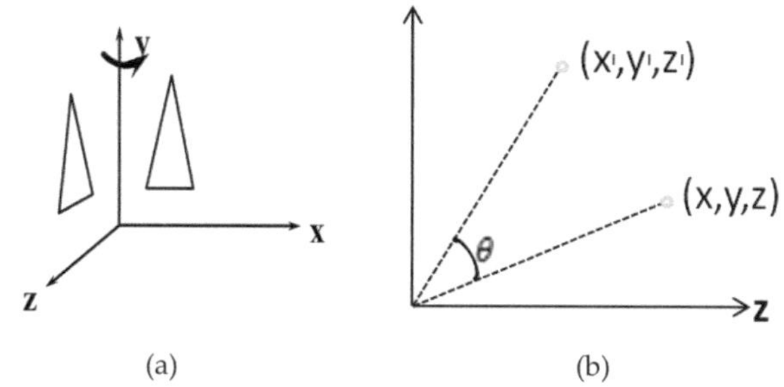

Fig. 17.6 Rotation about *y* axis

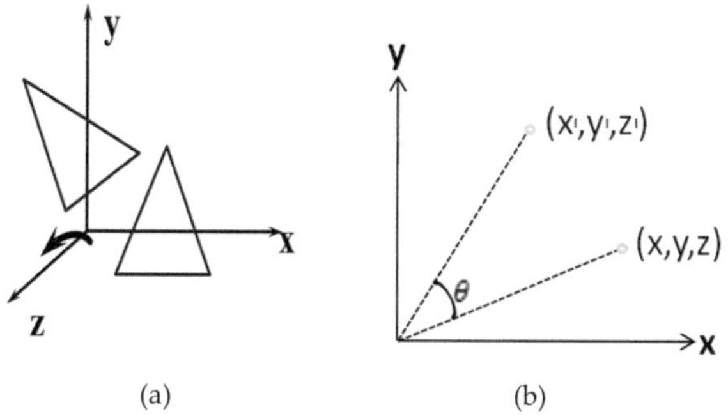

Fig. 17.7 Rotation about *z* axis

$$R_z(\theta)\colon x = x'\cos\theta - y\sin\theta$$
$$y' = x\sin\theta + y\cos\theta$$
$$z' = z$$

$$T_{R_z} = \begin{bmatrix} \cos\theta & -\sin\theta & 0 & 0 \\ \sin\theta & \cos\theta & 0 & 0 \\ 0 & 0 & 1 & 0 \\ 0 & 0 & 0 & 1 \end{bmatrix} \tag{17.37}$$

However, during the rotation of the object, we might need to transform the object from one coordinate axes to another, from $x - y - z$ to $x' - y' - z'$. In order to do this, we need to first derive a transformation matrix by assuming the following:

1. The unit x' axis makes the angle $\theta x'x$, $\theta x'y$, $\theta x'z$ around $x - y - z$ axis.
2. The unit y' axis makes the angle $\theta y'x$, $\theta y'y$, $\theta y'z$ around $x - y - z$ axis.
3. The unit z' axis makes the angle $\theta z'x$, $\theta z'y$, $\theta z'z$ around $x - y - z$ axis.

where $(\theta x'x, \theta x'y, \theta x'z)$, $(\theta y'x, \theta y'y, \theta y'z)$, *and* $(\theta z'x, \theta z'y, \theta z'z)$ are the directional cosines.

Hence, placing the directional cosines along the rows of the transformation matrix will produce Eq. (17.38) which is referred to as the resulting rotation matrix T_R:

$$T_R = \begin{bmatrix} \cos\theta x'x & \cos\theta x'y & \cos\theta x'z & 0 \\ \cos\theta y'x & \cos\theta y'y & \cos\theta y'z & 0 \\ \cos\theta z'x & \cos\theta z'y & \cos\theta z'z & 0 \\ 0 & 0 & 0 & 1 \end{bmatrix} \quad (17.38)$$

To *rotate* around the *center of the object*, which is usually more convenient, we must first *translate* from the *center of the object* to the *origin*, and then we apply *rotations* followed by *translating the object back to its center*. However, in order to achieve the *translation*, *rotation*, and *scaling* of the object using the transformation matrix, the order of the multiplication is important.

In *display coordinate system*, the coordinates are actual x, y pixel locations on the image plane, though display coordinate uses the same basis as view coordinates except it does not use -1, 1 range. The *view coordinates* determine *window size* and *view point*. *Display coordinates* determine how the negative one – to – one $(-1, 1)$ of *view coordinates* is *mapped* into *pixel locations of display*. With view port, it is possible to divide the port which ranges from 0, 1 for x and y axes and depth value representation with z axis. This is particularly useful in cases where one needs to render two different scenes but display them in the same window. The analysis and justifications for the preparation of datasets in its more suitable renderable form are also in line with the explanations of Schroeder et al. (2002).

17.3.2 Data Restructuring and Modeling

The prepared dataset has to be filtered thoroughly in order to enhance its pixel intensities. Similarly, the specified focused data should be geometrically mapped for better image quality. Modeling of the camera focus point in medical image analysis and visualization framework is likewise significant, often aligned with the physical laws of optics which could be better described by modeling the transport theory of light with specific attention on geometrical optics lights. Meanwhile, factors such as the wave character of light, possible light polarization states, diffraction, and interference are usually neglected.

If x is the radiant field at any point in the direction of the radiant energy n and around υ, the radiant energy could be defined as $R(x\,.\,n\,.\,\upsilon)$. Therefore, if θ is the angle between the direction n and the normal on da for time dt, the traveling radiant energy δE can be respresented in Eq. (39) provided that there is a specified frequency interval dυ around υ through a solid angle dΩ:

$$\delta E = R(x.n.\upsilon)\cos\theta da d\Omega d\upsilon dt \tag{17.39}$$

However, we can also define radiant energy using photon number density $\psi(x.n.\upsilon)$. If x denotes the position of the photons per unit volume, $d\upsilon$ represents the frequency interval around υ along the direction n and travels into an element of solid angle $d\Omega$. Then, the number of photon Ň per unit volume could be represented in Eq. (17.40).

$$\check{N} = \psi(x.n.\upsilon)d\Omega d\upsilon \tag{17.40}$$

Equation (17.40) could be extended for calculation of the number of photons Ň by representing surface da with time dt and traveling velocity c in Eq. (17.41):

$$\check{N} = \psi(\cos\theta da)(cdt)(d\Omega d\upsilon) \tag{17.41}$$

Nevertheless, if the energy carried by each photon is considered as $h\upsilon$ in accordance to the constant expressed in Planck – Einstein relation where h is the Planck's constant, hence, a new relationship could be established for radiant energy using photon number density:

$$\delta E = ch\upsilon\psi(x.n.\upsilon)\cos\theta da d\Omega d\upsilon dt \tag{17.42}$$

Apparently, since we have clearly defined radiant energy in Eqs. (17.39) and (17.40), we can therefore equate these equations:

$$R(x.n.\upsilon)\cos\theta da d\Omega d\upsilon dt = ch\upsilon\psi(x.n.\upsilon)(\cos\theta da d\Omega d\upsilon dt)$$

$$R(x.n.\upsilon) = ch\upsilon\psi(x.n.\upsilon) \tag{17.43}$$

Equation (17.43) shows clear similarity between radiance and photon number density as in Eq. (17.43). Therefore, in order to record all the focused points in an image, it becomes reasonable if we compute $R(x.n.\upsilon)$ for all the focused points. Some of these concepts are documented in Adeshina et al. (2012) and more elaborately in Hege et al. (1996).

17.3.3 Volume Segmentation and Classification

Volume or image segmentation entails partitioning of image or volume into meaningful region representation. Segmentation improves the analysis of an image establishing a reasonable correspondence between the image pixel properties and the type of tissue to facilitate the successful manipulation of data for medical visualization, while classification focuses on labeling the pixels of an image

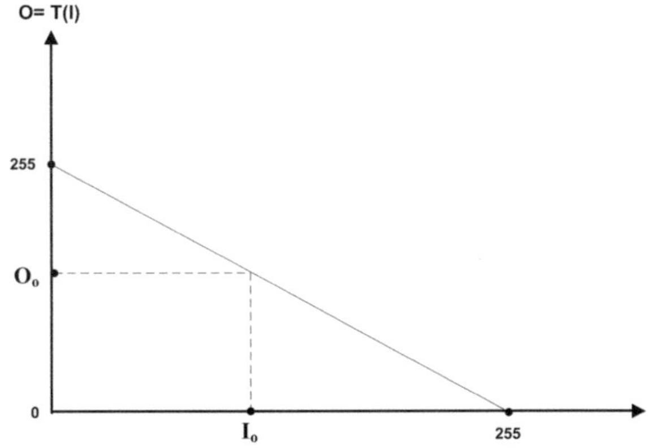

Fig. 17.8 Image point processing approach

corresponding to a specific type of tissue or anatomical structure usually with color and opacity. Apparently at the end of a successful volume segmentation and classification, specific objects within the image would be separated, and regions that have similar pixel properties would be identified along a specified predetermined boundaries. All these create rooms for a more detail image or volume analysis. In volume rendering, depicting region of interest based on color and transparency mappings of respective scalar values to the corresponding regions of volume is achieved using transfer function. Image point processing scale is presented in Fig. 17.8.

Transfer function could be established using image point ranging from 0 to 255 scale. *Point processing* image enhancement techniques is based on the intensity of individual pixels in the image. Hence, based on Eq. (17.44), intensity transfer function could be represented through *255 output pixels* and *255 input pixels* as in Fig. 17.8:

$$O = T(I) \tag{17.44}$$

where O represents the *output pixel*, T is the *transform*, and I is the *input pixel*.

Feature enhancement is extremely important in order to distinguish normal tissues distinctly from abnormal tissues especially when intensities of abnormal tissues match with the intensities of normal ones. Despite the fact that brain tumor might sometimes be large, space occupying, it could still exist in the same intensity as the normal tissues making it difficult to distinguish.

Transfer function was utilized in mapping data value to “renderable quantities” as the output value. The two (2) main transfer functions usually designed are the *opacity transfer function* and the *color transfer function*. The *opacity transfer function* maps intensities of volume elements (voxels) in the data sample to the corresponding opacity value based on the framework intensity scale and selectively makes some voxels transparent enough to be seen through the assigned opacity value in order to show the interior of the data sample. Meanwhile, *color transfer function* uses coloration for its classification procedures. It maps intensities of voxels to corresponding color values using lookup table and likewise does selective painting of voxels with different colors such that voxels of different intensity values are presented with appropriate corresponding color variances. However, in order to have better clarities of the output images, contrast enhancement transfer function, referred to as the *contrast transfer function (CTF), could be* applied in 3-D reconstruction procedures.

17.3.4 Shading and Gradient Computation

The *ambient coefficient*, the *diffuse coefficient*, and the *specular coefficient* are the three parameters that are usually modeled for illumination. Ambient lighting, the background illumination, is represented in equation (17.45):

$$R_c = L_c O_c \tag{17.45}$$

where R_c is the resulting intensity curve, L_c is the light intensity curve, and O_c is the color curve of object.

Ambient light has no direction and is independent of light position, orientation of the object, and observer's position. With this in mind, ambient is simply seen as the approximate contributions of light to the scene which is irrespective of the location of object and light. Figure 17.9 illustrates that.

Diffuse lighting is the non-shiny illumination and shadows. It has no dependence on camera angle. Diffuse lighting is illustrated in Fig. 17.10 and represented as Eq. (17.46). In order to determine diffuse's contribution to the surface, surface normal and the direction of the incoming rays are important:

$$R_c = L_c O_c \cos\theta \tag{17.46}$$

where L_c is the light color, O_c is the object color, and $\cos\theta$ is the product of the vector of light source (a negative value) and the vector of surface normal value to the object.

Specular lighting is the bright and shiny reflections which has no dependence on object color. Specular lighting is represented as Eq. (17.47) and illustrated in Fig. 17.11.

$$R_c = L_s K_s \cos{(\alpha)}^n \tag{17.47}$$

L_c represents the light color, K_s is the reflection constant, and R_c *is* the *color curve*. The product of the vector of light source, which is a negative value, and the vector of surface normal value to the object is $\cos\alpha$. However, specular power is denoted as n resulting from different n values of specular light. Equation (17.48) presents the integration of the three parameters that are usually modeled for illumination:

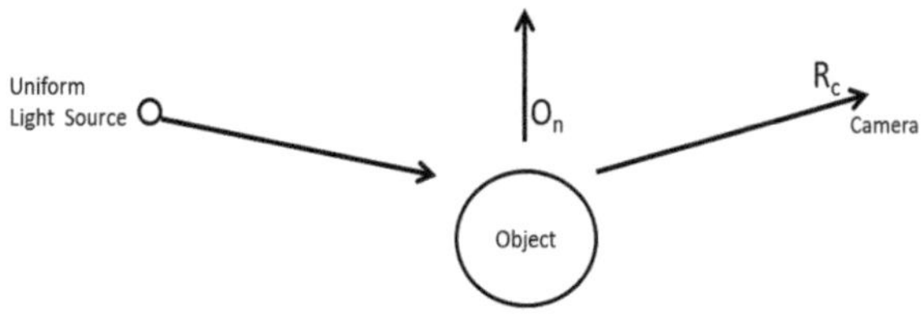

Fig. 17.9 Ambient lighting

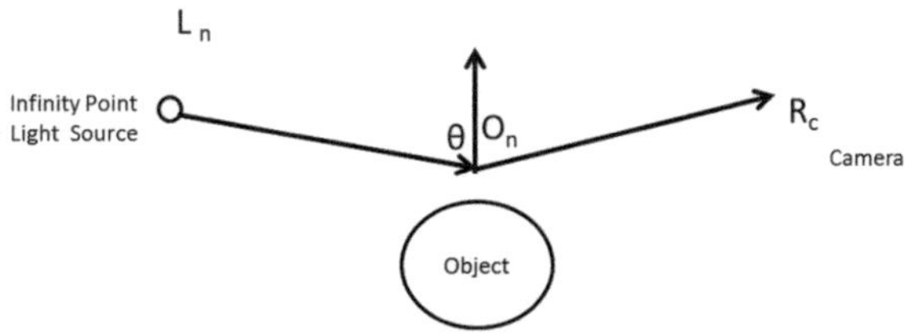

Fig. 17.10 Diffuse lighting

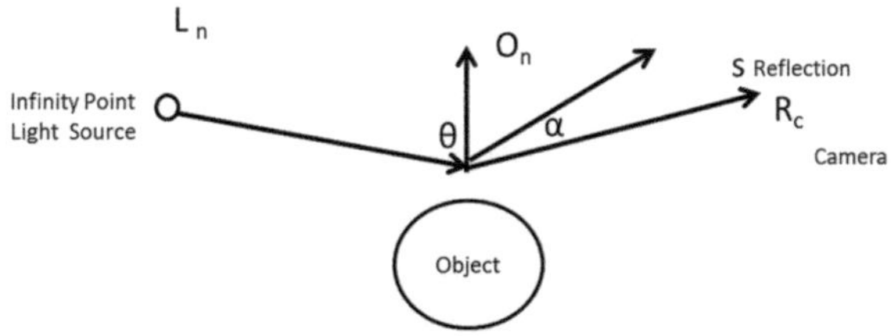

Fig. 17.11 Specular lighting

$$R_c = W_a + W_d + W_s \tag{17.48}$$

where W_a, W_d, and W_s are the relative weights of *ambient*, *diffuse*, and *specular*, respectively.

Moreover, in order to achieve quality image output, medical image analysis and visualization framework could be configured to select either *flat*, *Gouraud*, *Phong shading*, or their combination for better shading of images with respect to the level of pixels in the datasets. Flat shading is the earliest shading method which requires shading the polygons in the data samples with single color. However, because sometimes resulting interpolation color could be needed during shading to have a better image coloration, Gouraud shading was introduced. With Gouraud shading, polygons are shaded by interpolating color that are computed at the vertices of the image. Unfortunately, Gouraud shading usually produces *specular highlights*, a bright spot of light that appears on shining objects when illuminated. Phong shading produces better shading results compared to Gouraud shading by fixing the issue of specular highlights. However, despite the shortcomings in flat and Gouraud shading, using all in combination will contribute to obtaining better-shaded image.

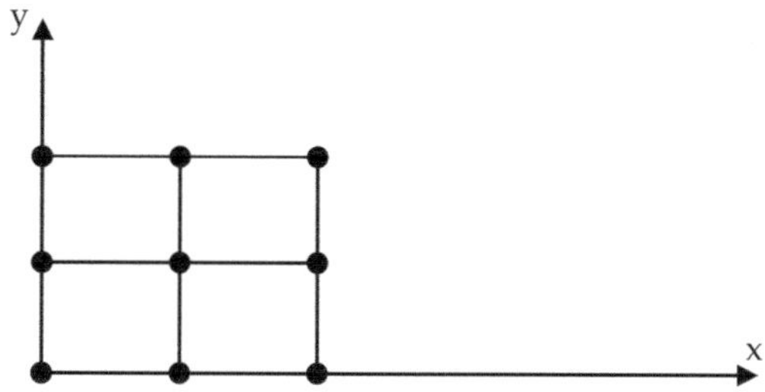

Fig. 17.12 A 3 by 3 image sample

17.3.5 Interpolation and Resampling

Interpolation is very important in medical visualization. Interpolation and resampling usually become necessary particularly whenever we perform scaling operations on digital images, for instance, a 3 by 3 image with its pixels represented in xy coordinates as illustrated in Fig. 17.12.

$$\begin{bmatrix} \widehat{x} \\ \widehat{y} \end{bmatrix} = \begin{bmatrix} s_x & 0 \\ 0 & s_y \end{bmatrix} \begin{bmatrix} x \\ y \end{bmatrix} \tag{17.49}$$

In the scaling operation presented in Eq. (17.49), $(S_{x,}\ 0)$ and $(0, S_y)$ respresent the transformation matrix, - xy is the coordinate of the pixel in the original image, and the $\widehat{x}$ and $\widehat{y}$ are - the coordinates of the pixels in the transform.The original 3 - *by* 3 image has 3 pixels in the horizontal direction- and 3 pixels in the vertical direction.- However, after performing scaling operation by - a factor of 3 in both axes, the final image size has 9 pixels in the horizontal and- 9 pixels in the vertical direction, leaving many pixels not- filled up. If we have a 1 dimensional signal- $f(t)$ and sampled signal as $f_s(t)$, we could therefore see the sample values $f_s(t)$ represented accordingly at discrete locations in the Fig. 17.13. However, since we do not have information of the positions at the intermediate locations as in Fig. 17.13, we need to do interpolation for all the values of t and subsequently do resampling in order to fill up those positions.

It is important to ensure that the interpolation operations follow the following properties:

1. The interpolation function should have a finite region of support, i.e., the interpolation operations should be carried out based on local information of the sample values and not on global information.
2. The interpolation function should be very smooth without introducing any discontinuity in the signal.
3. The interpolation function should be shift invariant; when the signal is shifted through operations such as translation, the same operation should be performed.

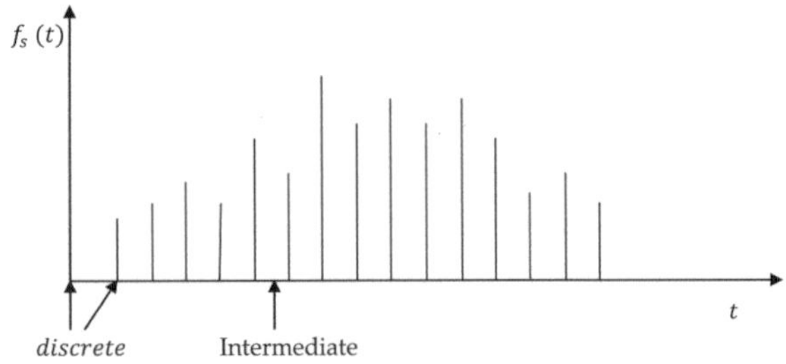

Fig. 17.13 Signals in discrete and intermediate locations

The stated properties are commonly satisfied with B-spline function (Prochazkova 2005). B-spline function is represented in Eq. (17.50):

$$x(t) = \sum_{i=0}^{n} P_i B_{i,k}(t) \tag{17.50}$$

where $n+1$ is the number of approximated samples, P_i are the control points that determine the smooth curve in B − spline functions, and $B_{i,\ k}$ is the normalized B − spline of order of k.

In order to produce smoother images with less artifacts, *trilinear interpolation* approach is usually being considered. Meanwhile this also comes with other overheads as computation of trilinear interpolation usually takes longer time. Therefore, optimization procedures should also be designed to reduce the computational overheads associated with interpolation and resampling procedures.

17.4 Compositing and Algorithm Performances

There are a number of notable volume rendering techniques (also referred to as the direct volume rendering) such as splatting, shear warping, texture mapping, and the ray casting, the Levoy's historic method of rendering. The commonly used compositing functions are the maximum intensity projection (MIP) and the local maximum intensity projection (LMIP). Apparently, LMIP is an extension of MIP. The image in MIP is created by selecting the maximum value along an optical ray that corresponds to each pixel of the 2-D MIP image, while the image of LMIP is created by tracing an optical ray traversing 3-D data from the viewpoint in the viewing direction and then selecting the first maximum value encountered that is larger than a preselected threshold value (Sato et al. 1998). Hence, MIP deals with maximum sampled values, while LMIP involves first local maximum above prescribed threshold and thus approximates occlusion. LMIP is considered faster and therefore better than MIP.

Due to the sensitivity nature and huge data cases in medical visualization, a robust, quality, high-fidelity, and high-performance rendering algorithm is important. Meanwhile, with the advent of high-performance computing architectures, integrating medical application into fast and parallel hardware has been seen as a viable alternative. Various acceleration approaches have been previously proposed for image composting and medical image visualization at large in order to reduce the usual associated computational cost. Apart from the huge number of graphic processing units (GPU) recently available, Compute Unified Device Architecture (CUDA) framework has been lately seen as a heel of high-performance computing which has been leveraged in many circumstances from the clinical data acquisition phase to the results analysis. With a firm design of algorithm, the computational complexities in some of the processes are handled by the high-performance graphic components. Compositing procedures and some of the previously proposed algorithms, although most of them mainly rely on different acceleration approaches, are intensively documented in Cabral et al. (1995), Fang and Chen (2000), Röttger et al. (2000), Engel et al. (2001), Aluru and Jammula (2014), and Leeser et al. (2014) and specifically with CUDA (Adeshina et al. (2012, 2013, 2014), Liu et al. (2014), Adeshina and Hashim (2015), and Kalms (2015). Sample *2-D* slices of brain MRI, *2-D* CT slices of human pelvic region, and the obtained *3-D* correspondents after a series of translational and visualization procedures are presented in Fig. 17.14.

17.5 Conclusion

Modern medicine is greatly benefiting from the fundamentals of mathematics and algorithmic approaches. The evolvement of high-performance algorithms also opened up more growth opportunities in traditional medicine, revolutionalizing the way medical image analysis and visualization are carried out for effective disease diagnosis and therapy management.

X-ray, CT, MRI, PET, SPECT, and other related techniques are used for acquiring morphological or functional information of patients. Apparently, each of the image modalities has its peculiar advantages over another making them somehow complementary rather than being a complete replacement; hence, each of the techniques may be used in various appropriate circumstances. Moreover, in certain decisions such as consideration on the level of exposure to radiation and in certain circumstances, acquisition time could also be considered by the physician while deciding the suitable image acquisition modalities to engage.

Data representation plays a significant role in achieving a reliable visualization results. In some cases, cross-sectional 2-D images in form of slices might need to be stacked. The stacked data needs to be properly enhanced in order to improve its pixel accuracy for effective segmentation and classifications. This chapter has presented some significant stages in data representation, data reconstruction, and modeling.

Compositing approaches and most of the stages in medical visualization could be accelerated using computational techniques such as CUDA, a parallel computing platform allowing programmers to have direct access to the GPU instruction and parallel computational elements. Such acceleration procedures drastically reduce

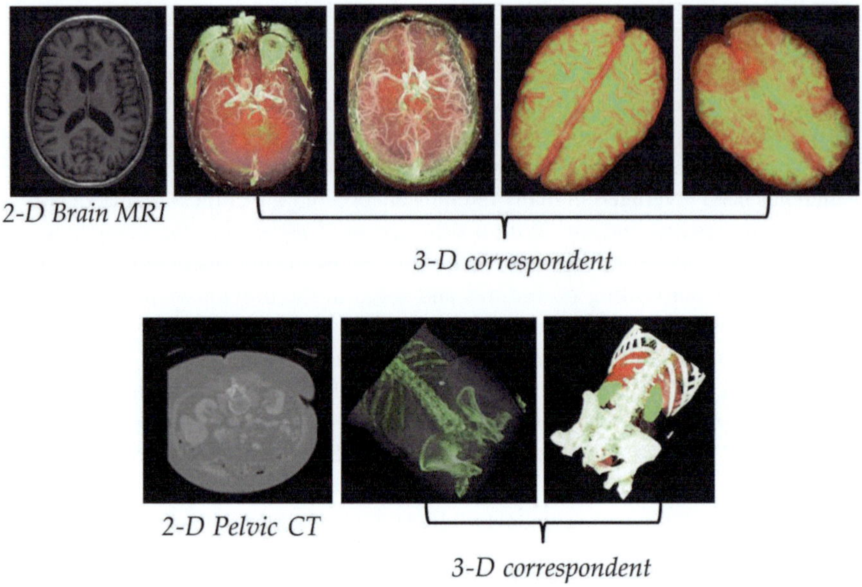

Fig. 17.14 Sample *2-D* slices, as being obtained from *MRI* and *CT* machines, and the correspondent *3-D* reconstructions (Adeshina et al. 2012, 2013)

computational overheads, thereby saving some of the associated computational cost. In the same vein, potential users (doctors) could spend less time in the disease and diagnosis procedures, thereby saving more lives.

References

Adeshina AM, Hashim R. ConnectViz: accelerated approach for brain structural connectivity using Delaunay triangulation. Interdisc Sci Comput Life Sci. 20151–13.

Adeshina AM, Hashim R, Khalid NEA, Abidin SZZ. Locating abnormalities in brain blood vessels using parallel computing architecture. Interdiscip Sci Comput Life Sci. 2012;4(3):161–72.

Adeshina AM, Hashim R, Khalid NEA, Abidin SZZ. Multimodal 3-D reconstruction of human anatomical structures using surlens visualization system. Interdiscip Sci Comput Life Sci. 2013;5(1):23–36.

Adeshina AM, Hashim R, Khalid NEA. CAHECA: Computer Aided Hepatocellular Carcinoma therapy planning. Interdiscip Sci Comput Life Sci. 2014;6(3):222–34.

Aldrich MB, Marshall MV, Sevick-Muraca EM, Lanza G, Kotyk J, Culver J, Wang LV, Uddin J, Crews BC, Marnett LJ, Liao JC, Contag C, Crawford JM, Wang K, Reisdorph B, Appelman H, Turgeon DK, Meyer C, Wang T. 2012. Biomedical optics express; 2012. p. 764–776.

Aluru S, Jammula N. A review of hardware acceleration for computational genomics. Hardware Acceleration in Computational Biology. IEEE Publications; 2014.

Bingham K. Mathematics of local X-ray tomography. Master's thesis. Helsinki University of Technology; 1998.

Butte AJ. Translational bioinformatics: coming of age. J Am Med Inform Assoc. 2008;15:709–14.

Cabral B, Cam N, Foran J. Accelerated volume rendering and tomographic reconstruction using texture mapping hardware. IEEEl; 1995. p. 91–131.
Chen J, Qian F, Yan W, Shen D. Translational biomedical informatics in the cloud: present and future. BioMed Res Int. 2013;2013:1–8.
Dhawan AP, Huang HK, Kim D-S. Principles and advanced methods in medical imaging and image analysis. Hackensack: World Scientific Publishing; 2008.
Engel K, Kraus M, Ertl, T. High-quality pre-integrated volume rendering Using hardware-accelerated pixel shading. ACM 2001 1-58113-407-X; 2001. p. 9–14.
Fang S, Chen H. Hardware accelerated voxelization. Comput Graph. 2000;24(3):433–42.
Faridani A. Introduction to the mathematics of computed tomography. Inside Out Inverse Prob Appl. 2003;47:1–46.
Faridani A, Ritman EL. High-resolution computed tomography from efficient sampling. Inverse Prob. 2000;16(3):635.
Hege HC, Höllerer T, Stalling D. Volume rendering – mathematicals models and algorithmic aspects. In: Nagel W (Hrsg.) PartielleDifferentialgleichungen, Numerik und Anwendungen. Konferenzen des ForschungszentrumsJülich GmbH, S; 1996. pp 227–255.
Hood L. Systems biology: integrating technology, biology, and computation. Mech Ageing Dev. 2003;124:9–16.
Hornak JP. The basics of NMR. Rochester: Department of Chemistry, Rochester Institute of Technology; 1997.
Jenkins R. X-ray techniques: overview. Encyclopedia of analytical chemistry; 2000. p. 13269–88.
Kalms M. High-performance particle simulation using CUDA. Sweden: Linköping University; 2015.
Leeser M, Mukherjee S, Brock J. Fast reconstruction of 3D volumes from 2D CT projection data with GPUs. BMC Res Notes. 2014;7:582.
Liu S, Chen G, Ma C, Han Y. GPGPU acceleration for skeletal animation-comparing Opencl with CUDA and GLSL. J Comput Inf Syst. 2014;10(16):7043–51.
Preim B, Bartz D. Visualization in medicine theory, algorithms, and applications. Amsterdam: Morgan Kaufmann Publishers, Elsevier Inc; 2007.
Prochazkova J. Derivative of B-Spline function. In: Proceedings of the 25th conference on geometry and computer graphics. Prague; 2005.
Radon J. Über die Bestimmung von FunktionendurchihreIntegralwerte längsgewisserMannigfaltigkeiten. Ber Verh Sächs Akad WissLeipzig Math Nat Kl. 1917;69 (1917):262–77.
Roentgen WC. On A New Kind of Rays. Ann Phys Chem. 1898;64:1–11.
Röttger S., Kraus M, Ertl T. Hardware-accelerated volume and isosurface Rendering based on cell-projection. In: Proceedings of the conference on visualization. IEEE Computer Society Press; 2000. p. 109–16.
Sato Y, Shiraga N, Nakajima S, Tamura S, Kikinis R. Local Maximum Intensity Projection (LMIP): a new rendering method for vascular visualization. J Comput Assist Tomogr. 1998;22(6):912–7.
Schroeder W, Martin K, Lorensen B. The visualization toolkit, an object-oriented approach to 3D graphics. 3rd ed. Clifton Park: Pearson Education, Inc.; 2002.

Index

© Shanghai Jiao Tong University Press, Shanghai and Springer Science+Business Media Dordrecht 2017

D.-Q. Wei et al. (eds.), *Translational Bioinformatics and Its Application*, Translational Medicine Research, DOI 10.1007/978-94-024-1045-7

MIX
Papier aus verantwortungsvollen Quellen
Paper from responsible sources
FSC® C105338

FSC
www.fsc.org

If you have any concerns about our products,
you can contact us on
ProductSafety@springernature.com

In case Publisher is established outside the EU,
the EU authorized representative is:
Springer Nature Customer Service Center GmbH
Europaplatz 3, 69115 Heidelberg, Germany

Printed by Libri Plureos GmbH
in Hamburg, Germany